Human Sperm Competition

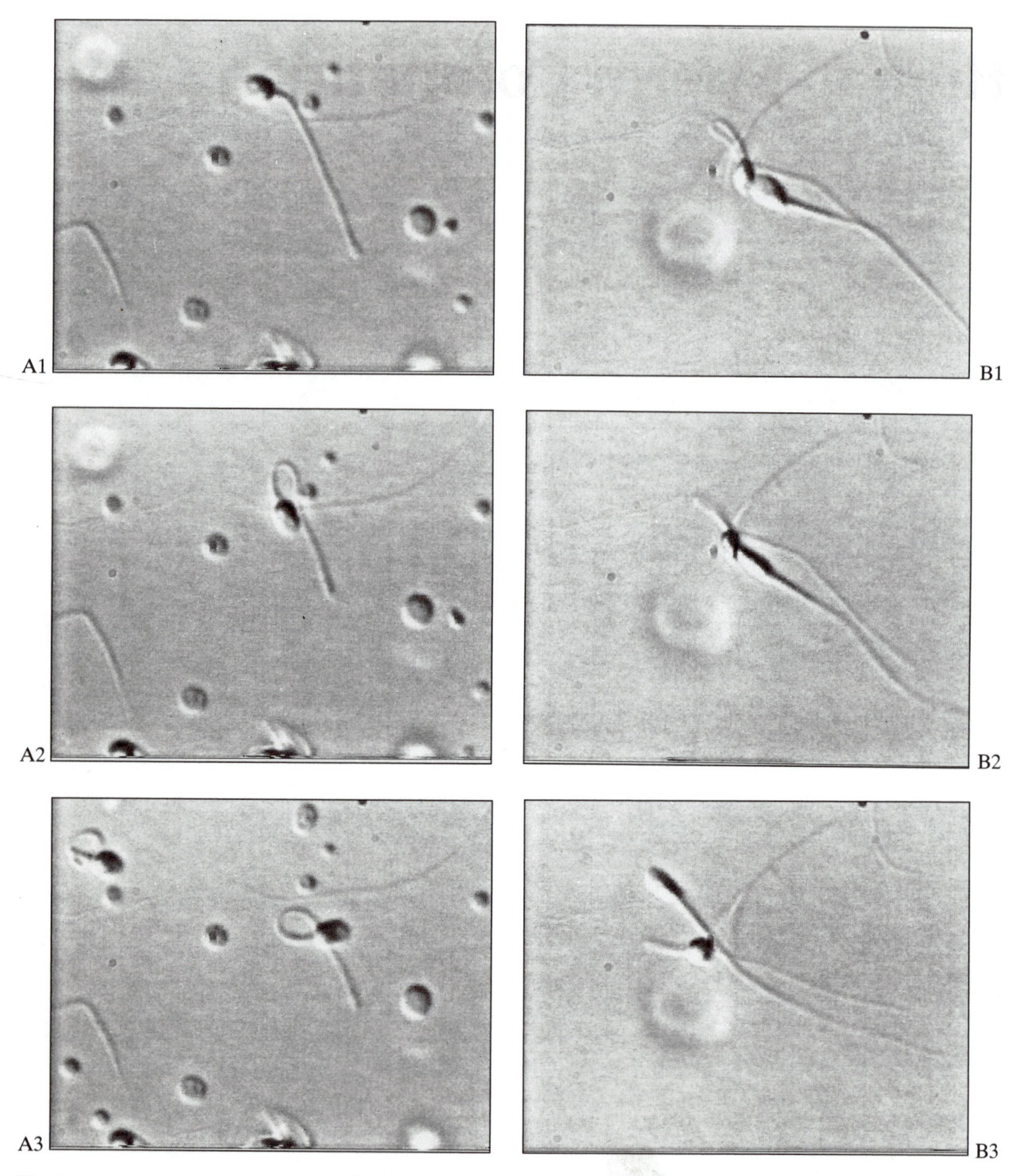

The two sequences show previously unfilmed facets of sperm behaviour.

(A1–A3) An apparently normal and still alive sperm 'metamorphoses' into a coiled-tail sperm in a single (unmixed) ejaculate. The sequence shown took about 20 seconds. In a further five minutes the sperm was as coiled as the sperm in the top left of A3. Both sperm were still moving.

(B1–3) Possible 'seek-and-destroy' behaviour by modal oval-headed sperm in a heterospermic mix of ejaculates from two different males. The pair of sperm (diad) were filmed for 30 minutes attached at the head. Both were active when first seen but by B1 the left-hand sperm was inactive, apparently dead. The surviving sperm, attached at the head tip, spent 15 minutes rotating on its long axis (B1–B2) before eventually breaking free and swimming off (B3).

Sequences taken from a video film made by Suzanne Jackson, Robin Baker, Chris Bainbridge and Karin Halliday, School of Biological Sciences, University of Manchester, using equipment provided by the British Broadcasting Corporation Natural History Unit, Bristol, and Nikon.

Human Sperm Competition

Copulation, masturbation and infidelity

R. Robin Baker and Mark A. Bellis

CHAPMAN & HALL
London · Glasgow · Weinheim · New York · Tokyo · Melbourne · Madras

Published by Chapman & Hall, 2–6 Boundary Row, London SE1 8HN, UK

Chapman & Hall, 2–6 Boundary Row, London SE1 8HN, UK

Blackie Academic & Professional, Wester Cleddens Road, Bishopbriggs, Glasgow G64 2NZ, UK

Chapman & Hall GmbH, Pappelallee 3, 69469 Weinheim, Germany

Chapman & Hall USA, One Penn Plaza, 41st Floor, New York NY 10119, USA

Chapman & Hall Japan, ITP-Japan, Kyowa Building, 3F, 2-2-1 Hirakawacho, Chiyoda-ku, Tokyo 102, Japan

Chapman & Hall Australia, Thomas Nelson Australia, 102 Dodds Street, South Melbourne, Victoria 3205, Australia

Chapman & Hall India, R. Seshadri, 32 Second Main Road, CIT East, Madras 600 035, India

First edition 1995

Typeset in 11/12 pt Times by Wilmaset Ltd, Birkenhead, Wirral
Printed at Alden Press Limited, Oxford and Northampton, Great Britain

ISBN 0 412 45430 0

A catalogue record for this book is available from the British Library

Library of Congress Catalog Card Number: 94-72449

∞

Printed on permanent acid-free text paper, manufactured in accordance with ANSI/NISO Z39.48-1992 and ANSI/NISO Z39.48-1984 (Permanence of Paper)

Dedications

RRB

To Liz, for being open-minded, and
to Nathanial, for being the first flowback baby

MAB

To my family, past, present and future, and
to Tammy, for being there

Contents

Preface

The romantic and traditional view of sexual reproduction is as the one phase of an animal's life when it cooperates totally with another individual (a member of the opposite sex) for their mutual benefit (the production of offspring). In contrast, behavioural ecologists see reproductive cooperation as a brief period when the biological interests of a male and female just happen to coincide. Even then, each sex is continuously vigilant for ways of promoting his own fitness at the expense of its partner's. According to this view, reproduction is a mating game; a subtle mixture of conflict and cooperation between the sexes.

By definition, most mating by monogamous species is in-pair copulation (IPC). However, an apparently universal feature of such species is that from time to time both sexes engage in extra-pair copulation (EPC). Although their own infidelity may be advantageous to both males and females, either may suffer through their partner's infidelity. Males risk cuckoldry or loss of the female to another male. Females risk reduced paternal care or loss of the male to another female. In response to these disadvantages, behavioural ploys have evolved in both sexes to attempt to reduce levels of partner infidelity.

The mating game does not only occur at the level of overt behaviour. Equally powerful aspects are played out cryptically within the female reproductive tract. For example, a special category of EPC is double-mating (the female mating with a second male while still containing competitive sperm from one or more previous males). The result is 'sperm competition' as the sperm from different males compete to fertilize the female's egg(s).

Models of sperm competition tend to view the female tract as a passive receptacle in which males play out their sperm competition games. Females have the potential, however, to influence the outcome of the contest in several different ways. Not least, females may physically eject more sperm from some inseminations than others.

Throughout most of his hominid evolution, the major part of the human population seems to have been 'monogamous' but with a range of subtle and sophisticated forms of infidelity. Many past and present populations show levels of EPC that indicate an evolutionary history of double-mating and sperm competition. We estimate that in Britain in the late 1980s, about 4% of children were conceived via sperm competition (i.e. were conceived while their mother contained within her reproductive tract competitive sperm from two or more different males).

Male humans respond to separation from a female partner by inseminating more sperm at their next copulation. Female humans, although not as blatant as some mammals, eject seminal fluid from their vagina after copulation. In view of the relative ease with which their ejaculates and relevant information can be collected, humans appear to be ideal subjects for the study of the cryptic mating game that takes place in the reproductive tract of female mammals. Yet, when we began our studies of human sperm competition in 1988, it was transparently clear that academically the area was virgin territory. We now know why.

The study of human sperm competition has led us into areas that many people consider to be taboo and to be unseemly as topics for academic consideration. An understanding of sperm competition demands information on patterns and levels of sexual behaviour, particularly levels of infidelity and double-mating. It also demands information on, and understanding of, such emotive and personal facets of human sexual behaviour as frequencies of copulation, rape, patterns of female orgasm, and levels of both male and female masturbation.

On the whole, our publications and lectures have been greeted worldwide with a gratifying level of interest. Most academics have responded with excitement and the question 'why has nobody done this before?'. Perhaps in answer to this question, we have also met our share of people who, from the perspective of their own particular hang-up, view us as nothing better than scientific (or actual) voyeurs who have violated others at a most intimate and personal level.

In carrying out our research, we have two major aims. Our first is a purely scientific understanding of the reproductive behaviour of humans and other animals. As our approach is novel, the type of understanding we are promoting is novel.

Our second aim is medical. About 10% of couples are, at some time, clinically defined as infertile and many expensive treatments for infertility have, and are being, devised. A 1985 estimate calculated the cost of infertility to the USA alone to be $64 billion for infertility investigations and treatment of around 2 million infertile couples with only 14% success. Many of these treatments are lacking information from basic research, particularly on the factors that influence the fertility of sperm. In all modesty, we feel our research could revolutionize the medical approach to infertility.

In writing this book we have but one objective. In the space of just six years we have seen in our research implications for a wide range of disciplines, from sociobiology on the one hand to medicine and veterinary science on the other. Although we shall continue to publish papers in appropriate journals, we could never give the overview that we feel is the main strength of this work. In a book, such an overview is possible.

Although this book is concerned primarily with the sexual behaviour of humans, throughout we take the opportunity of placing human behaviour in the context of that of other animals. At the same time, we point out the implications of the work for the medical understanding and treatment of infertility.

We make no apology for open discussion in these pages of the mechanics and consequences of human infidelity, copulation, rape, masturbation and orgasm. In our view, the result is a better understanding of the way males and females are programmed to behave in sexual matters. If, as a consequence, each sex manages better to understand the behaviour and motivations of the other, we shall feel that our whole project has been more than worthwhile.

Acknowledgements

Our studies have depended on the voluntary cooperation of many people who have had the courage to allow us to examine some of the most intimate aspects and products of their sexual behaviour. We are grateful to them all.

We also thank all those students who played a part in organizing and distributing questionnaires (Emma Creighton, Dominic Shaw, Viki Cook and Jo Moffitt), or in counting and typing sperm (Jo Bell, Jill Bownass, Jill Brundie, Jill Chew, Penny Cook, Tiffany Derwent, Cheryl Durkin, Katie Ellisdon, Helen Fisher, Kath Griffin, Debbie Haynes, Louise Heywood, Dave Johnson, Andy Lang, Steve Lockley, Naomi Matthews, Heather Newton, Alex Norman, Richard Oliver, Simon Pearson, Liz Shields, Mary Southall, Katy Turner and Paul Wilson).

In publishing our nationwide questionnaire, Gill Hudson and *Company* magazine gave us access to a mine of information on female sexual behaviour. Jo Bell, Kath Griffin, and Phil Wheater helped with the mammoth task of transferring the questionnaire data onto computer. Drs Keith and Anna Richardson provided considerable medical guidance and Andrew Burdett provided the same guidance in the field of veterinary medicine. A number of other friends and colleagues have also made invaluable contributions to various aspects of our work (Dr Matt Gage, Rachel Alcock, Jayne Burdett and Deborah and Sean Cochrane).

We are extremely grateful to St Mary's Hospital, Manchester for their help and collaboration over the past few years, particularly Dr Phil Matson and his staff at the IVF clinic and Anne Atkinson and her staff at the AID clinic.

Our work on sperm competition has been funded in part by the Science and Engineering Research Council; our work on menstrual synchrony by the Leverhulme Trust. We also thank Durex for their kind donation of condoms.

We have both benefited considerably over the years from discussions with the father of sperm competition, Professor Geoff. Parker (one of us, RRB, ever since 1962 when we began our undergraduate days together at the University of Bristol). If he had not started the whole study of sperm competition in 1970, we should not have enjoyed ourselves half as much since 1988 when we first decided to study the phenomenon in humans.

The manuscript was greatly improved by the typical and critical attention of Jack Cohen. More than anything, we benefited from his incredibly wide knowledge of all things reproductive and particularly through his eyes from seeing our book from the perspective of a non-behavioural ecologist.

Finally, we particularly thank both Dr

Tamsin Peachey, for advice on all matters biochemical and for her part in developing our model for the function of the female orgasm, and Elizabeth Oram, for drawing our attention to the potential value of studying flowbacks and for her role in developing the techniques of flowback collection. Without this single development, a major part of the work reported in this book might never have begun.

1 Introduction

It is probably fair to say that both traditional biological and current medical doctrines tend to interpret the characteristics of human ejaculates as features essential to the process of fertilization. However, such traditional doctrines find it difficult to explain why, for example, at each insemination a human male ejaculates around 350 million sperm (apparently enough to fertilize all of the women in America) to fertilize a single egg.

Over the past two decades, behavioural ecologists, one of the most recent subgroups of the biological sciences, have begun to question this traditional view. They suspect instead that most of the conspicuous aspects of animal enjaculates are more the evolutionary product of a phenomenon known as sperm competition than the result of selection for the efficiency of fertilization. Indeed, they have even begun to suspect that some characteristics of mammalian ejaculates evolved for the purpose of sperm competition may even be detrimental to the fertilization of eggs (Baker and Bellis, 1993a).

Sperm competition is the competition between sperm from different males for the 'prize' of the fertilization of the egg(s) produced by a single female. In internal fertilizers, such as humans, such competition naturally takes place within the reproductive tract of a single female. The risk of sperm competition is thought to have influenced many aspects of sexuality, not only concerning sperm and the ejaculate but innumerable other aspects of male and female anatomy and physiology. Thus, sperm competition may be argued to promote not only highly competitive ejaculates but also to shape anatomical devices such as the penis and vagina and to generate a whole array of copulatory behaviour.

The importance of sperm competition in the shaping of sexual behaviour and anatomy was first pointed out for insects by Parker (1970a). Since then, however, behavioural ecologists have come to realize that sperm competition is a virtually universal phenomenon that has shaped the sexuality of nearly all animal lineages (see collection of papers in Smith 1984a). They suspect that every male of nearly every animal that exists today, including every male human, is the descendant of many generations of male ancestors who have either successfully avoided contests or whose sperm have been successful in contests with sperm from other males. It is not surprising then that the stage has been reached when interpretation of human ejaculates as well as human sexual behaviour demands an understanding of the dynamics of human sperm competition.

Fertilization of an egg places relatively simple demands on the behaviour and physiology of both the sperm and the male who produces them. By contrast, sperm competition places enormous and complex pressures on males, their ejaculates, and on the females in whose reproductive tracts the contests take place. Thus, males do best who produce competitive ejaculates. Females do best who provide an internal environment that is optimally selective for the competitiveness of the sperm from different males.

As an analogy, let us compare a person (= DNA) leisurely driving a car (= a sperm) uncontested from A (= vagina) to B (= site of fertilization) with the same person in a rally car racing to arrive at B ahead of numerous other hostile competitors from one or more opposing teams (= sperm competition). In both cases only the car which arrives at B at precisely the right moment will receive a prize (= egg). However, the requirements, resources and strategies necessary to attain a prize in the second scenario are infinitely greater than in the first even though in the first the person may stand a better chance of completing the journey.

If this analogy seems whimsical, consider the bull whose semen is fully fertile when artificially inseminated on its own but fails to gain any fertilizations when mixed with semen from another bull (Beatty *et al.*, 1969). Little wonder, therefore, that behavioural ecologists suspect that when they look at ejaculates, or any other aspect of sexuality, they are seeing requirements for success in sperm competition rather than for success in simple fertilization. Although the latter is there, and obviously important, it is obscured by the former. The single tree of fertilization is difficult to find in the forest of sperm competition.

Since 1970, behavioural ecology has made tremendous progress in understanding the nature of animal sexuality. At the same time, other biological and medical disciplines have been unravelling the cellular and biochemical complexities of ejaculate structure. The complete picture, however, has been appreciated by only a limited and esoteric audience. By and large, the combined implications have escaped the attention of the wider interested audience of lay biologists, students in all branches of biological science, and medical and veterinary scientists.

This book is an attempt to correct that situation. Its message is simple and straightforward. Human sexuality, in all its anatomical, physiological and behavioural detail, owes more to sperm competition (both in the present and in the immediate and distant evolutionary past) than it does to simple fertilization. Acceptance of this fact revolutionizes one's view of all aspects of male and female sexuality and brings old enigmas into new perspective. As we hope this book will show, the medical and veterinary implications of this new perspective are enormous.

2 Sex, coyness and promiscuity: the evolution of sperm competition

2.1 Introduction

Each living human is the modern representative of a continuous evolutionary lineage stretching back from the present day through an unbroken sequence of generations to the origin of life on planet Earth some 4000 million years ago. During the course of these generations, the human lineage 'picked up' through evolution a variety of anatomical, physiological and behavioural characteristics that today constitute human sexuality.

In this and the next four chapters, we discuss the evolution of various facets of human sexuality. Throughout, we point out the pivotal role of sperm competition. Our approach is broadly chronological, considering each facet roughly in the order in which we think it appeared in the human lineage and since when it will have been a continuous feature of our ancestors' sexuality (Box 2.1)

Each section in this and the next four chapters deals with a particular facet of human sexuality and is organized according to the following convention. First, if necessary, we define the terms associated with the aspect of sexuality to be discussed. Next, we describe the characteristics of this aspect as manifest by humans. Finally, we briefly place the characteristic in the perspective of other animals before making a 'best guess' at its antiquity (i.e. when it first evolved in the human lineage).

We assume that many people who read this book will not have received formal training in behavioural ecology and thus will not be familiar with those basic hypotheses that have been formulated to explain the evolution of sexual behaviour. For those among such readers who would like a deeper understanding of the arguments in this book, we describe these hypotheses when they are appropriate. However, to avoid disrupting the flow of the main text, these descriptions have been separated into boxes.

2.2 Evolutionary inheritance: a programme for a lifetime

The earliest life-forms were probably simple, replicating molecules with relatively few heritable elements ('genes' etc.). Over a time scale of 4000 million years, however, the influence of mutation, breakage and recombination of heritable material has generated more, and more diverse, heritable elements.

Perhaps the first step in complexity was the greater reproductive success of individuals with a 'coat' that hindered other such molecules from 'stealing' their chemical building

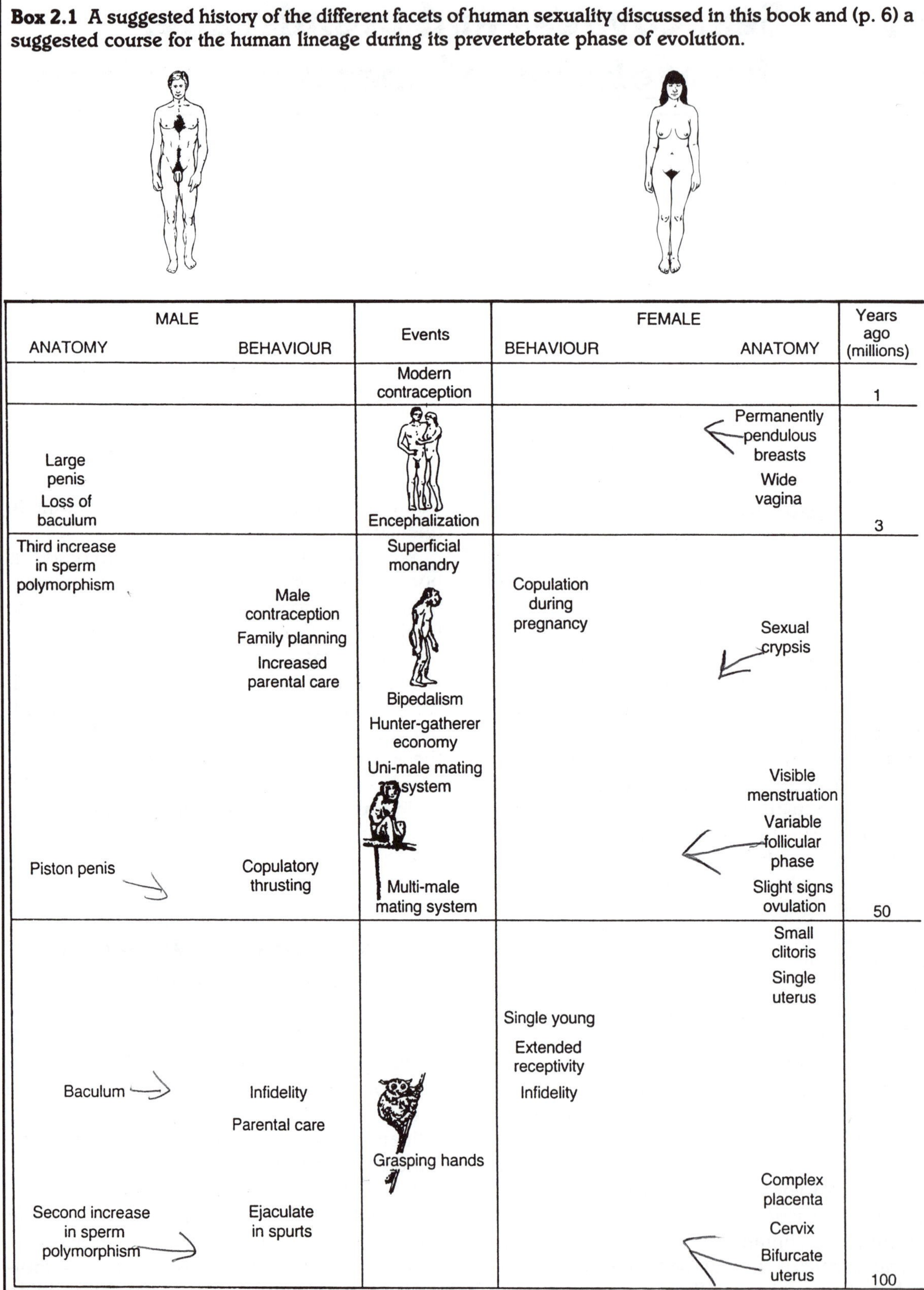

Box 2.1 A suggested history of the different facets of human sexuality discussed in this book and (p. 6) a suggested course for the human lineage during its prevertebrate phase of evolution.

MALE		Events	FEMALE		Years ago (millions)
ANATOMY	BEHAVIOUR		BEHAVIOUR	ANATOMY	
		Modern contraception			1
Large penis Loss of baculum		Encephalization		Permanently pendulous breasts Wide vagina	3
Third increase in sperm polymorphism Piston penis	Male contraception Family planning Increased parental care Copulatory thrusting	Superficial monandry Bipedalism Hunter-gatherer economy Uni-male mating system Multi-male mating system	Copulation during pregnancy	Sexual crypsis Visible menstruation Variable follicular phase Slight signs ovulation	50
Baculum Second increase in sperm polymorphism	Infidelity Parental care Ejaculate in spurts	Grasping hands	Single young Extended receptivity Infidelity	Small clitoris Single uterus Complex placenta Cervix Bifurcate uterus	100

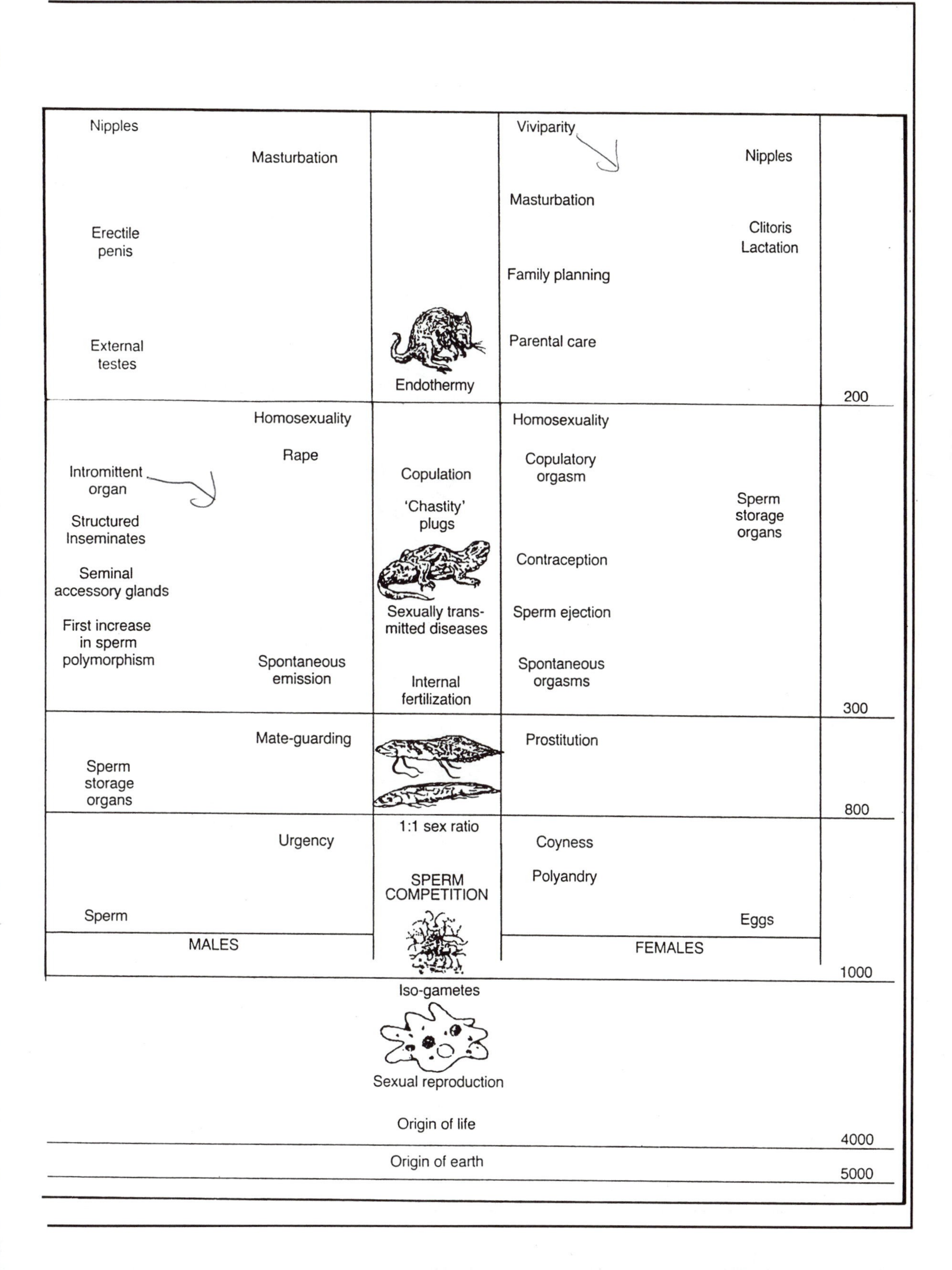

Nipples
Masturbation
Erectile penis
External testes
Endothermy
Viviparity
Nipples
Masturbation
Clitoris
Lactation
Family planning
Parental care
200
Homosexuality
Rape
Intromittent organ
Structured Inseminates
Seminal accessory glands
First increase in sperm polymorphism
Spontaneous emission
Copulation
'Chastity' plugs
Sexually transmitted diseases
Internal fertilization
Homosexuality
Copulatory orgasm
Sperm storage organs
Contraception
Sperm ejection
Spontaneous orgasms
300
Mate-guarding
Sperm storage organs
Prostitution
800
Urgency
Sperm
1:1 sex ratio
SPERM COMPETITION
Coyness
Polyandry
Eggs
MALES
FEMALES
1000
Iso-gametes
Sexual reproduction
Origin of life
4000
Origin of earth
5000

Box 2.1 (*continued*)
Adapted from Margolis (1981).

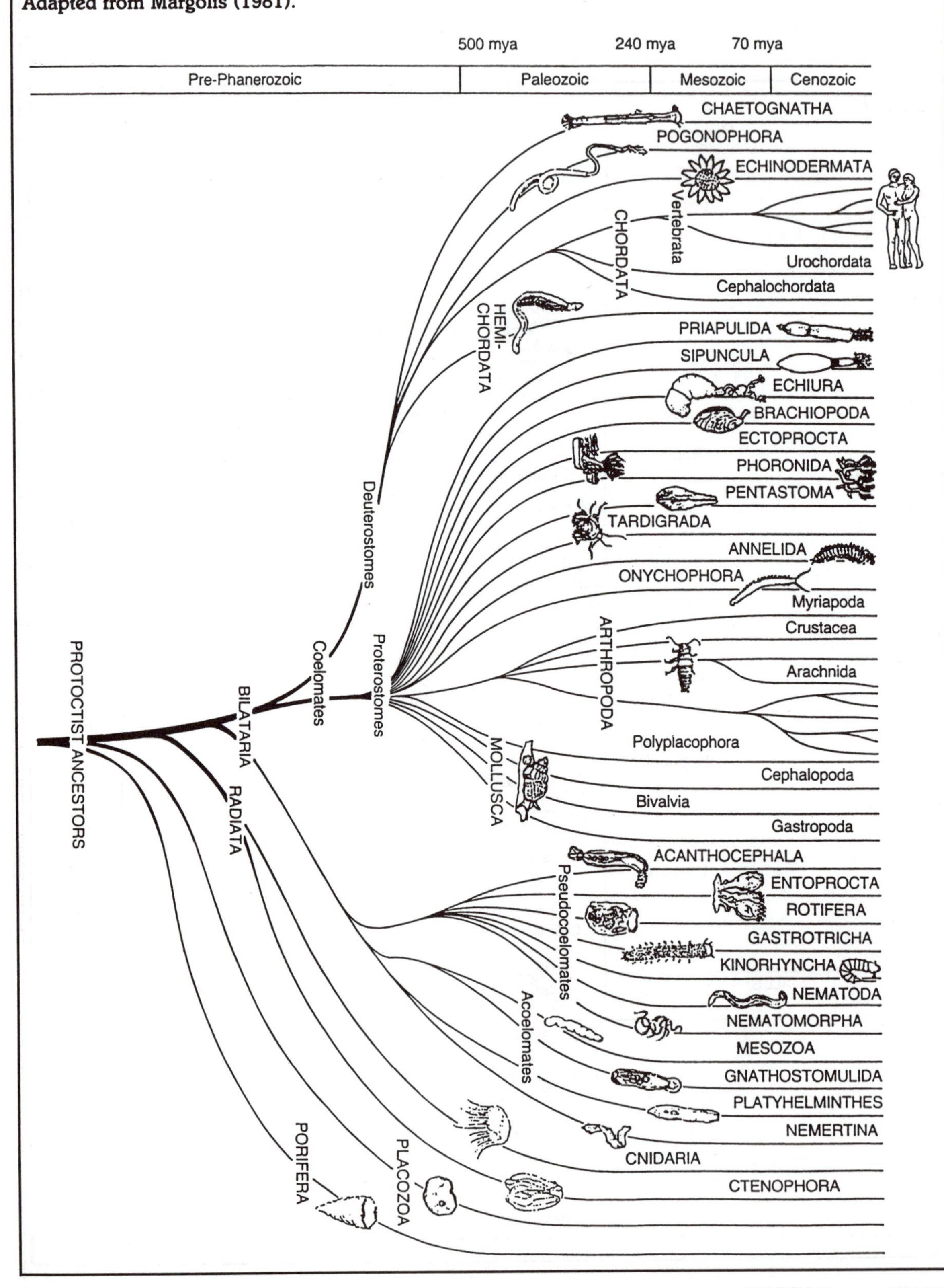

Box 2.2 Reproduction: the currency of natural selection

In many ways it is unfortunate that the phrase most widely associated with evolution through Darwinian natural section is 'survival of the fittest'. It is reproduction, not survival, that is the currency of natural selection. In so far as survival of the individual influences the time available, it may influence reproductive success. In its own right, however, survival without reproduction does not promote evolutionary success.

It is a mathematical inevitability that populations come to be dominated by those heritable characteristics that impart greatest multiplication power to the descendants of the lineage founder. Potential evolutionary success is not always obvious from the number of children produced by a lineage founder. Often potential success is not evident until the number of grandchildren or even of more distant descendants may be counted. Consider two examples. First, an individual may produce more grandchildren by producing fewer but better quality children. Second, an individual may produce more grandchildren by producing fewer children but expending time and energy in keeping them together and manipulating them into helping each other (Section 4.4.15).

Thus, when we come to examine the sexual behaviour of humans or other animals at the present time, we are seeing populations that are dominated numerically by heritable characteristics that imparted the greatest multiplication power on generations of past possessors. This statement has the certainty of all mathematical axioms and as such is immune to any further philosophical or ideological discussion.

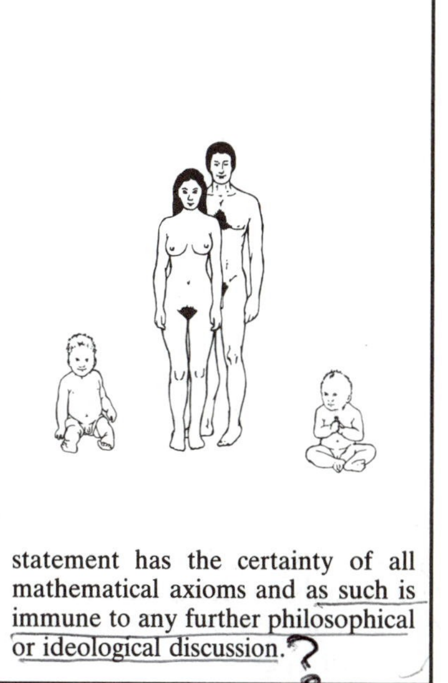

blocks (Dawkins, 1976). With the general increase in genetic complexity has come an increase in complexity of the coat. In a sense, the human body serves the same function; a 'coat' to protect its genes. It is the properties of this 'coat' that determine the chances of the germ cells meeting others in order to reproduce (Cohen, 1977).

So many heritable elements are now needed for the human (or rat, butterfly or tree) to function that the combination of elements are nearly infinite. However, despite the potential for infinite individual variation, populations are dominated numerically by heritable characteristics possessed by most individuals. For example, almost every human is programmed to reproduce sexually. With the certainty of mathematics, such widespread characteristics are those that over past generations have imparted the greatest production of offspring to their possessors.

The currency of natural selection is not survival but reproduction (Box 2.2) and the two are by no means always linked. For example, a dominant male gains no genetic benefit from his greater size, fitness and perhaps longevity if he is impotent (e.g. swine; Sumption, 1961). Inevitably, the genetic programmes that shape the behaviour of the majority of modern individuals are made up of a range of heritable characteristics, each of which in past generations was found in those ancestors who generated most descendants (i.e. had the greatest reproductive success).

Programmed behaviour does not equate to inflexible behaviour. The best artificial intelligence programmes are designed to change their response for the better as a result of their own experiences (Schank, 1990). However, even such dynamic intelligence is still simply a series of algorithms; a programme. Of course, some types of animals in some types of environment may gain greatest reproductive success from inflexible, unvarying behaviour though it is difficult to think of any unequivocal example. Without doubt, the majority gain from behavioural flexibility. Individuals whose behaviour changes with each change in circumstance, perhaps in part according to previous experience, will achieve greater reproductive success than those whose behaviour does not, or changes inappropriately. Nevertheless, which behaviour is shown in which circumstance and how, how quickly and to what extent behaviour changes with experience, will be part of the inherited programme. Humans, like other mammals, may be programmed to be extremely flexible. Nevertheless, the basis of the flexibility is still just a programme, a collection of interacting chemicals. Change the chemical balance inappropriately even in a human (for instance by alcoholic intake) and behaviour may become inappropriate and maladaptive (e.g. excitability, euphoria, aggression and unconsciousness; Littleton, 1981).

2.3 Human sexuality: the basics

2.3.1 Sex, gender and the sex ratio

Almost every modern human is programmed to reproduce sexually via the production of gametes which fuse to produce a zygote. Individuals are programmed to be either male or female producing, respectively, either sperm or eggs.

At a global level, the population consists of males and females in a roughly 1:1 sex ratio. However, at conception (primary sex ratio) males may out-number females 1.3:1 with much of this bias disappearing by birth (secondary sex ratio) and males faring worse than females in all subsequent stages (Singer, 1985). Perhaps most importantly, there is some evidence that a balanced 1:1 ratio may be reached at the age when male and female humans begin sexual exploration (Crew, 1952).

In these terms, basic human sexuality is indistinguishable from that of virtually all other mammals, the vast majority of other vertebrates and probably the majority of invertebrates. As far as sex ratio is concerned, a few animals do depart from a 1:1 ratio (Hamilton, 1967) and humans do show a bias toward producing male or female offspring depending on the mother's circumstance (Crew, 1952). However, further discussion of these subtle departures from a 1:1 ratio in humans is left until Chapter 4. For the moment, the important point is that sex ratios more often than not are nearly 1:1.

Theories explaining the evolution of these basic facets of sexuality are relatively well established (Boxes 2.3–2.5). Thus, it seems likely that sexual reproduction has been a continuous feature of the human lineage ever since the appearance of the simplest organisms, soon after the origin of life (3500–2500 million years ago; Box 2.1). Gametes and zygotes seem likely to have been a continuous feature since at least the appearance of the first eukaryotes (1200 million years ago; Box 2.1). The male:female phenomenon with a 1:1 sex ratio, however, although possibly orginating at about the same time may not have been firmly established until the explosive evolution of metazoans (around 800–600 million years ago; Box 2.1) (Parker *et al.*, 1972; Knowlton, 1974; Bell, 1978).

2.3.2 Urgent males and coy females

Surveys consistently show that human males claim to have inseminated more females than females claim to have been inseminated by males (Anderson *et al.*, 1991; Morris, 1993). Mathematically, of course, with a 1:1 sex ratio, mean (though not median) number of partners must be identical (Gurman, 1989). Recent analysis of the discrepancy suggests that almost the whole difference lies in the 'tail' of the distribution, in the claims made by the sexually most active males and females (Morris, 1993).

Two explanations for the discrepancy are normally given. Either: (1) more males than females have multiple partners (a few females being inseminated by many males) but most surveys are too small or too biased to contain a representative proportion of the few very active females; or (2) males tend to exaggerate, and/or females to underplay, their sexual activity, especially at the most active end of the spectrum.

The first of these two explanations has, in the past, generally been the more favoured (Johnson *et al.*, 1990). If this were the case, however, the larger the sample the greater the chance of including the putatively few very active females and hence the smaller the difference between male and female claims. Unfortunately, any evidence for this interpretation has recently been undermined by the findings of two very large surveys of 20 055 men and women in France (ACSF investigators, 1992) and 18 876 in England, Wales and Scotland (Johnson *et al.*, 1992, 1994). Even though both surveyed over 10 000

Box 2.3 Sexual reproduction

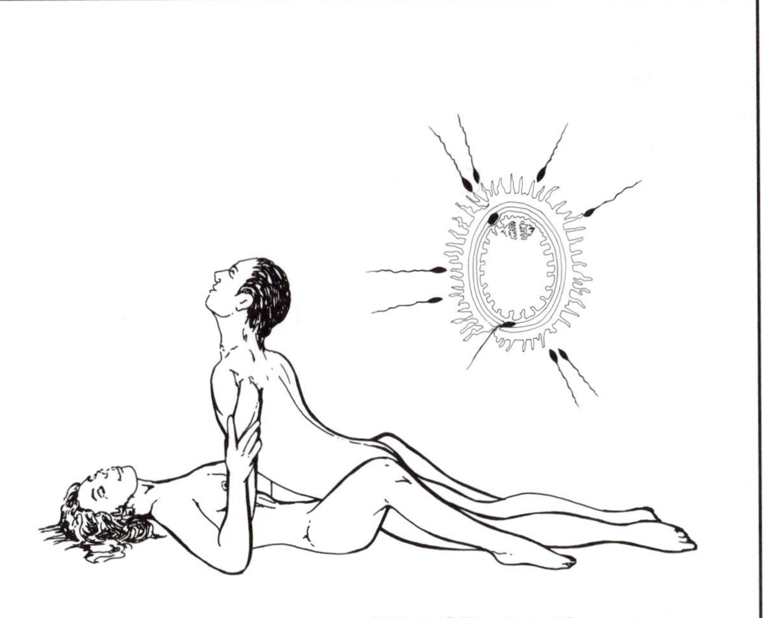

Reproduction is termed sexual when the mutual offspring of more than one (usually two) parents is first formed by the fusion of cells or parts of cells from each parent. The vast majority of living organisms, from the most simple bacteria and algae to the most complex fungi, plants and animals reproduce sexually. The alternative is to reproduce asexually. Reproduction is termed asexual when the offspring derive, without fusion, from a cell or cells of only one parent. In most lineages, asexually produced offspring are effectively clones, identical to the parent except when modified by mutation.

When sexual reproduction is via the fusion of single, whole cells, the cells are termed gametes and the product of fusion is termed a zygote. Most often, a gamete contains a replica of only part of the genetic material of each parent individual. Sexual reproduction is thus a mechanism that allows an individual to produce offspring which, while carrying a subset of their parent's genes, nevertheless vary in their characteristics. Asexual reproduction, in contrast, is a mechanism which allows an individual to produce offspring virtually identical to itself.

It is generally agreed that the preponderance of sexual reproduction among living organisms is an evolved result of the controlled variation it allows parents to generate among their offspring (Maynard Smith, 1978). Cloning may impart some short-term advantage to a lineage well-adapted to a stable environment but is less likely to lead to long-term evolutionary success in environments, such as those encountered by most mobile animals, that change quickly through time and/or space (Roughgarden, 1991).

females, the total numbers of claimed lifetime partners for males and females were, respectively, 11 and 3 in France and 10 and 3 in Britain. Both sets of figures are similar to the claims of 11 and 3 based on the first 1000 respondents in the UK survey (Maddox, 1989; Johnson *et al.*, 1990). There is no indication of a decrease in this anomaly with increase in sample size.

Aspects of the British and French 1992 surveys suggest that the second of the above interpretations is the more likely: males tend to exaggerate, and/or females to underplay, their sexual activity once number of partners becomes large (Morris, 1993). Of these two elements, the more potent seems to be the tendency for females to underestimate their past number of partners. For example, in the British survey (Johnson *et al.*, 1992, 1994), the mean number of partners claimed by females showed an unlikely decrease with age. Of course, such a trend could be attributable to age cohort differences. The past few decades could have seen such an increase in number of partners that young women now have more partners in their first eight years of sexual activity than older women did in their entire lifetime. However, the data for males shows the expected increase in number of partners with age. Even allowing for age differences in the selection of partners by males and females and an undoubted tendency for males to exaggerate, the most powerful factor still seems likely to be female understatement. Such understatement could be the result of a genuine lapse of memory or it could be due to subconscious deception (of self as well as others) and/or conscious secrecy. Self-deception as a female strategy is discussed in Chapter 6.

Both of the 1992 French and British surveys were carried out by interview, either by phone (in France) or by direct contact (Britain). Our own survey (females only) was by

Box 2.4 Anisogamy and the male:female phenomenon

Isogamous organisms produce populations of gametes with just one modal size. Anisogamous organisms produce populations of gametes with more than one modal size. When an anisogamous species produces gametes with just two modal sizes, the smaller of the two modes are termed sperm; the larger, eggs or ova.

Anisogamy evolved in lineages, such as metazoans, in which zygote survival was critically dependent on zygote size (Parker, Baker and Smith, 1972). There was then an advantage in producing (1) small gametes which were hence numerous but produced zygotes with reduced survival and (2) large gametes which were less numerous but produced zygotes with enhanced survival. Bimodality then arose because the producers of intermediate-size gametes produced too few to compete with small-gamete producers and zygotes with survival prospects that were too poor to compete with larger-gamete producers. Small- and large-gametes finally entered into an arms race (Dawkins and Krebs, 1979) over fusion with large gametes. Inevitably, as with most parasites, the arms race was won by the small-gametes. It is with the final appearance of large-gamete and small-gamete producers which produce gametes specialized to fuse with gametes of the other modal size, that the male–female phenomenon can be said to have arisen (small-gamete producers = males; large-gamete producers = females).

In Parker *et al.*'s (1972) original model for the evolution of anisogamy, the proposed initial advantage of small gametes was purely numerical. More recently, it has been suggested (Møller, unpublished) that an alternative or additional advantage of small gametes could be that they may be too small to contain disease organisms even as small as viruses. Males may thus function as a form of disease filter in each generation. It should be noted, however, that the small size of a gamete may not prevent it transmitting parasites on its outer surface (Charles and Larsen, 1987).

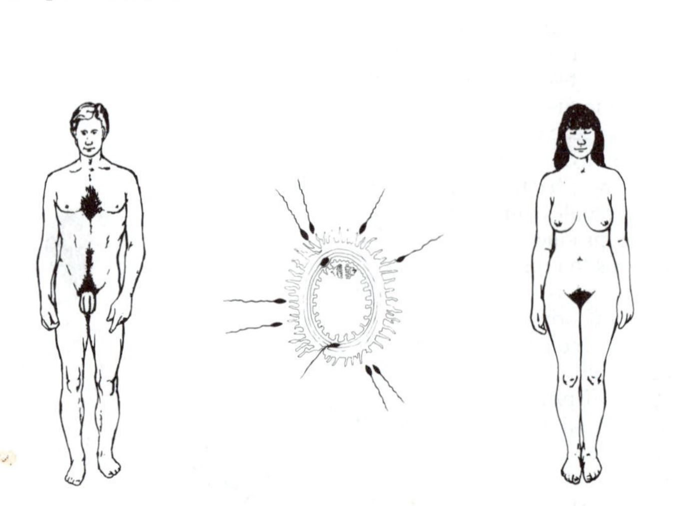

mail (Box 2.6) and produced claimed numbers of lifetime partners for females far more comparable to the number claimed by males in these other studies. Whether this higher claimed number of partners/female in our study reflects greater accuracy due to our less personal methodology or whether the women in other surveys really had fewer partners/female must remain a moot point. In favour of the former possibility, however, is the fact that in our study women did show the expected increase in number of partners with increased sexual experience.

Important though it is in its own right, the eventual detailed explanation for the male/female dichotomy in claimed number of partners does not matter in the present context. What matters here is that the discrepancy reflects an important difference between the sexual programming of human males and females: males are generally more urgent, indiscriminate and overt over sexual matters; females are generally more coy, cautious and secretive. This difference was graphically illustrated in a survey of Californian students in which males were significantly more likely than females to say that they would have sex with an anonymous partner (Symons and Ellis, 1989).

The greater urgency of males and their relative indiscrimination sometimes translates into enormous levels of reproductive success. For humans, the most offspring credited to a single male is 888 (achieved by an ex-

Box 2.5 Evolution and maintenance of a 1:1 sex ratio

Intuitive logic and female chauvinism concur that males are a waste of space. As males produce smaller, and thus more, gametes than females, each male is capable of fertilizing the gametes of several females. All of the females in a population could thus be fertilized by a smaller number of males. Yet, in most species, males and females exist in a roughly 1:1 ratio. From a population perspective, therefore, more males exist than seem necessary. Males really do seem to be a waste of biological resource. Why, then, has evolution produced such an apparently wasteful adaptation?

The generally accepted answer was produced as long ago as 1930 by the geneticist, R. A. Fisher. Refinements to Fisher's model were expounded by Trivers (1972) and Trivers and Willard (1973) (Section 4.4.13) and circumstances in which a female-based sex ratio would be favoured were detailed by Hamilton (1967).

In essence, Fisher's (1930) model shows that in a freely mixing population, if ever a population is dominated numerically by females, the most grandchildren are produced by parents who produce an excess of males in their offspring. Conversely, if ever a population is dominated by males, the most grandchildren are produced by parents who produce an excess of females in their offspring. Inevitably, therefore, if ever a population sex ratio begins to depart from unity, selective pressures are generated that cause that sex ratio to swing back towards unity. Whether males seem to be a waste of space or not, they are maintained by natural selection at a level comparable to that of females.

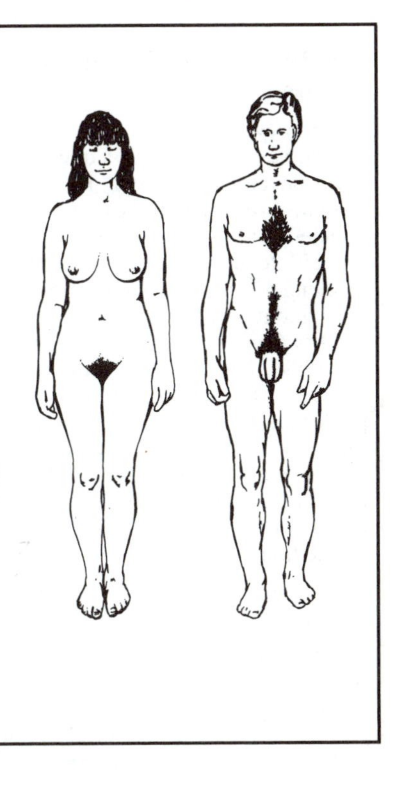

emperor of Morocco). By contrast, the most offspring ever credited to a single female is 67 (in 29 pregnancies) (*The Guinness Book of Records*). The tendency for males to be more indiscriminate and urgent than females over sexual matters is almost universal among animals and the theoretical basis of the phenomenon is relatively clear (Box 2.7).

The number of offspring produced by a population is determined by the females, not the males. It is the rate at which females convert time and energy into children that is critical, not their rate of sexual encounter with males. For males, however, it is the rate at which they fertilize eggs that is important to their individual reproductive success. In a classic experiment by Bateman (1948), equal numbers of male and female fruit flies, *Drosophila melanogaster* were placed in bottles. The number of matings and offspring produced by each individual was scored, using genetic markers to determine parentage. For males, reproductive success went up with number of matings, for females it did not (beyond the first mating).

The few animal lineages characterized by females who are more urgent and competitive than males are associated with ecological situations in which it is the male who limits reproductive output, not the female (Clutton-Brock and Vincent, 1991). Thus, in the Panamania poison-arrow frog, *Dendrobates auratus*, the male offers protection and transport to the tadpoles on his back. The male thus guards the offspring during their early development (Wells, 1981). In some birds (e.g. phalaropes), females are in a food-rich environment and are not limited in the number of eggs they produce. The limitation to their reproductive output is in finding males to incubate the eggs that they produce (Cramp, 1983). In both frog and phalarope, the females are the more urgent sex and males the more coy and cautious. Females compete (often aggressively) with each other for sexual access to males.

Urgent males and coy females have probably been a feature of the human lineage for as long as there have been mobile, metazoan males (around 600 million years; Box 2.1).

Box 2.6 **A nationwide survey (1989) of the level of polyandry and other sexual behaviour of British females**

The Survey

Our survey was based on a questionnaire developed between 1987 and 1989 in three pilot studies (two self-selected; one by interview), involving 250 females. The final version (seeking 57 answers) was distributed throughout Britain in March/April 1989 by *Company* magazine (Bellis *et al.*, 1989). Female readership of the relevant issue was estimated by the publishers to be 439 000. Our 3679 replies (excluding seven overtly spoilt) therefore represent 0.84% of the potential respondents who were themselves roughly 5% of the UK population of females of reproductive age. Of the respondents 92 claimed to be virgins leaving a sample of 3587 sexually experienced females aged between 13 and 72 years (mode = 21 years).

The major characteristics of our sample are shown as we present analyses from our survey at various points in this book. Any slight differences in sample size for different categories reported here and previously (Baker *et al.*, 1989; Bellis and Baker 1990; Baker and Bellis, 1993a,b) are the result of ongoing cleaning of this large data base of 209 703 data points.

Levels of Polyandry

Variation in the total number of different male partners so far at different stages of female sexual experience: Britain, 1989

No. of male partners	Total no. of lifetime copulations (so far)					
	1–50	51–200	201–500	501–1000	1001–3000	>3000
1	35.6%	19.4%	13.1%	7.4%	6.3%	5.6%
2–5	50.3%	45.4%	37.9%	35.0%	27.7%	22.2%
6–10	11.4%	19.8%	24.8%	25.9%	27.7%	25.0%
11–20	2.3%	11.6%	16.3%	17.9%	19.1%	15.7%
21–50	0.4%	3.3%	7.1%	11.9%	14.6%	21.3%
51–100	–	0.5%	0.7%	1.3%	4.0%	4.6%
>100	–	0.0%	0.1%	0.6%	0.6%	5.6%
n	481	796	896	632	477	108
Mean	3.2	6.1	8.5	12.7	14.9	36.2
± SE	0.1	0.3	0.3	1.3	1.0	10.5
Median	2	4	5	7	8	10
(IQR)	(1–4)	(2–7)	(3–10)	(3–13)	(4–17)	(4–30)

2.4 Polyandry and sperm competition: a prevertebrate legacy

2.4.1 Polyandrous females

A polyandrous female is one who, through her behaviour, exposes her lifetime production of eggs to the chance of fertilization by more than one male. Theories to explain the evolution of polyandry are well established (Box 2.8). Polyandry contrasts with monandry in which only one male has the opportunity to fertilize a female's lifetime production of eggs.

Promiscuity is a form of polyandry in which a female mates with more than one male in relatively rapid succession on an apparently random and casual basis. Designating a female as promiscuous rather than polyandrous involves an often arbitrary value judge-

Box 2.7 Urgent males; coy females

Urgent males

Males produce small gametes; females produce large. Each male is thus capable of fertilizing the gametes of more than one female. Yet males and females exist in equal numbers. Thus, for every male that fertilizes the eggs of more than one female, there will be a male or males that fails to fertilize any.

Evolutionarily, the most successful male in each generation will be the one that produces most descendants. In large part, this will be a function of the number of eggs a male fertilizes which in turn will be largely a function of the number of females he succeeds in persuading to allow his sperm access to their eggs. Human populations must now consist of the descendants of males (pre-human as well as human) who successfully pursued an urgent round of attempting to fertilize the eggs of as many females as possible. Each failure or even each delay in gaining access to a female's eggs experienced by males in past generations reduces the representation of their lineage in the modern population. An inevitable corollary of producing small gametes and a 1:1 sex ratio is that males will have an urgent approach to sexual reproduction and a predilection to fertilize as many females as possible.

Coy and cautious females

In contrast, females are limited in their potential reproductive success, not by how often they meet males, but by the time and energy they have at their disposal to reproduce. The most successful females are those that optimize their use of time and energy to produce offspring of such numbers and quality that the multiplication power of the female's lineage is maximized. As females produce fewer gametes than males, a female of most species requires only a single contact with a male to obtain enough sperm to fertilize all of her eggs, at least of the current 'batch'. In many species with internal fertilization, females have specialized sperm storage organs (section 3.5.1) so that a single insemination produces enough sperm to meet her long-term needs (even over a lifetime; e.g. the butterfly, *Papilio glaucus* (Levin, 1973)). Further contact and attention from males wastes the female's time and energy, those commodities that are most influential in affecting her reproductive success. As a result, the females of many species avoid contact with males as much as possible.

In general, female animals are programmed not only to reduce contact with males, they are also reluctant to accept sperm from any given male. They are certainly more reluctant to accept the male's sperm than the male is to ejaculate them. Although a male suffers some disadvantage through wasted time and energy if he ejaculates his sperm near to the eggs of a female that is in some way suboptimal (e.g. wrong species; infertile), his mistake is far less critical than a similar mistake by a female. If a female makes the mistake of allowing her eggs to be fertilized by a suboptimal male, her lifetime reproductive success is reduced (perhaps even to zero if she only has one sexual contact in her life and that contact is with a male who is infertile).

ment and the use of the term is rather limited. We prefer the term polyandry to promiscuity.

A prostitute is a female who allows a male to inseminate her in exchange for the donation of some resource. The term prostitution normally implies a high level of polyandry on the part of the female. At least biologically, however, even a monandrous female is a prostitute if she receives some resource from the male in exchange for sexual access.

(a) LEVELS OF POLYANDRY IN HUMANS

Human females may be more coy and cautious over sexual matters than human males. Nevertheless, in their lifetime, most human females are inseminated by more than one male. Our own nationwide survey (Box 2.6) of the sexual behaviour of females in Britain in 1989 is one of the largest ($n = 3679$) and in many ways the most detailed so far. Particularly striking is the difference between mean and median values for number of different males by which females claim to have been inseminated at different stages of their sexual life (Box 2.6). Such a difference between means and medians indicates a strong skew to the data; in this case a proportion of women who are highly polyandrous.

Relatively few females (7.4%) claimed to have been monandrous during their first thousand copulations. After 3000 copulations, the proportion claiming to have been inseminated by over 100 different males was as high as

Box 2.8 Polyandrous females: modes of sexual activity

Generally, females need relatively few matings, perhaps only one from one (fertile) male, to ensure fertilization and, on the whole, the females of most species are more cautious and less urgent than their conspecific males. As a general principle, coy females have maximal multiplication power because they make optimal use of time and energy and are less likely to commit their gametes to suboptimal males. Level of coyness, however, varies considerably between species and in most lineages female sexual activity seems to be greater than the level necessary for fertilization. Many circumstances have now been identified in which females benefit from an increase in level of sexual activity, both in terms of frequent mating with a particular male partner and polyandry.

In most cases the gain that females make from repeated copulation (with the same male or different males) is in time and energy. In effect, females exchange sexual access for resources. These resources may be obvious, such as food for herself or her offspring, or less obvious, such as being allowed to live near to the male, protected from predators, protected from harassment by other males, or even exposed to less risk that the male will behave aggressively toward her or her offspring.

In any particular lineage, we expect selection to fix the level of female sexual activity, both with the same and different males, at the optimum trade-off between: (1) savings in time and energy through the resources obtained from the males allowed sexual access; and (2) the loss of time and energy both directly through the excess copulations and indirectly through the aggressive interactions that may result from polyandry.

In some situations there may be more than one modal optimum. Such populations may consist of: (a) some individuals programmed for one level of polyandry, other individuals for the other; (b) all individuals programmed to respond to circumstance with the most appropriate polyandrous mode; or (c) a mixture of (a) and (b). As with balanced polymorphism (Chapter 5), the relative frequency of females showing the different modal levels of polyandry should stabilize evolutionarily at the level at which individuals at the different modes experience similar reproductive success.

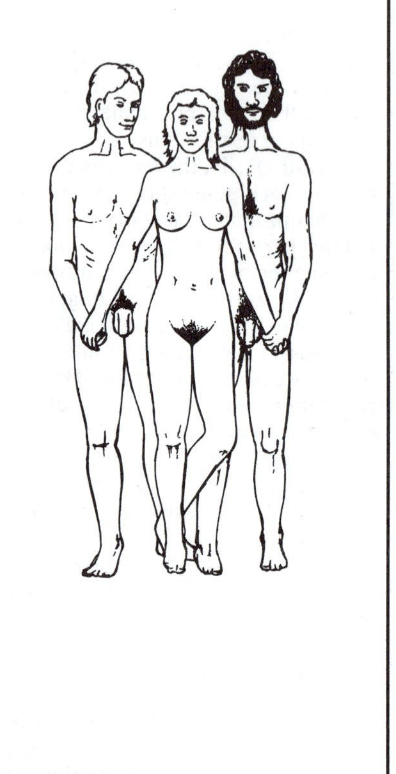

the proportion claiming monandry (5.6%). It is the presence in the population of this subset of very polyandrous females that generates the difference between means and medians. Of the 14 females (0.4% of the total sample of 3384 who provided all necessary information) who claimed to have had more than 100 lifetime partners, 12 claimed or implied that their high level of polyandry was due to prostitution.

(b) PROSTITUTION

The presence of female prostitutes is an almost universal feature of human societies. Out of 300 human societies coded in the Human Relations Area Files, only 12 (4%) were claimed not to contain any prostitutes (Smith, 1984b). The remaining 96% acknowledged their presence. Even so, it is difficult to gauge what proportion of females in any particular society ever allow a male to inseminate them purely in exchange for a resource. Estimates of the number of overt prostitutes at any one time range from our low figure of 0.4% in Britain in the late 1980s to a figure of 25% of women in Addis Ababa, Ethiopia in 1974 (Dirasse, 1978). However, such figures can never be wholly reliable and in most cases are probably underestimates, particularly of the proportion of women who ever engage in prostitution during their lifetime.

In part, the problem is in deciding what constitutes prostitution and what does not. In many ways, prostitution is simply the least ambiguous exchange of copulation for resources (or 'gifts') (Symons, 1979). Throughout human cultures, there are exam-

Box 2.9 Minimum time intervals between successive copulations with two different males by females at different levels of sexual experience

Sexual experience is categorized as number of lifetime copulations (so far). Sample sizes on histograms are the total number of females in each category of sexual experience. Monandrous females (at back of figure) are those who claim only ever to have been inseminated by one male in their life (so far). Data from UK Nationwide survey (Box 2.6).

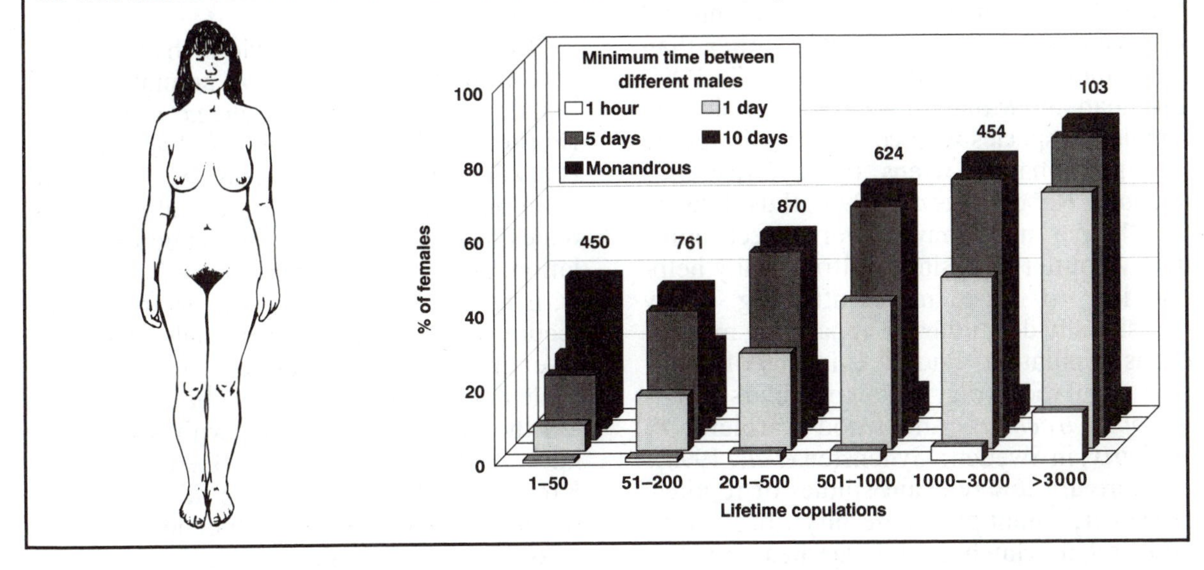

ples of males giving the female a 'gift' around the time of copulation without the female necessarily being designated a prostitute. Often the exchange is ritualized in some way, as during marriage ceremonies. Even on the wedding night money may be given to the bride before copulation can take place (Junod, 1962).

There is a strong sexual asymmetry in the giving of such gifts. Symons (1979) reviewed the relationship between gifts and sexual relationships across cultures. Out of 20 societies, only the male gave a gift in 15, the male gave a bigger gift in two, and male and female gave equal gifts in three. In no society did the female give the larger or only gift.

The overall impression is that there are degrees of prostitution and it is difficult to know where to draw the line between a traditional prostitute exchanging insemination for money and a female in a long-term partnership exchanging insemination for support, protection and 'gifts'.

As far as this book is concerned, the female's conscious or subconscious motivation for polyandry is relatively unimportant. The important factors are how many females have how many partners (as in Box 2.6), how long is the time interval between inseminations by different partners (Box 2.9) and what gain does the female obtain from her actions, conscious or otherwise.

(c) POLYANDRY IN OTHER ANIMALS

Given the difficulty encountered in trying to determine how many males inseminate a human female in her lifetime, it is not surprising that it is difficult to make any comparable assessment for other animals. The expectation is, however, that over the next few years, DNA fingerprinting studies will reveal a wealth of information on polyandry. For example, a preliminary study of Japanese macaques, *Macaca fuscata*, showed that between 1974 and 1988, only one female out of 13 gave birth to offspring sired by a single male (Inoue *et al.*, 1990). Each of the remaining 12 females produced offspring sired by two or more males. Studies of twins produced by noctule bats, *Nyctalus noctula*, detected several cases of half sibs (Mayer, 1993).

For other vertebrates, the data available

are only fragmentary. However, they also give the impression that very few species are dominated by monandrous females.

Superficially monogamous birds, which were once thought to pair for life and to be totally faithful, are now being found to be polyandrous, females cryptically seeking and achieving copulations with males other than their main partner (McKinney *et al.*, 1984; Birkhead and Møller, 1992). So, too, are the females of species with cooperative breeding (e.g. stripe-backed wrens, *Campylorhynchus nuchalis*; Rabenold *et al.*, 1990), thus showing that 'helper' males have a more direct reproductive outlet than simply altruistically helping kin to raise more offspring. Only occasionally do studies of apparently monandrous populations find no evidence of extra-pair copulation (e.g. willow warblers, *Phylloscopus trochilus*, and wood warblers, *P. sibilatrix*, in Sweden; Gyllensten *et al.*, 1990).

There are few relevant studies of reptiles. However, multiple paternity has been detected in clutches of loggerhead turtles, *Caretta caretta*, wood turtles, *Clemmys insculpta*, and snapping turtles, *Chelydra serpentina* (Galbraith, 1993).

The females of some insects seem often to be monandrous (e.g. speckled wood butterfly, *Pararge aegeria*, Svärd, 1985) but many, perhaps the majority, are highly polyandrous. Levels of polyandry are less well known in species for which fertilization occurs outside the female's body. However, there are probably few species in which monandry is absolute.

(d) THE ADVANTAGE OF POLYANDRY: RESOURCES

In many animals, the polyandry is clearly prostitution, food (for the female or her offspring) being the usual resource offered by the male in exchange for insemination. Thus, some female insects are given a meal by the male, either in the form of a prey item which the female eats while being inseminated by the male (e.g. hanging flies, *Hylobittacus apicalis*; Thornhill, 1976), or in the form of seminal fluids and excess sperm which the female digests after insemination (e.g. cave bat bugs, *Xylocoris maculipennis*; Carayon, 1974). The male mormon cricket, *Anabrus simplex*, produces a large spermatophore which the female eats. Females compete for the males and males pick the largest females which are also the most fecund (Gwynne, 1981).

Other resources less obvious than food are also traded by a range of animals. Thus, female dung flies, *Scatophaga stercoraria*, (Parker, 1970b) and baboons, *Papio anubis* (Packer, 1977) avoid excessive harassment from males by soliciting a particular male or males to protect them in exchange for copulation. A female dunnock, *Prunella modularis*, is more likely to receive assistance in the feeding of her fledglings from males who have inseminated her at least once (Davies, 1985). Male lions, *Panthera leo*, may be less likely to kill the offspring of females with whom they have copulated previously (Bertram, 1975; Krebs and Davies, 1993). In all such cases, the more polyandrous the female, the more resource donations she receives.

As in humans, prostitution offers the opportunity for females of other animals to exploit the male offering the resource. The male purple-throated carib hummingbird, *Eulamphis jugularis*, allows females to forage in its patch of nectar-producing flowers in exchange for copulation. Unfortunately for the male, these copulations do not often appear to lead to fertilization (Wolf, 1975). Male sepsid flies guard the female from harassment by other males while she is ovipositing (in dung) in exchange for copulation when she has finished. Often, however, the female escapes before the male has an opportunity to copulate (Parker, 1972).

(e) THE ADVANTAGE OF POLYANDRY: 'GOOD GENES'

The resource benefits accrued by female animals through being polyandrous are relatively clear-cut and generally accepted. Less clear-cut and far more contentious is the suggestion that females might gain some genetic benefits from copulating with several males (see the critical consideration by Parker, 1992).

Increased genetic variance in the female's

offspring as a result of polyandry might allow them to succeed in more varied habitats (but perhaps succeed less well in any single habitat). Alternatively, by having varied offspring sired by various males, a female might: (1) hedge against the risk of a disadvantage genetic 'match' between herself and her partner; and/or (2) increase the chances of a high quality, very successful genetic 'match' in at least a proportion of her offspring. On average, of course, females do no better by such 'bet-hedging' than by randomly selecting a partner. Any individual female, however, may through polyandry reduce the chances of her lineage becoming extinct in the next generation.

A female who 'bet-hedges' should never be as successful as a female who can identify and be inseminated by a male who is genetically 'the best'. However, if in each generation females all tend to select the same 'best' male, the residual genetic variation in the male characteristics used by the females for selection should soon disappear, and disappear more completely than seems to be the case (Parker, 1992). Perhaps, however, as long as: (1) there is selection on females to vary; (2) females select the male who is 'best for them' (i.e. genetically the most complementary; in other words best for their joint production of successful offspring); and/or (3) the male is 'best' only in terms of his genetic ability to avoid or resist biological entities, such as parasites, which can themselves adapt; then perhaps loss of variation is less of a problem.

Whatever the theoretical explanation, there does seem to be fairly clear evidence (including for humans; Chapter 6) that some elements of female polyandry are linked to genetic benefits. For birds, there is even some evidence that the females in an area tend to agree on which male is 'best'. In blue tits, *Parus caeruleus*, females paired to high quality males do not seek extra-pair copulations but females paired to lower quality males do so (Kempenaers *et al.*, 1992; Chapter 6).

Most paternity studies of birds seem to show that extra-pair males father only a proportion of the brood, the rest being fathered by the female's partner. Only rarely are there more than two fathers for a particular brood, suggesting that the female gives some paternity to her partner (to encourage paternal care) and the rest to the male in the vicinity she assesses to be of highest quality (Birkhead, 1993).

An exception is the aquatic warbler, *Acrocephalus paludicola* (Schulze-Hagen *et al.* in Birkhead, 1993). In this species, the male defends a territory in which the female and her young will nest and feed but provides no paternal care. Nearly half of 18 broods genetically fingerprinted were fathered by three or four different males, suggesting that in this species at least the female opts for variance in her offspring rather than simply for 'good genes'.

2.4.2 Sperm competition and double-mating

Sperm competition is the competition between sperm from different males for the 'prize' of the fertilization of the egg(s) produced by a single female (Parker, 1970a). It is, therefore, interejaculate competition, not intra-ejaculate (Parker, 1993). In species in which eggs are fertilized while still inside the female's body, sperm competition only occurs if the female shows a particular type of behaviour: double-mating. In the context of sperm competition, double-mating is a specific term meaning the mating of a female with one male while still containing in her reproductive tract competitive sperm from one (or more) different males. Double-mating thus results in the presence of competitive sperm from two (or more) different males inside the reproductive tract of a single female.

Modern female humans are polyandrous (Box 2.6) but, of course, not all polyandrous inseminations are double-matings. For sperm competition to occur, a female must be inseminated by two different males within the competitive lifetime of sperm from the first male (Chapter 8).

That some human females do from time to time contain sperm from more than one male is beyond question. The most graphic illustrations are provided by females who produce dizygotic (= fraternal) twins fathered by different males. Appreciation of such a phenomenon has a long history. According to Greek myth, Castor and Pollux were separately fathered by their mother's mortal husband and by Zeus (Bowen-Jones, 1992). The first real example to be documented, however, involved a woman who, in 1810, produced one black and one white (Archer, 1810). Since then a number of cases have been reported ranging from the obvious, as in this last example, to those in which the truth was only discovered by detailed tests (Shearer, 1978; Terasaki *et al.*, 1978; Bowen-Jones, 1992). In all such cases, assuming the female only contained sperm from two males, the result of sperm competition was a tie, 1:1!

Smith (1984b), in his classic paper, pointed out that a variety of human behaviour patterns would lead to sperm competition. He listed: communal sex; forced copulation; prostitution; adolescent promiscuity; and facultative polyandry (or infidelity). Of these, he considered infidelity to be the most important and discussed paternal discrepancy (children not sired by their putative father) and legal paternity suits as indicators of double-mating. Smith (1984b) concluded that sperm competition was a widespread feature of modern human populations and had been even more prevalent in the past.

Discussion of the possible levels of sperm competition in modern human populations is delayed until Chapter 8. Here we restrict ourselves to the results of our own survey (Box 2.6) on the shortest time interval between inseminations by different males reported by women at different stages of sexual experience (Box 2.9). For example, after 500 lifetime copulations, 1 in 200 females claim that on at least one occasion they have been inseminated by two different males within 30 minutes of each other. Around 30% claim to have been so inseminated within 24 hours. Median minimum time interval between inseminations by different males falls from 96 hours during the first 200–500 lifetime inseminations to 24 hours after 1000–3000.

The antiquity of sperm competition in the human lineage is probably as great as that of sperm themselves (say 800 million years). Parker (1982) points out that the evolutionary consolidation of anisogamy following the initial dichotomy into large and small gametes (Box 2.4) was itself a phase of sperm competition. As both small and large gametes competed for the fertilization of the larger gametes, it was the advantage of producing more, and hence smaller, gametes to outcompete rivals that provided the impetus for the evolution of sperm that are tiny relative to the egg.

Typically an external fertilizer such as a sea urchin has a sperm with a volume of 5 μm^3 and an egg with a volume of 1 000 000 μm^3 (a ratio of 1:200 000). Even the animal with perhaps the largest known sperm, the hemipteran insect, *Notonecta glauca*, with a sperm volume of 15 000 μm^3, has an egg 28 000 times larger with a volume of 420 000 000 μm^3 (Baccetti and Afzelius, 1976). In humans a sperm has an average volume of 17 μm^3 (Mann and Lutwak-Mann 1981) and an egg around 523 000 μm^3 (calculated from Alberts *et al.*, 1983). The ratio is thus 1:30 800.

Sperm competition is now recognized as a virtually universal feature of the reproductive behaviour of animals. In species in which the eggs are fertilized outside the female's body (e.g. many invertebrates, many fish, amphibians) sperm competition results from several males releasing their sperm in the vicinity of the spawning female (Breder and Rosen, 1966). In species in which the eggs are fertilized while inside the female's body, sperm competition results from polyandry and double-mating. Sperm competition was first recognized as a major feature of reproduction for insects (Parker, 1970a). Since then, its near universality has been recognized on many occasions (Smith, 1984a). The only lineages in which sperm competition between different males has ceased to be a factor in reproductive behaviour are those few species of self-fertilizing hermaphrodites, e.g. the nematode, *Caenorhabditis elegans* (Godfray and Harvey, 1991) though even in this species, hermaphrodites may mate with rare males which are <0.5% of the population, thus occasionally generating sperm competition (Van Voorhies, 1992).

The theoretical framework of sperm competition is described later in section 2.5.

2.4.3 Spermatogenesis and male sperm storage organs

In males, a sperm storage organ is a region specialized for the location and healthy maintenance of sperm between the time that they achieve maturity and ejaculation.

The reproductive tract of human males is shown in Box 2.10. Sperm are produced from the germ cells of the paired testes. One estimate of rate of production is about 84 million per day in mongoloids and about 185–253 million per day in caucasoids (Short, 1984). Higher rates (c. 300 million/day) have been reported by other authors (Johnson *et al.*, 1980; Neaves *et al.*, 1984). Spermatogenesis within the seminiferous tubules requires a sequence of mitotic and meiotic divisions followed by development around the genetic material of a transport system, the sperm (Box 2.10). This process requires 64 days in humans (Roosen-Runge, 1969; Johnson and Everitt, 1988) compared with around 3 days in a jelly fish, 20 days in a silk moth and 38 days in a rat (Roosen-Runge, 1969).

On completion of this process in humans, sperm are released into the lumen of the seminiferous tubules to begin their journey down about 40 cm of male tract. Fluid produced by the Sertoli cells (in the seminiferous tubules) washes the sperm through the rete testis. As sperm pass through the vas efferentia and epididymis, fluid is absorbed and the sperm mature and, in the epididymis, become increasingly capable of movement and fertilization. Passage takes 6–12 days, the sperm eventually arriving for storage primarily in the cauda epididymis and vas deferens (Johnson and Everitt, 1988). Here they remain until ejaculation. For humans, therefore, the total time from beginning of spermatogenesis to a sperm being ready for ejaculation is 70–76 days.

At least for conceptual convenience, the developing sperm may be envisaged as on a conveyor belt with younger immature sperm further back along the track and older mature sperm in the cauda epididymis and vas deferens waiting for ejaculation. Although some mixing may occur within the cauda and vas deferens, it seems likely that older sperm are nearer to the urethra waiting for ejaculation with increasingly young sperm in store back through to the epididymis.

The maximum number of sperm in the human cauda and vas deferens at any one time probably averages about 700 million sperm (Harvey and Mays 1989). The environments within the epididymis and vas deferens are hospitable and nutritive. Nevertheless, presumed older sperm are often shed in the urine (Mann and Lutwak-Mann, 1981; Johnson and Everitt, 1988) and others may be broken down and re-absorbed by lymphocytes via phagocytosis (Mann and Lutwak-Mann, 1981; Barratt and Cohen, 1986; Tomlinson *et al.*, 1992).

Male sperm storage organs of one form or another are virtually universal among animals and, to a large extent, probably owe their existence and size to the pressure of sperm competition. Despite some constraint, the risk of sperm competition has driven the evolution of males who ejaculate vast numbers of sperm at every opportunity of fertilization (Section 2.5). Ejaculation of huge numbers of sperm requires not only an appropriate rate of sperm manufacture, it also requires a site in which the sperm can accumulate in appropriate numbers until ejaculation. If the sperm are to remain viable during storage, the site must be hospitable. As long as the advantage via sperm competition of accumulating sperm for ejaculation is greater than the cost of developing and maintaining such a site, including any cost of maintaining the sperm, male sperm storage organs will evolve.

Males of the human lineage have probably had sperm storage organs of one form or another since at least early in the evolution of metazoans (say 800 million years). They probably did not assume their modern form, however, until the transition to mammalian (200 million years ago), even eutherian, characteristics (100 million years ago; Box 2.1).

2.4.4 Mate-guarding by males

Mate-guarding by males refers specifically to any mechanism by which one male attempts to prevent other males from having sexual

Box 2.10 **Sperm production and storage (In part adapted from Johnson and Everitt, 1988)**

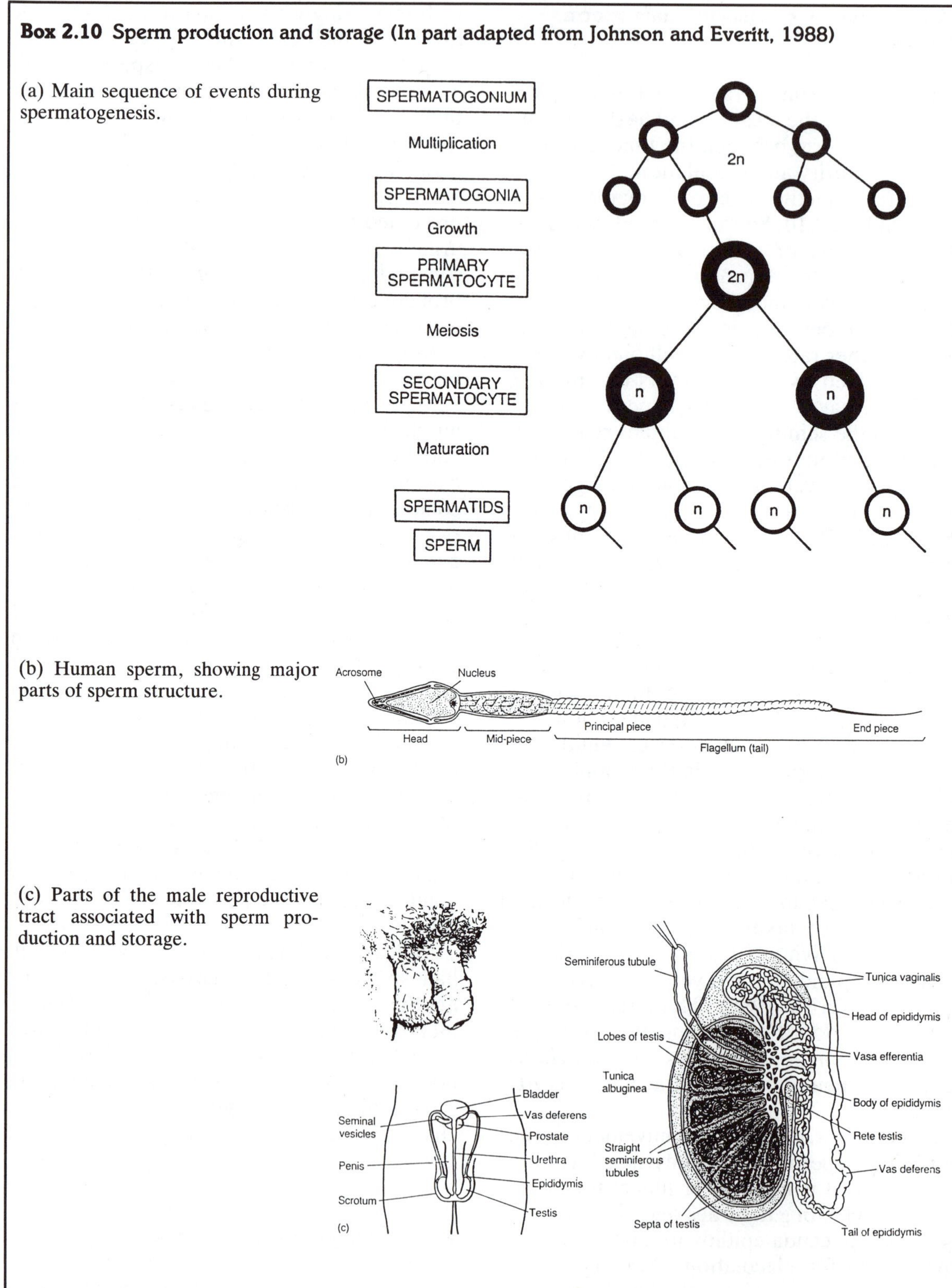

(a) Main sequence of events during spermatogenesis.

(b) Human sperm, showing major parts of sperm structure.

(c) Parts of the male reproductive tract associated with sperm production and storage.

Box 2.11 The efficiency and evolution of mate-guarding by males

Efficiency

Mate-guarding has been shown to reduce the probability of sperm competition in mammals (e.g. Idaho ground squirrels, *Spermophilus brunneus*; Sherman, 1989a). In humans, the less time that a male spends with his female partner between copulations, the more likely her last copulation is to have been with another male. Sample sizes are the number of females in each category.

(Data from UK nationwide survey, Box 2.6; drawn from tabulation in Baker and Bellis, 1993a.)

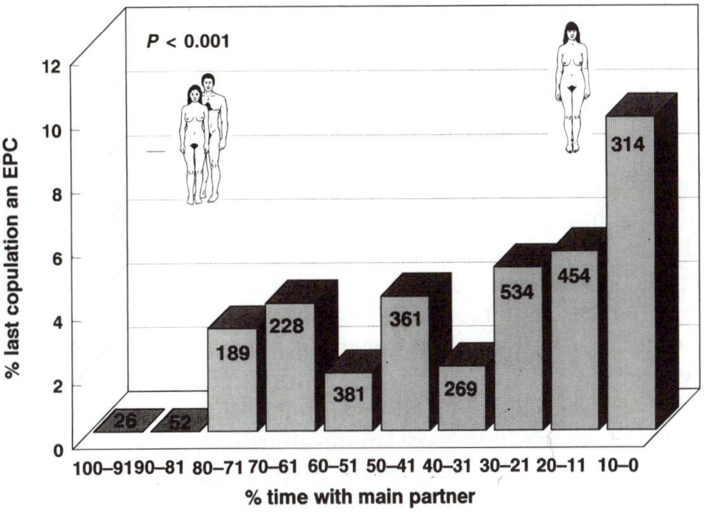

Evolution

Numerically, populations come to be dominated by males who have the same heritable characteristics that allowed their ancestors to fertilize more eggs than their contemporary male rivals. At first sight, the easiest way to achieve such success might seem to be to spend as little time with each female as possible before moving on to the next. However, although such a strategy invariably leads to a male ejaculating sperm in the vicinity of the eggs of as many females as possible, it does not necessarily lead to maximal fertilization.

This is especially so once populations contain many males pursuing the same urgent strategy. By spending minimal time with a female, males run the risk that another male will arrive and in some way supersede his sperm before the female's eggs have successfully been fertilized. In some circumstances, therefore, the male may fertilize more eggs in his lifetime if he stays longer with each female and attempts in some way (e.g. by guarding her from other males) to ensure that it is his sperm which fertilize her eggs. Although this may decrease the number of females he ever contacts, it will be advantageous if the extra attentiveness has a sufficiently great influence in increasing the number of eggs he fertilizes from each female (e.g. if he increases the number of eggs he fertilizes per female by 50% but decreases the number of females with which he mates by only 20%).

Any behaviour on the part of the male that restricts the access of other males to females may maximize the male's reproductive success. Thus, preventing males from access to particular living space containing females qualifies as mate-guarding in this context. Guarding the female from other males *before* the male ejaculates as well as after also qualifies.

access to a particular female. The reason for the evolution of such behaviour is relatively clear (Box 2.11). The behaviour is aimed at reducing the chances (ideally to zero) that the male's sperm will find themselves in competition with sperm from another male. As such, therefore, the behaviour is an evolutionary response to the risk of sperm competition.

The majority of male humans attempt to prevent other males from having sexual access to their female partner(s). They do this primarily by close association with, and observation of, their partner(s) and by behaving aggressively to the female(s) and/or the other male if there are any signs of sexual interest. In some societies, males were so attentive that they timed female absences, even when she was urinating or defaecating (Fortune, 1963). In a Caribbean village, mate guarding was most prevalent when females were most fertile (Flinn, 1988).

Of 849 human societies, only four (0.5%) did not show male mate-guarding (Murdock, 1967). More detailed analysis of 110 societies (Broude and Green, 1976) found intensive mate-guarding in 67 (60.9%) and essentially no mate-guarding in four (3.6%). In the remainder (35.5%), mate-guarding occurred but was relaxed on occasions when the male gained some benefit. Usually, relaxation of mate-guarding was part of a reciprocal arrangement with the other male (each having access to the other's female(s)). Often, such reciprocation was restricted to relatives (e.g. brothers).

In our UK nationwide survey (Box 2.6), we asked females: (1) 'Do you have a "main" male sexual partner at present?' (and, if yes) 'Roughly how many hours a week do you spend with him (include sleeping time)?'; and (2) 'If you have a main male partner, was . . . (your last sexual intercourse) . . . with him?' The probability that a female's last copulation was with a male other than her main partner was virtually zero if the male spent more than 80% of his time with her (Baker and Bellis, 1993a). At lower levels of mate-guarding, however, the more time the male spent away from his partner, the greater the probability that her last copulation was with some other male (Box 2.11). The correlation was significant whether the whole data set was used or analysis was restricted to copulations not involving contraception (Baker and Bellis, 1993a).

We conclude from the literature that some form of mate-guarding behaviour is a nearly universal feature of the sexual programming of modern male humans. Moreover, from our own study mate-guarding appears to be effective. The risk that a male's partner will be inseminated by another male is a direct function of the proportion of time that the male spends mate-guarding.

In other animals, mate-guarding by males is a common, but not universal, phenomenon. In many invertebrates, males release or inseminate their sperm near to a female's egg(s) and then move on immediately to search for another female with no guarding behaviour other than during gamete release. In other invertebrates, however, males show some degree of precopulatory (e.g. sepsid flies, *Sepsis cynipsea*, Parker, 1972) and/or postcopulatory (e.g. dung flies, Parker, 1970b; beetles *Tenebrio molitor*, Gage and Baker, 1991) guarding behaviour.

In vertebrates, mate-guarding in one form or another is the rule rather than the exception. Often (e.g. lizards, rodents) males guard a home range that overlaps the home ranges of one or several females and thus hinder access by other males. In others, males gain and/or retain access to a female or females by defending an area which is attractive to the female in that it offers resources for producing and/or raising young (Krebs and Davies, 1987). Again, at least while females are in that area, other males are denied sexual access. Examples are: male sticklebacks which build and defend a nest to which females are attracted and in which they lay eggs (Morris, 1972); song birds which defend a territory to which females are attracted (Krebs and Davies, 1993); and elephant seals which defend a stretch of beach onto which the females emerge to give birth (Le Boeuf, 1972; McCann, 1981).

Often in conjunction with the defence of living space, vertebrate males consort closely with a female at times of peak female fertility. Thus, male black bears, *Ursus americanus*, defend a home range that overlaps those of several females and associate with each female when she is most likely to conceive (Jonkel and Cowan, 1971). Males of many group living primates (e.g. lemurs) consort closely with the female at times that she appears to be most fertile (*Lemur catta*; Sauther, 1991).

It is difficult to guess at the antiquity of male mate-guarding in the human lineage. It has almost certainly been a facet of sexuality in the lineage since its transition to vertebrate characteristics (550 million years; Box 2.1) but is probably no older than the first mobile metazoans (800 million years). It probably did not take on its modern form, however, until the transition to primate (65 million years ago) or even hominid (5 million years ago) characteristics (Box 2.1).

2.5 Sperm competition theory

Sperm competition theory was first developed in a classical series of papers by Parker (1970a, 1982, 1984, 1990a,b) and is discussed in the context of Human Sperm Competition in Chapters 8–12. This section simply introduces the basic principles and main areas of contention.

2.5.1 The lottery principle

According to Parker, sperm competition is a lottery. The more sperm a male produces, the greater the chances that his sperm will outcompete the sperm from other males and thus have more chance of fertilizing, and/or will fertilize more of, a female's egg(s).

There is some direct evidence for the lottery hypothesis. Artificial insemination studies of chickens (using sperm from males genetically different for plumage) have shown that the more sperm are inseminated from a particular male, the greater his success in sperm competition (Martin *et al.*, 1974). However, even here it must be pointed out that males differ in their fertility (Martin and Dzuik, 1977; Chapter 12). Thus, it is true that, for any given male, 50 million sperm will do better than 25 million sperm when competing against 50 million sperm from another male. However, it is not necessarily true that with 50 million sperm from both males each will fertilize the same number of eggs (Parrish and Foote, 1985).

2.5.2 Sperm size and competitiveness

Parker (1982, 1984) argued that the lottery factor in sperm competition favoured the evolutionary maintenance of tiny sperm. The relative sizes of eggs and sperm in most animals are grossly different (Section 2.4.2). Thus, even if a male produced sperm twice the normal size, the sperm would still make very little contribution after fertilization to the size, energy reserves and hence survival and development, of the zygote. Yet such a male would, all else being equal, produce an ejaculate with half the number of sperm and therefore with half the chances of success in sperm competition. Males who produced small sperm would thus be more successful in sperm competition than males who produced larger sperm. In lineages with a high risk of sperm competition, therefore, males who produced more numerous, tiny sperm would have greater reproductive success than males who produced fewer, larger sperm.

In apparent contradiction to Parker's prediction, species of primates and rodents (Gomendio and Roldan, 1991) and butterflies (Gage, unpublished) with a higher risk of sperm competition have males which produce larger sperm (see Chapter 11). Gomendio and Roldan suggest that larger sperm swim faster and are therefore more competitive. Even if larger size reduces the number of sperm ejaculated, the larger size could evolve if the advantage of enhanced competitiveness outweighed any 'lottery' disadvantage.

2.5.3 Sperm warfare and the Kamikaze Sperm Hypothesis

The idea that sperm competition may take the form of strategic warfare rather than a lottery was first mooted for insects by Sivinski (1980) and for butterflies and moths by Silberglied *et al.* (1984). In elaborating our 'Kamikaze Sperm Hypothesis', we developed the idea of sperm warfare as a general principle for sperm in all animals (Baker and Bellis, 1988, 1989a).

Briefly, the Kamikaze Sperm Hypothesis suggests that animal ejaculates consist of different types of sperm each programmed to carry out a specific function. Some, often very few, are 'egg-getters', programmed to attempt to fertilize the female's eggs. The remainder, often the vast majority, are programmed for a 'kamikaze' role. Instead of attempting to find and fertilize eggs themselves, their role is to reduce the chances that the egg will be fertilized by sperm from any other male.

We envisage two primary categories of kamikaze sperm: 'blockers' and 'seek-and-destroy'. Blockers take up strategic positions en route to the egg, become relatively immotile, and bar passage to any later sperm. Seek-and-destroy sperm roam around appropriate areas of the female tract, seeking out and attempting to incapacitate and/or destroy any sperm from a different male probably using the highly destructive proteolytic enzymes produced by their acrosomal complex (Allen *et al.*, 1974).

In species in which the eggs are fertilized outside the female's body (usually in water), blockers can only really take up a position on the outer surface of the egg itself (Baccetti and Afzelius, 1976). Similarly, seek-and-destroy sperm are most likely to be effective in the water or other liquid immediately surrounding the egg. This may occur especially when mechanisms are used by external fertilizers to concentrate gametes (e.g. fish such as the European bitterling, *Rhodeus amarus*, laying their eggs within mussels; Breder and Rosen, 1966).

In species in which the eggs are fertilized inside the female's body, however, different blockers may take up positions in any suitable constriction in the female tract. Equally, different types of seek-and-destroy sperm may be programmed to locate and patrol any suitable region of the female tract. The more complex the female tract, the more scope for different types of kamikaze sperm, each type programmed to locate and operate within a particular region of the tract.

2.5.4 Restraint in sperm ejaculation

Whether sperm competition is a lottery, a race, or warfare, it is likely that the more sperm a male enters for the competition, the greater his chances of winning (Parker, 1990b). The selective pressures generated by the risk of sperm competition would seem only, therefore, to favour males who ejaculate as many sperm as possible on every occasion. Yet there is now abundant evidence, at both the species (Harcourt *et al.*, 1981; Short, 1981; Svärd and Wiklund, 1989) and individual (Baker and Bellis, 1989b; Bellis *et al.*, 1990a; Gage and Baker, 1991; Gage, 1991; Simmons *et al.*, 1993) levels, that males adjust the number of sperm they ejaculate according to the risk of sperm competition. When the risk of sperm competition is low, males apparently restrain themselves in the number of sperm ejaculated, reserving maximum ejaculation for occasions when the risk of sperm competition is highest.

Such restraint over the number of sperm ejaculated when the risk of sperm competition is low implies that males suffer some disadvantage if they ejaculate too many sperm on any given occasion. Two main disadvantages have been suggested: (1) that the sperm and other constituents in an ejaculate are costly to produce (Dewsbury, 1982); and (2) that, in the absence of sperm competition, the more sperm a male ejaculates, the lower his chances of fertilizing the egg(s) of the current female (Baker and Bellis, 1993a). The relative importance of these two potential disadvantages to overejaculation are discussed in detail in Chapter 9.

2.5.5. How many sperm should be ejaculated?

The precise format of sperm competition and the nature of any disadvantage in ejaculating too many sperm are still very much open to discussion and experimentation. Whatever the final conclusion, however, the major prediction of sperm competition theory remains the same. The number of sperm a male should ejaculate on any given occasion is the optimum trade-off between two opposing factors: (1) an advantage in ejaculating more sperm to increase the chances of winning any sperm competition that might occur; and (2) an advantage in ejaculating fewer sperm (either because of sperm cost or because of a lower chance of fertilization) if sperm competition does not occur. The result of these opposing pressures should be that more sperm are ejaculated when the risk of sperm competition is high than when it is low.

2.5.6 Females should promote sperm competition

In Section 2.4.1 it was argued that females gain a number of advantages from polyandry. Those direct advantages that derive from environmental benefits (food, energy and protection) are the most clear cut; those long-term advantages from genetic benefits the most contentious (Parker, 1992). It is possible, however, that not only does female polyandry, in the form of double-mating, generate sperm competition but that, in its turn, sperm competition generates a further advantage in polyandry. Such models of female benefit have been termed 'sexy sperm' models (Curtsinger, 1991).

If the competitiveness of a male's sperm is

Box 2.12 Sexy sons, sexy daughters and handicaps: the successful child principle in the evolution of female sexual behaviour

Sexy sons
Whenever males differ in heritable characteristics that impart different males with different potential reproductive success, females should prefer to mate with males with the greater potential. The advantage of such a preference to the female is that her sons (and/or grandsons, etc.) will be more successful at siring grandchildren (or great grandchildren, etc.) than if the female had preferred a male with some other characteristic. Inevitably, therefore, modern populations are dominated numerically by females with a preference to mate with males who will give them the most fecund sons.

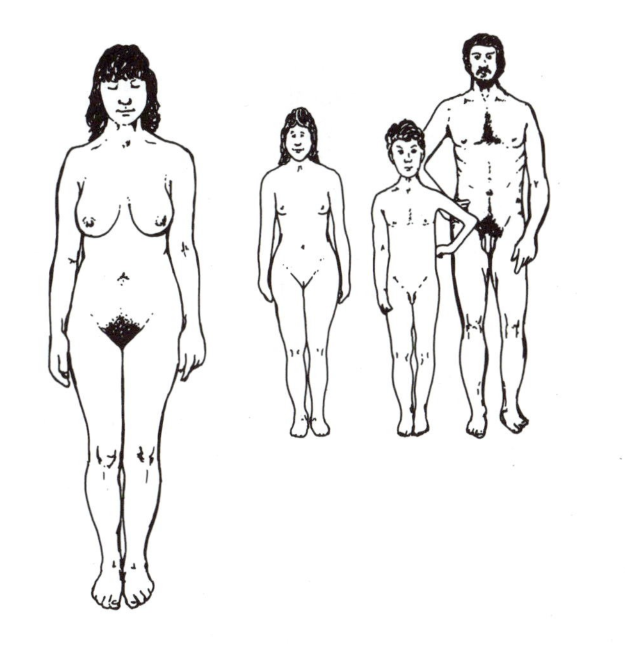

Handicaps
Perversely, one way that a male could advertise that he is a fit and desirable male with 'good genes' is to go through life with a handicap (e.g. long tail; bright plumage; Zahavi, 1975; Grafen, 1990; but see Baker and Parker, 1979). Only males who are otherwise very fit (both physically and genetically), perhaps free from parasites (Hamilton and Zuk, 1982), can afford to possess such a handicap.

Sexy daughters
A problem faced by the female when she chooses to mate with a male with a handicap is that her sexy sons inherit not only their father's good genes but also his handicap (Maynard Smith, 1976). If the handicap is sex-linked, however, the daughters of such fathers may gain the advantage of the good genes without the disadvantage of the handicap.

Both via sons (usually) and daughters (in handicap situations), females who mate with the fittest males should benefit through the greater success of their offspring.

heritable, females fertilized by the most competitive sperm will gain through the greater reproductive success of their sons (Box 2.12). If a female is unable to judge ejaculate quality from a male's appearance or behaviour, her best strategy is actively to promote sperm competition through double-mating (Smith, 1984b; Harvey and May, 1989; Bellis and Baker, 1990). Even an external fertilizer can promote sperm competition. By delaying the time from sperm arrival to sperm penetration, an egg can promote supernumeracy of sperm on the egg surface (e.g. sea urchins; Baccetti and Afzelius, 1976).

Promotion of sperm competition by females to select the males with the most competitive sperm has its parallel in veterinary science. A particularly sensitive method whereby breeders may select the most fertile males is, in effect, to allow the sperm to select themselves after artificial insemination of mixed sperm from different males (Robl and Dzuik, 1988).

There is some direct evidence that female animals can gain from promoting sperm competition. Thus, female adders, *Viperus berus*, like many other snakes, are able to store sperm throughout a breeding season or even longer. As a result, there should be competition between the sperm from all the males who inseminate a female over any given breeding season. Madsen *et al.* (1992) have found that female Swedish adders which copulate with more males during the breeding season give birth to a higher proportion of live young. According to Madsen *et al.*, female adders, by promoting sperm competition, are more likely to be fertilized by fitter sperm and will thus produce fitter offspring (though why more competitive sperm should produce more viable zygotes is not clear).

In this example, females were probably unable to judge the quality of each male's ejaculate without mating (unless, perhaps, they only allowed an additional male to mate if he was reliably assessed to have a 'better' ejaculate than previous males). For some other animals (e.g. sheep, Gibson and Jewell, 1982) there is direct evidence of female inability to judge the quality of a male's ejaculate. Such inability, however, need not be universal.

In the capercaillie, *Tetrao urogallus*, a large bird with a lek-mating system, females probably do mate preferentially with males with fitter ejaculates (Mjelstad, 1991). Males display on the leks. Females visit the leks to mate once or a few times, then depart to their nesting sites to raise the clutch without any further paternal involvement. Males who display most vigorously tend to achieve most matings and also tend to have a higher proportion of live and motile sperm. By mating with the most vigorous males, therefore, females should increase their chances of being inseminated with fertile sperm.

In this particular example, it is likely that the differences in ejaculate quality owe more to health than to genes. In humans, male subfertility is more often than not attributable to past or present infections (Chapters 3 and 7), and the same could be true for the capercaillie. Such infections can influence both the male's vitality and fertility, causing an association between the two that could be detectable to the female.

Differences in the competitiveness of sperm from different males (Chapter 12) may thus have environmental as well as genetic origins. Even so, a female who promotes sperm competition generally increases her chances of being fertilized by the male with the most competitive sperm. Whatever the origin of any difference in sperm competitiveness, the female can only benefit.

3 Legacies from the age of reptiles: copulation, flowback, and the female orgasm

3.1 Introduction

The earliest vertebrates in the human lineage (probably jawless fish some 500 million years ago; Box 3.1) were marine (Pough *et al.*, 1990). It is possible that the lineage next invaded freshwater before finally emerging onto the land, as protoamphibians, around 400 million years ago (Gee, 1991). As do modern amphibians, however, these earliest land-invading ancestors of modern humans were likely to have returned to water in order to reproduce. It was the advantage of a full-time life on land that stabilized one of the major changes in reproductive behaviour that characterizes humans today: internal fertilization.

As we argue in this chapter, the evolution of internal fertilization, and hence removal of the last barrier to the full-time invasion of land, may well have been driven by sperm competition. In its turn, internal fertilization itself changed the nature of sperm competition in the human lineage in a most dramatic way.

3.2 Internal fertilization and sexually transmitted diseases

Internal fertilization is the fertilization of eggs while still inside the female's body. It contrasts with external fertilization in which the eggs are fertilized after being shed by the female into the surrounding medium, usually water. The association of a male and female for external fertilization is referred to as spawning. When the male of external fertilizers clasps the female, the position is known as amplexus.

A sexually transmitted disease (STD) is a syndrome of symptoms that results from an infectious organism transmitted from one individual to another usually during a sexual encounter.

Humans are programmed physiologically and behaviourally for internal fertilization. With the exception of *in vitro* fertilization (Box 3.2), humans are known to develop only from fertilizations within the female's body. All mammals, birds and reptiles show internal fertilization. Most fish show external fertilization. The amphibians are intermediate between fish and reptiles. At one extreme, the vast majority of frogs are mainly external fertilizers (Van Tienhoven, 1968). At the other extreme, salamanders rely primarily on internal fertilization via the transfer of spermatophores (Pough *et al.*, 1990).

Even external fertilizers carry a range of infectious organisms which pass from one individual to another as males and females enter close proximity for spawning (e.g. the parasite *Mazocraes alosae* which lives on the

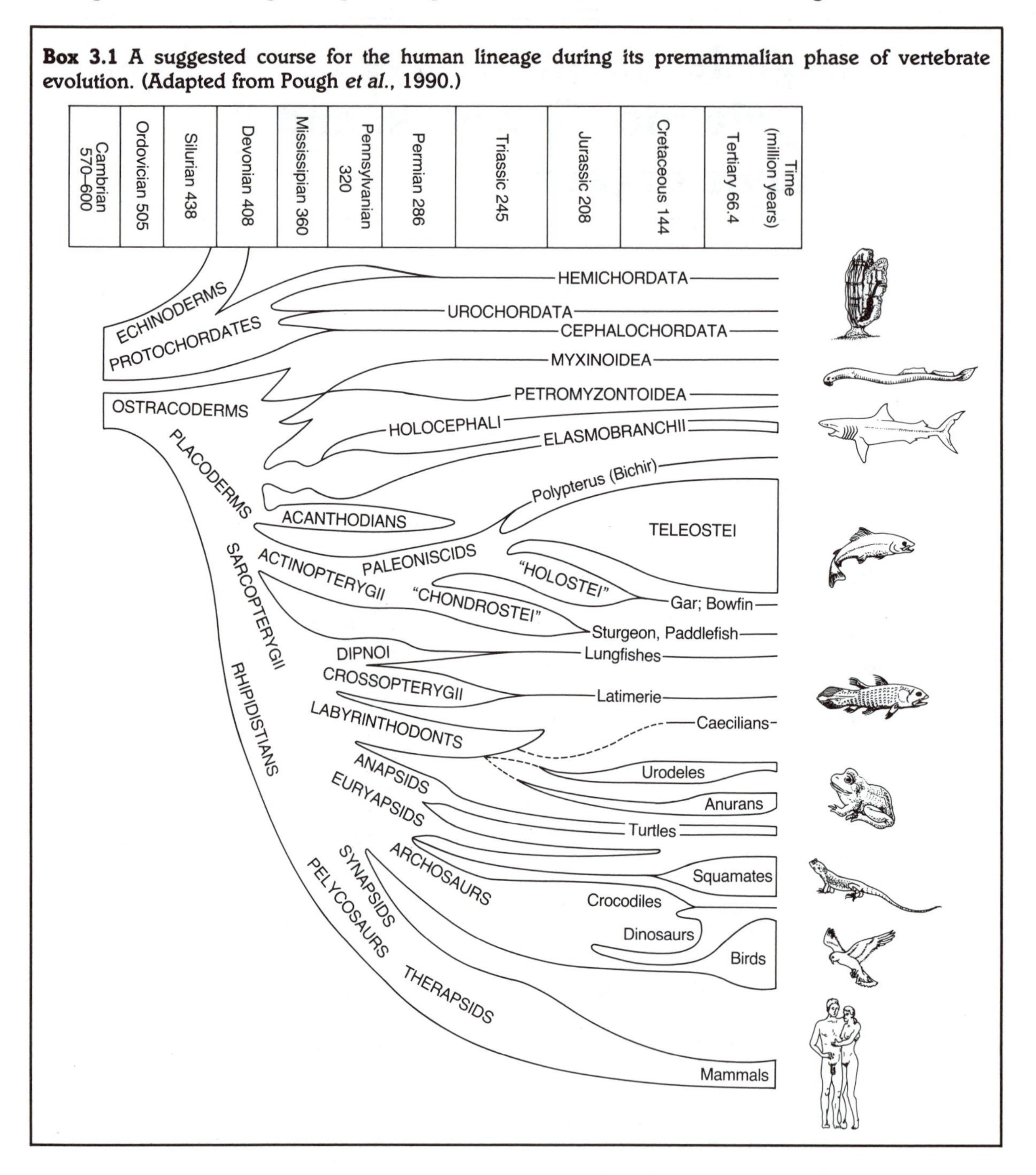

Box 3.1 A suggested course for the human lineage during its premammalian phase of vertebrate evolution. (Adapted from Pough *et al.*, 1990.)

gills of shad; Kennedy, 1975). Humans, in common with other internal fertilizers, suffer from a wide range of STDs caused by organisms as varied as viruses, bacteria, chlamydiae, mycoplasmas, fungi, protozoans and arthropods (Shanson, 1989).

Records of some of the STDs which inflict modern humans are as old as literature itself. There are references, attributed in the Old Testament to Moses, to the 'Botch of Egypt' (probably syphilis; Barlow, 1979) and 2400 years ago Hippocrates described 'strangury' (gonorrhoea; Parsons and Sommers, 1978). The first records of other STDs are more recent. The most notable of these, of course, is AIDS (acquired immune deficiency syn-

Box 3.2 ***In vitro* fertilization (IVF) and embryo transfer (ET)**

The foundations of current techniques were laid down in 1890 by Walter Heapes 'the patron saint of embryo transfer' (Betteridge, 1981). By transferring embryos from one rabbit to another, Heapes established that post-fertilization embryos could be placed in a female reproductive tract and still lead to pregnancy and birth. Successful mammalian *in vitro* fertilization (IVF; the fertilization of ova while outside the eventual mother) probably dates back to initial experiments with rabbit ova by Dauzier *et al.* (1954). The amalgamation of both IVF and embryo transfer to produce a successful birth from an embryo formed *in vitro* was finally accomplished, again in rabbits, by Chang (1959).

Human pregnancies from IVF techniques were first reported in the 1970s (de Kretzer *et al.*, 1973; Steptoe and Edwards, 1976, 1978). Although human IVF was initially developed for the treatment of female tubal infertility due to blocked fallopian tubes, the technique has now been shown to be useful in the treatment of male infertility (as in oligozoospermia) (Matson *et al.*, 1989).

Human IVF programmes often differ in methodology between clinics (Craft, 1984). Usually, however, females are maintained on a drug-induced ovulatory cycle while further drugs are used to promote superovulation, thus ensuring that a large number of ova will be available for fertilization. When ripe, eggs are removed from their follicles on the ovary surface by rupturing the follicle with a needle and sucking the egg into the needle's lumen. Eggs are usually fertilized in specially prepared culture media (not always in tubes; often in Petri dishes) using sperm which have been washed of all seminal fluids and resuspended in a similar medium (Craft, 1984).

Even if 10 or more eggs are harvested and fertilized, usually only about three are placed in the uterus. Although the extra eggs may increase the chance of successful pregnancy, they inescapably also dramatically increase the risk of multiple births.

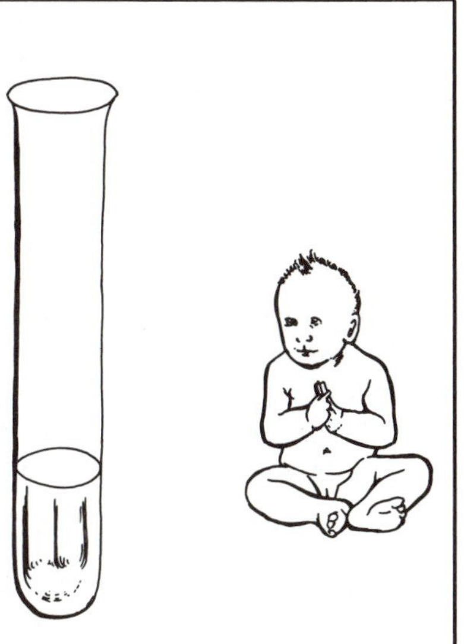

Multiple fetuses may in turn lead to premature delivery with associated mental and physical disabilities (Trounson, 1984). As many as 80–85% of patients may successfully have ova fertilized *in vitro*. However, only 20%–25% will achieve a clinical pregnancy and a number of these will suffer a spontaneous abortion (Craft, 1984).

drome), the first record for which is of a British sailor who died in Manchester, UK, in 1959.

Some STDs of humans are relatively benign, such as the external irritations caused by ectoparasites such as crablice, *Phthirus pubis*. Others are much more serious, potentially leading to infertility and/or death (e.g. genital warts, cervical cancer and AIDS (viral); gonorrhoea and syphilis (bacterial); and non-specific urethritis (often chlamydial)) (Parsons and Sommers, 1978).

The move to internal fertilization must have been a major factor in opening up a direct pathway between hosts for such arrays of parasitic organisms. STDs require neither independent mobility, a vector species, nor even a suitable external macroclimate to allow transmission. For example, the tropical disease yaws is caused by a close relative of syphilis. Transmission of yaws depends on skin contact between host individuals but needs a tropical macroenvironment. Its geographical spread has therefore been limited. Syphilis, on the other hand, experiences no such restriction for it has substituted a tropical macroclimate with a warm, wet microclimate, the male and female reproductive tract (Cameron, 1956).

The evolution of internal fertilization may well have been driven by sperm competition (Parker, 1982). Even in external fertilizers, males shed their sperm as near to the eggs as possible as they emerge from the female. In fish, males places their vent as near to the female's vent as possible. In some of the most primitive of fish (e.g. lampreys) the male wraps his body around the female in order to approximate extruded sperm to the eggs (Sharman, 1976). In many vertebrate species, small males exploit spawning pairs by rushing in and shedding their own sperm around the newly laid eggs (e.g. the salmon, *Salmo salar*, Jones, 1959; four-spine stickleback, *Apeltes quadracus*, Rowland, 1979). In some amphibians, such as frogs and toads, males position themselves on the female's back (in amplexus) so that they can shower the eggs with sperm as the eggs emerge (Halliday and Verrell, 1984). Again, however, other males may exploit the vulnerability of the spawning couple and place their sperm in competition with those of the paired male (e.g. the leaf frog, *Phyllomedusa callidryas*, Pyburn, 1970).

Amplexus, as in frogs and toads, may allow males to place their sperm at a competitive advantage over males who shed their sperm from further away. However, any male who could contrive for his sperm to enter the female would be at an even greater advantage, especially if he could fertilize the eggs prior to their emergence. Thus, the risk of sperm competition may have generated a continuous pressure on males for internal fertilization in no matter what environment they lived. At each step, however, sperm competition could only drive the male to increasingly intimate association with the female as long as the gains from greater success in sperm competition outweighed the cost of increased risk of STD.

The fact that relatively few aquatic organisms evolved internal fertilization may be less a reflection of a lack of selective pressure on males than of counterselection on females to resist invasion by male cells. Such invasion by non-self cells not only opens up the risk of an invasion of STD organisms but also raises immunological difficulties (Cohen, 1984). The problem faced by the female is to fight STDs while still allowing sperm to survive long enough for fertilization. On this theoretical scenario, internal fertilization evolves in those lineages the females of which, for some reason, gain an advantage from internal fertilization that outweighs primarily the disadvantage of invasion by non-self cells (sperm and pathogens).

The difficulty of evolving gametes capable of being fertilized while either osmotically protected (in a freshwater medium) or resistant to desiccation (on land) may well have generated for females the necessary advantage of internal fertilization (Sharman, 1976). Or rather, put another way, the advantage of being able to exploit the rich but empty freshwater and terrestrial universe was sufficient to favour those females who allowed males to place sperm inside their reproductive tract, thus protecting the act of fertilization from the rigours of the external environment.

One factor in this sequence may have been selection for the female to protect the egg, once outside the female, from osmotic stress or desiccation. The evolutionary solution appears to have been the cleidoic (enclosed) egg, now typical of reptiles, birds and (in a modified form) mammals, which protects the developing embryo. However, enclosing the egg necessitates fertilization prior to laying down a protective coat.

STD organisms are themselves exposed to selection. As are all parasites, those STD organisms are most successful which persist and multiply in hosts providing the best opportunities for onward infection. It is presumably for this reason that many non-STD parasites seem to manipulate the behaviour of their hosts to increase the chances of transmission (Kennedy, 1975). For example, some internal parasites of birds and mammals alter the behaviour of their intermediate hosts (e.g. snails, insects) so that the host is more likely to be eaten by their final target host. STD organisms might be expected similarly to manipulate their host's behaviour to increase chances of transmission.

The spread of human STDs is known to be favoured both by an increase in mating rate with the same partner (e.g. syphilis) and by an increase in the number of partners (e.g. *Chlamydia trachomatis*). Indeed, a correlation between the incidence of a disease and number of sexual partners is often taken to indicate that the disease is sexually transmitted (e.g. cervical cancer; Slattery *et al.*, 1989). It follows that natural selection should have favoured those STDs which change their host's behaviour to increase the host's rate of sexual contact, both with the same and other partners.

It is already known that some human STDs change their host's behaviour. Neurosyphilis, for example, causes delusions of great wealth and power and generates unusual social behaviour (Swartz, 1984). However, whether neurosyphilis or any other human STD manipulates its host's behaviour and physiology to increase rate of sexual contact must remain a moot point. The fact that the above studies of *Chlamydia*, syphilis and cervical cancer show that infected people have had more sexual partners could be due to either (or both) of two forms of host/parasite interaction. On the one hand, people with more sexual contacts should be more likely to encounter and contract an STD. On the other hand, people who contract an STD may be manipulated by the pathogen into greater sexual activity. The only way to disentangle the relative importance of these two host/ parasite interactions is to carry out longitudinal studies of individuals and accurately to monitor sexual activity before and after contracting an STD. At present, we are unaware of any such study.

An indirect but powerful mechanism by which an STD could manipulate its host's behaviour is to interfere with fertility. In humans, STDs reduce fertility, hinder fertilization and induce spontaneous abortion after fertilization. At least half of all human infertility in the USA is probably due to STDs (Khatamee, 1988). Such infertility in the female is often due to blocked fallopian tubes and, all too often, the only medical option is *in vitro* fertilization (Box 3.2).

These influences on the host's fertility could be selectively advantageous to the pathogen in that they reduce the period of time that the host is sexually inactive. In humans, encounter rate with novel sexual partners is likely to be reduced after fertilization and females could also have periods of sexual inactivity due to the carrying or production of offspring.

The human lineage almost certainly finally evolved internal fertilization during the transition from amphibian to reptilian characteristics (around 300 million years ago; Pough *et al.*, 1990). Although some level of association with STD organisms will be even older, the new level of intimacy between males and females opened up new opportunities for infection, particularly internal infection. A whole new range of STD organisms are likely to have evolved and to have speciated with their hosts.

3.3 The male: copulation, seminal fluids and copulatory plugs

3.3.1 Intromittent organs and copulation

When internal fertilization involves the male placing the sperm directly into the female tract, the coupling of the male and female is referred to as copulation. The act of placing the sperm inside the female tract is insemination and the organ used by the male for insemination is an intromittent organ, in mammals known as a penis.

Male humans are equipped with a single penis. They are programmed physiologically and behaviourally for penile erection in anticipation of a sexual encounter and to copulate by inserting the penis into the female reproductive tract before ejaculating sperm and seminal fluids into the vagina (Mann and Lutwak-Mann, 1981).

Not all animals with internal fertilization have evolved a penis. Many animals including molluscs (e.g. squid), arthropods (e.g. scorpions) and some amphibians (e.g. newts) produce their sperm in fluid-filled packages, known as spermatophores (Mann, 1975). In some amphibians (e.g. smooth-mouthed salamander, *Ambystoma texanum*) a large number of spermatophores are deposited relatively indiscriminately and these are taken up by passing females who insert the package into their own tract (Garton, 1972). However, in many other urodele amphibians (e.g. red-spotted newt, *Notophthalmus viridescens*; Verrel, 1982) and scorpions (e.g. *Opistophthalamus latimanus*; Alexander, 1957), males try to increase the probability that their spermatophores will be taken up by leading or pulling females to and onto the spermatophore. Many male insects, while continuing to transfer spermatophores, place the package inside the female tract. Such insects use a penis. The loss of the spermatophore and the insemination of free sperm in, for example, some Hymenoptera and Diptera were probably relatively recent evolutionary events in these insect lineages (Chapman, 1982).

It is possible that the vertebrate and thus human lineage passed through a sequence of events similar to that for insects. Labyrinthodonts, the likely ancestors of modern amphibians and reptiles, had aquatic larvae and presumably inherited external fertilization from their fish ancestors. A number of lineages, however, evolved internal fertilization. Among amphibians, the majority of modern urodeles (newts and salamanders) show internal fertilization via spermatophore deposition (Halliday and Verrell, 1984). Perhaps the lineage leading to reptiles also passed briefly through a phase of internal fertilization via spermatophore deposition on the ground or in water before taking the short step to deposition of the spermatophore in the external opening of the female's reproductive tract. Loss of the spermatophore and the consequent insemination of free sperm would lead to the modern form of insemination seen in all modern reptiles and mammals.

In reptiles, on perhaps a number of independent occasions, various, often cuticularized, penis-like structures have evolved to facilitate insemination. All modern reptiles (except the tuatara, *Sphenodon*) have an intromittent organ (a single penis in crocodiles and tortoises; paired hemi-penes in snakes and lizards; Tokarz and Slowinski, 1990). Most birds, however, do not have an intromittent organ for insemination (exceptions include the flightless ratites (King, 1981) and geese (Brooke and Birkhead, 1991)) but use instead their cloaca (the most distal part of their reproductive tract) which is everted into the female reproductive tract (Torrey, 1962; Birkhead and Møller, 1992).

3.3.2 Seminal fluids

Seminal fluid is any liquid medium with which the male surrounds the sperm in preparation for ejaculation.

While awaiting ejaculation, human sperm are stored in the cauda epididymis and vas deferens in a hospitable and nutritious fluid medium (section 2.4.3; Box 2.10 for male reproductive tract). Emission occurs prior to ejaculation. At emission sperm are moved from the vas deferens into the prostatic urethra (Harper, 1988), and ejaculation then follows. Sperm in the vas deferens are highly concentrated in just a small amount of fluid from the testis and epididymis (less than 10% of the final ejaculatory fluids). At ejaculation, contractile waves force the sperm from the urethra while secretions primarily from the prostate (responsible for 30% of the volume) and the seminal vesicles (for 60%) add volume to the semen (Lundquist, 1949).

The material produced by the prostate is a thin slightly opaque fluid with a pH near to neutral (Lundquist, 1949). Alkaline pH is essential to sperm movement (acid pH immobilizes sperm; El-Banna and Hafez, 1972) and therefore the prostatic secretions with the alkaline secretions from the seminal vesicles (Lundquist, 1949) provide a relatively alkaline buffer to help neutralize the acidic environment of the female vagina (Duerden *et al.*, 1987). Although originally believed to store sperm, the seminal vesicles of mammals, unlike similarly named areas in other groups (e.g. birds), are not sperm storage areas (Beams and King, 1933).

The volume of the human ejaculate varies from 2 to 6 ml (mean about 3 ml) (Belsey *et al.*, 1987). In contrast, the turkey ejaculates 0.2–0.8 ml (mean = 0.3 ml) and a boar up to 500 ml (Mann and Lutwak-Mann, 1981).

Even external fertilizers surround their sperm with a fluid medium for ejaculation. These fluids may influence the chances of fertilization in a variety of ways. The seminal fluids of mussels and oysters indicate to other individuals that sperm are present. These other individuals are then more likely to release eggs (Cohen, 1977). In the freshwater zebra fish, the seminal fluid may help to improve the osmotic environment around the eggs (Cohen, 1977).

In internal fertilizers, an enormous variety of male accessory glands have evolved which add volume and various chemical and physical characteristics to the ejaculate (Mann and Lutwak-Mann, 1981). The fact that sperm are inseminated surrounded by fluids originating in a male, opens up the possibility that these fluids could be used to influence the competitiveness of sperm from the other males. It would be surprising, therefore, if natural selection via sperm competition had not moulded seminal fluids beyond the form necesssary for simple passage to and in the female tract.

Surprisingly, the suggestion still persists that seminal fluids have no function other than to dilute sperm to a suitable concentration. In part, the confusion arises from the observation that sperm taken from the epididymis and placed in a relatively simple diluent can still achieve fertilization (Mann and Lutwak-Mann, 1981). However, such an observation tells us only that the seminal fluids do not play a vital role in fertilization. It does not tell us that the fluids have no role in sperm competition.

The large volume of seminal fluid in the boar ejaculate has been suggested to be a feature enabling the male to wash out the sperm inseminated by a previous male (Sumption, 1961). Such use of the seminal fluid could give the sperm from the second male more of a competitive chance against sperm from the first than if the seminal volume were less.

In the dogfish ejaculate, serotonin stimulates muscular contractions and sperm uptake by the female (Mann, 1960). As explained in section 2.5, enhanced sperm uptake could be an advantage to the male in sperm competition. Prostaglandins in human semen have been suggested to play a similar role (Mann and Lutwak-Mann, 1981) but in normal vaginal insemination they are probably ineffective.

In some insects (e.g. the butterfly, *Heliconius erata*; Gilbert, 1976) males are less likely to copulate if they detect the presence in the female of seminal fluids from a previous male. The influence is so strong that such chemicals have been termed anti-aphrodisiac. The presence of such chemicals could give the first male to inseminate a female more of an advantage in sperm competition than would otherwise be the case. Some human males find the smell of non-self semen samples in a medical laboratory so repulsive that they are unable to work there (Cohen, personal communication). Whether human males are ever discouraged from copulation on detecting semen from another male in a female's tract appears to be untested. As is clear later, however, any such effect is unlikely to be absolute.

Seminal fluids not only contain sperm-immobilizing compounds (Luterman *et al.*, 1991) but the final fraction of the human ejaculate is spermicidal (Eliasson and Lindholmer, 1973). Once again, both of these facets could give the first male to inseminate a female more of an advantage in sperm competition than would otherwise be the case.

The influence of seminal fluids from one male on the inseminate of another appears to be unstudied *in vivo* for healthy individuals. However, in certain immunological conditions (usually pathological), antibodies in the ejaculate from one male are known to prevent fertilization by another male (Quinlavan and Sullivan, 1977).

Taken together, such observations strongly indicate that seminal fluids now represent adaptations to sperm competition as well as to the protection and passage of sperm during insemination and while in the female. The different potential involvements of seminal fluid in sperm competition are discussed at various points in the sections and chapters that follow. However, one major such adaptation, the use of seminal fluids to form a copulatory plug, is discussed in the next section.

3.3.3 Copulatory plugs

A copulatory plug is a structure, usually formed from a mixture of sperm and seminal fluids, which when positioned by the male at a suitable location in the female tract, interferes in some way with the next copulation. Such plugs are found in a wide range of animals and are sometimes termed chastity plugs.

Within seconds of insemination into the upper vagina, most of the human ejaculate coagulates to form a soft, gel-like structure (Mandal and Bhattacharyya, 1985). After about 15–20 minutes *in vitro* and perhaps faster in the vagina, the structure decoagulates and, at least in part, is eventually ejected in the flowback (section 3.5.2). Coagulation is achieved by an enzyme from the seminal vesicles and decoagulation by one from the prostate (Mann and Lutwak-Mann, 1981).

This means that the first part of the ejaculate often does not coagulate or decoagulate very quickly, the last part may stay coagulated for an extended period (Tauber and Zaneveld, 1976).

Coagulation may retain the inseminate in position while sperm leave to travel further up the track. However, an additional function could be that the structure as a whole could interfere with the passage of sperm from any second insemination within the next 15 minutes or so and thus could function as a soft, short-term plug. (In our UK survey, Box 2.6, between 1/100 and 1/200 of 1181 women who had had >500 lifetime copulations had been inseminated by two different males within <30 min.) Such a weak, though not necessarily insignificant, physical barrier may be enhanced by the sperm immobilizing compounds in the seminal fluids (Luterman *et al.*, 1991) and the spermicidal final fraction to the ejaculate (Eliasson and Lindholmer, 1973) mentioned earlier.

Soft, spongy copulatory plugs of similar texture to the one found in humans are widespread among animals. Even within primates, a wide range of plug consistencies and decoagulation times are found (Roussel and Austin, 1967). Birds do not have copulatory plugs (Birkhead and Møller, 1992) but many other vertebrates (e.g. snakes, Devine, 1975; mammals, Martan and Shepherd, 1976; Mosig and Dewsbury, 1970; Baker and Bellis 1988; Koprowski, 1992) have seminal fluids which rapidly harden to form a relatively solid, hard plug in the female tract.

Vertebrates are not the only lineage of internal fertilizers to have evolved a copulatory plug. Various invertebrates (e.g. acanthocephalan worms; Abele and Gilchrist, 1977), but most notably the insects (Parker, 1970a) have evolved a variety of plugs. In some insects, the plug is formed not by hardened seminal fluids, but by the detached genitalia from the male. Thus male honey bees (*Apis mellifera*) wrench off their genitalia after copulation, leaving the penis inserted (Eberhard, 1985). This hinders (but does not prevent) insemination by later males, thus usually giving the first male to mate with the female the greatest share of fertilizations (Parker, 1970a). Male praying mantids (*Mantis religiosa*) are often eaten by the female during copulation (under natural conditions as well as in captivity; Lawrence, 1992) but their genitalia remain in the female.

Whereas most of the above animals use copulatory plugs to keep the sperm from other males out, males of the spider crab, *Inachus phalangium*, use them to keep other males' sperm in (Diesel, 1990). The last male to mate displaces the ejaculate of his predecessors dorsally into the apex of the female's seminal receptacle. Sperm of previous matings are sealed in with hardening seminal plasma and are thus prevented from being used by the female to fertilize eggs.

In the majority of animals (e.g. rats, Orbach *et al.*, 1967), including humans, the consistency of the plug (hard or soft) seems to be dictated by the male, being more or less the same whether ejaculated into the female or an artificial substitute. However, in some species, secretions and cells from the female tract may form a substantial part of the plug (e.g. the opossum, Hartman, 1924).

Copulatory plugs have been found successfully to block spermatozoa inseminated by subsequent males (e.g. guinea pigs; Martan and Shepherd, 1976) and may be interpreted as a male mechanism for extended mate-guarding (Parker, 1970a). One function may be a backup to direct mate-guarding by the male (i.e. a form of rearguard defence when more direct methods fail and a second male succeeds in attempting to inseminate the female). Another function may be as an attempt to gain the advantage of mate-guarding without incurring the costs of continuing to associate directly with the female (section 2.4.4).

Copulatory plugs thus provide the male with a means of increasing his chances of fertilizing the current female's eggs without reducing the rate at which he can encounter new females. For both functions to evolve it must be assumed that any potential gain in terms of an increased rate of fertilizing eggs outweighs any increased cost incurred through plug production (e.g. increasing the volume of seminal fluid and/or changing their chemical composition to facilitate plug formation).

There is discussion of the factors influencing whether the optimum plug for a particular lineage is hard or soft in Chapter 6.

3.3.4 The antiquity of copulation and insemination in the human lineage

Copulation seems likely finally to have evolved in the human lineage in association with the full-time invasion of the terrestrial environment, about 300 million years ago. Cloacal accessories for sperm transfer from male to female are first found in reptiles and a cloaca is found in the embryos of birds and mammals (Torrey, 1962). However, in all mammals except the prototherians (in which the ventral part of the cloaca forms the penis; van Tienhoven, 1968) the cloaca is replaced in adults by the specialized vagina/penis structures which have a non-cloacal embryonic origin (Torrey, 1962).

The initial appearance of major accessory glands probably dates back to early metazoans and relatively complex arrangements of accessory glands are found in fish (Mann and Lutwak-Mann, 1981). With the evolution and establishment of internal fertilization, seminal fluids are adapted first to protect and transport gametes both while outside (e.g. the deposited spermatophores of newts and salamanders) and once inside the female tract. Immediate pressure from sperm competition, however, seems likely to have led to the increased complexity associated with the formation of copulatory plugs. The widespread occurrence of hard plugs among reptiles raises the possibility that the earliest plugs in the human were of this type and that the ephemeral soft plugs found in modern humans and some other primates are a secondary adaptation (Chapter 6).

Although intromittent organs, accessory glands and copulatory plugs may first have been shaped about 300 million years ago, their subsequent evolution will have tracked changes in the configuration and chemistry of the female tract. The modern form of the human penis and accessory glands seems likely to have been shaped primarily by the changes in the female tract that occurred following the early evolution of eutherian mammals about 200 million years later. Further discussion of these changes is thus delayed until Chapters 4 and 6.

We suspect that two other notable features of male human sexuality also owe their origins to events during the age of reptiles: spontaneous emission and homosexuality (Box 2.1). However, these will also have been modified considerably during the early mammalian phase of human evolution and so discussion is again delayed, in this case until Chapter 5.

3.4 Forced copulation

When a male forces copulation on an apparently resistant female despite the latter's continuing rejection responses, the copulation is described as rape or, for the purposes of this book, forced copulation. When a female is forced to copulate with more than one male acting in cooperation, each inseminating the female in turn, the event is described as gang rape.

As gang rape results in the female containing sperm from more than one male in her reproductive tract at the same time, such behaviour invariably leads to sperm competition.

Theories of the adaptive significance of forced copulation are well established (Box 3.3) and have been discussed in detail by Thornhill and Thornhill (1983) and Ellis (1989). The latter author distinguishes between predatory rape, in which a male forces copulation on a previously unknown female, and date rape, in which a male forces a female to copulate after a period of amicable association.

A recurrent problem in the legal determination of forced copulation is that female sexual behaviour, particularly with a novel partner, often involves an element of coyness, elusiveness and physical resistance before final consent. Again, the theoretical interpretation of such behaviour seems straightforward (Box 3.3) and even accommodates otherwise unexpected facets of female response. For example, a USA study showed that 39% of date-rape victims continued to date their assailant after the assault as opposed to only 12% of females who success-

Box 3.3 The behavioural ecology of rape

With few exceptions, animal populations are numerically dominated by relatively urgent, competitive males and relatively cautious, coy females (section 2.3.2). When a male meets a female, the male benefits most from mating as soon as possible, the female from either not mating at all or not mating until the male has satisfied particular critiera. Essentially, the male has to pass certain tests of appraisal for suitability before being acceptable to the cautious female.

One way in which the male can reduce the time from meeting to mating to near his optimum (and thus away from the female's optimum) is to force copulation by raping the female.

Rape and the male

Rape aids a male's reproductive success as long as three conditions are satisfied: (1) the time and energy required to force mating is not greater than the time and energy involved in waiting for the female to acquiesce to an unforced mating or involved in abandoning the female and searching for another; (2) the risk is not too great of the male being wounded or killed by the female (or an individual or individuals guarding the female); and (3) the risk is not too great of the female being wounded to the point that she is unable to produce and perhaps raise the offspring sired by the male.

Numerically, populations should now be dominated by males with heritable characteristics that allow them accurately to assess whether forced mating is an advantageous strategy in any particular circumstance (i.e. whether the benefit of rape in terms of increased reproductive success via the current female is greater than the combined costs of (1), (2) and (3)).

In some animals (e.g. butterflies; birds) males are physically limited in their ability to force copulation on a 1:1 basis. Gang rape is thus the only means by which any given male can reduce costs (1) and (2). Gang rape, however, increases the risk of cost (3) and dilutes the benefit of rape because of sperm competition. Put crudely, the number of males participating in a gang rape must increase the chances of copulation per male by more than the number in the gang. Thus a male would benefit from being a member of a gang of three if he achieved more than three times the number of copulations he would achieve on his own. This simple equation, however, must be modified by changes in the risk of the male being damaged through being a member of a gang and changes in the risk of the female being damaged through being gang raped.

Rape and the female

If males that force mating under appropriate circumstances have greater reproductive success than males who do not force mating, females will produce more grandchildren (via their sons – Box 2.12) if they mate with males who have the ability both to force copulation and to assess when forced copulation is advantageous. Females with a preference for such males will on average leave more descendants than females with no such preference.

The problem faced by females, however, is how to recognize that a male has the preferred ability. The obvious solution is for the female to test the male for his ability not only to force copulation but also to assess when doing so is advantageous. The technique would be to resist the male until he attempts to force copulation, at least up to the level of aggression at which she begins to risk physical damage. There would seem, therefore, to be an advantage to females of elaborating on their coyness by exposing males to some degree of elusiveness and/or physical resistance.

fully thwarted the attempt to force copulation (Wilson and Durrenberger, 1982).

Such female behaviour makes the definition and recognition of forced copulation a difficult procedure. On any one occasion, the categorization of a copulation as forced requires an often delicate legal decision over whether female resistance was token or real and, therefore, over what constitutes consent (Symons, 1979).

In a review of anthropologically indexed human societies, Broude and Green (1976) found forced copulation to be common in 41.1% of societies, relatively uncommon in a further 35.3%, and reportedly absent in 23.5%. In the USA, it is estimated that 600 000 forced copulations occur each year and that about 10% (60 000) are reported (Green, 1980). Russel (1984) estimates that over a lifetime in San Francisco 24% of women will be forced to copulate and another 20% will experience attempted forced copulation. Some figures put gang rape as high as 70% of all forced copulations (Steen and Price, 1977).

Forced copulation is particularly common during warfare, whether at the local or global level (Brownmiller, 1975). For example, it is estimated that during the Pakistani occupation of Bangladesh in 1971, about 300 000 forced copulations took place in nine months. Allowing for population size, this is roughly double the (peacetime) rate estimated for the United States.

A contentious and emotive issue concerns the lifetime fertility of rapists and the fertility of each forced copulation event. A common view is that rapists are individuals who have difficulty in gaining sexual access to females by less forceful means (Thornhill and Thornhill, 1983). For example, rapists are three-times more likely to have a facial deformity than the average male (though not more likely than the average male in the prison population from which the rapist sample was taken) (Ellis, 1989). Rapists are also more likely to be young and poor (Thornhill and Thornhill, 1983).

Even so, there is no indication that rapists are less likely than matched non-rapists to have a partner and children (Ellis, 1989). If a rapist's reproductive success through non-rape copulations is potentially at least average, the implication is that forced copulation is an additional mating tactic rather than a substitute (Ellis, 1989).

Most often, forced copulation seems to represent a sexual intent on the part of the male (Symons, 1979). Forced copulation victims are a non-random subset of the female population and are most likely to be in their peak reproductive years (Thornhill and Thornhill, 1983). The age distribution of forced copulation victims is also very different from that of female murder victims (Thornhill and Thornhill, 1983). The two factors together suggest that in the majority of forced copulations reproduction, not violence and/or subjugation (cv Brownmiller, 1975), is the ultimate (though not necessarily conscious; Chapter 7) function. Physical injury to the female is perhaps then a by-product of the male forcing copulation rather than the main function. On occasion, however, rapists show such disregard for the female's life and safety that copulation and insemination can only be an incidental and effectively non-reproductive element in the behaviour.

As far as the fertility of individual forced copulation events is concerned, data seem to show that forced copulations are more likely to lead to conception than unforced, a trend that has been attributed to coitus-induced ovulation (Clark and Zarrow, 1971; Jöchle, 1973, 1975). However, it is impossible fully to control such data for the possibility that females are more likely to report forced copulations that do or may lead to conception (Singer, 1973).

All of the elements of forced copulation and associated behaviour found in humans are also found in other animals. Males who respond to particular circumstances by attempting to force copulation on a resistant female are characteristic of a wide range of species, including insects (e.g. scorpion flies, *Panorpa* spp., Thornhill, 1980), fish (e.g. guppies, *Poecilia reticulata*, Farr, 1980), birds (e.g. mallard, *Anas platyrhynchos*, Cheng *et al.*, 1982) and mammals (big horn sheep, *Ovis canadensis*, Hogg, 1984; chimpanzee, *Pan troglodytes*, Tutin and McGinnis, 1981). Moreover, gang rape is not uncommon (e.g. house sparrows, *Passer domesticus*, Summers-Smith, 1955; mallard, *Anas platyrhynchos*, Cheng *et al.*, 1982) and is found in at least one non-human primate, the woolly spider monkey, *Brachyteles arachnoides* (Milton, 1985).

It is a common feature of courtship in animals that the female responds elusively and/or aggressively to male sexual advances and only allows mating after a chase and/or physical struggle. In the mink, *Mustela vison*, for example, males and females fight before copulation and some level of physical trauma (often biting from the male) is actually necessary for the female to ovulate (Ford and Beach, 1952). However, as in humans, levels of aggression normal for the species can sometimes escalate to such an extent that females may be wounded or even killed (e.g. mallard; Huxley, 1912).

In many animals (e.g. birds: bank swallow, Beecher and Beecher, 1979; lesser scaup, Afton, 1985; white-fronted bee-eater, Emlen and Wrege, 1986), as in humans, rapists often have female partners with whom they achieve most of their reproduction. These partners are actively defended against forced copulation by other males. Also as in humans, gang rape is common and forced copulation victims are predominantly females at peak fertility. In such species, successful rapists may experience greater reproductive success than other males.

In some other animals, however, forced copulation is clearly a suboptimal strategy and rapists are males who are physically unable to adopt more profitable mating tactics. In *Panorpa* scorpionflies, for example, rapists are the smallest males in the community, are unable to gain reproductive opportunities by unforced means, and have relatively low reproductive success (Thornhill, 1981). If the larger males are removed from the community, the smaller males abandon forced copulation attempts and adopt more successful reproductive tactics. In this example, males seem to assess which strategy best suits their physical ability and circumstances and change appropriately from one tactic to another with each change in situation.

Elements of forced copulation behaviour may well have been present in prereptilian males of the human lineage, throughout the long phase of external fertilization. Forced copulation in its human form, however, seems likely to have evolved hand-in-hand with copulation as a mechanism for internal fertilization (say 300 million years ago) and to have been a feature of the human lineage ever since. Whether forced copulation and gang rape by male humans are now more or less common than earlier in human or prehuman evolution cannot presently be guessed.

3.5 The female: sperm storage organs, flowbacks and copulatory orgasms

3.5.1 Sperm storage organs in the female

The main regions of the reproductive tract of female humans are shown in Box 3.4.

A female sperm storage organ is a specialized site or sites in which sperm are located in the female for a time between insemination and fertilization. Sperm storage organs should, by definition, provide conditions that are at least more hospitable to sperm than other regions of the tract. They may also provide nutrients. The theoretical advantages of sperm storage organs to females are discussed in Box 3.5.

Human females have blind, branched crypts in the wall of the cervix (Box 3.6) which receive sperm (Koch, 1980). Diagonal channels through the cervical mucus (Box 3.5) may actually direct some of the inseminated sperm to these crypts. While in these cervical crypts, human sperm become immotile or slow ('hibernating'; Höglund and Odeblad, 1977) and are relatively safe from phagocytosis (Hafez, 1973). There are up to ten thousand crypts in the walls of the human cervix (Bernstein *et al.*, 1977). In all, these crypts are potentially capable of accommodating millions of sperm (Insler *et al.*, 1980) though current estimates are that they receive less than 1% of the inseminated sperm.

The expulsion of sperm from each crypt is a sudden phenomenon occurring at different times from each crypt (Höglund and Odeblad, 1977). One hypothesis is that this staggered ejection results in a steady traffic of sperm out of the cervix over a period of from 2–3 (Harper, 1988) to 8–10 days (Koch, 1980; Baker, Bellis, Creighton and Penny, unpublished) after insemination.

Sperm are also relatively safe in, and may be stored in, the isthmus of the oviduct (Hunter, 1987a). How many sperm travel directly to storage in the oviduct after insemination, how many spend some time in the cervical crypts before travelling to secondary storage in the oviduct, and how many fail to

Box 3.4 The reproductive tract of the human female

Ovaries and oviducts
In humans, the ovaries are paired, located at each lateral side of the uterus, and project into the pelvic cavity.

When six months old, while the young female is still inside her mother, the fetus' ovaries contain about 7 million primary follicles capable of becoming eggs (Linkie, 1982). At birth, this number has reduced to about 400 000. Only about 400 of these can ever develop into eggs and be shed during the female's lifetime. The number that actually does so is usually less than about 60.

Near to, but separated from, the ovaries are the ciliated fimbriae at the ends of the paired oviducts which carry eggs from the ovaries to the uterus. The human oviduct is about 10 cm long (Snell, 1986), and the final 1–2 cm as they pass through the wall of the uterus are so convoluted and tortuous that a probe cannot be passed through. This intramural part has a diameter ranging from 0.1 to 1.0 mm (Harper, 1988).

The uterus
The oviducts open into the uterus at the uterotubal junctions. In humans, there is a single uterus which receives both oviducts. The uterus is a muscular organ which, in its virginal state, is the shape of a flattened pear measuring roughly 8 × 5 × 3 cm (Last, 1978). The walls are thick, about 3 cm at the fundus, and normally adpressed leaving a tiny internal volume to the lumen of no more than 4 ml. During pregnancy, the volume increases to about 4000 ml (Symonds, 1992).

The cervix
The lower part of the uterus (which leads to, and opens into, the vagina) is the cervix. The human cervix has a straight, narrow, spindle-shaped lumen, 2–3 cm long (Dobson, 1988). The cervical walls of elastic fibres are capable of stretching during birth to the circumference of a baby's head. At other times, however, the cervical lumen is lightly adpressed. The opening of the human cervix into the vagina (= os cervix) is a circular dimple in a nulliparous woman but often a transverse slit after childbirth (Last, 1978). During female orgasm, the os is thought to gape (Masters and Johnson, 1966).

The complexity and tortuosity of the cervix varies considerably between species and must have a major influence on the nature of sperm competition.

The vagina
The cervix opens, and extends by about 2.5 cm, into the upper part of the vagina (Goldenson and Anderson, 1986), an elastic passage, lined with stratified squamous epithelium, leading to the outside of the female. The external opening of the vagina is the vestibule and is surrounded by the vulva (i.e. all of the external female genitalia, primarily the mons pubis (or veneris) labia major and labia minor, clitoris, and Bartholin's glands; Snell, 1986).

When unstretched, the vagina of an adult female is roughly 10 cm long and less than 5 cm wide (external diameter). The internal surface is marked by ridges (Goldenson and Anderson, 1986). Despite the way it is normally depicted in diagrams, the lower vagina has virtually no internal lumen when not distended. Over most of its internal surface it is non-secretory though on either side of the entrance to the vestibule are a number of secretory cells (Bartholin's glands) that produce a mucus-like lubricant during sexual arousal. The pH of the vagina is maintained at an acidic 3.5–5.8 probably with the aid of bacteria (e.g lactobacilli; Duerden *et al.*, 1987), in part as a mechanism to resist infection.

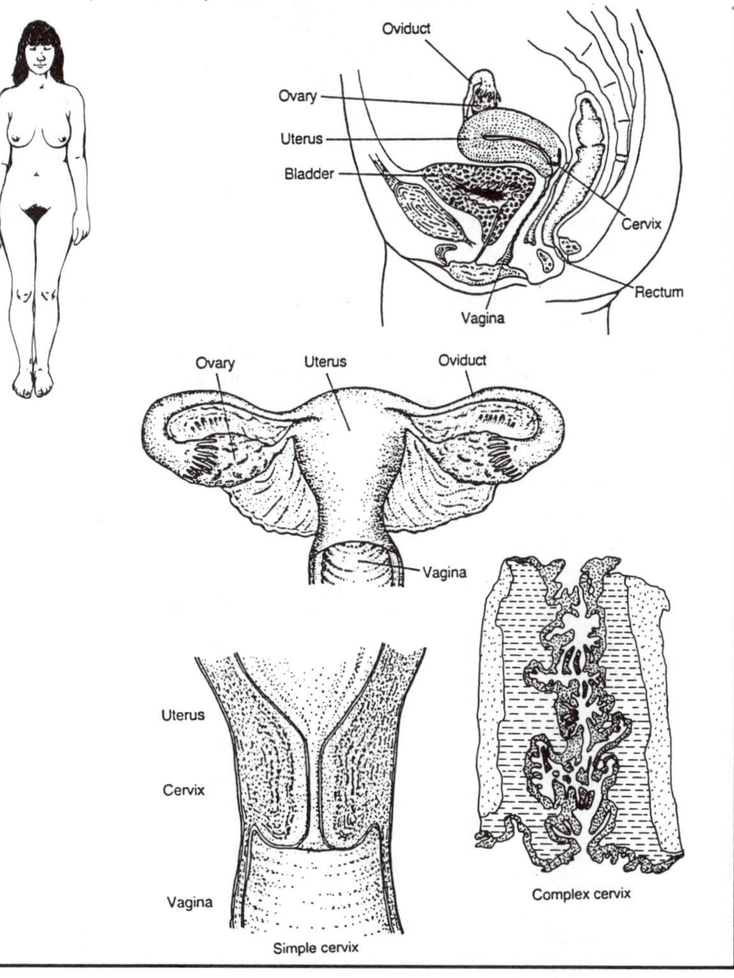

Box 3.5 Cervical mucus

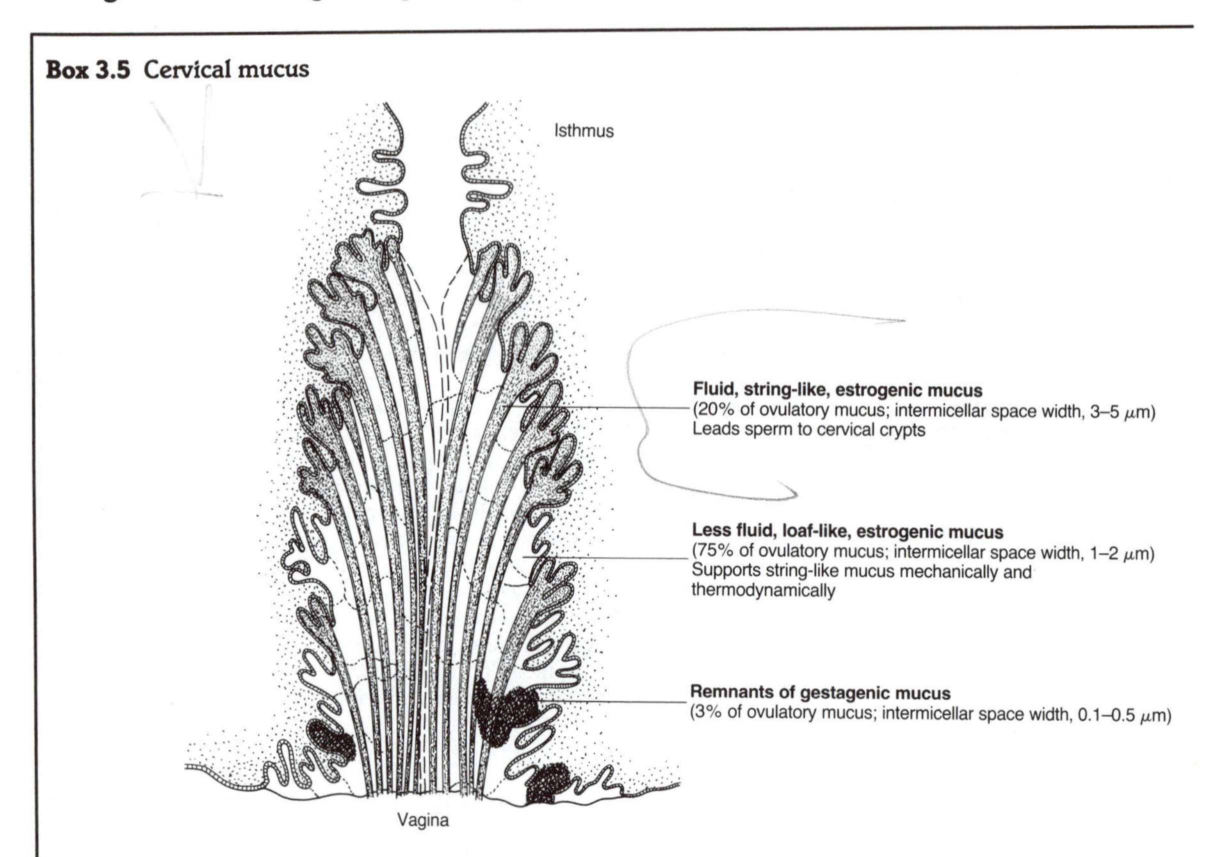

Mucus types and structure
The cervix is filled with a column of mucus which is secreted partly by the endocervical glands (Linkie, 1982). The action of cilia on epithelial cells on the cervical walls moves the mucus column in glacier-like fashion slowly down through the cervix (Austin, 1975). Eventually, the mucus drips into the upper vagina (Jaszczak and Hafez, 1973), where it is resorbed and/or is eventually lost through the vaginal opening (Hafez, 1973).

Cervical mucus has a complex macromolecular structure (Harper, 1988). Macromolecular fibrils are arranged to form parallel chains with spaces between them which thus form channels (Elstein and Daunter, 1976). Sperm migrate through the mucus by way of these channels, perhaps having to push aside or cut through small interstices between mucus macromolecules.

There appear to be two types of ovulatory mucus (Höglund and Odeblad, 1977): loaf-like mucus which provides mechanical and thermodynamic support for the diagonal mucus which is string-like and runs from the cervical crypts to the lower (vaginal) end of the mucus column. Compared with a sperm, which is about 3 μm wide (Perloff and Steinburger, 1964), the channels though the ovulatory mucus are adequate (3–5 μm) whereas those through the gestagenic (post-ovulatory) mucus are relatively narrow (0.1–0.5 μm).

As plug material ages and travels down the cervix (Linkie, 1982), it may become cluttered with leucocytes and other cells and debris from the female as well as sperm and other male products from a previous copulation. Leucocytes invade the cervical mucus within minutes of the arrival of sperm (Pandya and Cohens 1985) and at their peak may outnumber sperm by up to 3:1 (Cohen, 1984). These leucocytes remove dead, dying and healthy sperm (Mattner, 1969) and other male-produced cells, as well as infectious organisms. The presence of leucocytes and other cells and debris in the mucus renders it less penetrable to sperm (Parsons and Sommers, 1978; Belsey *et al.*, 1987).

Rate of mucus production varies with phase of the menstrual cycle (Box 4.6). It is greatest (700 mg/d; Pommerenke, 1946; Moghissi and Syner, 1976; Koch, 1980) during the fertile phase a few days before ovulation and least (20–60 mg/day; Moghissi, 1977) during the infertile post-ovulatory phase. Not only the quantity but also the nature of the cervical mucus varies during the menstrual cycle. During the preovulatory phase, under oestrogenic influence, the cervical canal begins to fill with thin, clear mucus (Kremer and Jeger, 1988).

The faster production rate and lower viscosity of preovulatory mucus is associated with much lower densities of leucocytes and other cells during the fertile phase. It is as if one of the functions of the increased production of mucus is to 'flush-out' cells and debris from the cervical column. In contrast, during pregnancy, the column is heavily populated by leucocytes and epithelial cells (from the vagina) and forms a plug that is considered to be virtually impenetrable to sperm (Davey, 1986).

Cervical mucus, copulation and sperm

Before and during copulation, the vagina becomes lubricated and stretches to accommodate the erect penis. The lubricating material is probably derived from several sources. The vagina 'sweats' during sexual arousal and this may provide some of the lubrication (Masters and Johnson, 1966). Bartholin's gland may also produce a mucus-like secretion but the most likely source of the major part of the sexual lubricant is the cervical mucus (Kinsey *et al.*, 1953). One estimate (Dickinson, unpublished; in Kinsey *et al.*, 1953) is that a female human secretes between 1.0 and 4.0 ml of cervical mucus during sexual arousal.

The end result of copulation is the deposition of the ejaculate as a 'seminal pool' in the upper part of the vagina adjacent to the cervical opening. The cervix dips into the seminal pool (perhaps with a gaping opening if the female orgasms (Box 3.4)), and establish an interface between the semen and cervical mucus. This interface consists of a series of finger-like projections of semen (= semen phalanges) into the cervical mucus (Moghissi, 1977) perhaps primarily into the string-like mucus (Högland and Odeblad, 1977). Swimming sperm then travel either slowly through the narrow spaces of the loaf-like mucus, perhaps directly into the uterus, or along the diagonal strings of mucus into the cervical crypts (Mortimer, 1983). Once the vanguard sperm have passed through the mucus, however, the macromolecular structure of the mucus seems to be distrupted and any following sperm make much slower progress (Katz *et al.*, 1990). The result seems to be the establishment of a gradient of sperm density from ectocervix to endocervix in both the cervical crypts and the cervical canal by one hour after copulation (see Figure for rhesus monkey).

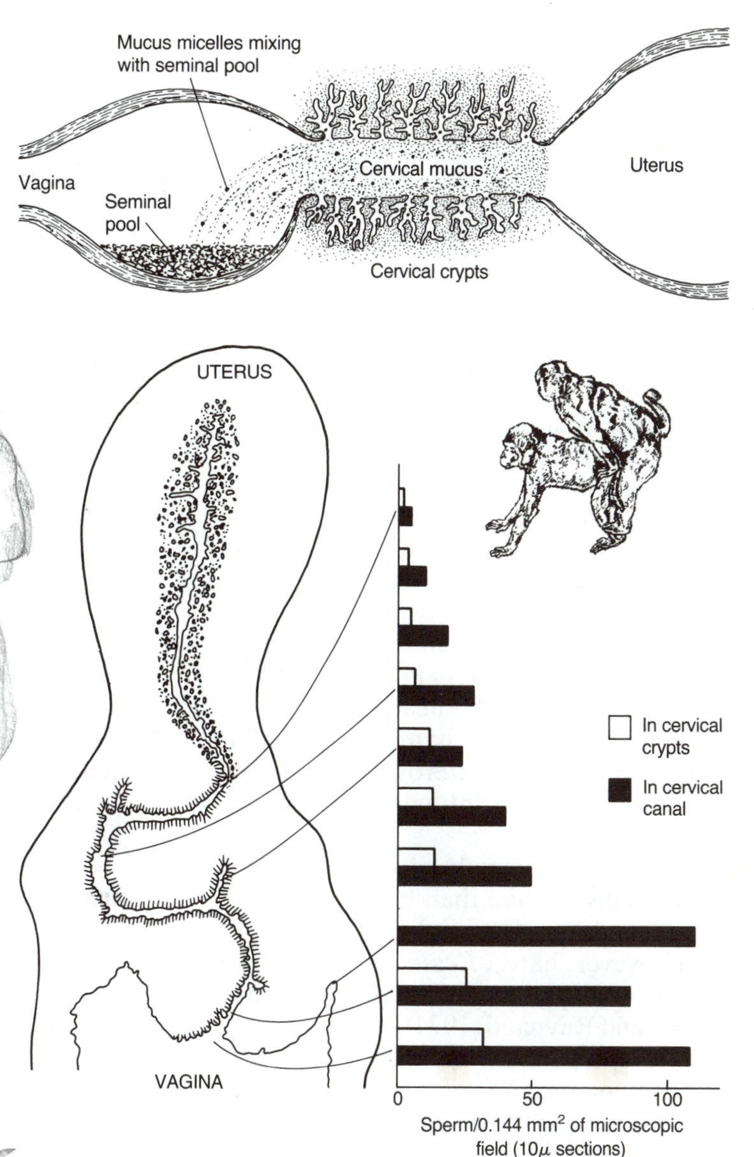

Sperm are immediately immobilized in acidic environments at pH levels below 6.5. Even at a pH of 7.0 the ability of sperm to penetrate cervical mucus is minimal. Penetration is 'normal' at pH 7.5 and above normal at pH 8.25 (El-Banna and Hafez, 1972). The pH of seminal fluid is usually in the range of 7.0–7.8 (Raboch and Skachova, 1965) and buffers the sperm from the acidic (pH 3.5–5.8; Duerden *et al.*, 1987), and thus very hostile, vaginal environment. The pH of the cervical mucus varies considerably from a favourable 7.4 to a hostile 4.0 (Kroeks and Kremer, 1977).

Once sperm are no longer buffered by seminal fluid, they cannot live for more than a maximum of 10–12 h at the hostile pH levels found in the vagina (Vander Vliet and Hafez, 1974). Most medically significant infections of the cervix also prefer a more-alkaline environment (Duerden *et al.*, 1987) and any increase in acidity could have an antibiotic effect. Thus, any physiological ability to lower the pH of the cervical mucus would endow the female with some influence over both the viability of any infectious organisms and sperm motility and survival (see Chapter 10).

(Cervical mucus diagram compiled and modified from drawings and data in Högland and Odeblad, 1977; seminal pool and sperm distribution diagrams inspired by Hafez, 1973.)

enter any form of storage is unknown (Chapter 8).

All of the major animal lineages with internal fertilization have sperm storage organs or areas in the female. Thus insects have a cuticular storage organ called a spermatheca (Happ, 1984) where sperm may be stored for weeks (e.g. large white butterfly, *Pieris brassicae*, David and Gardiner, 1962) or even years (e.g. honey bee, *Apis mellifera*, Wilson, 1971). In vertebrates, some viviparous teleost fishes (e.g. *Heterandria formosa*) may store sperm sometimes for nearly a year in the ovaries (Turner, 1947). Some reptiles possess tubules which act as sperm stores (Fox, 1956, 1963) such that some snakes (e.g. *Acrochordus javanicus*) may be able to store sperm for 7 years (Magnusson, 1979).

In most birds, sperm are stored in tubules at the uterovaginal junction. Sperm storage durations range from 6 days in the quail to 42 days in the Turkey (Birkhead and Møller, 1992). As in humans, less than 1% of inseminated sperm enter these uterovaginal storage tubules in birds (Birkhead, personal communication).

On average, mammals store sperm for shorter lengths of time than birds (Birkhead and Møller, 1992; Gomendio and Roldan, 1993). However, hares, *Lepus* spp. may store sperm for 42 days, even through pregnancy (Martinet and Raynaud, 1974), and some bats may store sperm for even longer (e.g. 198 days in the noctule, *Nyctalus noctula*, Racey, 1973). Discussion of the length of time that humans store sperm is delayed until Chapter 8.

Mammals have four potential sites and/or mechanisms for sperm storage and any one, or combination, of these may be found in any given species. Sperm-receiving structures, which are more appropriately termed crypts than tubules, may be found: (1) as in birds, at the junction of the vagina and uterus (e.g. rabbits, ruminants and primates, Hunter, 1973); and/or (2) in the isthmus of the oviduct (e.g. opossum, *Didelphis virginian*, Rodger and Bedford, 1982). The remaining mechanisms are simply to maintain sperm in close association with the tract epithelium in either the uterus (e.g. bats; Racey, 1979) or the isthmus of the oviduct (e.g. pig, Hunter, 1973, 1981).

In mammals, the vagina–uterus junction is anatomically specialized (the cervix; Box 3.4), has walls with cervical crypts and in most species has a mucus plug (Box 3.5). Cervical crypts are pockets of columnar epithelium of the cervical mucosa which extend in many directions and can even be branched (Elstein *et al.*, 1972). The general case for the cervix, and particularly the cervical crypts, being a sperm storage area in mammals seems very strong (Morton and Glover, 1974a; Harper, 1988). Cervical crypts occur primarily in species, such as humans, in which sperm are inseminated into the vagina. Such mammals may also have oviductal sperm storage.

All mammalian storage sites provide a relative haven from phagocytosis by the female leucocytes (cervical crypts, Mattner, 1968; oviduct, Hunter, 1981) as well as providing varying levels of nutrition for the sperm (Moghissi, 1972). In many mammals, as in humans, stored sperm are relatively inactive and dormant (e.g. rabbit, Overstreet *et al.*, 1980).

The evolution of sperm storage organs in females provides a new focus for the process of sperm competition. No longer is this process simply a race to fertilize the egg(s) as it seems to be in external fertilizers. Now sperm must seek in the first instance to gain access to the hospitality provided by the storage organs. Then they must attempt, while within the storage organs, to attain a position that allows them a quick and easy exit when departing for the site of fertilization. If, at either of these stages, they are in the presence of sperm from another male, both manoeuvres will be subject to competition.

Sperm already present in the storage organs should attempt to deny access to these organs for any sperm from a later male. Sperm from the later male must attempt to gain access to these organs or else suffer the consequences of remaining in a relatively inhospitable environment outside of the organs.Once in the organs, the sperm must jostle for position ready for a quick and easy exit.

Box 3.6 The evolution of sperm storage organs in females

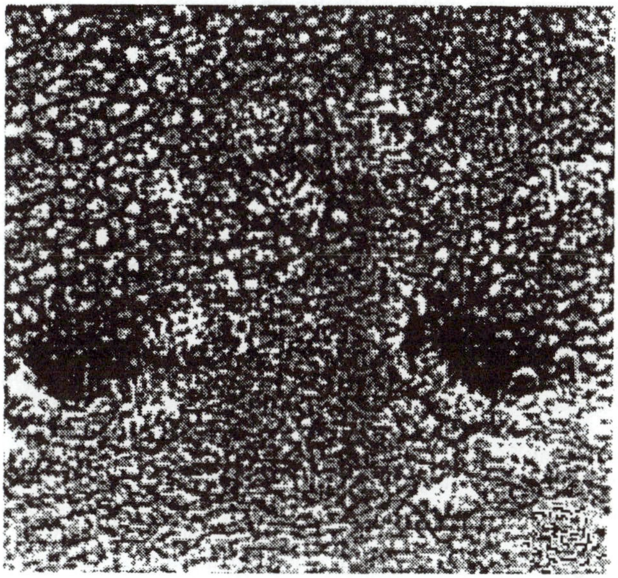

Females gain less advantage from repeated copulation than males (section 2.3.2) and in most cases there is an optimum rate for sexual interaction with males (section 2.4.1). One constraint on the female which forces her to contact males more often than might otherwise be optimum is the need to have viable sperm for the fertilization of her eggs. External fertilizers thus need to encounter a male whenever it is optimum for them to reproduce. One advantage to the female of internal fertilization (section 3.2) is that sperm received to fertilize one egg or batch of eggs could still be available for the next egg or batch. This renders the female virtually hermaphrodite, able to fertilize her own eggs (Cohen, 1977) and free from the need to contact a male every time she produces eggs.

Critical pressures in this trade-off are: (1) the savings in time, energy and perhaps risk of injury in being less constrained over when to have sexual encounters with males; and (2) the increased costs of using sperm that have been inside her body for extended periods. At least two factors could be important in determining these costs: (a) older sperm may be less fertile than younger sperm; and (b) the extended presence of non-self (i.e. from another individual) cells in the body may raise immunological problems in the recognition and attack of infectious organisms.

One way of ameliorating many effects of sperm age is to offer as hospitable an environment as possible or even to provide nutrients. The best way of reducing the problems associated with the presence of non-self cells is to collect the sperm together into discrete sites. Here, at least until they are needed for fertilization, the sperm can be isolated from the immunological war against infection that the female is waging in the main thoroughfare of her reproductive tract.

Sperm storage, once evolved, endows the female with more control over fertilization, effectively separating the acts of insemination and fertilization in both space and time (Birkhead and Møller, 1992).

Human cervical crypts

0.001 cm

Drawing of two human cervical crypts among inactive secretory cells from a photograph in Chretein, 1989.)

There is some indication that the cervix, with its storage crypts, may be an important site for sperm competition in primates, a possibility already suggested for the analogous storage tubules in birds (Birkhead and Møller, 1992). Within primates, the least developed cervices appear in the monogamous gibbons, *Hylobates* sp. In contrast, the highly polyandrous macaques (*Macaca* spp.) have a vast and complex cervix (Hafez, 1973).

3.5.2 The flowback: female ejection of sperm

The flowback is a mixture of seminal fluid, female secretions, sperm and other cells (originating in both the male and female) that emerges from the female tract and is lost to the outside after insemination.

To our knowledge, since the loss of sperm by female humans was described by Perloff and Steinberger (1964), there has been only

Box 3.7 Written instructions for the collection of flowback samples

The following instructions were given to volunteers participating in the study by Baker and Bellis (1993a,b) (Box 3.8). They are here given verbatim in the hope that they will make it easier for other workers to repeat and extend our research.

Instructions for collecting flowbacks
'You are provided with a large glass beaker and a smaller, screwtop jar containing fixative.

Flowback samples cannot be collected if copulation involves: (a) barrier contraceptives (condoms, caps); (b) any form of spermicidal cream or even other barriers (lubricants such as Vaseline); or (c) penis withdrawal before or during ejaculation.

When the penis is first withdrawn after copulation, there is invariably a certain amount of white material around the entrance to the vagina and in surrounding hair, etc. THIS IS NOT THE FLOWBACK. Ideally, the female should remain lying down for at least 15 minutes after the male has ejaculated. Collection becomes easier with time from ejaculation (it is really easy after an hour). It is a good idea for the female to try to ensure she will not need to urinate for perhaps 30–60 minutes after copulation.

Two methods of collection may be tried:

Method (a) works nearly every time: 35 minutes or more after the male ejaculated, the female stands up, then crouches down, legs apart, so as to compress the vagina and open the vaginal opening. Coughing followed (if able) by contraction of the vaginal muscles, repeated a few times, produces the flowback within about a minute. The flowback appears as a series of white drips/globules which should be collected in the LARGE BEAKER provided. DO NOT ATTEMPT TO COLLECT THE FLOWBACK DIRECTLY IN THE JAR OF FIXATIVE! To ensure the entire flowback has been collected, continue until coughing and contraction fail to produce any further flow. Finally, collect the last remnants by pressing the spout of the beaker against the vaginal opening. At this stage, any material sticking to the surrounding hair can also be collected.

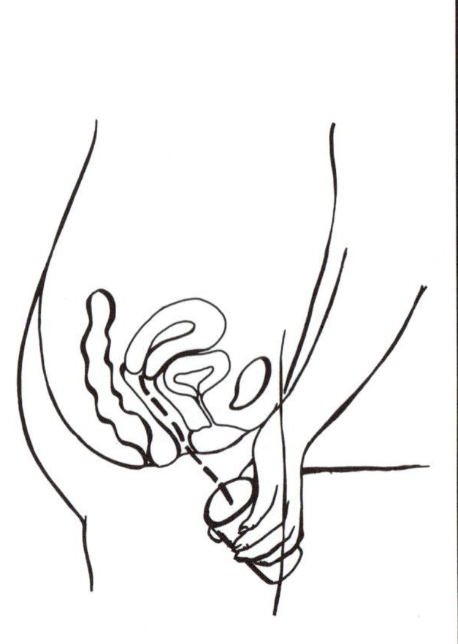

Method (b). Don't be discouraged if method (a) does not work the first time. It takes a little practice but then rarely fails. However, on any occasion that a flowback does not appear using method (a), use can be made of the fact that the flowback will emerge at urination. Collection, however, is tricky and requires some dexterity. Unfortunately, WCs are not well designed for flowback collection, a more-open situation being needed. A bath (empty) is ideal. The problem is to separate the urine flow from the flowback. Crouch in the bath and separate the lips around the urethra and vagina so that none of the urine drips backwards. Attempt to begin urination with as much force as possible so that the urine stream projects forward. A few seconds after urination has begun, the flowback emerges with some force. Careful positioning of the beaker is necessary in advance to catch the flowback while missing the urine. It does not matter if a few drops of urine go into the beaker but minimize the amount as much as possible. Again, some practice is necessary before this method becomes reliable.

Both methods (a) and (b) can be carried out by the female alone but are easier if a partner holds and positions the beaker. As soon as possible after the flowback has been collected (particularly if some urine contamination occurs), push any material sticking to the rim of the beaker down into the beaker with a finger, then pour the fixative into the beaker. Gently shake and swirl the fixative around in the beaker (taking care not to lose any) until all collected material is in the liquid, then deftly pour the fixative back into the screwtop container, screw on the top, and shake gently. As soon as possible, thoroughly wash the beaker with washing-up liquid and hot water so that there is no contamination of the next sample to be collected.'

one detailed study of flowback in humans (Baker and Bellis, 1993b). Between May 1988 and July 1993 we collected 150 flowbacks from 11 female volunteers (Box 3.7) and counted the number of sperm (and other cells) being ejected by the female (Box 3.8). In a separate study (Box 3.9), nine females estimated the volume of flowback without collection.

The human flowback emerges from the vagina as a discrete series of 3–7 white globules and measures normally up to 3 ml (Baker and Bellis, 1993b). Flowback occurs either while the female is still horizontal after copulation, when she next begins to walk, or, perhaps most often and most spectacularly, when she next urinates. Indeed, when the

Box 3.8 A direct study of whole ejaculates and flowbacks

The following methods were the basis of the study reported by Baker and Bellis (1993a,b), the results of which are presented at various points throughout this book. Volunteers were recruited through staff and students in the School of Biological Sciences, University of Manchester, UK.

Whole ejaculates were collected in condoms during copulation. Flowbacks were collected in a 250 ml glass beaker (Box 3.7). All samples were fixed in 2% glutaraldehyde in a phosphate buffer, pH 7.2; Pursel and Johnson, 1974).

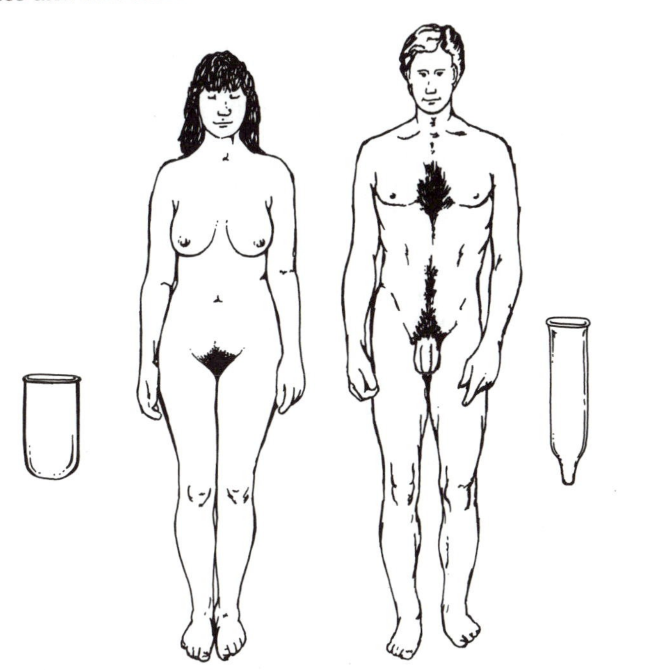

The male–female pair completed a questionnaire for each sample. Among the information requested was: whether the ejaculate was collected during masturbation or copulation; time since last ejaculation; time since last copulation; percentage time together (including sleeping time) since last copulation; day of the female's menstrual cycle; whether the female was taking an oral contraceptive; and whether the female experienced an orgasm associated with copulation and, if so, the timing of the orgasm relative to copulation and ejaculation. The female was also asked whether she had experienced an orgasm between the present and previous IPC and, if so, how many, under what circumstances and how many hours before the present IPC. The completed questionnaire was placed in an envelope which was then sealed.

Samples were returned for sperm counts to Bellis (MAB); questionnaires to Baker (RRB). Estimates of the number of sperm per flowback were based on procedures in the Human Semen Manual (Belsey *et al.*, 1987). All standard errors are within the range 5–25% and are not considered further. All counts were performed 'blind', the counter being unaware of any details of the sample being counted (except that flowbacks were visibly different from whole ejaculates).

Each screw-top jar contained 52 ml of fixative and weighed ≈75 g when first given to the volunteers. By the time the fixative was used, however, evaporation had often reduced the volume by up to 3 ml and the weight by up to 3 g. Although this weight and volume loss did not affect estimation of sperm numbers, it did prevent any reliable estimate of ejaculate volume.

Pairs were encouraged to return as many samples as they found socially acceptable. Details of the couples taking part in the study and a complete list of the number of whole ejaculates and flowbacks donated by each couple are given in Baker and Bellis (1993a).

Analyses in this book use the total number of samples available when that particular section of the book was being written. Total sample sizes therefore vary from section to section and are usually greater than those in Baker and Bellis (1993a,b).

All analyses in this book use Meddis' non-parametric rank-sum tests, either specific or non-specific (Meddis, 1984). The reasons for adopting these tests are discussed in detail in Baker and Bellis (1993a).

female urinates, the flowback is sometimes ejected with surprising force.

Median time to emergence of the flowback after male ejaculation is 30 minutes (IQR = 15–44) with a range of 5–120 min (Baker and Bellis, 1993b). All collected flowbacks have contained sperm and 94% of copulations monitored subjectively for flowback volume have been followed by noticeable flowbacks.

The range of sperm numbers in collected flowbacks has been from 7 million to 443 million (Box 3.10). Median number of sperm in these flowbacks was 131 million. This represents a median ejection of sperm that is about 35% of the number inseminated (Baker and Bellis, 1993b). We calculated that 12% of copulations were followed by virtually 100% ejection of sperm (i.e. <1% retained).

Box 3.9 The subjective estimation of flowback volume

Seven females, while willing to record details of their copulations and orgasms, preferred not to collect flowbacks directly but volunteered instead to estimate the volume of flowback subjectively. In addition, two pairs who collected flowbacks also made some subjective estimates of volume (Table I in Baker and Bellis, 1993a). All nine pairs recorded flowback volume as either 'normal' (for that female) (=2), heavier than normal (=3), lighter than normal (=1), or none (=0). Crudely, these estimates approximate to the volume of the flowback in ml. None of the subjects knew the results of the investigation of flowback samples that simultaneously was being carried out.

Although 115 copulations were monitored for flowback volume, one female did not classify six flowbacks that emerged at night.

Box 3.10 Frequency distribution of number of sperm in whole inseminates (collected by condom) and ejected in flowbacks

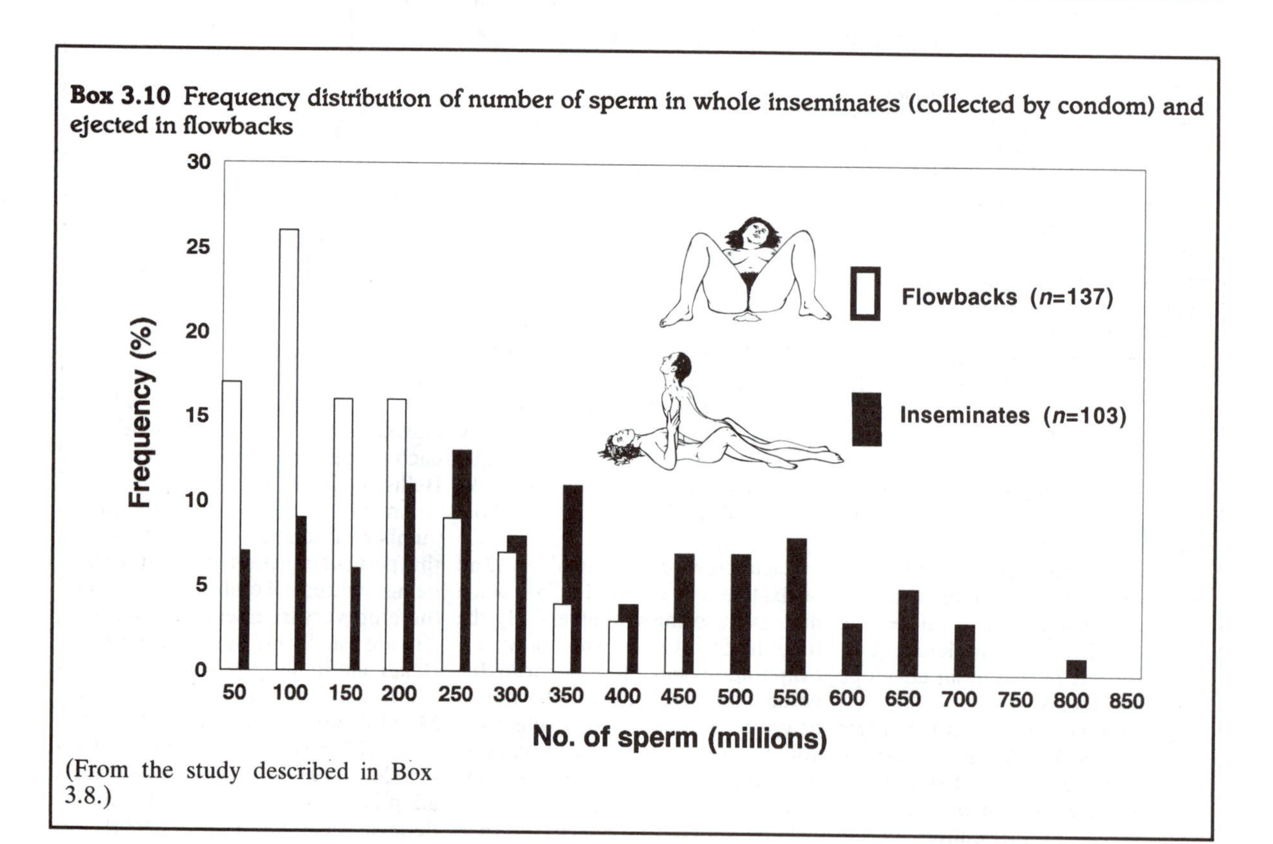

(From the study described in Box 3.8.)

Females appeared capable of near-total ejection of inseminates (Chapter 10).

Sperm ejection by the females of other animals has rarely been investigated. However, among the invertebrates, a few cases are known for insects. For example, the boll weevil, *Anthonomus grandis*, actively pumps sperm from the spermatheca (Villavaso, 1975). Surgical cutting of the nerves to the spermathecal muscles appears to prevent sperm ejection. If a male of the carrion fly, *Dryomyza anilis*, strokes the female with his antennae before copulation he can induce her to eject a drop of fluid containing sperm from a previous mating. Finally, males of the bush cricket, *Metaplastes ornatus*, have specialized plates with which, before copulation, they can open the female's genitalia. The female then everts the distal part of her reproductive tract and grooms away any sperm and other material from a previous mating (von Helversen and von Helversen, 1991).

Box 3.11 Ejection of flowback by female Grevy's zebra

(Photo © J. R. Ginsberg 1984.)

As far as vertebrates are concerned, there appears to be no information on sperm ejection by reptiles but there are relevant observations of birds and non-human mammals.

Female birds eject portions of inseminates, both spontaneously and in response to a male pecking their cloaca (Davies, 1983, 1985). Less than 10% of the inseminated sperm are retained in birds (Brillard and Bakst, 1990).

Ejection of the flowback by female Grevy's zebra (*Equus grevyi*) after mating is spectacular (Box 3.11). The volume of fluid emanating from the vagina is about 300 ml (Ginsberg and Huck, 1989). Loss of seminal fluid after mating has also been reported in swine (Sumption, 1961), rabbits (Morton and Glover, 1974a), and sheep (Tilbrook and Pierce, 1986).

Ginsberg and Rubenstein (1990) imply that the plains zebra (*Equus burcheli*) has no flowback but this may only be relative to the dramatic flowback of Grevy's zebra rather than absolutely. In rabbits, about 80% of sperm from an ejaculate are lost in the flowback within minutes of insemination (Morton and Glover, 1974a).

Our description of sperm competition theory in section 2.5 examined sperm competition from the perspective of the male. We gave the impression that the female tract is simply a passive receptable in which males play out their sperm competition games. The phenomenon of flowback warns (as have Thornhill, 1983; Eberhard, 1990, 1991; Baker and Bellis, 1993b; and Birkhead and Møller, 1993a) that this scenario may be misleading. By no means all of the sperm inseminated by the male ever attain a position in the sperm storage organs. In the rabbit, at least, only the minority stay in the female longer than a few minutes.

Even more importantly, the number and proportion of sperm in the flowback varies from insemination to insemination (Box 3.10). The possibility emerges that females may influence the number of sperm placed into store from any particular male on any particular occasion. The phenomena of sperm storage and ejection could give the female some element of control over the outcome of sperm competition.

By definition, flowback is confined to species in which the seminal fluids form soft, liquid or no copulatory plugs (sections 3.3.3 and 6.8). The females of at least some species with hard plugs, however, may show behaviour that is functionally comparable to ejection of the flowback. Thus, female fox squirrels, *Sciurus niger*, and eastern grey squirrels, *S. carolinensis*, often remove the plug with their incisors while grooming their genitalia shortly after copulation (Koprowski, 1992). Such variable behaviour could again give the female some element of control over the outcome of sperm competition.

We have argued (section 2.5.6) that females may double-mate and promote sperm competition in order to ensure that their sons inherit the ability to produce competitive sperm. However, competitiveness of sperm is not the only male characteristic that may be favoured by the female. Other features, such as behaviour, size or appearance may also be favoured (Darwin, 1871). Females with advantageous preferences will leave more descendants via their successful sons (Box 2.12). A female, while allowing the sperm to compete, may gain through biassing the outcome of sperm competition in favour of a particular male. One way in which she may achieve such a bias would be to retain and store more sperm from that male than from others.

The ejection of sperm with all of its consequences for the probability of conception from any given copulation is, in effect, a form of contraception (i.e. copulating while reducing the chances of conception). This and other forms of female contraceptive behaviour are discussed in more detail in Chapter 4.

3.5.3 Female copulatory orgasm

The female orgasm may be loosely defined as the sudden climactic release of sexual tension, release being manifest in the form of rapid contractions of various sections of the reproductive tract but primarily the uterus and vagina. We define a copulatory orgasm (= CO) as an orgasm while the male's penis is inserted in the female's vagina. This definition is different from that used previously (Baker and Bellis, 1993b) in which a copulatory orgasm was defined as any orgasm associated with copulation, including those during foreplay and postplay.

In humans, the female sexual response may be divided into three phases (Box 3.12): excitation; plateau; and orgasm (Masters and Johnson, 1966). These phases are associated with changes in heart and breathing rate, muscle tone, contractions of various parts of the reproductive tract and vocalizations. Often parts of the body (chest, neck and vulva) become red just before climax.

Not all copulations lead to female orgasm (Symons, 1979). Indeed, the uterine and other contractions associated with copulation are notoriously variable (Kelly, 1962). According to Hite (1976) only about 30% of women in the United States orgasm regularly during copulation. In Finland only 12% of 66 women reported always having an orgasm during copulation compared with 8% who reported never having an orgasm during copulation. The remaining 80% sometimes experienced copulatory orgasm and sometimes did not.

In our UK survey (Box 2.6), only 7% of females experienced orgasm at their first copulation. Only 53% experienced a copulatory orgasm during their first 50 copulations but 84% had done so on at least one occasion after 500 copulations. In an analysis of the most recent copulation by 3555 females, 65% of copulations involved a copulatory orgasm. Of 314 copulations monitored by 22 females as part of our study of whole ejaculates and flowbacks (Boxes 3.8 and 3.9), 46% involved copulatory orgasm. This is similar to the 49% of 1432 copulations reported to be orgasmic by 66 women in Finland.

Even when copulatory orgasms do occur, their timing relative to male ejaculation is very variable (Box 3.12). Approximately half (49%) of female copulatory orgasms climax more than 30 s before the male ejaculates, compared with 25% which climax more than 30 s after the male ejaculates and approximately 26% within ±30 s of male ejaculation.

Female copulatory orgasms are not, as often assumed, peculiar to humans and the behavioural and physiological changes associated with orgasm in humans have been reported in a wide range of non-human mammals (e.g. dogs, Evans, 1933; cats, Ford and Beach, 1953; cattle, Hartman, 1957; other primates, Kinsey *et al.*, 1953; Allen and Lemmon, 1981). As in humans, not all copu-

Box 3.12 Course and timing of female copulatory orgasms in humans

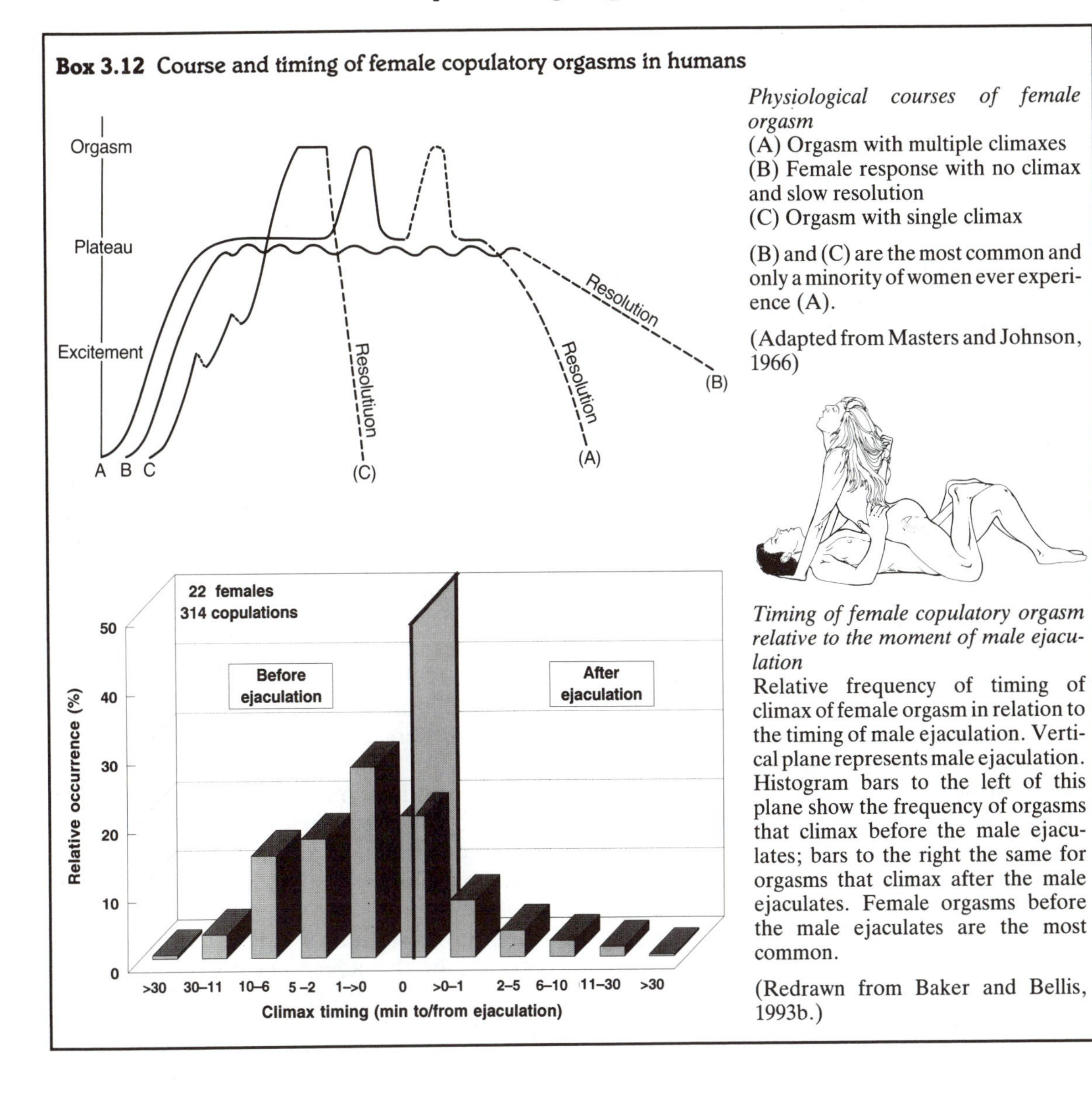

Physiological courses of female orgasm
(A) Orgasm with multiple climaxes
(B) Female response with no climax and slow resolution
(C) Orgasm with single climax

(B) and (C) are the most common and only a minority of women ever experience (A).

(Adapted from Masters and Johnson, 1966)

Timing of female copulatory orgasm relative to the moment of male ejaculation
Relative frequency of timing of climax of female orgasm in relation to the timing of male ejaculation. Vertical plane represents male ejaculation. Histogram bars to the left of this plane show the frequency of orgasms that climax before the male ejaculates; bars to the right the same for orgasms that climax after the male ejaculates. Female orgasms before the male ejaculates are the most common.

(Redrawn from Baker and Bellis, 1993b.)

lations in mammals lead to female orgasm (Kelly, 1962; Symons, 1979).

Allen and Lemmon (1981) reported orgasm in several adult, adolescent and juvenile female chimpanzees, *Pan troglodytes*, in response to manual stimulation of the circumclitoral area and vagina by the experimenter. One of the females allowed stimulation to continue to orgasm on ten separate occasions. The orgasms were manifest as 2–9 rhythmic perivaginal muscular contractions, after which further stimulation was rejected. Relatively few blatant emotional responses (e.g. vocalizations) were manifest by the chimpanzees, even during the peak of intense vaginal contractions.

The possibility that female birds and reptiles might also experience orgasm during copulation has not so far been investigated. Anticipating the results and discussions in Chapter 10, however, it seems likely that a comparable phenomenon will be found to exist in all animals with internal fertilization via copulation and in which the female has sperm storage organs and exhibits flowback.

Currently, there are two favoured hypotheses concerning the function of the female copulatory orgasm: (1) the ‘poleaxe’; and (2) the 'upsuck'. The poleaxe hypothesis proposes that, as humans are bidpedal, it is important for the female to lie down after copulation in order to reduce sperm loss (Morris, 1967; Levin, 1981). The orgasm thus functions to induce fatigue and sleep. The upsuck hypothesis proposes that the orgasm functions to suck up sperm during copulation (Fox *et al.*, 1970). These two hypotheses are tested empirically in Chapter 10.

Whatever the relationship between female copulatory orgasms and sperm uptake and loss, there is no indication that such orgasms are necessary for conception. In Chapter 10 we propose instead that female orgasm is concerned primarily with sperm competition. We suggest that the variability in the occurrence and timing of female copulatory orgasm is a primary strategy on the part of the female to exert some control over sperm retention and hence over the outcome of sperm competition.

3.5.4 The antiquity of sperm storage and manipulation by females of the human lineage

Sperm storage organs probably evolved in the human lineage soon after the evolution of internal fertilization (say 300 million years ago). For most of the time since, there have probably been storage organs, in the form of tubules or crypts, at the junction of the vagina and uterus with perhaps secondary storage sites or areas in the oviducts.

Sperm ejection (= flowback) seems widespread among mammals and is likely to have been a characteristic of the human lineage for at least the last 130 million years. In all probability, however, it too dates back to the dawn of the age of reptiles (say 300 million years).

The antiquity of the female copulatory orgasm is less clear than most of the facets of human sexual behaviour discussed so far in this book. In view of its apparently widespread occurrence in mammals, it seems certain to have been a feature of the human lineage for at least 100 million years. Most likely, however, for the reasons given below, it has been a feature for as long as copulation itself (i.e. since early in the age of reptiles).

From the very earliest of stages in the evolution of internal fertilization, as opposed to copulation, females seem likely to have evolved neural and muscular responses to help and hinder sperm in their travels through the female tract. The pre-adaptation for this evolutionary development was probably the series of contractions previously used by the female to move and expel eggs during the long evolutionary phase of external fertilization. If the human lineage passed through a phase of internal fertilization via external spermatophore deposition, as in modern urodele amphibians, neuromuscular contractions of the female tract to move around sperm may even have evolved before copulation (but after internal fertilization).

Some of these contractions may have been slow and drawn-out, moving sperm around slowly. Others, particularly if they functioned through the build up and sudden release of pressure in the female tract (as in the sudden expulsion of eggs by external fertilizers during spawning), may have been orgasmic. Such non-copulatory, spontaneous orgasms may have been particularly important during and after the evolution of sperm storage organs because of the advantage to the female of controlling how many sperm and from whom gained access to the storage sites.

We suspect, therefore, that spontaneous orgasms had their origins in contractions of the female tract which evolved hand in hand with the evolution of internal fertilization (say 300 million years ago; Box 2.1). Such neuromuscular systems for contraction will have been the raw material on which selection could have acted during the evolution of copulation to produce the copulatory orgasms discussed above. Perhaps even more directly, they could also have been the raw material for the evolution of those non-copulatory orgasms discussed in Chapter 5 in relation to female masturbation and homosexuality.

4 The mammalian inheritance: maternal care, family planning, and sperm polymorphism

4.1 Introduction

Lurking among the numerous and diverse populations of land vertebrates during the age of reptiles was a relatively insignificant group eventually destined, about 65 million years ago, to radiate and produce a wide diversity of mammals (including, 60 million years later, humans) (see Boxes 3.1 and 4.1). These mammal-like reptiles will undoubtedly have shown internal fertilization and in all probability will have had females which laid eggs, stored sperm, ejected flowbacks and sometimes experienced copulatory orgasms. Males will have shown some form of mate-guarding, produced some form of copulatory plug and on occasion employed forced copulation as a sexual strategy.

Relatively little has changed in the sexual behaviour of modern reptiles since that time. In the mammalian lineage, however, an event took place that, perhaps surprisingly, had far-reaching consequences in shaping the sexual programming of modern humans. Yet again, the major force to rebound from the event and generate new facets of sexual behaviour was sperm competition.

The event in question was the development of endothermy (a relatively constant body temperature maintained above that of the environment through internally generated thermal energy). Endothermy seems to have triggered (Boxes 4.2 and 4.3) an evolutionary sequence which began with selection for extended parental care by females and ended with those major features of female human reproduction, viviparity and pregnancy. The result was the evolution of a female reproductive tract that, as an arena for sperm competition, became of catacomb-like complexity.

4.2 Maternal care, viviparity and lactation

4.2.1 Maternal care and lactation

Female humans are typical mammals in that they show maternal care of their offspring. After conception (fertilization of the egg), the developing embryo is nourished and carried inside the female (section 4.2.2). In humans, the average time interval from conception to giving birth is 270 days (range 231–329 days; Kenneth and Richie, 1953). Thereafter, for up to 3 or more years (e.g. Banyarwanda, !Kung, Hartmann *et al.*, 1984), the female human lactates and nourishes the child with milk produced by ventral, thoracic mammary glands (Box 4.4). Ducts from these glands

Box 4.1 **A suggested course for the human lineage during its mammalian phase of evolution**

(Redrawn and adapted from Novacek, 1992).

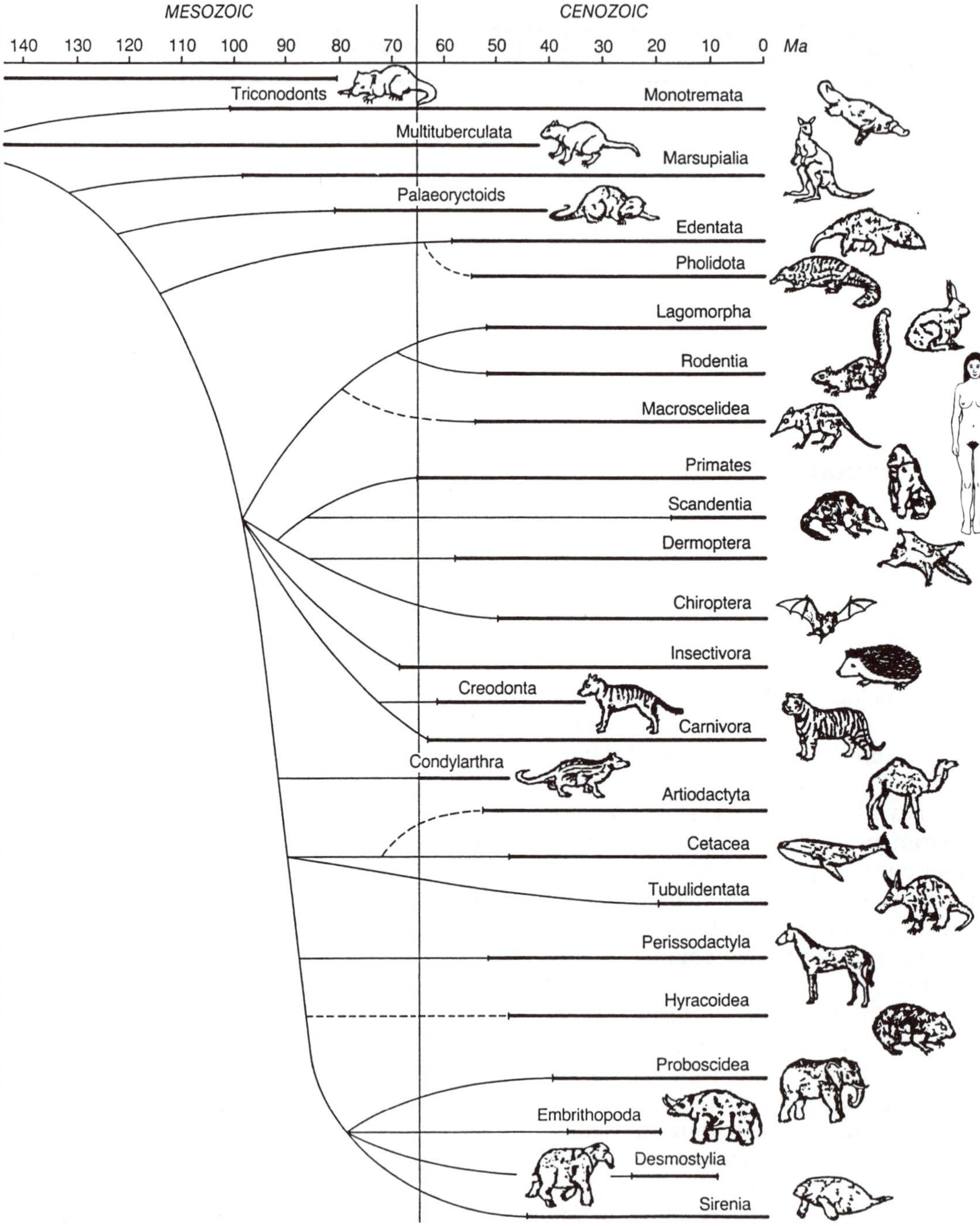

MESOZOIC
CENOZOIC
140 130 120 110 100 90 80 70 60 50 40 30 20 10 0 Ma
Triconodonts
Monotremata
Multituberculata
Marsupialia
Palaeoryctoids
Edentata
Pholidota
Lagomorpha
Rodentia
Macroscelidea
Primates
Scandentia
Dermoptera
Chiroptera
Insectivora
Creodonta
Carnivora
Condylarthra
Artiodactyta
Cetacea
Tubulidentata
Perissodactyla
Hyracoidea
Proboscidea
Embrithopoda
Desmostylia
Sirenia

Box 4.2 The behavioural ecology of parental care

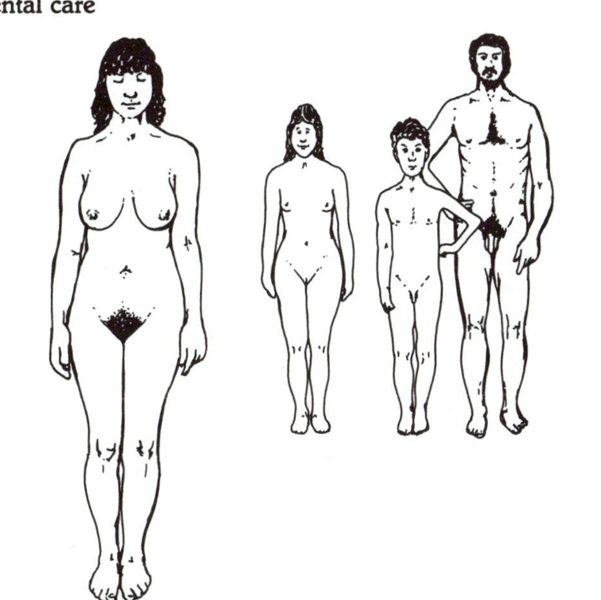

Parental care is uncommon among animals. Males rarely stay with their shed gametes beyond their fertilization of eggs; females rarely stay with their gametes beyond the point that the eggs are either fertilized and/or shed as zygotes. By minimizing the time and energy invested in each set of shed gametes, both males and females free more time and energy for potential future reproduction. We assume that individuals of species without parental care maximize their number of grandchildren by maximizing their number of children.

In a few lineages, one or other sex stays with the developing zygote/embryo until it has reached a particular level of independence. Thus, males of some species not only guard the female to the point of fertilization (section 2.4.4) but also guard the resulting young. Similarly, females of some species guard the young after fertilization, either inside or outside their body. Parental care from one or both parents usually protects the young against the ravages of climate and predation and/or increases the young's quality. We assume that individuals of species with parental care maximize their number of grandchildren by maximizing the survival and/or quality of fewer children (Box 2.2). We further assume that parental care only evolves in lineages in which: (1) adults are physically capable of protecting their young; and (2) the species' environment and ecology allows such protection.

In lineages that have evolved parental care, both parents benefit from the maximization of the number of grandchildren. Yet, in most such lineages, only one parent shows parental care. In many cases, this is because the ecology of the species militates against parental care by both parents. Thus, suppose male black bears were programmed to stay with the female instead of deserting her after fertilization. There may be less food for each, the female may lay down less fat in autumn, be less able to suckle the young over winter, and fewer of the pair's young may survive (see study by Jonkel and Cowan, 1971). Similarly, in many ducks, the young feed themselves from hatching. Consequently, the male may not help by staying, the presence of two adults may attract more predators, and fewer of the pair's young may survive. It is not difficult to see how, in such ecological conditions, both parents might benefit if one of them leaves.

Two factors may influence which parent leaves: (1) the parent with the first opportunity to do so; and, much more importantly, (2) the parent with least probability that they are the genetic parent of the offspring in their vicinity. Consequently, in external fertilizers (e.g. many fish; a few amphibians), the female is more likely to desert their young than in internal fertilizers (e.g. most birds and mammals). Among bony fish (teleosts), for example, only 3% of families with male parental care are internal fertilizers compared with 37% with female parental care (Gross and Shine, 1981). Most of the species with male parental care show male territoriality, other males being excluded from the area which females visit to have their eggs fertilized.

The parent who deserts conserves time and energy and has the opportunity to reproduce again with another partner. The partner who does not desert is left in the 'cruel bind' of having either to: (a) raise the offspring single-handedly; or (b) abandon the young (thus wasting the current attempt at reproduction) and trying again with a new partner.

open to the outside via a pair of protuberant nipples. The nipples are carried on enlarged pendulous breasts which contain much of the mammary gland tissue.

Humans usually have a single pair of nipples (the number of nipples possessed by a female mammal broadly correlates with normal litter size). Supernumerary nipples, however, are not uncommon in humans. These additional nipples occur along a line from the armpit to the groin. However, rarely are they associated with mammary tissue and are usually mistaken for moles (Snell, 1986).

The process of decreasing the child's

Box 4.3 Maternal care in the animal kingdom

Maternal care is uncommon in the animal kingdom. In most species, females do not stay with their gametes beyond the point that the eggs or young become separated from the mother's body.

Relatively few invertebrates have evolved any level of such maternal care. In a few insects, the mother stays to guard and protect her eggs, and sometimes her young (e.g. earwigs, De Geer, 1758 in Imms, 1951; some species of bugs and some cockroaches). Mothers in a few species of crickets not only guard their young but also feed them, first with small, infertile eggs that are layed especially and later with food brought in from outside (Evans, 1984).

Among vertebrates, maternal care is rare in fish and amphibians and even more rare in reptiles (Pough *et al.*, 1990). It is, however, the norm in birds (Birkhead and Møller, 1992), and in effect universal in mammals.

Only about 33% of fish families show any parental care and in only about half of these is that care maternal (Ridley, 1978). Many amphibians show no parental care at all (Pough *et al.*, 1990) and again, only about half of those that do (mainly frogs and toads) show maternal care (Ridley, 1978). In dendrobatid frogs, the eggs are laid on land and are transported by a parent to water. Usually it is the male who does the transporting but the females of at least seven species have been observed to show such maternal care (Silverstone, 1976). Female *Dendrobates pumilio* feed the tadpoles with unfertilized eggs (Weygold, 1980).

Apart from some guarding of young by various crocodiles, alligators and caiman, maternal care by reptiles is unknown (Pough *et al.*, 1990). In contrast, among birds, maternal care is absent in only a few species (e.g. obligate nest parasites, such as cuckoos; the megapodes of Australasia; and a few waders; Brooke and Birkhead, 1991). Even those female waders which lay eggs in clutches incubated by a series of different males usually eventually lay a clutch which they incubate themselves (Brooke and Birkhead, 1991) All together, the absence of maternal care in birds has been reported in only 8% of families (Ridley, 1978). All female mammals show maternal care, at least for the duration of lactation.

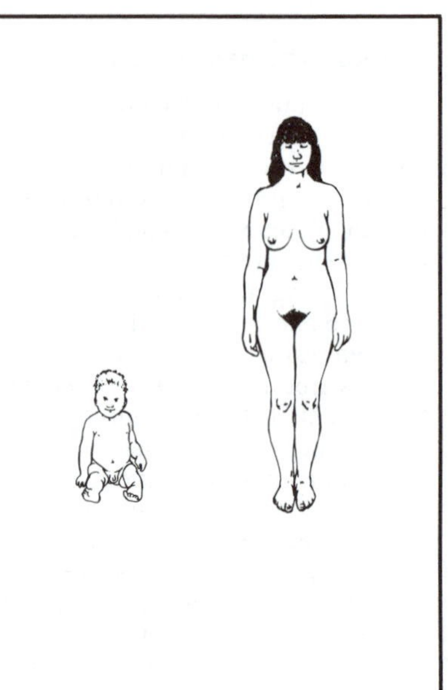

Maternal care was apparently a much more unequivocal advantage in bird and mammal lineages than in other vertebrates with internal fertilization. One possibility is that the adaptation was associated with the evolution in these two lineages of endothermy (a relatively constant body temperature maintained above that of the environment through internally generated thermal energy).

The advantages of endothermy are that: (1) chemical processes are faster at higher temperatures (e.g. for any given body size, the metabolic rate of birds and mammals is about 6 times that of an ectotherm; Pough *et al.*, 1990); and (2) enzyme and other chemical interactions can occur over a more limited range of temperatures. Fewer enzymes are thus needed and these can be more specialized and can function at maximum efficiency. Endothermy also removes the animal's dependence on external sources of heat (e.g. sunshine).

Endothermy is much less costly for larger than smaller animals due to volume:surface area ratios. Until young animals are large enough for endothermy to be efficient, some means of external temperature regulation, such as close association with an adult's body, could maximize the growth and survival of the young. Endothermy thus provides an adult with a means of influencing their offspring's quality and survival. As both bird and mammal lineages are internal fertilizers, the presssure to have such an influence was greater on females than on males.

dependence on mother's milk, eventually to zero, is termed weaning. After weaning, the human mother continues to show maternal care until her child becomes capable of independent survival. Although in some societies a child may be providing 50% of its own food by the age of five years (e.g. Hadza; Blurton Jones, 1989), full dietary independence may not occur until some time after puberty (the phase of ontogenetic transition from a non-reproductive to a potentially reproductive individual). Some mothers may maintain close association with their children, perhaps even throughout their lives. Others may show no further association beyond the age of puberty.

In terms of behaviour, maternal care is uncommon in the animal kingdom (Box 4.3). Its virtual absence in the early vertebrates and the principles outlined in Box 4.2 suggest that

Box 4.4 Lactation, breasts and nipples

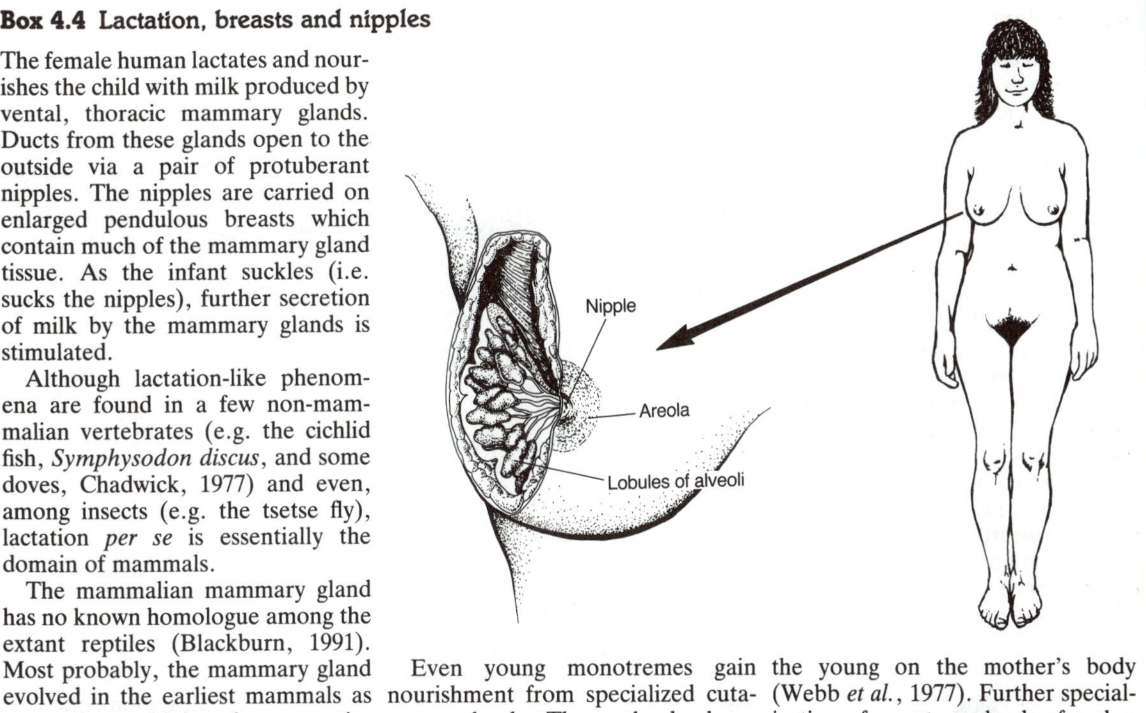

The female human lactates and nourishes the child with milk produced by vental, thoracic mammary glands. Ducts from these glands open to the outside via a pair of protuberant nipples. The nipples are carried on enlarged pendulous breasts which contain much of the mammary gland tissue. As the infant suckles (i.e. sucks the nipples), further secretion of milk by the mammary glands is stimulated.

Although lactation-like phenomena are found in a few non-mammalian vertebrates (e.g. the cichlid fish, *Symphysodon discus*, and some doves, Chadwick, 1977) and even, among insects (e.g. the tsetse fly), lactation *per se* is essentially the domain of mammals.

The mammalian mammary gland has no known homologue among the extant reptiles (Blackburn, 1991). Most probably, the mammary gland evolved in the earliest mammals as a mosaic combining the properties of other already extant cutaneous glands, most likely the apocrine and sebaceous glands (Blackburn, 1991).

Even young monotremes gain nourishment from specialized cutaneous glands. These glands, however, do not form well-defined teats but instead form milk areolae and open into a pair of longitudinal depressions in a temporary pouch for the young on the mother's body (Webb *et al.*, 1977). Further specialization of secretory glands of such a type to form discrete organs opening onto a nipple must have occurred during the early evolution of marsupials and placentals.

the human lineage has shown maternal care only since the origin of endothermy (say 180 million years). Lactation via nipples seems likely to have appeared soon after, during the early origins of therian mammals (say 135 million years ago; Pond, 1984).

Internal fertilization is thought to be an important factor in the evolution of maternal care. The usual explanation (Box 4.2) is that, in lineages which evolve internal fertilization, males have a reduced certainty of paternity because of the risk of sperm competition. In lineages which evolve both internal fertilization and parental care, therefore, the male is less likely than the female to respond to selection. Thus, in part, even maternal care owes its evolution in the human lineage to sperm competition.

Further features of human maternal care are viviparity and pregnancy. The evolution of these features had violent repercussions on the evolution of the female reproductive tract, the amphitheatre of sperm competition.

4.2.2 Viviparity and pregnancy

When the females of a species with internal fertilization shed their offspring to the outside as an egg, the species is said to be oviparous. If the young are born 'live' (i.e. not enclosed in an egg), the species is said to be viviparous. In some viviparous species the developing young obtains energy (through some form of filter) directly from the mother's blood supply, rather than from a supply of food (or 'yolk') in the egg. The females of such species are said to be pregnant during the period the young are inside them. The site of association

Box 4.5 Viviparity, pregnancy and the placenta

Humans, in common with all eutherian mammals, show viviparity and pregnancy and have a placenta.

Viviparity has evolved on a number of occasions in a number of lineages (e.g. the tsetse fly, *Glossina*, in insects; some sharks (in fish); and on a number of occasions in reptiles). Among mammals, monotremes are oviparous but marsupials and placentals are universally viviparous with a variable period of pregnancy (Kenneth and Richie, 1953). A placenta is universal among eutherian mammals but is also found in some sharks (e.g. the hammerhead, *Sphyrna tiburo*) and even in the occasional insect (e.g. the earwig, *Hemimerus telpoides*) (Hogarth, 1976).

The duration of pregnancy (= gestation period) varies greatly among mammals (see table).

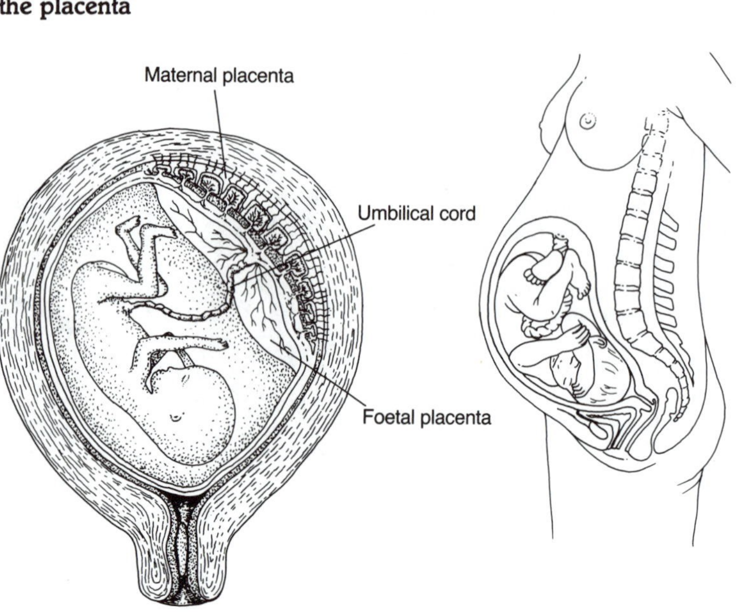

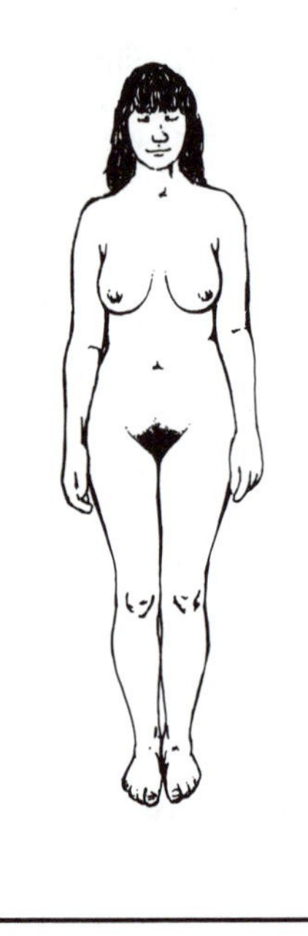

Gestation period in various mammals

Species	Gestation period (days)
Opossum (*Didelphis virginianus*)	13
Shrew (*Sorex araneus*)	20
Rabbit (*Oryctolagus cuniculus*)	31
Cat (*Felix cattus*)	63
Dog (*Canis familiaris*)	63
Guinea pig (*Cavia porcellus*)	68
Tiger (*Panthera tigris*)	105
Pig (*Sus scrofa*)	114
Sheep (*Ovis aries*)	150
Rhesus monkey (*Macaca mulatta*)	163*
Chimpanzee (*Pan troglodytes*)	227*
Human (*Homo sapiens*)	280*
Cow (*Bos taurus*)	280
Blue whale (*Balaenoptera musculus*)	365
Camel (*Camelus dromedarius*)	390
Black rhinoceros (*Diceros bicornis*)	540
Indian elephant (*Elephas maximus*)	623

* Taken from first day of last menstruation

with the mother's blood supply is the placenta.

Humans, in common with all eutherian mammals and various other lineages, show viviparity and pregnancy and have a placenta (Box 4.5).

The human lineage has probably shown viviparity since the earliest origins of therian mammals (say 135 million years; Pond, 1984) and shown pregnancy with a placenta since the origin of placental mammals (say 100 million years). This is the antiquity, therefore, of the broad shape of the arena for human sperm competition.

4.3 The female tract: the arena for sperm competition

The reproductive tract (Box 3.4) of female humans is the arena within which sperm compete and its structure is of critical importance to the nature and course of human sperm competition. Conditions within the tract are not constant but vary in a regular way with the female's menstrual cycle (Box 4.6).

Specialization of sections of the female tract to form the uterus and cervix was clearly an evolutionary innovation of early mammals, presumably concomitant with the evolution of viviparity and pregnancy (section 4.2.2). The uterus provides the appropriate environment and the cervix both holds the developing fetus in position within the uterus and, in combination with its own column of mucus (Box 3.5) and a highly acidic vagina (Box 3.4), reduces the risk of invasion of the area by infectious organisms.

Many such organisms are likely to be introduced during copulation, either on the penis or in the ejaculate (e.g. the bacterium, *Escherichia coli*; Aroux *et al.*, 1991). Rapidly, however, a secondary function of the cervix was probably to become a site for influencing the progress of sperm up the female tract. In particular, the cervix could endow the female with influence over whether and which sperm reach the sperm storage organs (section 3.5.1). This control is more potent if the sperm storage organs become concentrated beyond the main site for sperm selection, the cervical mucus (Box 3.5).

Viewed in this light, facets of the cervix and cervical mucus may have been the target of selective pressures generated by sperm competition. Further up the female tract, the oviducts could also have been the target of selective pressures generated by sperm competition. These tubes vary considerably in straightness and tortuosity even in primates (Hill, 1970b) All these regions of the reproductive tract could endow the female with an ability to be much more than a passive receptacle in which males play out their sperm competition games. Rather, in association with orgasms (Chapter 10), they allow the female a real opportunity to have some direct influence over the outcome of such games.

The earliest mammals seem likely to have had not only paired ovaries and oviducts as in humans but also paired uteri, cervices and even vaginae (though with a single vaginal opening), such as found in modern marsupials such as the oppossum and kangaroo (van Tienhoven, 1968). In the lineage leading to modern placentals, however, the vaginae fused to form a single vagina but with the upper parts of the tract remaining paired, as in modern rats and rabbits.

Fusion of the cervices to form a single cervix, while retaining two 'horns' to a rather small uterine body, as in modern pigs, dogs, cats and lemurs (Eckstein and Zuckerman, 1956; van Tienhoven, 1968), occurred in the lineages leading to ungulates, carnivores and primates. The uterine horns became relatively short in the lineages leading to modern horses and cows and more or less disappeared completely in the lineage leading to the anthropoid primates through to humans. Evolutionarily, the development of a single uterus with no, or only short, uterine horns seems to have been associated with a reduction in litter size to two or one, as occurred in ungulates and primates.

The major regions of the female tract were almost certainly shaped in the earliest therian (130 million years ago) and placental (100 million years ago) mammals and are likely to have been exposed to selection via sperm competition ever since. However, as even prosimians have distinct uterine horns (Eckstein and Zuckerman, 1956), the final general form of the human tract, with a single uterus and cervix, was probably not shaped until the early anthropoid members of the human lineage (say 50 million years ago).

4.4 Family planning and contraception: the female perspective

Any consideration of sperm competition in modern humans has to take place within the framework of the array of forms of family planning available to the people being dis-

Box 4.6 Ovulation and menstruation: the menstrual cycle

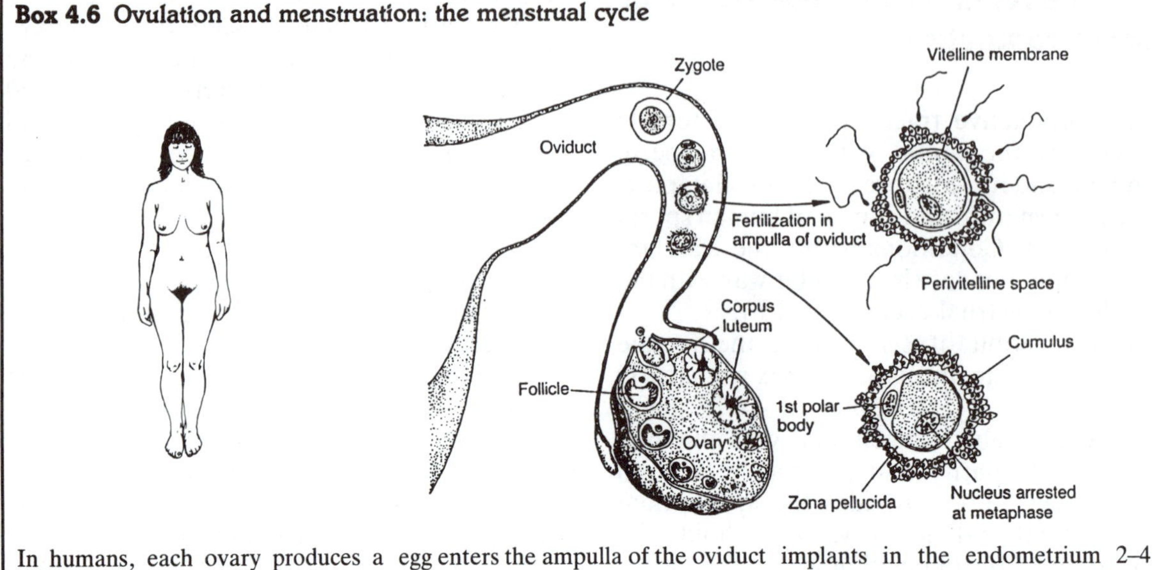

In humans, each ovary produces a mature egg at roughly two-monthly intervals. Through alternation, the female ovulates roughly once a month (but see Box 4.10). In advance of each ovulation, 6–12 primary follicles of the thousands present in the ovary (Box 3.4) begin to grow. After about a week of growth, however, one of the follicles begins to outgrow the remainder which subsequently break down (Guyton, 1991). The surviving follicle eventually grows to 1.0–1.5 cm just before ovulation at which time the wall of the ovary ruptures and the egg is expelled with fluid from the follicle.

The egg is shed into the body cavity and is then drawn into the ampullae of the oviducts on a current of body cavity fluid. The current is created by the beating cilia of the fimbriae at the ends of the oviducts (Harper, 1988). If fertilization occurs, it does so, as with most mammals (Ahlgren, 1975), soon after the egg enters the ampulla of the oviduct (Guyton, 1991).

Passage of the egg or developing zygote down the oviduct to the uterus takes about 80 h (Harper, 1988), speed of travel thus being just over 1 mm/h. Transport is affected mainly by a feeble fluid current in the tube resulting from epithelial secretion plus action of the ciliated epithelium that lines the tube, the cilia always beating towards the uterus (Guyton, 1991). Weak contractions of the oviduct may also help. If the egg has been fertilized while in the ampulla, the zygote divides during its journey and by the time it reaches the uterus is a blastocyst of about 100 cells (Guyton, 1991).

Each month, changes take place in the inner lining (= endometrium) of the uterus in preparation for the arrival of the egg (Guyton, 1991). Primarily, the walls become thicker (increasing from 1 to 3–4 mm) and more highly vascularized. If the egg has been fertilized, the blastocyst implants in the endometrium 2–4 days after entering the uterus. While waiting to implant, the blastocyst receives nutrition from endometrial secretions known as uterine milk and develops special trophoblast cells on its surface. These cells secrete enzymes which digest and liquefy nearby cells in the endometrium. The trophoblasts then multiply rapidly, and invade, digest and imbibe further endometrial cells en route to achieving implantation and eventually developing into the placenta and various other membranes (Guyton, 1991).

If the egg was not fertilized or if a fertilized egg fails to implant or if an implanted egg/embryo is spontaneously aborted, the endometrium shrinks, breaks down, is sloughed off and is shed (= menstruation). In humans and many other primates (Hrdy and Whitten, 1987) the products of menstruation emerge from the vagina as a bloody discharge. In part, menstruation may be an

cussed. It is important to realize, however, that family planning is not a modern, or even a human, invention but that it is part of humankind's mammalian inheritance.

In its modern sense, family planning is the regulation of the timing, spacing and number of children to match some individually perceived optimum. In its adaptive sense, it is the optimization of the timing, spacing, number, paternity/maternity and sex, etc. of children in order to maximize the individual parent's reproductive success.

It is unfortunate, but typically human, that the phrase 'family planning' has such strong overtones of conscious strategy. In this book, to avoid confusion, we persist with the use of this familiar phrase despite arguing that over the course of human evolution the major part

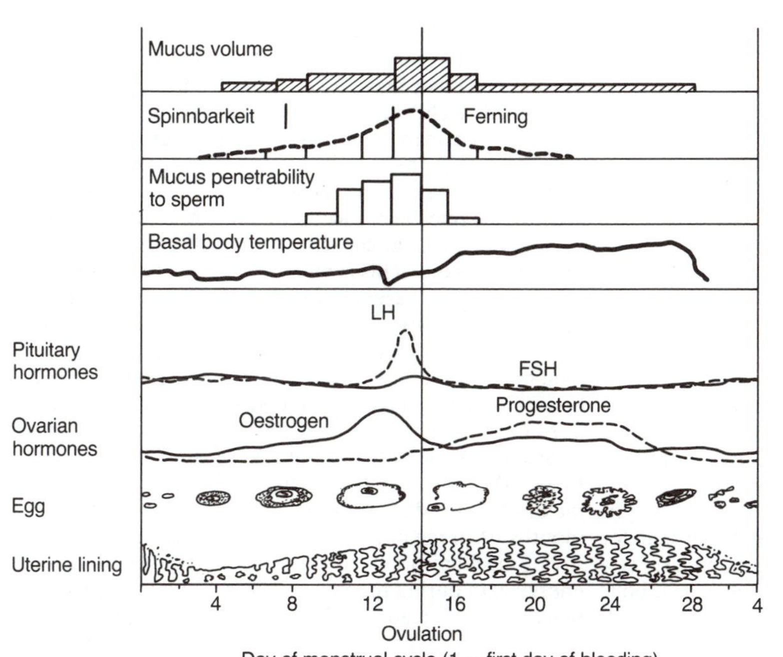

adaptation for defence against pathogens carried into the uterus by sperm (e.g. *Escherichia coli*; Auroux *et al.*, 1991) (Profet, 1993).

In the case of an unfertilized egg, menstruation typically begins at a fairly fixed 14 days after ovulation (Clark and Zarrow, 1971) and continues for about 5 days. In contrast, the time interval from the initiation of menstruation to the next ovulation is highly variable. The result is a range of lengths of the menstrual cycle in normal, healthy women of from 21–42 days (Symonds, 1992).

Such variation is found not only between women but from cycle to cycle in the same woman (Chapter 6). The traditional gynaecological view that an intraindividual variation in the length of the menstrual cycle greater than five days is an indication of hormonal disturbance (Weingold, 1990), or even suggestive of disease, has now been re-appraised (Münster *et al.*, 1992). In a thorough longitudinal study of nearly 4000 women in Denmark, over two-thirds of women aged 15–44 years showed more than a 5-day variation in cycle length between cycles over the course of a year (Münster *et al.*, 1992).

Conventionally, studies of the menstrual cycle of human females use conversion to a standardized 28-day cycle (McCance *et al.*, 1937) as a means of identifying patterns in the face of such variation. This standardized cycle may be divided into three major hormonal phases (Hawker, 1984): I (days 1–5, menses); II (6–14, proliferative or follicular); and III (15–28, secretory or luteal). In this standardized cycle, ovulation occurs on day 14 and copulations are most fertile on days 9–14 (peak fertility = day 12; Barrett and Marshall, 1969). The rank-order of the three phases for fertility is II>I>III (i.e. phase II is the most fertile). In cases of rape, however, there is a greater spread of conceptions through the cycle (Box 6.10).

The typical sequence and timing of hormonal and other events in a standardized human menstrual cycle is shown in the figure and discussed further in Chapter 6.

of family planning has derived from subconscious mechanisms. Here, we briefly introduce from the female's perspective those natural methods of family planning that are part of humankind's mammalian inheritance. In Chapter 7, we consider the interplay of natural and modern forms of family planning and discuss the extent to which the latter is also adaptive.

At first sight, the best method of family planning would seem to be to avoid copulation except at those times that conception is advantageous. However, as discussed in section 2.4.1, copulation can increase a female's reproductive success in many and diverse ways in addition to being the preliminary necessary for fertilization. Yet females who conceive and attempt to carry and raise a

Box 4.7 Energetics of female reproduction: fat distribution, body weight and nutrition

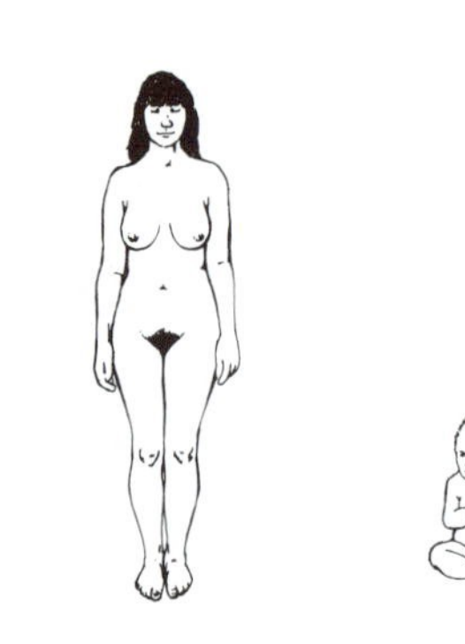

The energetics of reproduction
Humans, like most primates, have a single offspring with very slow fetal and postnatal growth rates. Consequently, the energetic costs of primate reproduction are lower than for any other group of mammals (as a percentage of the mother's metabolic turnover; Prentice and Whitehead, 1987).

Pregnancy and lactation should require an increase in caloric intake of 14% and 24% respectively but even affluent women, eating to appetite, do not meet these levels (Prentice and Whitehead, 1987). Even so, there is no apparent impairment of reproductive performance.

If energy intake is restricted, fetal growth rate may be reduced by about 10% but lactational capacity is relatively unaffected. This is due to utilization of the fat reserves which thus serve as an important buffer (Prentice and Whitehead, 1987). Basal metabolic rate is lowered during the first 6 months of pregnancy and again during lactation. There may also be reduced female activity at these times. Despite an increase in basal metabolic rate during the last three months of pregnancy, these energy-saving mechanisms may lead to a nett energy gain during pregnancy (Prentice and Whitehead, 1987).

The role of fat stores
The main function of adipose tissue is the uptake, storage and controlled release of lipids; processes involving large and rapid changes in tissue mass. In most invertebrates as well as ectothermic vertebrates, the storage tissues are intra-abdominal, near the centre of gravity (Pond, 1992a,b). Mammals and birds differ from ectothermic vertebrates in that adipose tissue occurs in a dozen or more discrete depots, associated with several different organs, including viscera and skeletal muscle.

Nutrition and other energy-flux factors seem to be a major determinant of female reproductive success in mammals (deer, Clutton Brock *et al.*, 1982; monkeys, Whitten, 1983; apes, Wrangham, 1979). It is generally accepted that food restriction or prolonged exercise can suppress pubertal development and ovulation while simultaneously depleting fat reserves, suppressing growth and decreasing lean body mass (Bronson and Manning, 1991). Cause and

Adipose tissue distribution in some mammals

Deer
Tiger
Human
Guinea pig

child when environmental (either ecological and/or social) circumstances are suboptimal may actually reduce their reproductive success.

There are two main factors: (1) the stress and damage that a female may suffer through an untimely attempt to raise a child may reduce her chances or ability to reproduce when circumstances improve: and/or (2) the child may suffer to the point of having

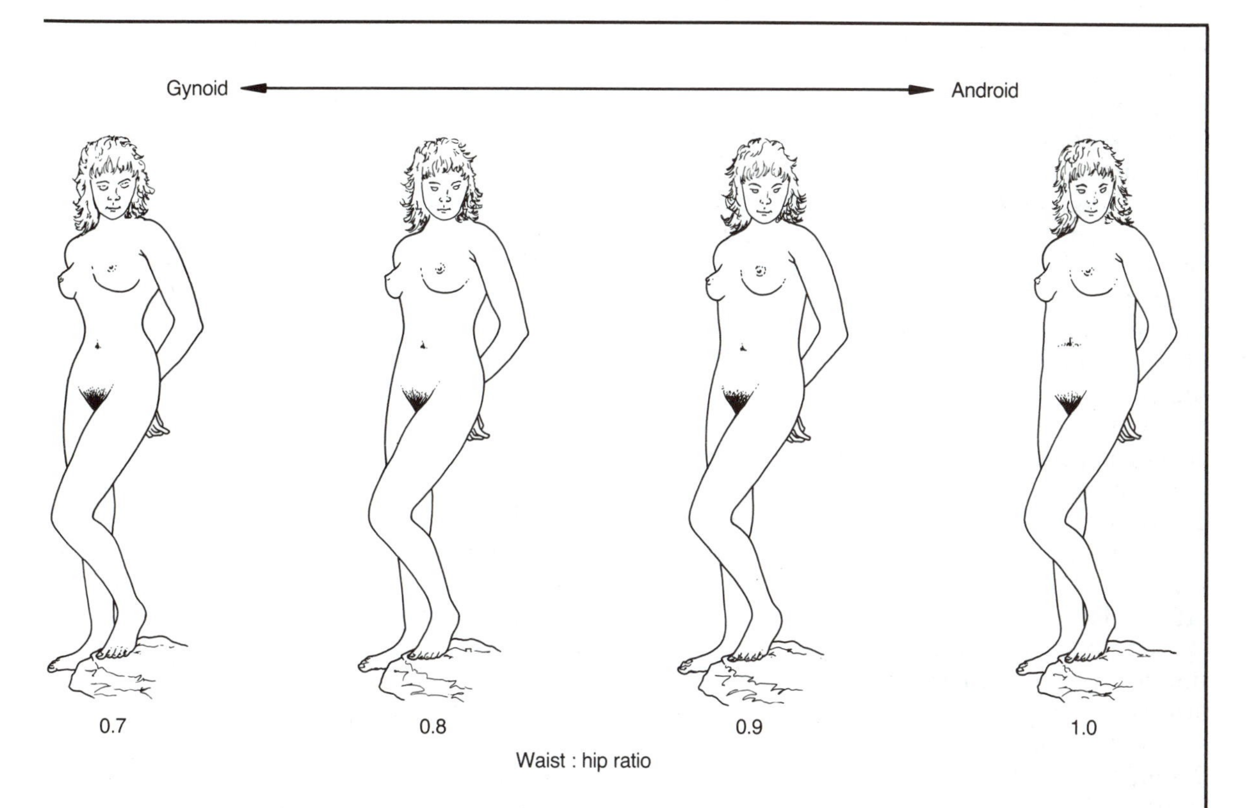

effect in this sequence, however, is less clear and it now seems (Bronson and Manning, 1991) that body fat does not play as central a role in the energetic regulation of reproduction as was first envisioned by Frisch and McArthur (1974). For example, acute starvation blocks ovulation in the Syrian hamster, but this can be countered simply by adding glucose to the drinking water (Morin, 1986). Blood sugar levels may thus be the central factor and any involvement of fat stores could derive from their influence on these levels.

Fat distribution in female humans: android and gynoid shapes and the waist-hip ratio

The distribution of adipose tissue in mammals does not yet form a clear evolutionary picture, there being no obvious evidence for the previously expected role in insulation (Pond, 1992a,b).

Humans show considerable sex differences in fat distribution and human females are unusual (though not unique) among mammals in having large subcutaneous fat stores. The most striking gender-specific differences are seen in the regions of the abdomen and the buttocks and thighs (gluteofemoral) (for review, see Björntorp, 1991). Testosterone stimulates fat deposits in the abdominal region and inhibits fat deposits in the gluteofemoral region. The oestrogens, in contrast, inhibit fat deposits in the abdominal region and maximally stimulate fat deposits in the gluteofemoral region more than in any other region of the body. These hormonal influences produce a gynoid (female-like) or an android (male-like) body fat distribution. The best index of these body shapes is the waist:hip ratio (WHR), low WHR ratios indicating a gynoid shape, high ratios an android shape.

The fat deposit from the gluteofemoral region is almost exclusively used by females during late pregnancy and lactation (Björntorp, 1987; Rebuffé-Scrive, 1987). Body fat distribution in women of reproductive age seems to have more impact on fertility than age or obesity. A high WHR correlated with irregular menstruation (Hartz *et al.*, 1984) and a 0.1 unit increase in WHR led to a 30% decrease in probability of conception per cycle during artificial insemination (Zaadstra *et al.*, 1993).

(Fat distribution in mammals modified from Pond, 1992a; WHR drawings inspired by Singh, 1993).

reduced survival and reproductive prospects. Selection should thus favour females who time their attempts to raise children such that they maximize their number of grandchildren. Critical factors will be the age of first reproduction and the spacing of subsequent reproductions.

Human litter size is relatively invariable, being rarely more than one (twins = 1.25% of births; triplets = 0.01%; quadruplets =

0.0003%). This implies that the lineage has been exposed to strong selection for females not to attempt to raise too many offspring at once. In contrast, age of production of the first litter varies considerably from society to society (e.g. 18 years, 21 years and 28 years for the Ache, UK and Andamanese respectively: Hill and Kaplan, 1988; Baker, 1978; Cappieri, 1970). So, too, does the spacing of subsequent litters. For example, in the Ache of Paraguay (who, until the 1970s, were a society of forest-living nomadic hunter–gatherers), mean interbirth interval was 3.2 years (range 1.2–9.0 years; mean for individual women, 1.9–5.0 years) (Hill and Kaplan, 1988). Intervals were shorter for, say, the Hutterites (Tietze, 1957) and longer (about 4 years) for the desert living !Kung San of Africa (Howell, 1979).

Attempts to determine if the observed timing of first, and spacing of subsequent, attempts at reproduction in different human societies are optimal are at an early stage. In a detailed study of the !Kung hunter–gatherers, Blurton Jones (1989) concluded that the observed interbirth interval of 4 years could well be optimum. A suggested critical factor was the energetic (and hence mortality) costs to the mother of having to carry both the child and often heavy food on foraging trips. Another potential critical factor, to some extent supported for the Ache, is the time taken for females to rebuild their body fat reserves after each reproduction (Hill and Kaplan, 1988). One factor hindering all such attempts at evaluating adaptiveness is a continuing level of confusion over the precise interaction of fat distribution, body weight, and nutrition in the energetics of female reproduction (Box 4.7).

However far we may be from understanding the ecological details of female family planning, the general principle seems clear. Females are faced continuously with conflicting pressures from the non-reproductive advantages of copulation and the reproduction advantages of not attempting to raise new offspring at suboptimal times. These conflicting pressures seem to have generated in females the evolution of contraceptive physiology and other family planning behaviour. Strategically, the function of evolved mechanisms for family planning is to continue to gain advantage from copulation while reducing the costs of the untimely production of offspring. Not surprisingly, with such a powerful function, family planning is a widespread feature of mammalian physiology and behaviour.

Although a female's reproductive success (say, number of grandchildren) may be primarily a function of the number and timing of her attempts to have and raise children, other more subtle factors may also be influential. Among these are the sequence and final ratio of her children's sex, paternity and behavioural type (e.g. stay-at-homes v migrants; see analysis of the reproductive consequences of human migration in 19th century Sweden, Clarke, in press). Any mechanism that allows the female to influence the array of these characteristics in her offspring could therefore be a target for selection.

Female humans share with the females of other mammals a number of natural family planning techniques: (1) contraception (i.e. any physiological or behavioural mechanism that reduces the probability that sperm will fertilize the egg(s)); (2) prevention of implantation of the fertilized egg; (3) spontaneous abortion of the developing fetus (= miscarriage); and (4) infanticide. There is not a stage of the reproductive process at which females do not have some facility for family planning.

We assume that the evolution of such a set of backup forms of family planning in mammals is a response to the risk that circumstances may change rapidly. An environment that may be conducive to reproduction one month may deteriorate rapidly the next, thus favouring females who abandon their current attempt.

In the next few sections, we discuss in turn each of the methods of family planning available to females. Many of these mechanisms could also be used by the female to influence the outcome of sperm competition. In our discussion, we use the term stress to describe the physiological syndrome by which suboptimal external (e.g. lack of food; lack of social support) or internal (e.g. infection) events are mediated internally to produce an appropriate response.

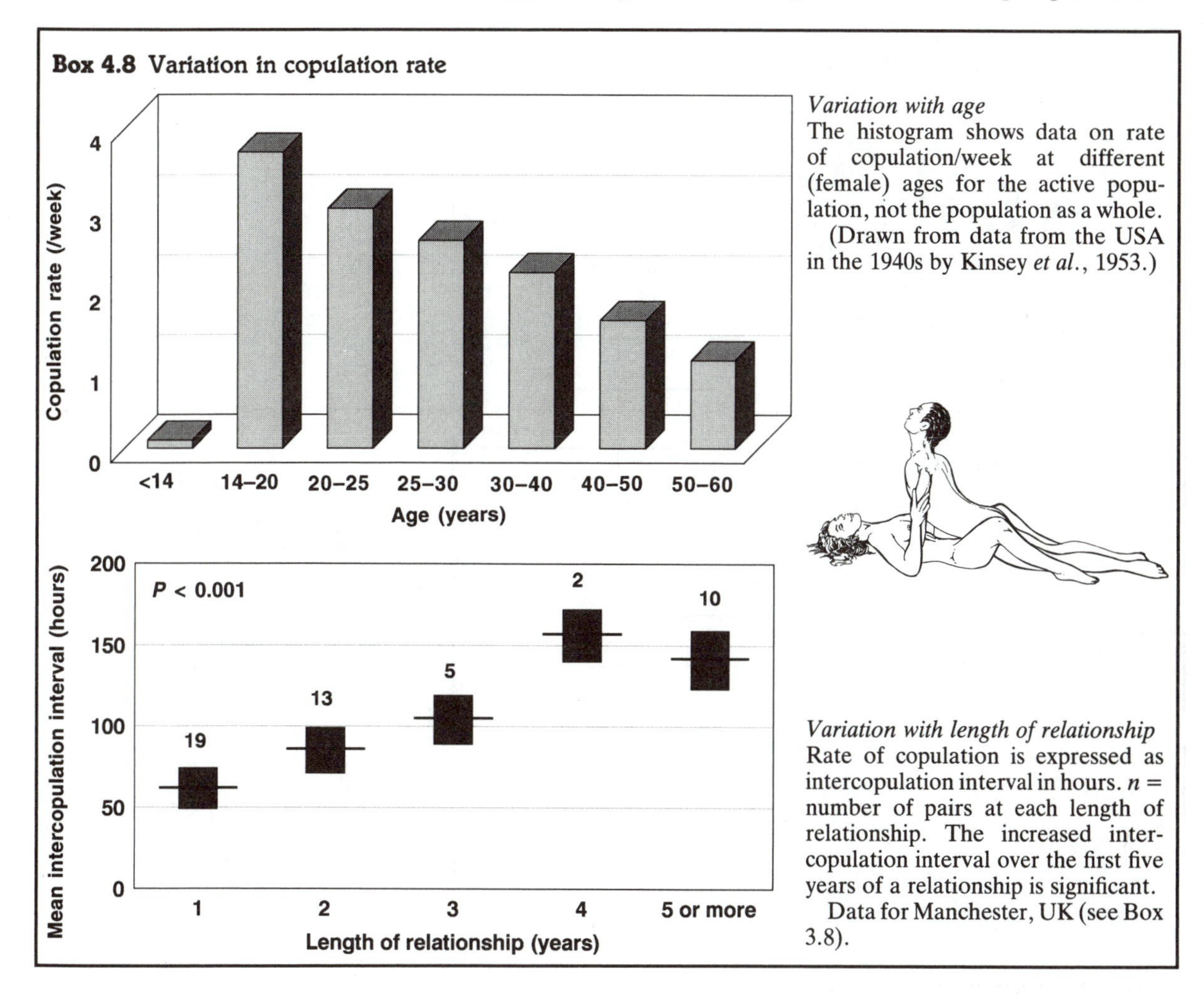

Box 4.8 Variation in copulation rate

Variation with age
The histogram shows data on rate of copulation/week at different (female) ages for the active population, not the population as a whole.
(Drawn from data from the USA in the 1940s by Kinsey *et al.*, 1953.)

Variation with length of relationship
Rate of copulation is expressed as intercopulation interval in hours. n = number of pairs at each length of relationship. The increased intercopulation interval over the first five years of a relationship is significant.
Data for Manchester, UK (see Box 3.8).

4.4.1 Contraception: avoiding copulation

Most of the methods of family planning discussed in this section assume that the female continues to copulate as part of a strategy to gain the various indirect benefits of copulation (section 2.4.1). However, the most direct method of avoiding conception is to avoid copulation.

The majority of female animals, not least mammals, avoid or resist copulation (section 2.3.2) unless they are approached in a suitable way by a suitable male at a suitable time. Female copulation may thus vary considerably in space and time depending on circumstance. Mammalian females may only seek or allow copulation during particular phases of their cycle but then may copulate many times in a short space of time. Female lions, for example, in the few days preceding ovulation, may copulate every 15 minutes, day and night, and are inseminated by all (usually 2–3) males within the pride (Bertram, 1975). Within monogamous pairs (e.g. many birds and a few ungulates and primates), the couple tend to settle into a characteristic pattern and rate of population, which may also have peaks and troughs linked to female cycles of receptivity.

Most surveys of industrialized human societies suggest an average copulation rate of 1–3/week within couples (Smith, 1984a,b). Anthropological records suggest a much more varied level of copulation rate. Thus, although the Keraki (1/week), Lesu (1–2/week), Chiricahua and Trukese (2–3/week) and even the Hopi Indians (3–4/week) claim levels comparable to industrialized societies, most (e.g. Siriono of eastern Bolivia) claim rates of 1/day or more (Ford and Beach, 1952). The reliability of the more extreme

Box 4.9 The duration of maternal care and the evolution of puberty

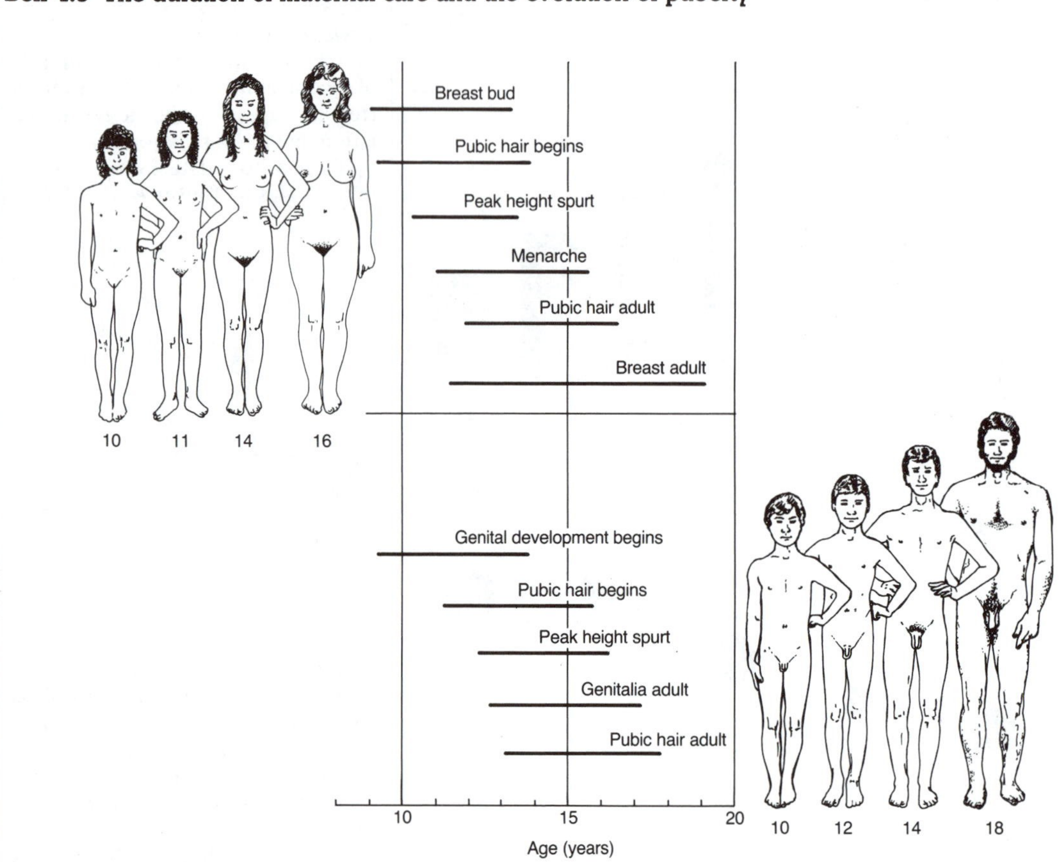

Populations come to be dominated by hereditary characteristics that cause their possessors to multiply (via descendants) faster than their contemporaries (Box 2.2). Mathematically, selection thus favours lineages in which parents produce more offspring which survive to reproduce and/or produce young which begin to reproduce as early in their lives as possible. Biologically, however, attempting to breed before the individual's location, body and experience is optimal for the animal's ecology may lead to reduced or zero lifetime reproductive success (e.g. the body may have insufficient energy reserves for both reproduction and survival; in competing too soon for access to females, a small, young male may be damaged by larger, older males). The timing of puberty for any given species should reflect the optimum compromise, in the context of the animal's ecology, between these two opposing (mathematical vs biological) pressures.

During prepuberty, time and energy is channelled, not into reproduction, but into body growth and the gaining of experience appropriate to each stage of body development. As such, the existence of the phase of prepuberty seems to be an adaptation to shorten the time to first reproduction.

The final phase of the transformation from a prereproductive to reproductive individual is the process of finding a suitable place in which to reproduce and a partner or partners with which to do so. In almost all

reports, however, is difficult to assess. In many cases, they probably reflect targets rather than performance.

Even societies with extreme claims (e.g. Lepcha, 5–6/night; Ford and Beach, 1952) acknowledge that copulation rate declines with age (to about l/night at 30 years and eventually to once every 2–4 nights). The more reliable estimates for industrialized societies show clearly a decline in copulation rate with age (Box 4.8). There is also a decline with length of relationship (Box. 4.8).

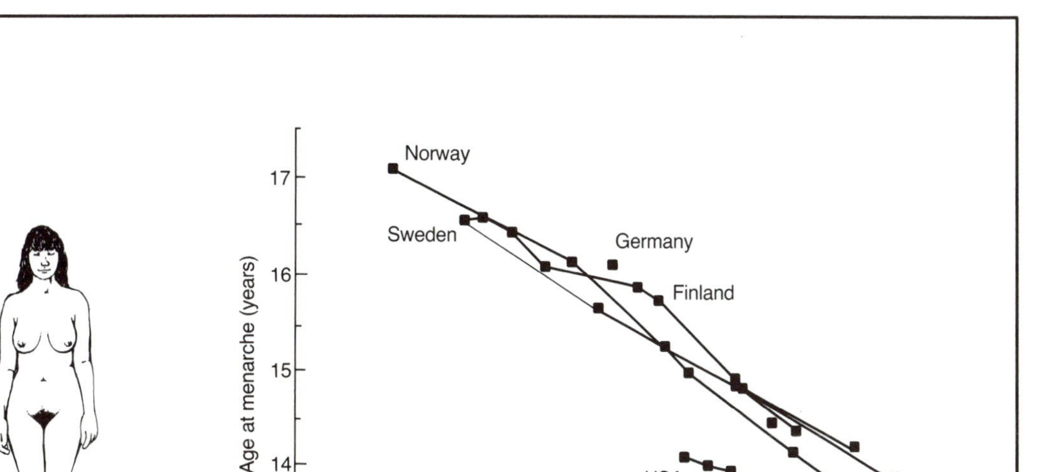

vertebrates, this phase is a phase of exploration and, depending on the animal's ecology, may occur just before, during or after sexual maturation (Baker, 1978).

Discovery of a suitable location and/or partner(s) may speed the final stages of puberty. Thus, in many mammals, both females and males show delayed maturation if surrounded by a high proportion of individuals of the same sex as themselves (e.g. female mice; Vandenburgh *et al.*, 1975). Often, the critical factor may be the balance of male and female pheromones in the immediate environment.

The timing of puberty in modern humans appears on average to be delayed relative to most mammals (Johnson and Everitt, 1988) but to be about typical for primates, occurring at an age that is about 20% of longevity (Dietze, personal communication). In the macaque, *Macaca nigra*, puberty occurs at an age that is about 23% of longevity, in the gibbon, *Hylobates lar*, 28%, the chimpanzee, *Pan troglodytes*, 22%, and the gorilla, *Gorilla gorilla*, 17% (calculated from data in Harvey *et al.*, 1987).

Age at puberty in humans is variable and sensitive to environmental conditions. It tends to decrease when diet, workload and morbidity patterns improve (Eveleth and Tanner, 1976) and rise when they get worse. Average age at puberty in females (measured as age of menarche, i.e. first menstrual bleeding) ranges from a minimum of 12.5–13 years in urban post-industrial settings to a maximum of 18 and perhaps even 20 years (Wood *et al.*, 1985) in some communities in highland Papua New Guinea. Over the past century, age at puberty has decreased considerably in industrialized societies (see figure).

Although 'ecological' stress seems to delay puberty in humans, 'family' stress may have the opposite effect. A Canadian study of 1247 women (M. K. Surbey, personal communication) has shown that girls who experience high levels of stress in the home matured significantly earlier than those experiencing less stress. In the absence of their putative genetic father, girls experienced menarche 4.5–6 months earlier than girls living with both parents.

Evidently, there is no single effect of stress. Rather, humans become ready for reproduction at a time that is most adaptive for the environment in which they find themselves.

(Puberty figure inspired by Diagram Group, 1981).

These declines strongly suggest that avoidance of copulation at particular times is at least in part a mechanism to avoid conception as the female ages. However, we could find no indication that, when controlled for age, females reduced their copulation rate as number of children increased ($P>0.5$).

4.4.2 Contraception: prepuberty and menopause

Female humans begin to menstruate (a girl's first menstruation is termed menarche) and ovulate at puberty (Box 4.9) and thereafter ovulate at intervals throughout their repro-

Box 4.10 Anovulatory menstrual cycles

Externally, the menstrual cycle of female humans and other primates that void the products of menstruation to the outside through the vagina begins and ends with the first appearance of blood from the vagina. Often, between these two events, the female has ovulated. Often, however, she has not. Instead, the cycle has been anovulatory. Basal body temperature curves for individual women often show sequences of ovulatory and anovulatory cycles. In the example illustrated (simplified from Döring, 1969), the woman, of known fertility, showed a clear tendency to ovulate in the winter months but not in the summer.

Variation with age

The histogram shows the incidence of fertile ovulatory cycles, infertile ovulatory cycles with a shortened luteal phase (<10 day) and infertile anovulatory cycles at different ages. The decline in proportion of ovulatory cycles from 30 years onwards and rapidly after 40 years is reflected in a decline in birth rate after age 40 years.

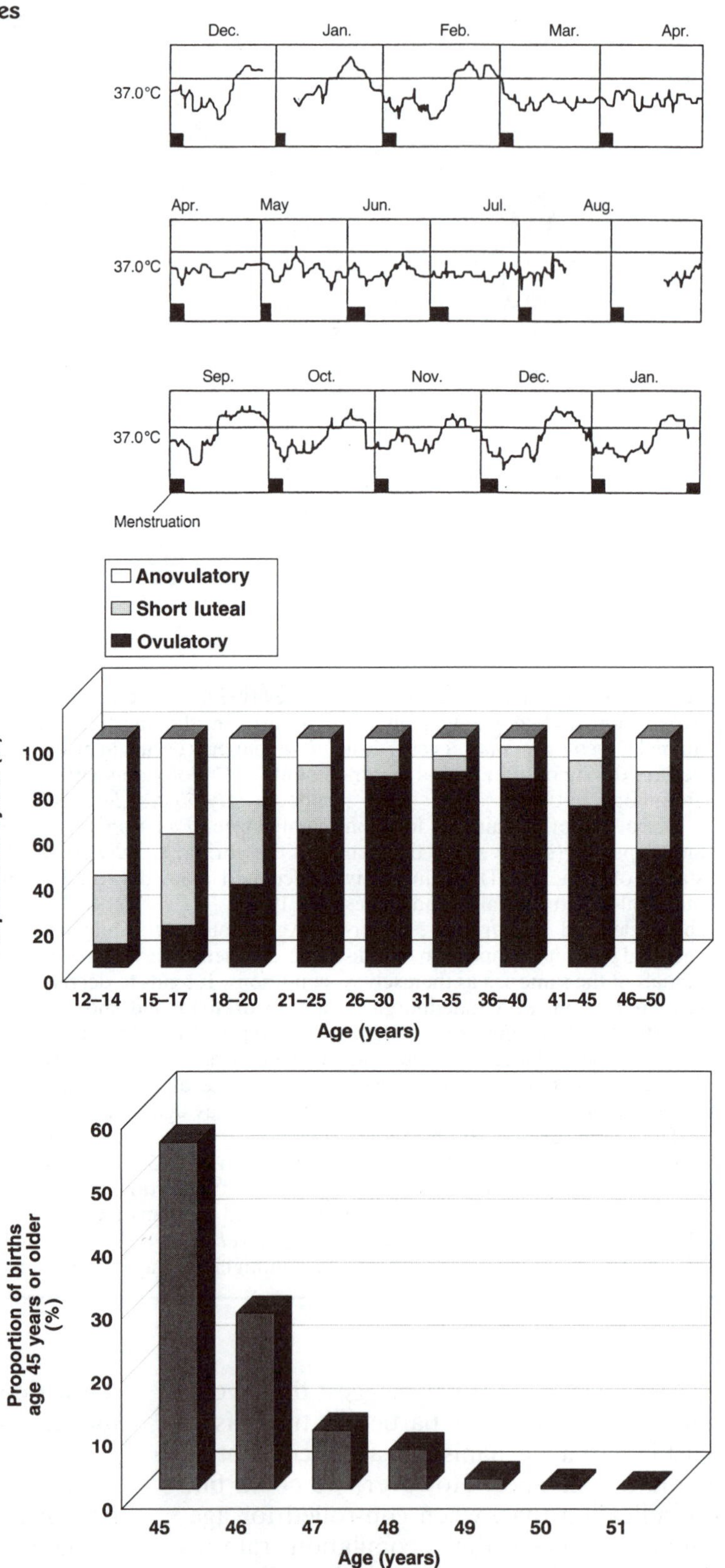

Seasonal variation
Although humans throughout the world may give birth, and hence conceive, at any season, in most locations there are characteristic peaks in the incidence of births, and hence conception. In the USA, peak birth season varies with latitude and longitude (months of birth shown at the top of columns; months of conception at the bottom; peaks are grey, troughs black) (data from Huntingdon, 1938). At higher latitudes, the births peak in February/March and trough in October/December. At middle latitudes a second peak apears in September and at lower latitudes a trough appears in the hottest part of the year. Cycles are reversed in the Southern Hemisphere (Huntingdon 1938).

Similar latitudinal variation is found in Europe and Asia (Cowgill, 1969). In the UK, births peak in February/March and again in September (Cowgill, 1969), corresponding to conception peaks in May/June and December. We have analysed the dates of 10 000 births in the Manchester region and have shown that they follow the national pattern apart from a slight local extension of the late spring/summer peak of conception into August.

There is no apparent survival benefit to children in being born at peak times (Cowgill, 1969). The phenomenon may well be a balanced polymorphism. Original seasonal advantages to being born at particular times of year lead to an increase in numbers being born at those times. Density dependent influences, however, could then act more strongly against individuals born at peak times. Birth distribution stabilizes when benefits and costs just balance.

Our own pilot study of the putative December ovulation peak in Manchester, UK, monitoring Basal Body Temperature and LH surge (Box 4.6), shows a distribution of probability of ovulation among women aged 19–29 years that is a significant ($P<0.02$) function of the distribution of probability of conception in Manchester. Sample sizes are number of cycles monitored in each month.

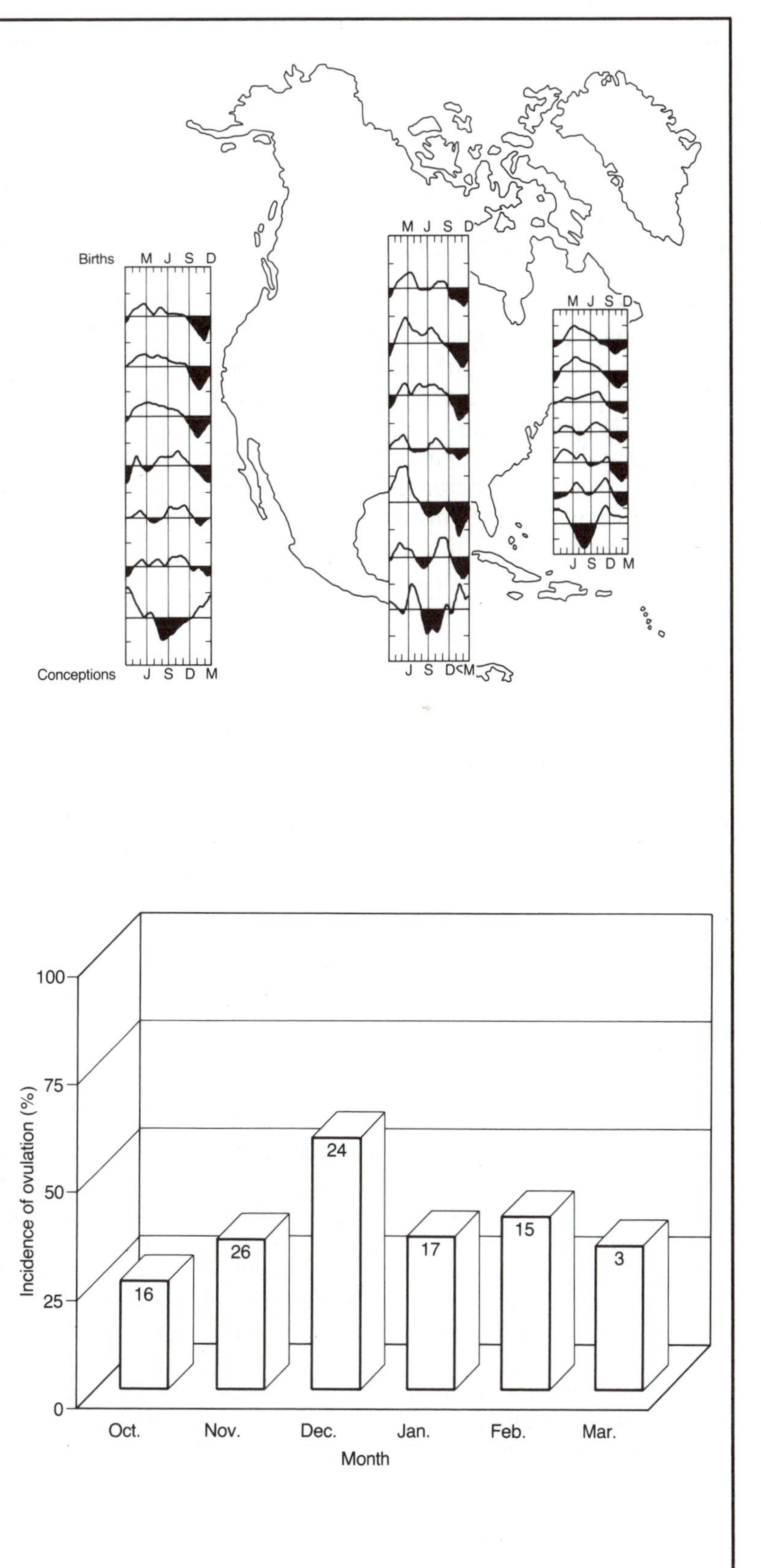

ductive life. Eventually, however, ovulation and menstruation become irregular and finally cease altogether. Females then begin a post-fertile phase of life during which most of their reproductive effort is directed toward assisting their children to raise grandchildren. The physiological syndrome associated with this change from a fertile to post-fertile state is menopause.

The sequence of events at female puberty is: (1) breast and nipple development; (2) thickening of vaginal epithelium with a drop in vaginal pH (i.e. vagina becomes more acidic); (3) pubic hair development; (4) menarche and subsequent menstruations; and finally (5) gradually increased incidence of ovulation (Box 4.10) (Symonds, 1992). This sequence itself suggests that females are programmed to begin to copulate some time, perhaps several years, before they are programmed to begin to produce offspring.

Female age at puberty varies considerably from population to population in relation to the change in age at first reproduction noted above. Undernourished girls reach puberty about 2 years later than well-nourished girls (Dreizen *et al.*, 1967). Body weight at menarche is about 46 kg in several Caucasian populations but lower among Japanese (Tanner, 1962).

At the other end of their reproductive lives, females in a wide variety of human populations rarely give birth after 50 years and menopause is now around 48–51 years (Talbert, 1977). The Hutterites, for example, who do not use modern methods of contraception, continue to bear children until a mean age of 49 years (Tietze, 1957). Characteristically, however, human females are increasingly less likely to produce children beyond the age of 40 years (Box 4.10).

Age at puberty has decreased over the past 100 years (Box 4.9) and age at menopause has increased (Frommer, 1964; but see McKinlay *et al.*, 1972).

All mammals pass through an infertile prepubertal phase and many have an, albeit often brief, post-fertile phase. True menopause (i.e. after which the female's ovary contains no normal oocytes) is primarily but not entirely associated with species in which the female lives in social groups with her daughters and grandchildren (e.g. many carnivores; ungulates and primates; Talbert, 1977). In the macaque, *Macaca mulatta*, menopause occurs when females are 25–30 years old (Van Wagenen, 1970). Chimpanzees, however, are often still menstruating at 44 years (Guilloud, 1968). Cows cease to reproduce at 15–21 years but may live to be 30 years and similar postmenopausal phases have been reported for horses, Indian Hariana cattle, and Indian buffalo (see references in Talbert, 1977).

Females of species that do not have such a tight, female-based, social structure (i.e. females living in a group with mothers, aunts, daughters, granddaughters etc.) may also cease to reproduce but tend to retain normal oocytes in their ovaries. Thus, rats cease to reproduce at 15–18 months but live for 3 years and hamsters, guinea pigs, rabbits and gerbils are also capable of living for significant lengths of time after reproduction ceases (references in Talbert, 1977).

The selective advantage of a prepubertal phase seems to be that the young female can channel energy into fuctions other than reproduction until the optimum time of first reproduction (Box 4.9). In nomadic hunter–gatherers, females do not reproduce until they reach the mean weight for females in the population, e.g. the Ache of Paraguay (Hill and Kaplan, 1988). In the Ache, body weight, particularly fat reserves, of the mother is such a major determinant of the survival prospects of the offspring that females who attempt to reproduce too early in life have a higher risk of losing the offspring and of damaging future attempts (Hill and Kaplan, 1988).

The selective advantage of the postreproductive phase seems likely to be that beyond a certain age, a female maximizes her reproductive success either by putting more effort into child care than child production and/or by assisting her children to raise grandchildren (Rogers, 1993).

Copulation may occur both during prepuberty and during and after menopause, albeit at a reduced level (Box 4.8). In all human societies, at least some females first copulate during prepuberty and in many societies prepubertal sexual experience and experimentation is at least tolerated if not openly encouraged (Ford and Beach, 1952). Such prepubertal experience should speed the development of female sexual ability (Chapter 5), aid in mate choice, and confuse future paternity (Chapter 6). Post-fertile copulation allows the female to continue to gain resources from males (Chapter 2).

Despite the strongly contraceptive nature of prepuberty and postmenopause, males gain some, albeit slight, advantage in copulating with apparently prepubertal and postmenopausal females. Female humans have given birth at all ages between 7 years and at least 57 years with some unconfirmed reports of birth to mothers over 70 years (*The Guiness Book of Records*). Thus, there is always some chance of fertilization. Moreover, there is never a reliable sign available to the male that fertilization is impossible. Young girls may ovulate before first menses and apparently postmenopausal women can conceive even up to 18 months after their last menstruation (Snaith and Williamson, 1947).

Prepuberty and menopause, plus a level of reproductive crypsis, thus allow the female to gain the indirect benefits of copulation while avoiding the disadvantages of reproduction at suboptimal phases of their life. As such, therefore, the two syndromes could well be evolved forms of family planning.

4.4.3 Contraception: avoiding ovulation from puberty to menopause

Externally, the beginning and end of the menstrual cycle of female humans and other primates who void the product of menstruation to the outside is marked by the appearance of blood from the vagina. Often, between these two events, the female has ovulated. Often, however, she has not. Instead, the cycle has been anovulatory.

Anovulatory cycles were first recorded in rhesus monkeys in the 19th century (Heape, 1898 in Döring, 1969). Their occurrence in women was first realized and described in the 1930s (Mazer and Ziserman, 1932). Now it is clear that, even between the ages of puberty and menopause, by no means all of a female's menstrual cycles are ovulatory, even though externally the cycles may appear perfectly normal.

Anovulatory cycles are just one of the forms of infertile departure from the normal fertile cycle shown by women. A more extreme form is amenorrhoea, the cessation not only of ovulation but also of menstruation. A less extreme form is to ovulate but to shorten the postovulatory (= luteal; Box 4.6) phase from its normally invariable 12–15 days. Luteal phases shorter than 10 days (as judged from basal body temperature curves, Box 8.1) are generally infertile (Döring, 1969).

On average, only about half of the menstrual cycles of perfectly normal, healthy female humans are fertile, though the proportion varies with age and season (Box 4.10). In addition, fertile cycles are much less common during lactation (e.g. Díaz *et al.*, 1991) and during periods of nutritional or other stress.

The first menstrual cycle after giving birth is rarely ovulatory as are only 46% of the first three cycles thereafter (Döring, 1969). The primary stimulus to suppress ovulation while a woman is lactating is the suckling stimulus from the infant. However, maternal nutrition also influences the length of the anovulatory period after giving birth, undernourishment increasing the time from giving birth to first postpartum ovulation.

In humans and in other mammals there is a programmed loss of fat reserves after giving birth, irrespective of whether lactation occurs. However, as shown by studies of Gambian women, actual gain or loss of weight is the nett product of this programmed loss, the demands of lactation, and food intake (Prentice and Whitehead, 1987).

The onset and incidence of ovulation in humans may be a function of attainment of a minimum weight:height ratio representing a critical fat store (Frisch and McArthur, 1974). Undernourished women have delayed menarche, a reduced incidence of fertile cycles, and early menopause. Female athletes and ballet dancers as well as slimmers also have a reduced incidence of fertile cycles which appears to be caused by the loss of 10–15% of body weight (i.e. about 30% loss of body fat).

The behavioural syndrome anorexia nervosa (weight phobia; Crisp, 1980) involves self-induced weight-loss. As a result, anorexics have few or no fertile cycles. Such behaviour is most common in young females (1% of 16–18 year olds compared with 0.1% for the population as a whole). Thus, anorexia usually starts in early or mid-adolescence. Thereafter, 75% of women return to 'normality', 15–20% continue their behaviour throughout their lives and 5–10% die of their behaviour.

Anorexia is often associated with an unsatisfactory sexual situation (Raboch and Faltus, 1991), though cause and effect are often difficult to separate. Nevertheless, anorexics continue to show sexual activity, albeit sometimes at a reduced level (Raboch and Faltus, 1991). The characteristics of anorexia nervosa are such that it is tempting to interpret the syndrome as a form of contraceptive behaviour which, except in extreme cases, may well be adaptive.

Nutrition and absolute weight are not the only 'energetic' factors to influence the incidence of fertile cycles. Fat distribution may also do so. There is evidence (Zaadstra *et al.*, 1993) that females with a more gynoid body fat distribution (Box 4.7) are more likely to conceive. A 0.1 unit decrease in waist:hip ratio led to a 30% increase in probability of conception per cycle during artificial insemination (Zaadstra *et al.*, 1993) both before and after adjustment for confounding variables. Absolute level of leanness or obesity, however, was not a significant factor.

In Zaadstra *et al.*'s study, older females were also less likely to conceive, presumably due to reduced probability of their having a fertile cycle (Box 4.10). As a high waist:hip ratio, like obesity, is known to be associated with irregular menstruation (Hartz *et al.*, 1984), the likelihood is that the influence of fat distribution on chances of conception also acts through the probability of a fertile ovulation. However, in our own pilot study in Manchester, UK, in 1992–93, females were less, not more, likely to ovulate if their shape was more gynoid. Clearly, more data on this point are needed.

Forms of stress other than those associated with nutrition also influence the probability of ovulation. Amenorrhoea is a known response both to major psychiatric disorders and to simple stress due to external catastrophe, separation from partner, and fear of becoming pregnant. In Japanese and German concentration camps, 50–70% of women became amenorrhoeic as have 100% of women awaiting execution in prison.

Anovulatory cycles are as much a feature of non-human primates and other mammals (e.g. baboons, *Papio* spp; Hendrickx and Kraemer, 1969) as they are of humans. Many of the factors influencing the probability of ovulation are the same. Thus, it is generally accepted that food restriction or prolonged exercise can suppress pubertal development and ovulation while simultaneously depleting fat reserves, suppressing growth and decreasing lean body mass (Bronson and Manning, 1991). For example, starvation blocks ovulation immediately in lean Syrian hamsters but not until after two or three cycles in fat individuals (Schneider and Wade, 1989). In female pigs, food limitation reduces ovulation rate at both the current and next oestrous cycle (Reid, 1960).

The existence of seasonal variation in the proportion of ovulatory cycles giving rise to peak seasons of birth is well established for some primates (e.g. squirrel monkey, *Saimiri sciureus*; macaque, *Macaca mulatta*; Dukelow, 1978). In rhesus macaques, the main breeding season is from September to March. With lengthening photoperiods, however, anovulatory cycles become more common.

Females of humans and other mammals continue to copulate during anovulatory cycles which thus seem to be an adaptation for contraception. The variation with age and season (Box 4.10) and the links with lactation, nutrition and stress seem to be clear adaptations for family planning. All help to avoid conception at suboptimal times of life or season, at suboptimal times after the birth of a previous child, and at suboptimal times when the environment is unfavourable and hence stressful.

4.4.4 Contraception: copulation at infertile stages of fertile menstrual cycles

Copulations during anovulatory cycles clearly do not lead to fertilization. Copulations during ovulatory cycles also have a low probability of leading to conception. This is because there is only a relatively short phase in the ovulatory cycle of women when fertilization can occur.

The typical, ovulatory human menstrual cycle of about 28 days is described in Box 4.6. In the standardized menstrual cycle, ovulation occurs on day 14 and the most fertile period extends from day 9 to day 14. This means that, on average, human females are fertile, even during a standard ovulatory cycle, for less than 25% of the time.

Human females copulate at all stages of the menstrual cycle. Many other mammalian females also show receptivity at all stages of the cycle and the vast majority on occasion will copulate during infertile phases (see full discussion in Chapter 6). Such females thus continue to gain the non-reproductive benefits of copulation while avoiding the risk of conception. The existence of long periods of time that the female cannot conceive thus appears to be a contraceptive strategy.

In humans, the 'contraception window' (i.e. the ratio of lengths of the infertile to fertile phases) is >3. We suggest that this window is an evolved optimum for females that gives them the optimum trade-off between gaining the non-reproductive benefits of copulation (both with their partner and other males) while retaining some opportunity for conception in any given cycle. The contraception window is increased by the occurrence of anovulatory cycles and amenorrhoea. It is decreased by any tendency to concentrate copulations into the fertile phase of the cycle and by any tendency towards induced ovulation (section 4.4.7).

The fact that the contraception window for humans is >3 is due to three factors: (1) the fertile life of the egg; (2) the fertile life of the sperm; and (3) humans usually being spontaneous, rather than induced, ovulators. If we assume that the human contraception window of >3 is an optimum trade-off for females, we should consider the possibility that the fertile life of eggs (section 4.4.5) and sperm (section 4.4.6) and the timing of ovulation (section 4.4.7) have themselves evolved as adaptations to produce this trade-off.

4.4.5 Contraception: short-lived eggs

The human egg is fertile for only about 12–36 hours after release from the ovary (Austin, 1982). Eggs fertilized early in this period, however, are more likely to produce embryos that survive for a full-term pregnancy and to produce normal young (Austin, 1982). Different mammals have slightly different egg longevities but in all the fertile life of the egg is relatively short (usually <24 h except in a few species, such as the horse, in which it may be 2–3 days; Austin, 1975).

Female mammals, in common with most internal fertilizers, seem adapted to a strategy of first collecting sperm through copulation, then ovulating. When the egg appears, sperm are already waiting to fertilize. Such a system endows the female with maximum control over fertilization. In advance of ovulation she can collect sperm from one or several of the potentially most suitable males available to her. She can test males for their ability to guard or to sneak copulations. She also has time to bias the numbers from individual males via differential phagocytosis and storage (section 4.4.14) and then to allow the sperm to compete via numerical mixing, warfare or jostling/racing for position (Chapter 2). The egg is then produced to be fertilized by the most competitive sperm from among the males the female has preselected.

The actual longevity of the egg should thus be the optimum trade-off between a time: (1) long enough to ensure fertilization; but (2) not so long as to allow less competitive sperm to arrive at the site of fertilization. In most mammals, the evolved result seems to be an egg with a fertile life that is relatively short. Such short-lived eggs are, in effect, contraceptive adaptations that allow the female to minimize the risk of being fertilized by less competitive sperm and/or sperm from males who are assessed to be suboptimal in other respects.

4.4.6 Contraception: sperm longevity in the female tract

Human sperm live for up to 10 days in the female tract but remain fertile for only 5 days

Box 4.11 Longevity of mammalian sperm

Most estimates are probably underestimates (see discussion in Chapter 8). The significance of the sperm drawings is explained in Chapter 11. In humans, at least 27% of sperm coil their tails and become immobile some time before they die.

Species	Fertile life
Rodents	
Mouse	6–12 h
Rat	14 h
Hamster	13–15 h
Guinea Pig	22 h
Lagomorphs	
Rabbit	30–36 h
Bats	
Myotis lucifugus	138 days
Eptesicus fuscus	156 days
Carnivora	
Ferret	126 h
Dog	72–96 h
Cat	50 h
Ungulates	
Pig	36–48 h
Sheep	36–60 h
Horses	96–144 h
Cow	28–51 h
Primates	
Humans	120 h

or so (see discussion in Chapter 8). Different mammalian species have characteristically different sperm longevities (Box 4.11).

There is clearly no absolute constraint on how long mammalian sperm can live in the female, some species having females who store sperm for up to several months. The fact, therefore, that so many species have relatively short-lived sperm suggests that sperm longevity in any given species is optimum for that species. It then becomes a moot point whether the optimum is primarily adaptive for the male or the female (Chang, 1965).

In principle, males should always gain from having sperm live longer in the female as long as the cost of programming and provisioning them to do so does not outweight the benefit of the greater longevity. Restricting the fertile life of sperm from any given copulation, however, could be a clear contraceptive strategy on the part of the female. Some evidence that the fertile life of sperm once inseminated is largely determined by the female is provided by the observation that sperm tend to live longer in the reproductive tract of females of a different species. For example, rabbit sperm keep their fertility twice as long in the rat uterus as in the rabbit uterus (Bedford, 1966).

Human sperm are inseminated into the upper vagina. A third or more, on average, are ejected in the flowback. Most of the remaining two-thirds lodge in the cervical mucus and perhaps, for a while, in the upper vagina (Box 3.5). Less than 1% gain a safe haven in the cervical crypts and only about a millionth of the original ejaculate get anywhere near the point of fertilization in the oviduct.

Except in the seminal coagulum, from the moment of insemination both live and dead sperm are systematically annihilated by phagocytosis (Mattner, 1969). Only those that enter the cervical crypts are initially safe but even these are eventually expelled by the female to run the gauntlet of phagocytosis.

The only sperm to reach the vicinity of the egg are those spared by the phagocytes. Eventually, all sperm are destroyed.

It is difficult to escape the conclusion that sperm survival in the female tract is determined by the female. In consequence, males seem likely to have programmed and provisioned their sperm to function only for as long as they can, on average, 'expect' to be allowed to survive. In which case, the observed correlation between sperm longevity and the length of the menstrual and oestrous cycles in mammals (Parker, 1984), could well be the result of selection on females for an optimum contraception window.

This possibility and its link with sperm competition is discussed further in the context of cryptic ovulation (Chapter 6).

4.4.7 Contraception: the timing of ovulation

Female mammals as a whole may be broadly categorized as either spontaneous or induced ovulators (Box 4.12).

Female humans are usually considered to be spontaneous ovulators. They ovulate at a particular phase of the menstrual cycle as the end product of several days of hormonal changes (Box 4.6). The female may copulate several times in the days preceding ovulation. In contrast, induced ovulators are defined as species which ovulate soon after, and in response to, the female's first copulation in any given cycle.

In fact, the distinction is not as clear-cut as the normal classification suggests. In both types, ovulation is preceded by a characteristic hormonal phase of finite length. In both types, the timing of ovulation may be influenced by copulation during this phase. Even in humans, the preovulatory follicular phase can be interpreted as a period during which the female is 'ready and waiting' for ovulation, the precise timing of which can then be influenced by a particular copulation (Clark and Zarrow, 1971; Jöchle, 1975).

In effect, therefore, in both types hormonal changes induce a state of readiness for ovulation the precise timing of which is influenced, to a greater or lesser degree, by the occurrence and timing of copulation. Induced ovulators respond more quickly and more reliably to copulation and are less likely to ovulate if copulation does not occur. Spontaneous ovulators, like humans, respond more slowly and less reliably to copulation and are more likely to ovulate even if copulation does not occur. However, even spontaneous ovulators may cease to ovulate if copulation also ceases (e.g. horses, Mann and Lutwak-Mann, 1981; humans, McClintock, 1971).

The main difference between the two types of female is in the time window they offer to males to copulate and to sperm to compete. In this respect, induced and spontaneous ovulators are simply two ends of a continuous spectrum. Induced ovulators effectively promote intense male–male competition for copulation during a limited period and offer an advantage to the male with the fastest sperm (Gomendio and Roldan, 1993). Spontaneous ovulators have more time and hence opportunity to escape and deceive guarding males. They also offer less advantage to males with the fastest sperm but more to males with sperm who compete in other ways throughout the female tract (Chapter 11).

Of the two extremes, spontaneous ovulation appears to be more a part of a contraceptive strategy than induced ovulation. Much depends, however, on the extent to which the females of each type will copulate at other times in the cycle. On the whole, induced ovulators copulate less than spontaneous ovulators during infertile parts of the menstrual cycle. They are often also more vigorous in resisting copulation.

4.4.8 Contraception: control of sperm numbers

Human sperm swim through cervical mucus and along the oviduct. Passage from the vagina to the cervix and from the cervix to the oviduct (through the uterus), however, is sometimes assisted by contractions of the female tract. En route, some sperm are given safe haven in the cervical crypts. Elsewhere, however, they are subject to massive attack from the female's leucocytes (Box 3.5).

Box 4.12 Spontaneous and induced ovulation in mammals

Induced ovulators		Spontaneous ovulators
	Carnivora	
Northern fur seal		Dog
Large brown weasel		Fox
Ferret		
Mink		
Raccoon		
Domestic cat		
	Rodents	
Tree mouse		Guinea pig
California meadow mouse		Golden hamster
Asian vole		Mouse
13-lined ground squirrel		Rat
	Lagomorphs	
Common hare		
Jack rabbit		
Eastern cottontail		
Domestic rabbit		
	Bats	
Lump-nosed bat		
	Insectivores	
Hedgehog		
Mole shrew		
Common shrew		
Madagascar hedgehog		
	Ungulates	
		Goat
		Sheep
		Pig
		Cow
		Horse
	Primates	
		Rhesus monkey
		Human

Compiled from van Tienhoven (1968).

In principle, the combination of copulatory orgasm (section 3.5.3), sperm storage organs (section 3.5.1), cervical mucus (Box 3.5) and leucocytes (Box 3.5) could endow the female human and other mammals with some, if not considerable, control over the number of sperm that eventually arrive at the site of fertilization in the oviduct.

Leucocytes are thought to be selective over which and how many sperm they phagocytose. In rabbits (Tyler, 1978) and humans (Pandya and Cohen, 1985; Thompson *et al.*, 1992), the release of leucocytes in the female tract is elicited by the presence of sperm, not simply by copulation or even, in the rabbit, seminal donation from a vasectomized male (Tyler, 1977). The release of leucocytes is accompanied by the transudation of IgG (immunogammaglobulin) from the blood (Smallcombe and Tyler, 1980). The IgG coats most, but not all, sperm and the leucocytes seem selectively to target for phagocytosis those sperm which are IgG coated; sperm recovered from the oviduct do not coat with IgG (Cohen and Werrett, 1975; Cohen and Tyler, 1980).

Level of selectivity, number of phagocytes, and speed of phagocytosis could combine to influence the number of, or even prevent, sperm from reaching the oviduct and the fertile egg. In addition, females do not inevitably assist any sperm on their passage to the oviduct. For example, conception is enhanced following artificial insemination with sperm from a fertile donor if the female copulates with her infertile partner within 4 hours after artificial insemination (Kesserü, 1984). Stressed female humans are less likely to conceive and may completely inhibit sperm passing through the cervix and uterus, apparently as a result of unfavourable myometrial contractions. If the nerves to the uterus are cut, however, the inhibition seems to disappear.

An alternative to allowing few or no sperm to reach the egg is to allow too many. Sperm release chemicals, probably from the acrosome, that dissolve the egg. Even as low as a tenfold increase in the number of sperm around the egg can result in the egg being killed and dissolved, in part, by the sperm. In the rabbit, increasing the number of sperm inseminated peritoneally from 10 million to 390 million decreased the proportion of eggs subsequently recovered from 95% to 28% and the proportion of eggs fertilized from 79% to 20% (Adams, 1969).

On the whole, the usual concentration of sperm at the site of fertilization in rodents is considered to be optimal. It is just high enough to give all eggs a good chance of being penetrated before the end of their fertile life but not so high that there is a risk of polyspermy (Braden and Austin, 1954a,b) or of the eggs being damaged by the sperm. The frequency of successful sperm–egg collisions is estimated to be about every 2 min in the rabbit and about every 10 min in the rat. In rats the female tract becomes less restrictive towards the end of the eggs' fertile lives. The number of sperm per oviduct increases from 24 at about the time of ovulation to 50–60 about 3 h later (Braden and Austen, 1954a,b). Rabbits, however, show no such increase with time after ovulation.

4.4.9 Avoiding pregnancy: failure to implant

Even when fertilized, eggs do not always implant in the uterus as they pass down the female tract. Obtaining data on fetal loss before pregnancy can be diagnosed in humans is extremely difficult. Yet the loss of fertilized eggs could be at its maximum during this period.

On the basis of observations of hysterectomy specimens, it has been estimated that whereas 90% of fertilized human eggs survive to reach the uterus, only 58% implant and only 42% survive to the twelfth day of pregnancy (Hertig and Rock, 1959; Hertig *et al.*, 1959).

Factors affecting the likelihood of implantation are unknown. Again, however, the possibility exists that implantation is a function of stress levels and other female reactions to her environment and circumstance, including the chromosomal and other characteristics of the foetus.

Perhaps the best-known example of implantation failure in other mammals is found in mice (the Bruce effect; Bruce, 1960). A female, having been inseminated by one male (the stud male), fails to implant if the stud male is removed and replaced by another (the alien) male. Failure to implant is less likely if the female is exposed to the alien male in the presence of the stud male and does not occur at all if she is re-exposed to the stud male after a separation of 24 h. The manifestation of the Bruce effect is also a function of the relative social status of the stud and alien males, dominant males having more influence than subordinate (Huck, 1982). As the males of several strains of mice kill unrelated young (Labov, 1980), the young of females who lose the presence of the stud male may have reduced prospects.

4.4.10 Avoiding development: polyspermy

Polyspermy is the fertilization of an egg by more than one sperm. It is said to be physiological polyspermy if the zygote is nevertheless capable of full development into a normal individual and pathological polyspermy if not (Austin, 1975).

In humans and other mammals, polyspermy seems usually to be pathological. There are no data for humans on the number of polyspermic eggs which fail to implant. However, 10% of all spontaneous abortions in the first three months of pregnancy are polyploid (Simpson *et al.*, 1982) and a retrospective study of 1499 spontaneously aborted fetuses showed that 39% had abnormal karyotypes of which 25% were polyploid (Boue *et al.*, 1975; Wolf *et al.*, 1984). In other mammals, the level of polyspermy *in vivo* is around 2% (Austin, 1961).

Of course, polyspermy is usually interpreted as a malfunction rather than as a female contraceptive strategy. Nevertheless, there are at least some grounds for considering the latter possibility.

Polyspermy need not be pathological nor need it occur. Other animals successfully cope with, or avoid, polyspermy. In some, it is even an obvious part of an adaptive female strategy. For example, in the invertebrate ctenophore, *Beroe*, several sperm may penetrate the egg. The egg nucleus then appears to 'inspect' a succession of sperm nuclei before 'choosing' one (the first above threshold; they do not return to previously inspected nuclei) with which to fuse (Carre and Sardet, 1984). Physiological polyspermy is also common in birds and reptiles. The extra sperm are allowed in but only one is allowed to pair with the female pronucleus. The other male pronuclei (often 50 or more) are bundled out of the way and play no part in fertilization (Austin, 1975). Finally, both fish (Alberts *et al.*, 1983) and amphibians (Englert *et al.*, 1986) have evolved a fast block to polyspermy, the egg preventing access to other sperm once one has entered beyond a certain point.

There seems no immediate reason why polyspermy should be pathological so generally to mammals nor why, given that it is pathological, a fast block should not have evolved. An alternative approach is to consider the possibility that polyspermy in mammals is a positive female strategy rather than an evolutionarily unavoidable pathological condition.

In rats and rabbits, higher concentrations of sperm at the site of fertilization are associated with larger numbers of sperm in eggs (Braden and Austin, 1954a,b). In humans, the incidence of polyspermy *in vitro* is directly related to the concentration of sperm (Simpson *et al.*, 1982; Wolf *et al.*, 1984). Possibly, therefore, females can increase the chances of polyspermy by allowing above average numbers of sperm to reach the vicinity of the site of fertilization (section 4.4.8).

If the probability of polyspermy ever proves to be a function of stress levels and/or other female reactions to her environment and circumstance, the possibility that the phenomenon is part of female contraceptive strategy would increase.

4.4.11 Avoiding birth: spontaneous abortion

Even once an embryo has implanted in the uterus, the female retains the ability to discard the embryo in a spontaneous abortion. Estimates of the incidence of spontaneous abortion suggest that about 20% of all pregnancies recognized by the end of the first lunar month end in spontaneous abortion (Box 4.13).

The frequency of spontaneous abortion is greatest during the first three months of pregnancy (Shapiro *et al.*, 1971). It is most common when the mother is very young (<15 years) and least when she is 15–24 years old (Potts *et al.*, 1977). Thereafter, it increases with age and is a function of the number of children the mother has already had (Box 4.13).

The probability of spontaneous abortion is greatest when the foetus has reduced viability or prospects, either due to features of the foetus itself or due to circumstances affecting the mother. Rarely, diseases or infections generating a high temperature (e.g. malaria) trigger spontaneous abortion (Potts *et al.*, 1977). Death of the mother's male partner, sudden social disturbance (e.g. outbreak of war), sudden discovery of the male partner's infidelity may all be followed within hours by spontaneous abortion. It is also more likely if the embryo has reduced prospects, either through polyspermy (section 4.4.10) or chromosomal abnormality (Boxes 4.13 and 4.15).

Estimates of early fetal loss in cattle, horse, sheep and swine range from 10% to 60% (Corner, 1923; Perry, 1954a). Similar levels are suggested for rodents and a wild population of rabbits (Frazer, 1955). The same link is found as for humans between maternal stress and spontaneous abortion. Thus, severe nutritional deficiencies and overcrowding raises the incidence of spontaneous abortion in a range of mammals (e.g. mice; Potts *et al.*, 1977). Horses mated soon after foaling miscarry four times as often as those mated later and in monkey colonies the probability of spontaneous abortion also decreases with time between last birth and next conception (French and Bierman, 1962).

4.4.12 Avoiding maternal care: infanticide

As is well known to anybody who has kept a pet hamster, gerbil or rabbit, female mammals are very prone to abandon, kill or eat their young soon after birth, particularly if exposed to any form of environmental stress (Elwood, 1992). Some rodents (e.g. hamsters, Elwood, 1991) often cull their litter to a particular size even without overt stress.

Box 4.13 Spontaneous abortion

Incidence
Estimates of the proportion of spontaneous abortions as a percentage of all pregnancies recognized by the end of the first lunar month.

Place	Time	Proportion (%)
UK	1940s	10
Hawaii	1950s	24
Ireland	1950s	14
USA	1950s	11
India	1960s	11
Sweden	1960s	12
Thailand	1960s	16

Influence of age

Age (years)	Pregnancies ending in spontaneous abortions (%)
15–24	5
35–39	11
40–44	32

Influence of the number of children the mother has already had

First child,	18%
Second,	6%
Third,	14%
Up to sixth plus,	21%

Abnormal development and spontaneous abortion
Evidence going back to the 1880s for continental Europe has shown that fetus' spontaneously aborted during the first 10–12 weeks of pregnancy were likely to show signs of abnormal development, for example:

Europe	1880s	37.5%
USA	late 1930s	61.7%
Kuwait	late 1960s	50%

Particular chromosomal abnormalities are found in only 0.2% of the population but are found in 10–22% (e.g. Canada; Carr, 1967, 1972) of spontaneous abortions.

(Compiled from Potts *et al.*, 1977)

The implication is that the mothers use infanticide to optimize the litter size, often at zero, to the level that is suited to their current situation and environment.

Humans are typical mammals in that the female also has a period shortly after giving birth when she has a low threshold for killing or abandoning the new-born baby. The physiological and behavioural syndrome associated with this reduced threshold is often manifest as postnatal depression. Many cultures recognize and are sympathetic to this aspect of female behaviour and many legal systems accept that the female may experience such an uncontrollable disposition at this time.

Postnatal infanticide was widespread among hunter–gatherers (e.g. the Ache of Paraguay, 6.9% of all children born are killed before the age of 2 years; Hill and Kaplan, 1988) and is clearly evident in figures for modern western societies (Daly and Wilson, 1984). The association between the incidence of postnatal infanticide and female stress levels and other reactions to the environment and circumstance is also well established.

Cross-culturally, lack of a father's support is the most common factor cited by infanticidal mothers (Daly and Wilson, 1984). In the Mangaia of the southern Cook Islands, females may stop feeding their male partner's children until they die if the male is discovered to be unfaithful (Marshall, 1971).

Infanticide may also occur later in a child's life, again usually when the female is in a suboptimal situation. In the Ache, for example, a child has only a 0.6% chance of being killed before age 15 years if its father survives to this time but a 9.1% chance of being killed if he does not. With older children, the mother never actually does the killing herself but it is not clear to what extent she is compliant (Hill and Kaplan, 1988).

4.4.13 Optimizing family structure: influencing the sex ratio

It is now well established that the sex ratio of offspring in humans and other mammals (e.g. red deer, Clutton-Brock *et al.*, 1984) departs from the population average in a way that is a function of the mother's circumstances (Box 4.14). Females who are paired to high-status males are significantly more likely to give birth to boys than females paired to low status males. Females without a long-term partner are more likely to give birth to girls (Crew, 1952). The behavioural ecology of this phenomenon is now well established (the Trivers–Willard hypothesis; Box 4.14) but the mechanism by which it is achieved is unknown.

It is unlikely simply that the X/Y ratio of a male's sperm changes with the male's social status, otherwise females without a partner would not show a bias towards producing daughters. It is much more likely that the adjustment is achieved by the female. This could be done either by: (1) selective transport or access of X and Y sperm to the egg before fertilization; (2) selective passage of X and Y sperm through the outer egg layers during fertilization; (3) selective implantation of male and female embryos; and/or (4) selective spontaneous abortion of the embryo-foetus during pregnancy.

After birth, females have another opportunity to alter the sex ratio of their offspring via selective infanticide, abuse or neglect. Daughter infanticide was common in the highest caste or class in British Colonial India and traditional China (Dickemann, 1979). Daughter neglect and abuse leading to death are reported for prosperous families living in German Schleswig-Holstein from 1770–1869 compared with male-biased mortality in the less prosperous families (Voland, 1984). Finally, on Ifaluk, the more prosperous parents spent about twice as much time with sons as with daughters while among less prosperous parents the pattern was reversed (Betzig and Turke, 1986).

Optimum sex ratio is not just a function of social status. The sequence with which sons and daughters are produced may also influence the mother's reproductive success. In Micronesia, for example, Ifalukese women have more children (c. 9 children) if their first two children are girls than if their first two children are sons (c. 5 children).

Female mammals are not the only animals with the ability to influence the sex ratio of their offspring. Perhaps the best known examples are among insects in the social

Box 4.14 Sex ratio of offspring as a function of social status: the Trivers–Willard hypothesis

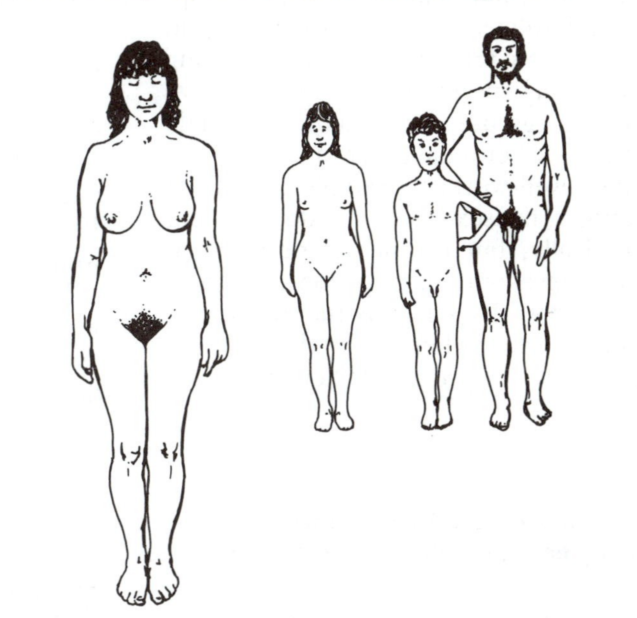

The Trivers–Willard hypothesis (Trivers and Willard. 1973)

1. Male reproductive success has a greater variance than female (section 2.3.2). As a result, more males than females should have more than the median number of children and more males than females should be childless.
2. Parents' access to resources is affected by the parents' status and positively affects offspring survival and status (here, offspring status is relative to the mean for the population within which the offspring will reproduce).
3. As males have greater variance in reproductive success, high status male offspring should be reproductively more successful than high status female offspring and low status male offspring should be less successful than low status female offspring.
4. Parents should maximize their reproductive success by biasing the sex ratio of their offspring towards males if they (the parents) are of high status and towards females if they are of low status.

Social status of parents and sex ratio of offspring in humans

Analysis of the sex ratio of offspring born to male Americans (born 1860–1939; $n = 1014$) and male and female (born 1860–1945; $n = 1628$) Germans listed in their national *Who's Who* show a very significant bias towards males (Mueller, 1993). The sex ratio of eminent British industrialists (born 1789–1925; $n = 1179$) is also significantly male biased (Mueller, 1993). The sex ratio of offspring born to single-parent women, however, is female biased (Crew, 1952).

Sex ratio biases in other mammals

Among primates, several studies on macaques have also shown that high-ranking females produce more sons than daughters in accord with the Trivers–Willard hypothesis (e.g. Meikle *et al.*, 1984; Paul and Kuester, 1990). Similar results have also been obtained for ungulates (red deer, *Cervus elaphus*, Clutton-Brock *et al.*, 1984; bighorn sheep, *Ovis canadensis*, Hogg *et al.*, 1992). In these species, sons are more costly to produce than daughters and so subordinate females in bad condition may not be able to afford to produce a son. However, a number of primates (galagos, Clark, 1978; macaques, Simpson and Simpson, 1982; Silk, 1983; baboons, Altmann *et al.*, 1988) produce sex ratio biases opposite to that predicted by the Trivers–Willard hypothesis: high ranking females produce more daughters. Two alternative hypothesis have been proposed:

1. *The advantaged daughter* (Simpson and Simpson, 1982): Maternal rank is passed on to daughters (who stay in the natal group) but not to sons (who migrate and eventually settle in a new social group). The daughters of high-ranking mothers thus may have higher than average reproductive success whereas the sons may have average reproductive success.
2. *Local resource competition: the costs of daughters* (Clark, 1978; Silk, 1983): When females stay in the social group, they compete for resources. Established females attempt to prevent recruitment of new females. Subordinate mothers are less able to protect their daughters from group harrassment and thus do better to produce sons to migrate to new groups.

Summary

Many mammals clearly show a bias in the sex ratio of their offspring according to their circumstance. The Trivers–Willard hypothesis alone cannot explain all observed variation. It has been suggested that both the Trivers–Willard process and local resource competition are working, exerting selective pressures in opposite directions in social mammals, such as many primates, in which daughters are less likely than sons to migrate from the natal group. Where local resource competition is low, the Trivers–Willard effect will prevail and *vice versa*. However, more recently, Hiraiwa-Hasegawa (1993) has argued that the Trivers–Willard effect is unlikely to be important among primates. Humans, however, show a much less clear-cut difference than most primates over which sex leaves the natal group (Baker, 1978) or over which sex will suffer most from local resource competition. As a result, and as supported by the above analyses, the Trivers–Willard hypothesis seems likely to apply, at least to some extent.

Hymenoptera (wasps, ants, bees; Wilson, 1971). In these, males are haploid, developing from unfertilized eggs; females are diploid and develop from fertilized eggs. By releasing or not releasing sperm into her reproductive tract as each egg passes through, the female can produce either a daughter or a son.

Another well-worked example is that of some reptiles, the sex of which is determined, not genetically, but by the temperature at which they develop (e.g. crocodiles; Magnusson *et al.*, 1989). By selecting the location at which to lay eggs, the female can thus bias the sex ratio of her offspring in either direction.

4.4.14 Optimizing family structure: influencing paternity after insemination

If the female has such post-insemination mechanisms for selectively influencing the sex of her offspring, there seems no reason in principle why she may not also show similar selectivity to influence the paternity of her offspring (Birkhead and Møller, 1993a,b). Evidence of selective paternity is not yet available for mammals but is now available for birds.

In the tree swallow (*Tachycineta bicolor*), females containing sperm from two different males seem selectively to favour one male's sperm over another for fertilization (Lifjeld and Robertson, 1992). To some extent, which male's sperm they will prefer can be predicted from advance behaviour with the two males. Moreover, the preference can remain for several days after the preferred male has departed, despite the female continuing to copulate with the less preferred male.

4.4.15 Optimizing family structure: producing helpers

Many animals, ranging from eusocial insects (wasps, ants, bees; e.g. Wilson, 1971), to birds (e.g. acorn woodpeckers, *Melanerpes formicivorus*; Koenig *et al.*, 1984) and a few mammals (naked mole rat, *Heterocephalus glaber*; Jarvis, 1981) produce offspring which are physically or behaviourally sterile. Instead of leaving the parental home range to reproduce, they remain and help their parents to raise their (the offspring's) siblings (Box 4.15).

Clarke (in press) has analysed the reproductive consequences of human migration in 19th century Sweden. She has shown that women who stayed in their village of birth had a higher likelihood of remaining childless whereas those who stayed but had children appear to have had, on average, more children than those who migrated from their village. Some women who remained at home may thus have assumed the role of sterile helpers to their parents and siblings. Such a possibility has also been explored for other societies (e.g. Kipsigis, Borgerhoff Mulder and Milton, 1985).

Human females progress from a phase in which they achieve reproductive success through the production of children to a phase (via menopause; section 4.4.2), in which they abandon further attempts to produce children and concentrate instead on helping their children raise their own children. There is an intermediate phase, around the time of menopause, during which human females have an increased probability of producing genetically sterile offspring (Down's syndrome; Box 4.16). Chimpanzees show a similar syndrome (Barnard, 1983).

Down's syndrome individuals have reduced longevity but those that survive a childhood often needing high parental investment are characteristically philopatric (= stay at home), affectionate, caring and helpful, their social skills exceeding their other abilities (Box 4.16). They are usually sterile but, when allowed, they can play a major role in caring for parents, siblings and nephews and nieces.

It is at least a hypothesis worth testing that, by producing such a helper towards the end of her reproductive life, a female may have a greater reproductive success via the earlier production, higher survival and/or better quality of grandchildren than if she had produced a competitive and fully reproductive child.

Box 4.15 Parental manipulation

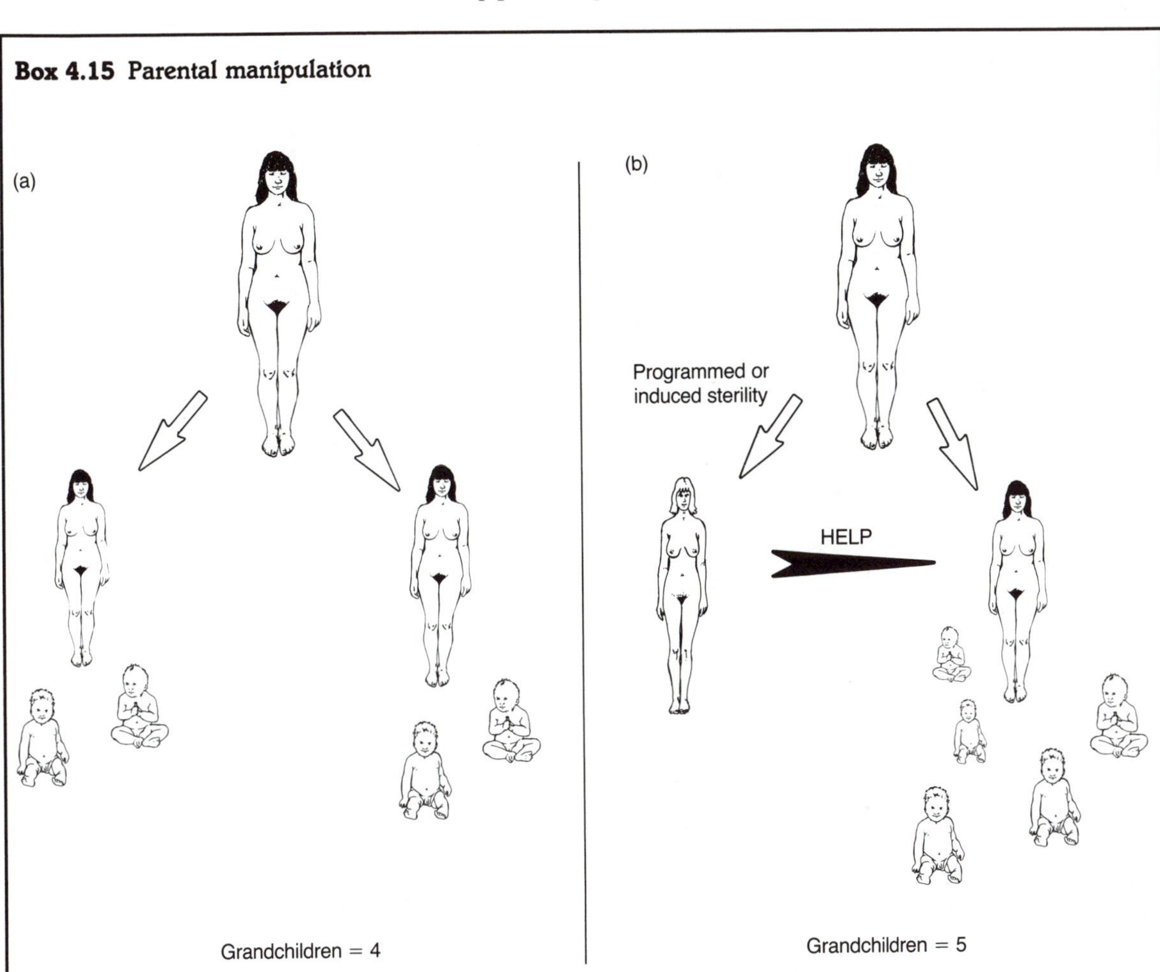

(a) Most often, parents should maximize their reproductive success (Box 2.2) by producing offspring who each attempt to maximize their own reproductive success.

(b) In some ecological and/or social circumstances, however, parents may maximize their reproductive success (e.g. number of grandchildren) by producing some offspring programmed to avoid reproduction themselves but instead to help their siblings to produce more offspring, and hence more grandchildren for their parents.

Most behavioural ecologists (e.g. Krebs and Davies, 1993) prefer to interpret such cases in terms of inclusive fitness (Hamilton, 1964) and kin selection (Maynard Smith, 1964). They address their question to the sterile offspring by asking 'why are you behaving altruistically towards your kin?'. We prefer to step back a generation (or more) and address our question to the parent by asking 'why do you programme or induce some of your offspring to behave altruistically to your other offspring?' The mathematics of the two answers may have many similarities but philosophically the latter approach raises fewer problems. It also allows the satisfactory interpretation of some situations in biology (e.g. level of relatedness of social Hymenoptera workers with multiple fathers) that sit uncomfortably, or not at all, within kin selection.

4.4.16 Antiquity of female family planning

Here, we have discussed female family planning from a mammalian perspective. However, females from a wide range of animal groups may gain non-reproductive advantages from mating and reproductive advantages from regulating reproduction. Thus, elements of female family planning may well have existed in the human lineage long before the transition to mammalian characteristics. Nevertheless, the importance of controlling

Box 4.16 Down's syndrome: sterile helpers?

Humans with Down's syndrome have characteristically shaped heads with distinctive eyelids and faces, are short in stature, and frequently have malformations of the heart (Singer, 1985).

Down's syndrome seems to be a universal and constant characteristic of human populations (Bell, 1991) but was not described scientifically until 1866 (Singer, 1985). On 95% of occasions, it is genetically simply the result of an extra chromosome (47 instead of 46). The extra chromosome is a rather small autosome, by convention designated chromosome 21. Down's syndrome individuals are thus trisomic (i.e. have three copies instead of the normal two) of chromosome 21. On 2% of occasions, individuals are mosaic, being trisomic only in some cells (Hull and Johnston, 1993).

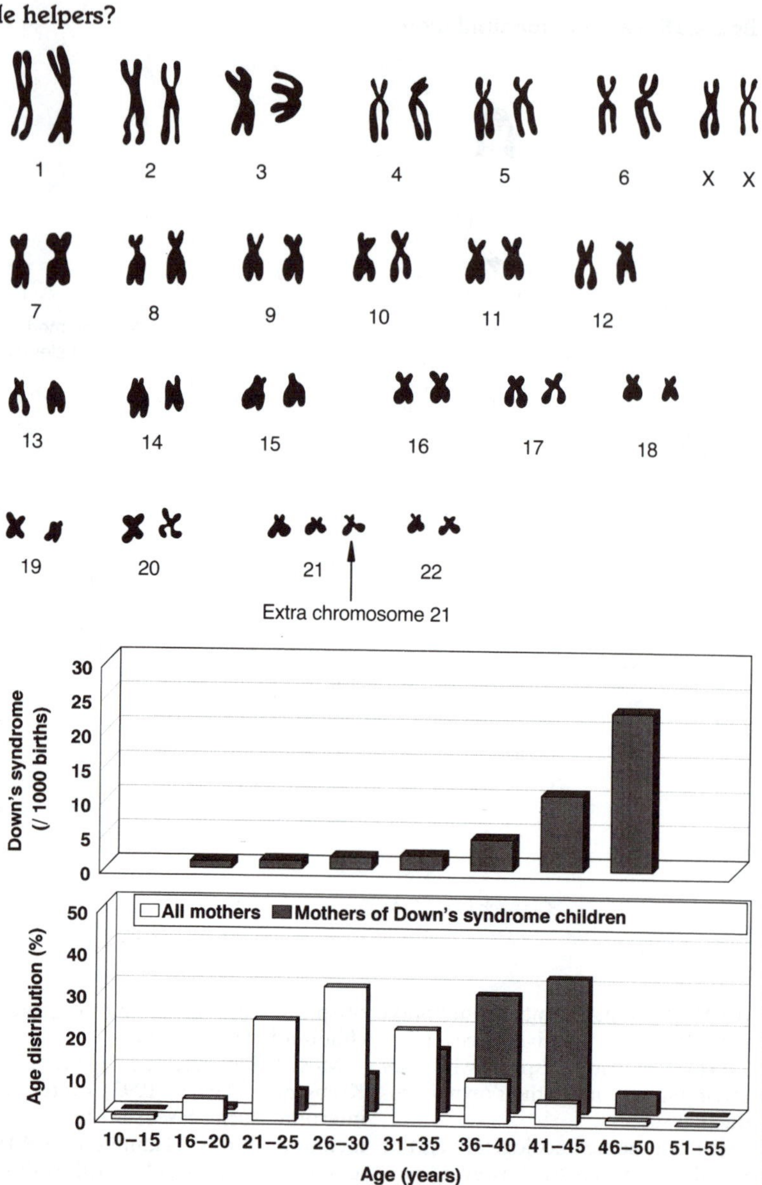

Trisomy of chromosome 21 is most often (on 80% of occasions) the result of a cell division in the mother's egg-cell line in which, instead of each cell receiving one copy of chromosome 21, one cell receives two and one receives none (Singer, 1985; Hull and Johnston, 1993). Eggs with no chromosome 21, if ever fertilized, are spontaneously aborted (section 4.4.11). Eggs with two copies of chromosome 21 (and therefore trisomic after fertilization) are also more likely to be spontaneously aborted but produce Down's syndrome individuals if they survive to be born.

About 60% are spontaneously aborted and at least 20% are stillborn (Connor and Ferguson-Smith, 1984). Of those not stillborn, 15–20% die before the age of 5 years, usually as a result of severe inoperable congenital heart disease. Life expectancy for the remainder is well into adult life. By the age of 40 years, however, almost all develop Alzheimer's disease (Hull and Johnston, 1993). Social skills often exceed other intellectual parameters. They have characteristic mental aptitudes, being slow or incapable of learning much of the behaviour normally necessary for an independent existence. On the other hand, they are particularly happy, affectionate and caring and, in most cases, when adult are quite capable of assisting parents and siblings with routine activities. Occasionally, Down's syndrome females reproduce. Then, half of their offspring are 'normal'; half are also Down's syndrome (Clegg and Clegg, 1975).

Non-Down's females are more likely to produce offspring with Down's syndrome as they age (see figure), irrespective of the age of the child's father. About 2.2% of births to women over 45 years are Down's syndrome. Relative to women under 20 years, this represents about 50 times the incidence (compared with c. 2× for women 25–30 years old; 3× for 30–35 years old; 8× for 35–40 years; and 23× for 40–45 years) (Colman and Stoller, in Singer, 1985).

This increased incidence with age could be simply due to a decreased efficiency in female cell division with age. However, Down's syndrome births to older women are so common (>2%) and such a characteristic feature of humans that an adaptive explanation could be considered necessary. Perhaps females strategically ovulate eggs with duplicated chromosome 21 around menopause when a sterile helper is an advantageous addition to the family.

the timing of attempts to raise offspring becomes orders of magnitude greater with the evolution of extended female parental care (Boxes 4.2 and 4.3). We assume, therefore, that family planning behaviour was exposed to even greater selection during mammalian evolution and most of the mechanisms discussed here are closely linked to mammalian physiology.

Contraceptive behaviour is most obvious in mammals in which females are more or less continuously receptive at all stages of their menstrual cycle (Chapter 6). Such behaviour is widespread among primates, including humans, and probably arose in the human lineage about 60 million years ago. Since that time, the minority of copulations by females have been for the purpose of conception. For example, it has been calculated for lions that 3000 copulations are needed to produce a lion who survives to maturity (Bertram, 1975). For comparison, the 3679 human subjects in our UK survey (Box 2.6) reported a total of 2.5 million copulations and 800 children (i.e. 3200 copulations/child).

It follows that neither modern women, nor their female progenitors over the past 60 million years, should closely associate copulation with conception. In fact, on the contrary, most copulations have been concerned with avoiding conception. The recent availability of even more reliable female contraceptive techniques (e.g. oral contraceptives; intrauterine devices; barrier devices) simply offers an extention of pre-existing physiobehavioural mechanisms and requires no shift in normal female psychology (Chapter 7).

4.5 Male genitalia: size and location of testes

Male humans are programmed to develop genitalia (i.e. a single penis and paired testes) that are permanently external (i.e. outside of the body). The testes are contained in a scrotum, a thin pouch of skin and subcutaneous tissue (Boxes 4.17 and 4.18).

Virtually all aspects of the size, structure and location of the genitalia of human males have been influenced at some stage of their evolution by the risk of sperm competition. The influence of sperm competition on the size, structure and function of the penis is considered in Chapter 6. This section is concerned with the possible influence of sperm competition on the size and location of the testes (the production and storage of sperm are described in section 2.4.3.).

4.5.1 Testis size

The influence of the risk of sperm competition on the size of animal testes is fairly clear. In general, males of species in which risk of sperm competition is higher have testes proportionally larger than males of species in which the risk is lower. Such a trend has now been demonstrated in butterflies (Gage, unpublished), birds (Møller, 1988a), and mammals (Kenagy and Trombulak, 1986; Møller, 1988c). Within mammals, the trend has also been shown separately for rodents (Kenagy and Trombulak, 1986), voles (Heske and Ostfeld, 1990), equines (Ginsberg and Rubenstein, 1990), baleen whales (Brownell and Ralls, 1986) and primates (Harcourt *et al.*, 1981; Møller, 1988b).

It is generally assumed that the link between high risks of sperm competition and relatively large testes resides in the fact that, all else being equal, larger testes produce more sperm at a faster rate (e.g. primates; Møller, 1988b). As a result, in lineages exposed to more intensive sperm competition, males with larger testes, because they have more sperm to inseminate, are favoured by selection more intensively than in lineages exposed to less-intensive sperm competition.

In humans, the right testis is on average 5–6% heavier than the left (see data in Diamond, 1986). Moreover, the size of the testes varies considerably from male to male and from population to population. For example, at autopsy the left testis of 100 Hong Kong Chinese was 9.4 g compared with 20.4 g for 140 Danes. The man testis volume was 24 ml for American males and 19 ml for Korean males (Kim and Lee, 1982). Rate of sperm production varies with testis size, not only

Box 4.17 **Male–male variation in testis size, rate of sperm production and involvement in sperm competition**

Individual variation in daily sperm production relative to size of testis
In our study of the number of sperm ejaculated (Box 3.8), males measured the size of their left testis using callipers. Daily sperm production rate was estimated by summing the number of sperm ejaculated over 2–40 ejaculates (copulatory and masturbatory) by each male and dividing by the sum of interejaculation intervals for those ejaculates.

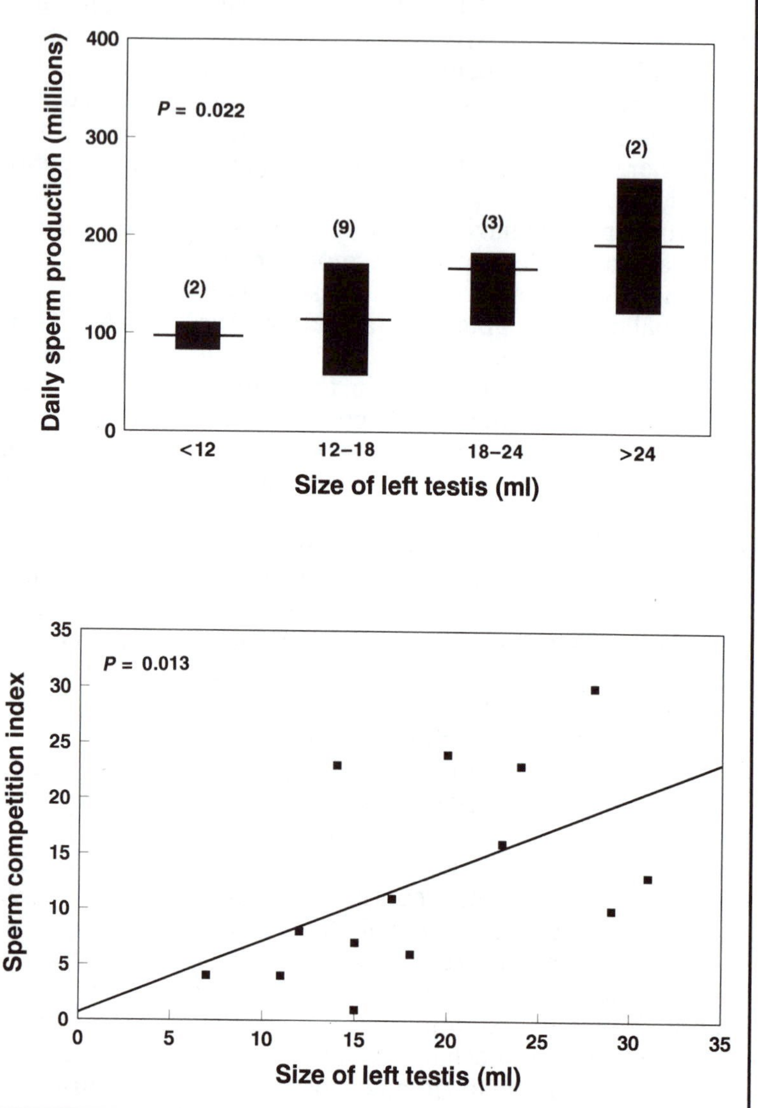

Size of testes and involvement in sperm competition
Males were ranked by 20 independent judges (subjectively but blind with respect to the size of their testes) according to their predisposition to pursue or not to pursue a 'sperm competition' strategy. There was a strong rank-order correlation with testis size.

between races (section 2.5.3) but also between individuals (Box 4.17).

Rushton (1988) reports that negroids have larger genitalia than caucasoids who in turn have larger genitalia than mongoloids. According to Rushton, this rank-order is associated with average behavioural differences between races which, if correct, would also indicate a positive link with risk of sperm competition.

Whether individual differences in testis size within a population are linked to different sexual strategies by males has received no attention. The prediction would be (section 2.5) that males more predisposed to pursue a sexual strategy involving a greater risk of sperm competition should have larger testes. Such males should be those predisposed to invest less in mate-guarding and to be more active in attempting to copulate with females likely to contain sperm from other males. In contrast, males with smaller testes should be predisposed to invest more in mate-guarding and to be less active in pursuing females associated with other males.

In our study of the number of sperm

inseminated by males during copulation (Box 3.8), some males measured the size of their left testis and were ranked according to their predisposition to pursue or not to pursue a 'sperm competition' strategy. There was a strong rank-order correlation between involvement in sperm competition and testis size (Box. 4.17).

This potential male polymorphism for sexual strategy is discussed further in Chapter 5.

4.5.2 Testis location

Although there is a clear association between the size of testes and level of sperm competition across and perhaps even within species, a link between sperm competition and the location of the testes in different lineages is more open to debate.

Human testes are external and scrotal (Box 4.18). Actual distance from the abdomen may be adjusted by contraction and relaxation of certain muscles. Muscles of the scrotal sheath (the dartos muscles) allow the scrotum to be pendulous when the ambient temperature is high but to be retracted into the lower pubic hair and adpressed to the abdomen when the ambient temperature is low. Other (cremasteric) musculature around the spermatic cords raise the testes for short periods when the male is excited (van Tienhoven, 1968) such as during copulation or when ready to fight or flee (Setchell, 1978).

Even though the scrotal muscles allow an element of temperature regulation, one result of an external, scrotal location is that the testes are normally cooler than the remainder of the body. In humans, mean scrotal temperature is about 31.0 °C, about 6.0 °C cooler than mean internal body temperture. Wearing clothes increases scrotal temperature by about 3.0 °C, thus reducing the difference to about 3 °C. In other mammals with external testes the difference between scrotal and body temperatures ranges from 2 to 6 °C (Carrick and Setchell, 1977).

More or less external (though not necessarily fully scrotal) testes are characteristic of all higher primates and so have been a feature of the human lineage for at least the past 50 million years. Throughout that time, therefore, sperm in the human lineage will have been exposed to selection for development at temperatures characteristic of the testes rather than the general body. It should be no surprise, therefore, if, when sperm are forced to develop at higher temperatures, infertility and/or testicular damage can occur (Moore, 1926). Such problems, however, may be the result of testes normally being external and cooler, not the cause.

In an evolutionary sense, testes do not have to be external for sperm to develop normally. Many animals with body temperatures higher than humans nevertheless have testes that are permanently inside their body. Thus, all birds and many mammals (e.g. monotremes, whales, elephants; Eckstein and Zuckerman, 1956; Box 4.18) have internal testes without any obvious deleterious influence on fertility and sperm performance. Among the apes, the chimpanzee, like humans, has scrotal testes but the orangutan and gorilla have not. Rather, the orangutan has a postpenial bulge of bare black skin under which the testes reside. A similar arrangement is found in the gorilla, but the postpenial bulge is covered with hair (Smith, 1984a,b).

So far, there is no general consensus on the adaptive significance of permanently external, scrotal testes. Any association with factors such as optimum temperature for sperm development and use of the testes in visual signalling to females (Portmann, 1952) is likely to be a secondary result of external testes rather than a primary cause.

The most likely explanation involves sperm competition and is based on the suggestion that external testes are adaptive for sperm storage, rather than sperm development (Bedford, 1977). Scrotal position and morphology is such as to ensure cooling of the sperm-storage area rather than of the testes themselves. A typical pattern across a wide range of mammals is for the hair on and around the scrotum to end abruptly at the lower border of the scrotum adjacent to the cauda epididymis where the mature sperm are stored (Bedford, 1977).

Stored sperm may not only use up energy reserves, their genetic materials may slowly undergo spontaneous mutation (Ehrenberg

Box 4.18 Testis location

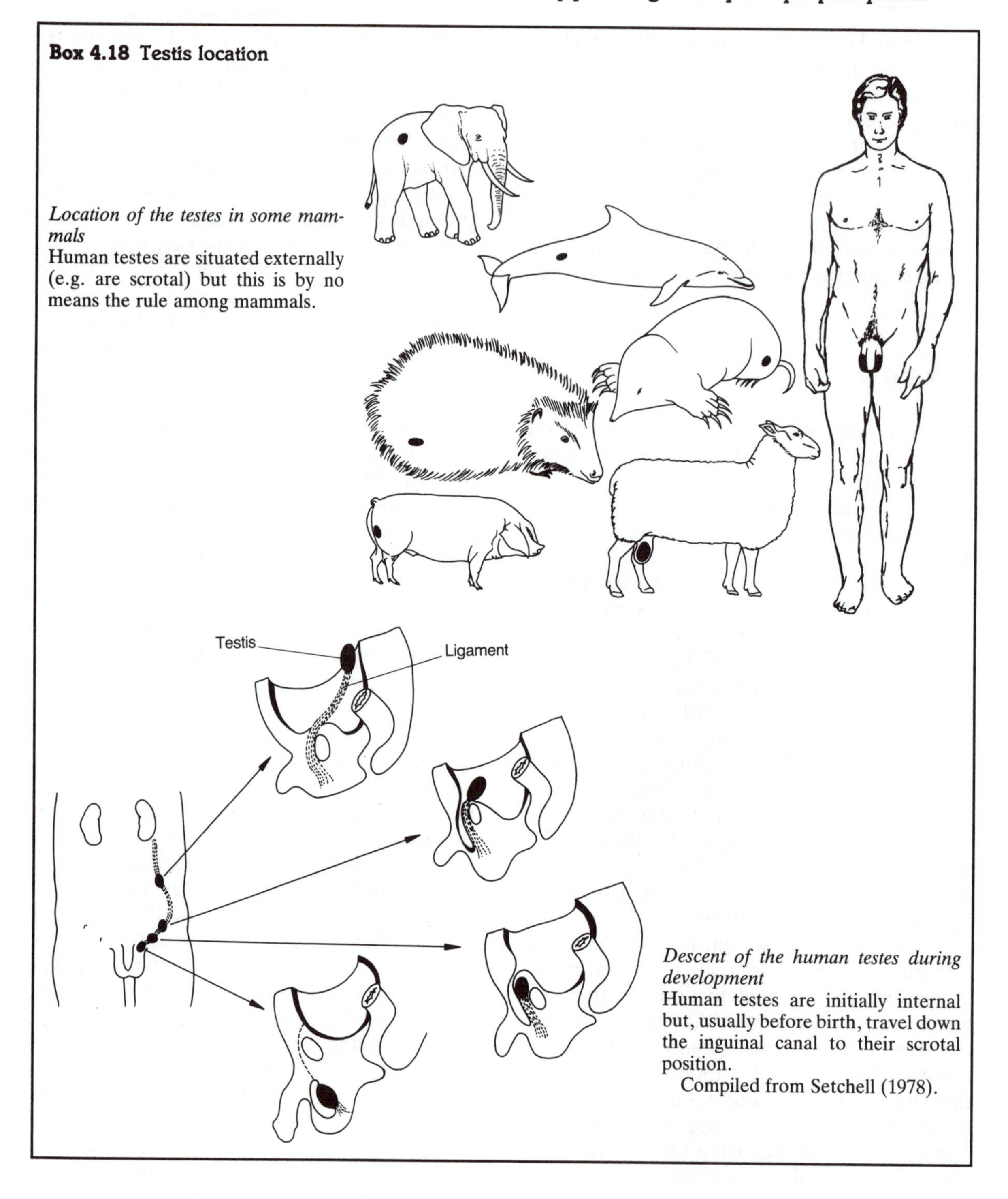

Location of the testes in some mammals
Human testes are situated externally (e.g. are scrotal) but this is by no means the rule among mammals.

Descent of the human testes during development
Human testes are initially internal but, usually before birth, travel down the inguinal canal to their scrotal position.
Compiled from Setchell (1978).

and von Ehrenstein, 1957). In this respect, sperm are probably no different from any other cell but, being so small and independent, they possess neither the apparatus nor the energy reserves to repair themselves. Energetically, it may be cheaper for the male to replace rather than repair old and/or damaged sperm (but see also Chapter 11).

The lower the temperature at which sperm are stored the longer they are likely to retain their viability. This will be true no matter what the optimum temperature for development. Thus, human sperm survive more than twice as long at a temperature of 21 °C than at body temperature (37 °C; Bellis and Baker, unpublished data) and may survive indefinitely when frozen (Mann and Lutwak Mann, 1981).

Experiments with rats and rabbits, neither of which have abdominal testes, surgically repositioned the epididymis in the abdomen. There was no influence on sperm maturation but a dramatic influence on sperm longevity (Bedford, 1977). Stored rat sperm in the cauda epididymis remained viable for 21 days in the scrotum but only 3–4 days in the abdomen; stored rabbit sperm for 35 days in the scrotum but only 8–10 days in the abdomen.

Bedford suggests that the scrotal state should be found in mammal species in which the male must be able to produce fertile ejaculates repeatedly over a relatively short period, as in polygynous species (Chapter 6). However, these are species which store sperm for shorter periods and which may, therefore, be less influenced by storage temperature. Nevertheless, Smith (1984b) proposes that scrotal testes should be found in species with intense sperm competition. On this basis, he explains the difference between the well-developed scrota of humans and chimpanzees and the poorly developed scrota of male orangutans and gorillas.

An alternative, while still accepting Bedford's sperm-storage hypothesis, is that the differences between species may depend as much on differences in the disadvantage of scrotal testes as on differences in the advantage. Whether a species has scrotal testes or not then depends on the trade-off between the advantage(s) and disadvantage(s), rather than simply on differences in the advantage.

Even if scrotal testes were always advantageous in terms of sperm viability during storage and hence in terms of any subsequent sperm competition, there are obvious disadvantages that may have militated against their evolution in certain lineages. External testes could be disadvantageous to animals for which aerodynamic (e.g. birds, bats) or hydrodynamic (e.g. seals, sea lions, whales) performance is a priority. All of these groups have internal testes, though many bats descend their testes into a scrotum during the season that they produce sperm and retract them again at other times (Racey, 1973).

External testes could also be disadvantageous to lineages living in environments in which any external organ would have a high risk of being damaged. Most male humans can identify with the vulnerability of scrotal testes. In addition, scrotal testes are inevitably more prone to torsion, one wrapping around the other and reducing blood supply, leading often to death of the testicular tissue in one of the testes (Hull and Johnston, 1993). There must also be a possibility that scrotal testes are less protected from background radiation damage.

Species which live in dense undergrowth and/or spend a great deal of time in trees are much more likely to damage scrotal testes than species living in open, relatively clear, terrain on the ground. This seems as, if not more, likely an explanation than differences in the level of sperm competition for the differences in scrotal arrangement between the fully scrotal humans and chimps and the less scrotal orangutans and gorillas. Both risk of damage and risk of sperm competition, however, should be involved in the selective trade-off. In principle, scrotal testes should evolve only in lineages in which the sperm competition benefits of longer sperm viability while stored in the male exceed the potential costs due to risk of damage to the scrotum or testes.

4.6 Sperm polymorphism

The sperm in a typical human ejaculate are polymorphic (i.e. of several different morphologies or 'morphs'). In this book, we follow the World Health Organization classification of sperm types (Belsey *et al.*, 1987). Thus, we recognize eight types of sperm head, four types of sperm tail, and two types of mid-piece. In addition, we recognize that some sperm have a cytoplasmic droplet attached whereas others do not. The different sperm types and their relative proportions in an average human ejaculate are shown in Box 4.19.

Even the least common visually recognizable morph, the bicephalous, is still numerous, about 300 000 being present in the typical ejaculate. The most common sperm morph, with an oval head, straight mid-piece, single long tail and no cytoplasmic droplet makes up about 60% (i.e. about 180 million) of a typical ejaculate. This most common morph is often referred to as the 'normal' sperm (and all other morphs, collectively, as 'abnormal') but we avoid this terminology through most of this book.

One, rather whimsical, interpretation of sperm polymorphism is that it is an artefact of the lifestyle of modern humans, abnormal sperm being the result of exposure to high temperatures through wearing tight trousers or having hot baths. However, we may rule out this interpretation. Many mammals, none of which have hot baths or wear tight trousers, have sperm equally as polymorphic as humans (Box 4.19) and, to a greater or lesser extent, all mammals have polymorphic sperm.

Sperm polymorphism also occurs in animals other than mammals. The phenomenon has received less attention for external than internal fertilizers. The general impression is that there is less sperm polymorphism among external fertilizers, but this impression may be more a reflection of a lack of data than of reality.

Among insects, sperm polymorphism is largely one of size (e.g. fruit flies, *Drosophila* spp.; Joly *et al.*, 1989). In the Lepidoptera (butterflies and moths), however, there are two distinct sperm types: (1) fertile eupyrene sperm; and (2) infertile apyrene sperm, which lack a nucleus (Meves, 1902; Silberglied *et al.*, 1984; Sonnenschein and Hauser, 1991).

The traditional interpretation of sperm polymorphism is that 'abnormal' sperm are deformed and in some way mistakes (Cohen, 1967). Sperm are considered to be difficult to manufacture and natural selection to be insufficiently powerful to make the process 100% efficient. The evidence in support of Cohen's 'development error' hypothesis is an apparent across-species correlation between chiasmata frequency and the redundancy of sperm (i.e. number of sperm – number of eggs to be fertilized) inseminated by the male (Cohen, 1973). The logic was that the probability of developmental error is a function of the number of chiasmata and that the greater the risk of error the more sperm a male needed to produce to have the necessary number of fit sperm. However, although subsequent authors have sometimes assumed to the contrary (e.g. Harcourt, 1991), Cohen never suggested for mammals that these development errors correlated with gross morphology and visible sperm polymorphism.

The correlation between number of chiasmata and sperm redundancy has since been criticized statistically (Mather, 1974; Wallace, 1974) and seems unlikely to be real. A further criticism of Cohen's hypothesis is biological. Females successfully produce and mature gametes with a high degree of efficiency, relatively few errors being detected compared with the level postulated for males. Cohen (1975) has suggested that strict comparison between eggs and sperm is unreasonable. Unlike males, females reabsorb the vast majority of the gametes that they initiate (Box 3.4) and those that are reabsorbed may well represent the meiotic failures. However, at least in insects, as acknowledged by Cohen, which cells are to become eggs and which are to be reabsorbed (as nurse cells) is determined before meiosis. Egg-line redundancy, therefore, is unlikely to be the result of errors during meiosis and this casts doubts on whether errors during meiosis can also explain sperm redundancy.

Some authors (e.g. Harcourt, 1989, 1991) still use Cohen's developmental error hypothesis to explain sperm polymorphism and clinically, in the treatment of human infertility, 'abnormal' sperm are considered to be undesirable passengers in the ejaculate. To an evolutionary biologist, however, it seems highly unlikely that sperm polymorphism would persist at such a high level if the 'abnormal' sperm were really undesirable deformities. An adaptive explanation, ascribing a real function to the different sperm morphologies, would be much more acceptable.

Our Kamikaze Sperm Hypothesis (section 2.5.3) is an attempt at an adaptive interpretation of sperm polymorphism. We argue that the observed polymorphism reflects a division

Box 4.19 Sperm shape and polymorphism (pp. 90–91) in humans and other mammals

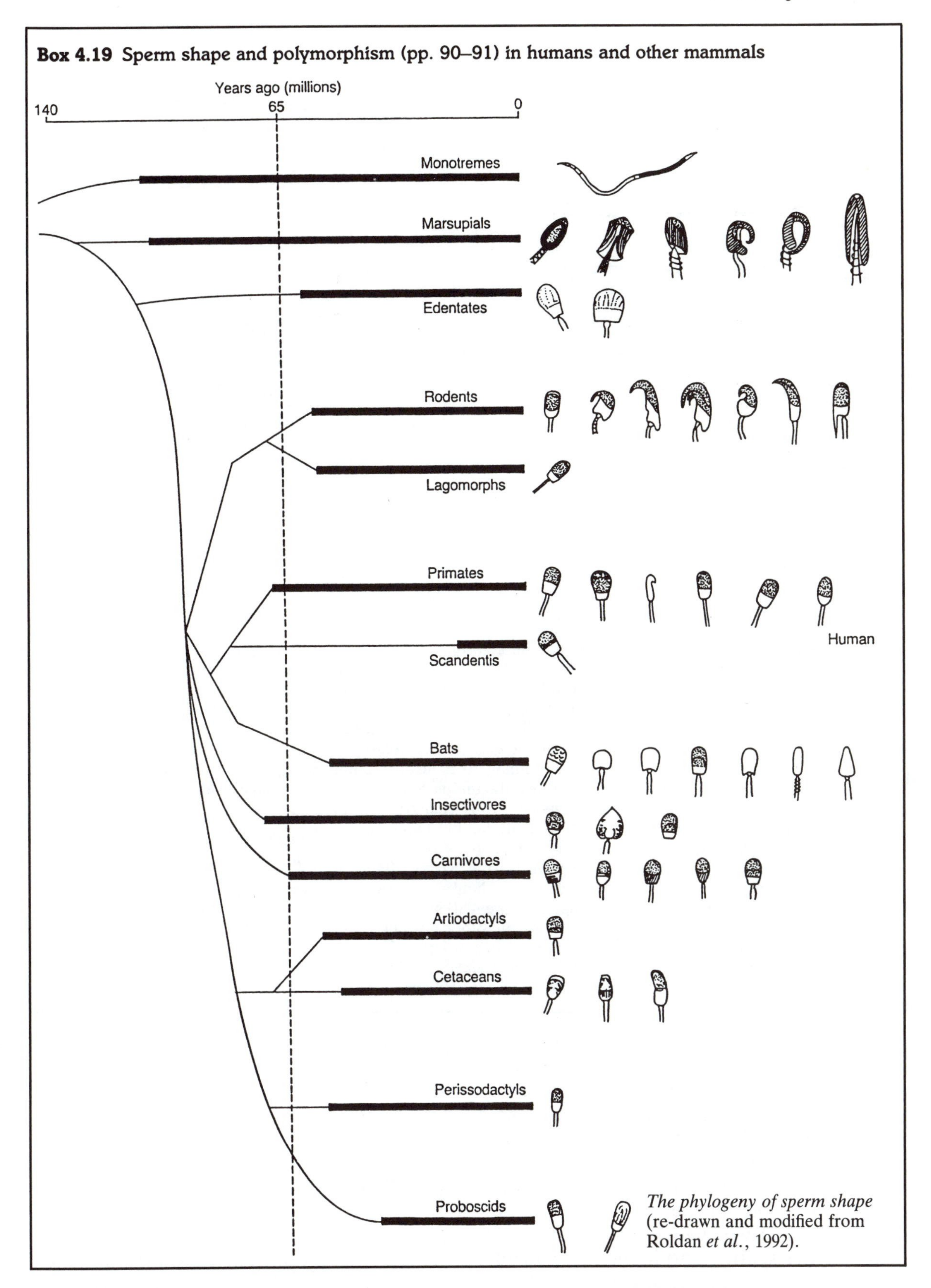

The phylogeny of sperm shape (re-drawn and modified from Roldan *et al.*, 1992).

Box 4.19 (*continued*)

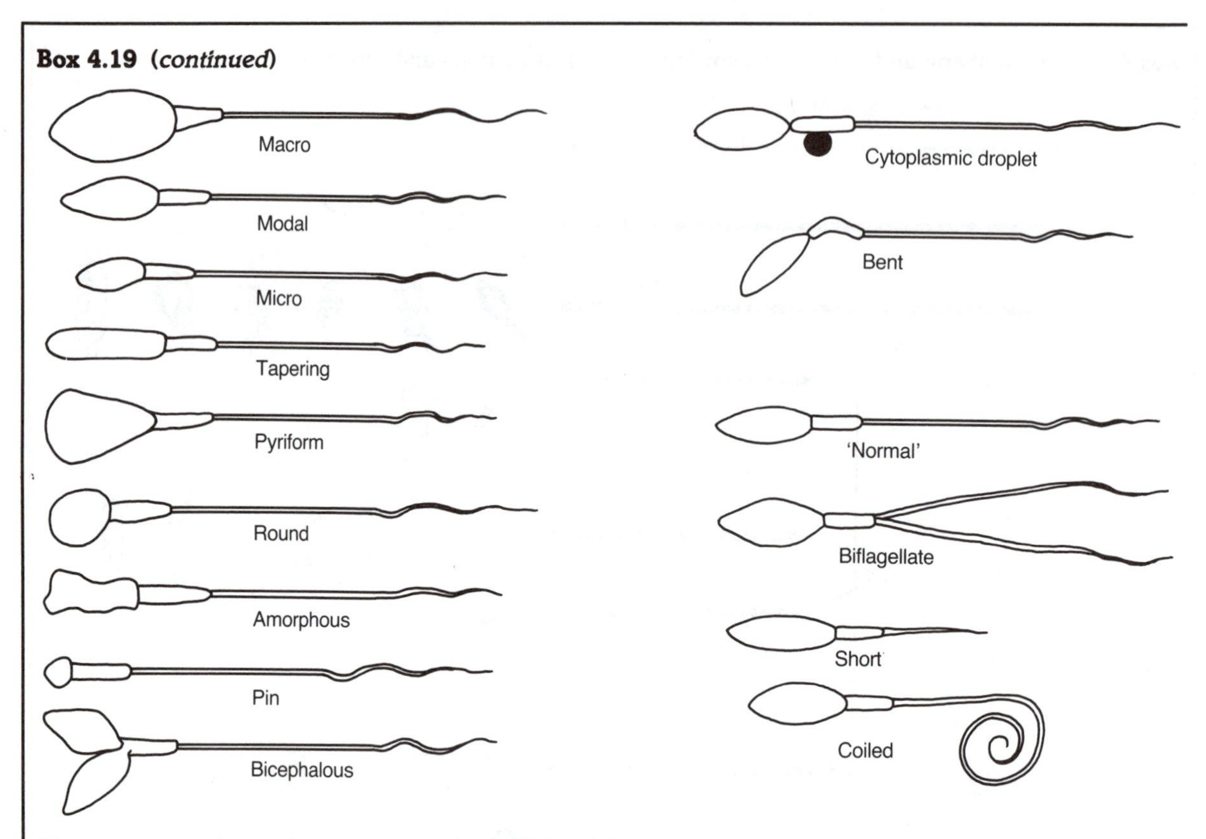

Human sperm polymorphism

Head morphs

Oval-headed sperm

Modal sperm: Regular oval-shaped head, 3–5 μm long and 2–3 μm wide. The head is rounded at the junction to the mid-piece.

Macrocephalous: Oval-shaped but >5 μm long and >3 μm wide.

Microcephalous: Oval shaped but <3 μm long and <2 μm wide.

Other head morphs

Pin: Sperm with head too small to contain a nucleus.

Tapering: Cigar-shaped heads with parallel sides which may or may not come to a point at the mid-piece. The acrosome may or may not be pointed. The most common forms have heads >7 μm long and <3 μm wide.

Pyriform: Regardless of size, have an obvious or exaggerated tear-drop shape coming to a distinct point just above the midpiece region.

Bicephalous (two-headed): Sperm with a single tail but two heads of any size or shape.

Amorphous: Any bizarre shape of head (e.g. dumb-bell, twisted, bulging, depressed) not classifiable as any of the other morphs. Usually normal size or larger.

Round (marblehead): Spherical heads with no visible acrosome.

Tail morphs

Uniflagellate: single and flagellate, at least 45 μm long.

Coiled: normal length tail, usually coiled in a spiral.

Short: tail shorter than 45 μm.

Biflagellate: single-head (of any size or shape) with two tails

Midpiece morphs

Straight: 7–8 μm long and 1 μm wide

Bent: not straight

Other morph

Cytoplasmic droplet: Sperm with a residual cytoplasmic attachment at least one-half the size of the head and still attached to the head, mid-piece or upper tail region.

of labour between the sperm in an ejaculate, each sperm morph having a different role to play in sperm warfare between males. The more complex the female tract and the greater the intensity of sperm competition in any particular animal, the more different roles are available for sperm to play.

In Chapter 11 we analyse the dynamics of many of the sperm morphs in the human ejaculate and deduce the most likely role for each in the context of the Kamikaze Sperm Hypothesis.

4.7 Ordered ejaculates

Human males inseminate the female by ejaculating a series of from 3–9 spurts of semen,

Some estimates of morph frequency in humans and other mammals

Proportion (%) of sperm of different head morphs in various mammals

Species	Modal	Macro	Micro	Tapering	Pyriform	Round	Amorphous	2-headed
Human (cop.)[a]	78	5	4	4	2	3	3	0.3
Human (mas.)[a]	72	5	5	6	4	3	5	0.3
Human (mas.)[b]	73	2	<1	<1	–	–	18	0.5
Gorilla[b]	71	2	1	1	–	–	16	0.5
Chimpanzee[b]	95	<1	–	<1	–	–	<1	–
Pigmy chimp[b]	98	–	1	–	–	–	–	–
Orangutan[b]	99	–	–	–	–	–	1	–
Horse[c]	85	1	1	–	<1	–	<1	0.1
Cheetah[d]	98?	1	<1	–	–	–	–	0.03
Clouded leopard[e]	98?		(2)	–	–	–	–	–
Domestic cat[f]	99?	<1	<1	–	–	–	–	–
Lion[g]	98?	1	<1	–	–	–	–	0.2

Proportion (%) of sperm of different tail and other morphs in various mammals

Species	Uniflagellate tail	Short tail	Coiled tail	2-tailed	Bent midpiece	Cytoplasmic droplet
Human (cop.)[a]	86	1	12	0.6	4	6
Human (mas.)[a]	91	1	7	0.7	4	7
Gorilla[b]	–	–	–	–	1	<1
Chimpanzee[b]	–	–	–	–	<1	–
Horse[c]	96	<1	3	0	<1	7
Cheetah[d]	68?	–	32	0.04	22	13
Clouded leopard[e]	90?	–	10	–	9	9
Domestic cat[f]	89?	–	11	–	6	12
Lion[g]	98?	–	2	0.04	4	15
Domestic ferret			4		30	
Black-footed ferret			23		11	

[a] Manchester, UK, survey (Box 3.8; $n = 119$ copulatory ejaculates, 92 masturbatory); [b] Seuánez (1980); [c] Bielański *et al.* (1982); [d] Free ranging, East Africa, Wildt *et al.* (1987a); [e] Wildt *et al.* (1986); [f] Wildt *et al.* (1983); [g] Serengeti, Africa, Wildt *et al.* (1987b).

Authors often include round-headed and pyriform sperm in the amorphous category. Note the difference in proportion of tapering + amorphous (including pyriform and round) heads in humans and gorillas on the one hand and chimpanzees and orangutans on the other (section 11.3.4).

often with a copulatory thrust between each spurt. Many other mammals also ejaculate in spurts though for some (e.g. the pig; Mann and Lutwak-Mann, 1981) the ejaculatory flow seems to be more or less continuous.

The seminal fluid in the human ejaculate has long been known to differ between spurts. Differences in the appearance, smell and taste of the different parts of the ejaculate were described as long ago as 1746 (John Hunter in Mann and Lutwak-Mann, 1981).

The first part of the human ejaculate is dominated by fluids from the prostate and Cowper's gland (Tauber *et al.*, 1976) and the later parts by fluids from the seminal vesicle (Tauber *et al.*, 1976). The concentration of sperm (Amelar and Hotchkiss, 1965), IgG,

Box 4.20 **A study of sperm in the different spurts of the human ejaculate**

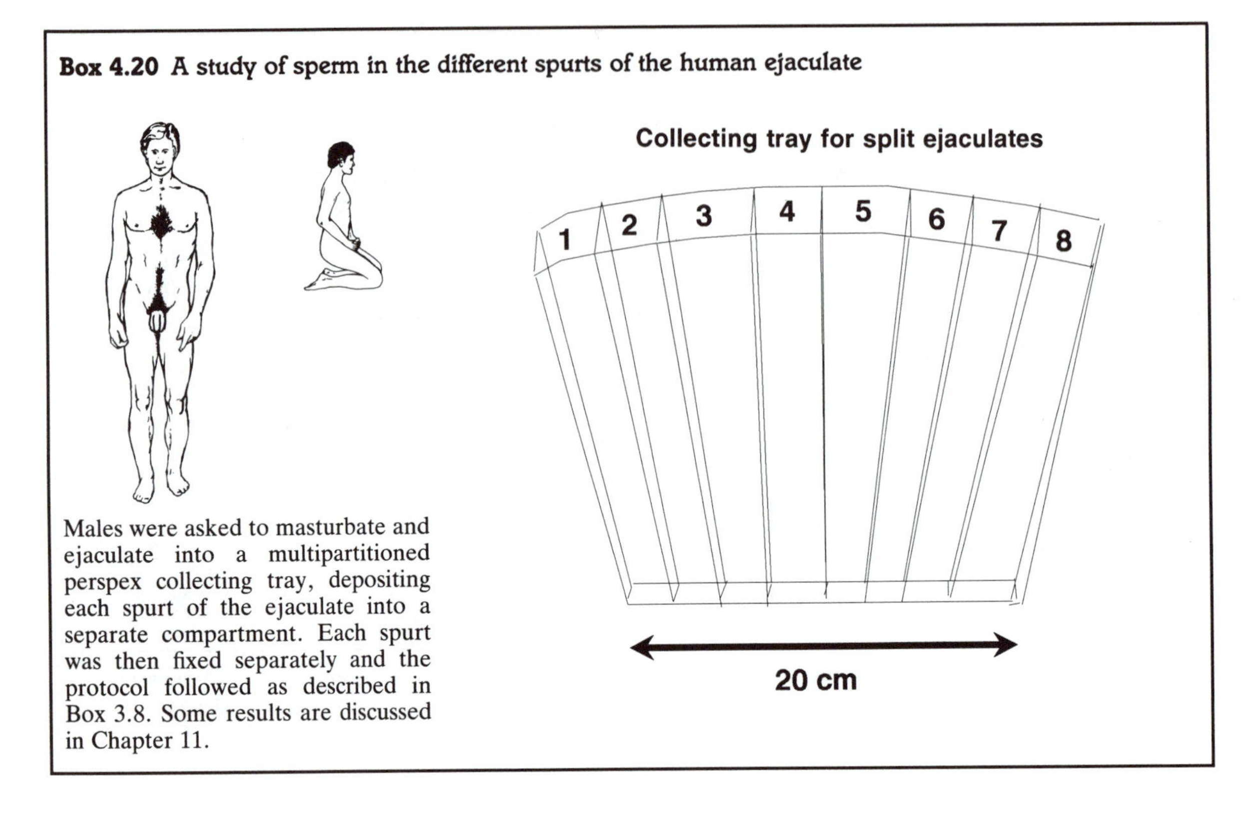

Males were asked to masturbate and ejaculate into a multipartitioned perspex collecting tray, depositing each spurt of the ejaculate into a separate compartment. Each spurt was then fixed separately and the protocol followed as described in Box 3.8. Some results are discussed in Chapter 11.

and albumin decrease from the first part of the ejaculate to the last (Tauber *et al.*, 1976). In contrast, the greatest amounts of fructose and lactoferrin are present in the final portion from the seminal vesicle (Tauber *et al.*, 1976).

The very last spurt of seminal fluid is strongly spermicidal. Sperm show better survival, detectable after only 40 minutes, in the first than the last half of the ejaculate (Lindholmer, 1973). The reduced motility and faster death rate of sperm in the second fraction is due to the presence of a suppression factor (Lindholmer, 1973), perhaps involving fructose. Such a spermicidal function could be an adaptation to sperm competition, the last part of the seminal fluids from one male potentially decreasing the motility and survival prospects of sperm in the first part of the ejaculate from any male who inseminates the female within the next 30 minutes or so.

On the other hand, the prostate secretion in the first half of the ejaculate provides the sperm with some protection from the seminal vesicle secretion in the second (Lindholmer, 1973). Again this is consistent with a role in sperm competition; the seminal fluid in the first part of a male's ejaculate protecting the contained sperm from the spermicidal influence of the second part of the ejaculate from a (very) recent previous male.

In humans, the first half of the ejaculate supports a higher proportion of motile sperm than the second (Farris and Murphy, 1960; Eliason and Lindholmer, 1972; also Marmar *et al.*, 1979a,b) but analyses of sperm morphology have so far been relatively lacking in detail. We have carried out our own, more detailed, analysis of the variation in sperm morphology from early to late spurts in the masturbatory ejaculates of humans (Box 4.20) and some of the results are discussed in Chapter 11.

Not only do the spurts differ in seminal chemistry, they are inseminated with different force by a penis that may change in length and turgor from first spurt to last. Exactly where each spurt is deposited in the vagina will also be influenced by whether or not there is a preceding copulatory thrust. Equally, each spurt may have different properties of viscosity, its chemical constituents causing different levels of seminal coagulation and subsequent decoagulation (section 3.3.2). The result must be that the seminal pool deposited in the upper vagina (Box 3.5) will not be homogeneous but will be ordered and structured. This in turn will inevitably influence the interaction between cervical mucus and semen, both in the presence and absence of a copulatory orgasm by the female.

Unfortunately, we are not yet in a position to be more specific about just how the different spurts will position themselves in the vagina in relation to the cervix and cervical mucus.

Even mammals that ejaculate in a single spurt, such as the boar, have a chemically ordered ejaculate (Mann and Lutwak-Mann, 1981). The ejaculate of such species will also have an ordered interaction with the female tract. The only exceptions may be those species with such a small volume to their ejaculate (e.g. cat, 51 million sperm in 0.03 ml and bat, 300 million in 0.05 ml compared with the human with 300 million in 3 ml; Mann and Lutwak-Mann, 1981), that there may be little opportunity to structure the ejaculate.

Contrary to the usual rule, Harvey (1956) found that six men out of 243 studied had a higher concentration of sperm and better motility in the second part of their ejaculate. These males, may have had a reversal of the normal prostate–seminal vesicle sequence as subsequently found for a few males by Kvist (1991). In such men, the normal ejaculation sequence could have been reversed either through swelling of the prostate or neuromuscular changes in the ejaculation process (Kvist, 1991). Whatever the cause, such reversal of the seminal fluid sequence renders the ejaculate infertile. A similar situation has been found in horses which normally also produce a last part to the ejaculate with a low concentration of sperm (Mann *et al.*, 1957). Stallions with reversed ejaculates were subfertile (T. Mann in Amelar and Hotchkiss, 1965).

Ordered ejaculates seem to be a mammalian characteristic and we thus assume they have been a feature of the human lineage for at least 100 million years. Not enough is known about the taxonomic distribution of ejaculation in spurts to make any deduction about the antiquity of this aspect of sexual programming in humans.

5 The mammalian inheritance: masturbation, homosexuality and push buttons

5.1 Introduction

During the early mammalian phase of human evolution, sperm competition made an impact on some fairly obvious features of human sexuality. Examples are the form and function of different parts of the reproductive tracts and genitalia of both males and females (Chapter 4). However, there is one feature of humankind's mammalian inheritance that at first sight seems very strange. This is the apparent predilection of both males and females to seek, trigger and experience orgasms that have nothing at all to do with copulation.

For both sexes, there are four main types of non-copulatory orgasm. These are: (1) spontaneous (usually during sleep); (2) stimulated by self-masturbation; (3) stimulated by a heterosexual partner (but without copulation); and (4) stimulated by a homosexual partner.

At first sight, such events may appear to have little positive connection with sperm competition, at best seeming irrelevant, at worst actually seeming to make nonsense of sperm competition theory. For example, in principle, sperm competition should favour males who are very precious over their allocation of sperm (section 2.5.4). Yet, by virtue of their non-copulatory ejaculations, males on the contrary appear to be extremely wasteful. Such apparent wastefulness, if it is not to be considered maladaptive or to run counter to sperm competition theory, demands evaluation.

Historically, few aspects of human sexuality have been as misunderstood and as subject to prejudice as the various forms of non-copulatory orgams shown by males and females. These features are the most secretive, have often seemed the most enigmatic and futile, and have certainly been the most open to prejudice and victimization. Nevertheless, they are part of humankind's mammalian inheritance. So, too, is their secretive nature and even the suspicion, prejudice and victimization they elicit from other individuals who in all probability are themselves secretly showing the same behaviour.

In this chapter, we first introduce and describe the different types of non-copulatory orgasm and then consider their interrelationships and patterns of occurrence. Finally, we discuss their possible functions and antiquity in the human lineage.

In attempting to synthesize a picture of the incidence and patterns of these orgasms, we make considerable use of the only single survey known to us that provides information on most of the factors we require. By using a single source we avoid any temptation to be selective over the figures we use. The source we adopt is the survey for the USA in the 1940s by Kinsey *et al.* (1948, 1953). Although this survey has received some criticism over the years, most controversy has concerned levels of behaviour rather than the patterns that are the main concern here.

In addition, and for comparison, we present when relevant the results of our own nationwide, UK (Box 2.6) survey of the sexual behaviour of women and our Manchester, UK (Box 3.8) survey associated with our collection of ejaculates and flowbacks.

5.2 Orgasms in a lifetime

There are very few human males who never ejaculate over an entire lifetime (0.4%, USA). Indeed, the majority of males have ejaculated by age 15 years and then have very rapidly established an adult rate of emission which persists at about a more or less constant 3/week until nearly 35 years of age (Box 5.1). Thereafter, rate slowly declines.

Box 5.1 Variation with age in incidence of ejaculation (males) and orgasm (females) (USA, 1940s)

Frequency of ejaculation and orgasm per week is divided into copulatory and non-copulatory events.

(Compiled from data in Kinsey *et al.*, 1948, 1953.)

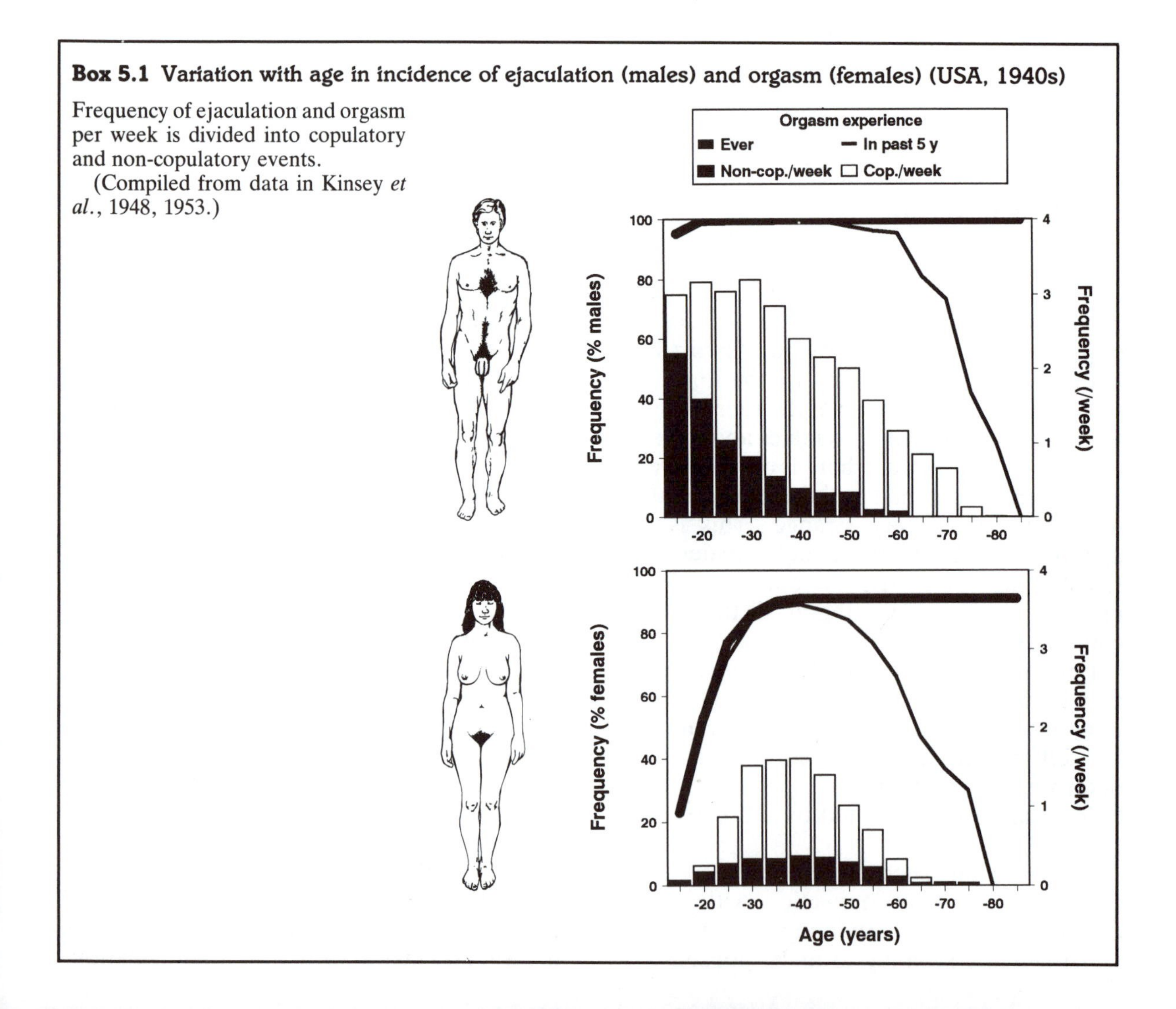

Box 5.2 Frequency with which first lifetime orgasm is non-copulatory or copulatory (USA, 1940s; UK 1980s)

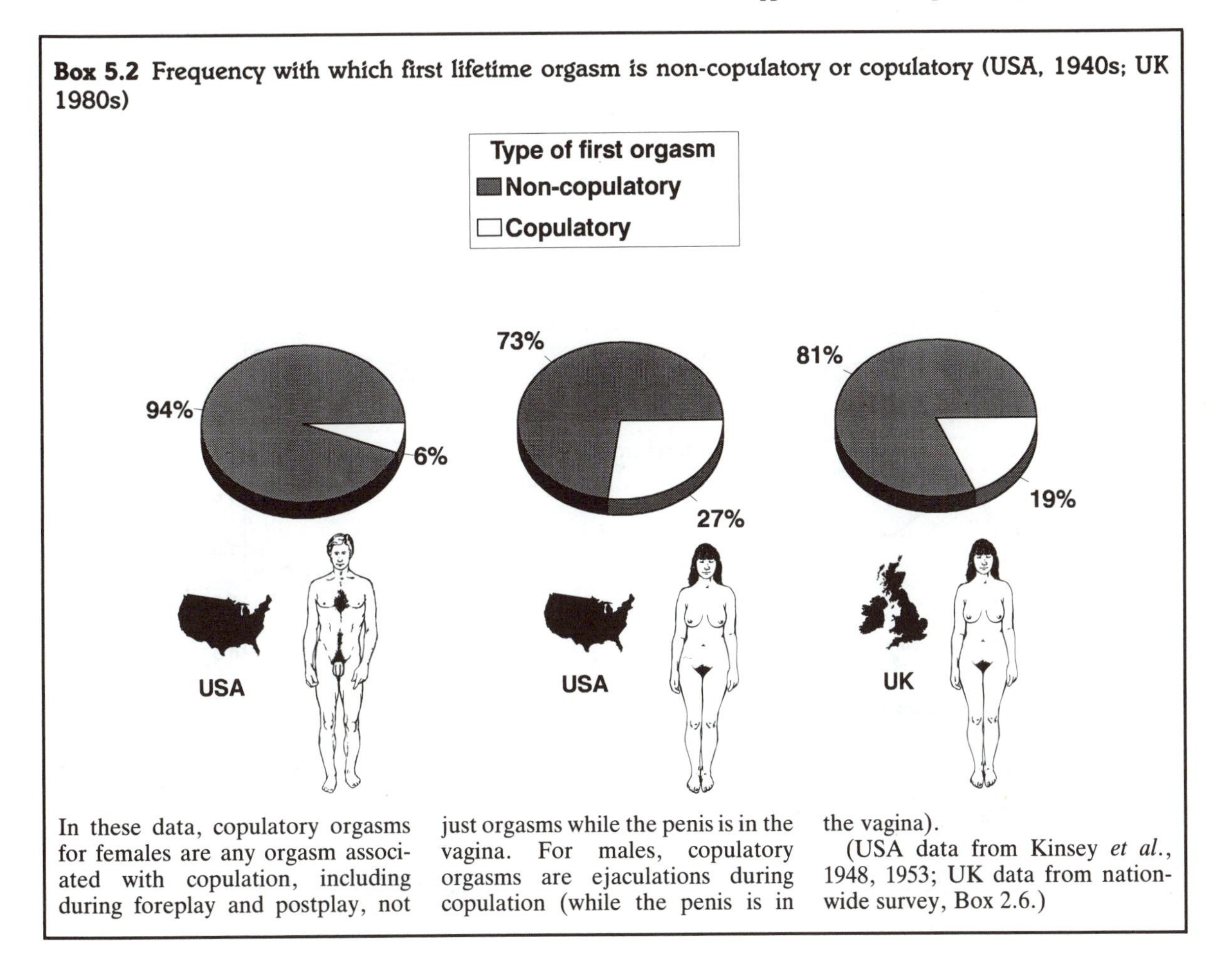

In these data, copulatory orgasms for females are any orgasm associated with copulation, including during foreplay and postplay, not just orgasms while the penis is in the vagina. For males, copulatory orgasms are ejaculations during copulation (while the penis is in the vagina).

(USA data from Kinsey *et al.*, 1948, 1953; UK data from nationwide survey, Box 2.6.)

About ten times as many females as males never orgasm in their lives (9%, USA; 2%, UK). Moreover, even those females who do experience orgasm have their first climax at a much more variable age than males and do not reach their peak rate (about 1.5/week) until 30–40 years of age (Box 5.1). On average, females also cease to orgasm earlier than males.

For the vast majority of people the first orgasm is non-copulatory, usually more or less totally disassociated from copulation (94% of males; 73–81% of females; Box 5.2). Moreover, such non-copulatory events continue to make a significant contribution to total orgasms throughout reproductive life. In males, non-copulatory orgasms decline in frequency with age but in females they actually increase in frequency throughout the fertile years (Box 5.1).

5.3 Types of non-copulatory orgasms

5.3.1 Spontaneous (nocturnal) orgasms

For a few males (14%, USA) and females (9%, USA), first orgasm occurs spontaneously, without any direct physical stimulation. Mostly, this happens at night while asleep but for a few males (1%) it occurs while awake (e.g. while climbing a tree; reading a book; watching a film).

Occasionally (about 1% of people; USA), humans can experience a nocturnal orgasm without any recollection of an associated dream. Usually, however, there is recollection of such a dream. Even so, cause and effect is not clear. Some of our own Manchester, UK, female subjects reported that the

Box 5.3 Variation with age in the incidence and frequency of homosexual, nocturnal and self-masturbatory ejaculations (males) and orgasms (females) (USA, 1940s)

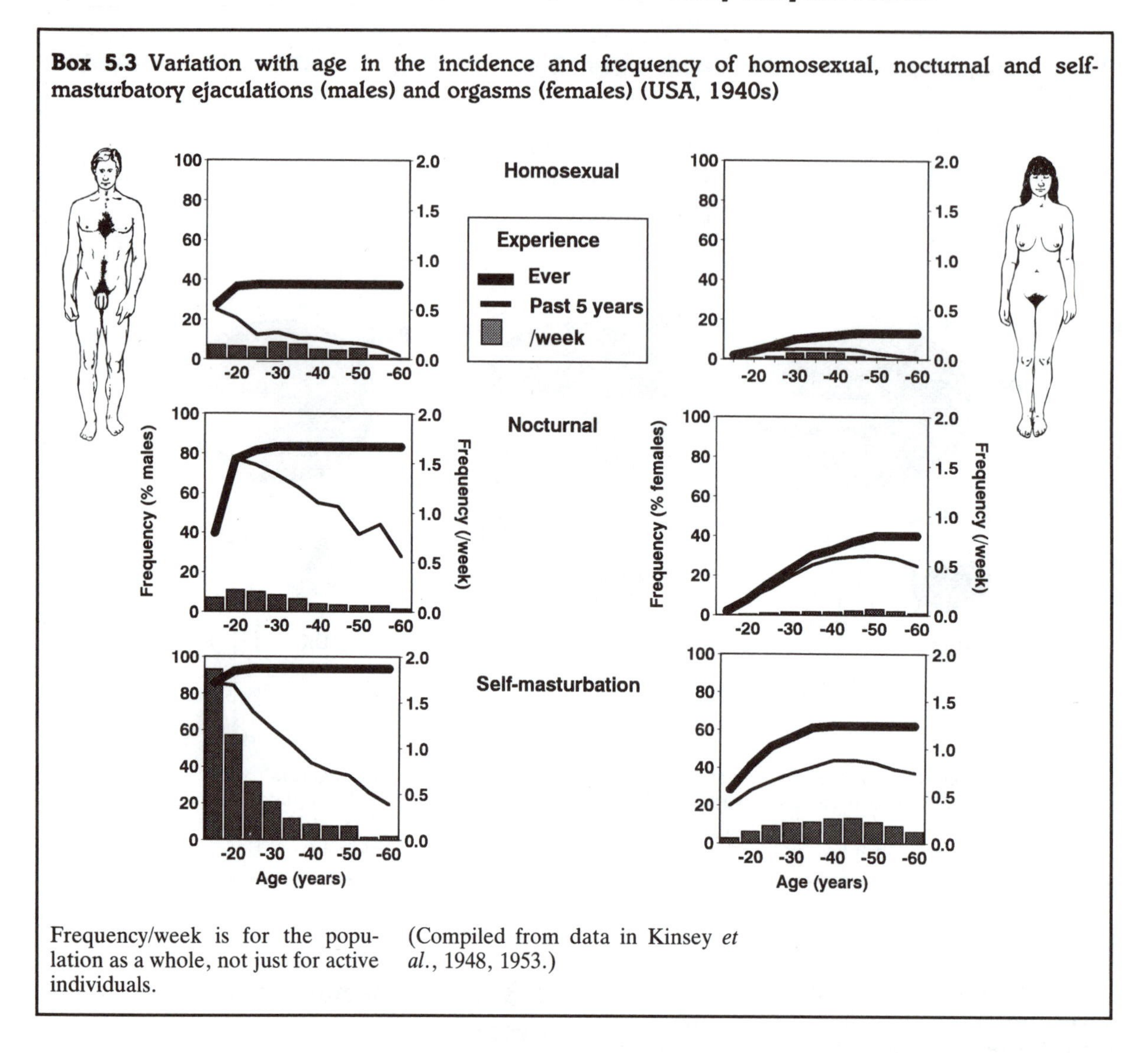

Frequency/week is for the population as a whole, not just for active individuals.

(Compiled from data in Kinsey *et al.*, 1948, 1953.)

orgasm often came first but seemed to trigger the sexual dream that followed almost simultaneously.

Over a lifetime, the majority (83%, USA) of males experience at least one nocturnal emission and most have done so by age 20 years (Box 5.3). After this age, the proportion experiencing such events declines to about 30% by age 60 years. Rate of nocturnal emission peaks (about once every 5 weeks) between the ages of 15 and 20 years and then declines. Even among adolescent males, however, nocturnal emissions account for only 6–10% of all ejaculations.

Although a roughly similar proportion of females (9%, USA) to males (13%) experience their first orgasm spontaneously at night, at least in some societies more than half of females may never experience a nocturnal orgasm. In the USA, for example, the proportion who have ever experienced such an event increases only slowly with age and first experience may occur at any time up to about age 50 years (Box 5.3). By their mid-20s, about 12% of women have experienced at least one nocturnal orgasm (Russia, Weissenberg, 1922; USA, Box 5.3). By their mid-40s, the proportion rises to from 37% (USA: Hamilton, 1929; see also Box 5.3) to 50% (Germany: Heyn, 1924). Perhaps in some societies, however, virtually all women experience nocturnal orgasms from time to time (e.g. Melanesia, Davenport, 1965).

Even women who do report nocturnal orgasms experience them only infrequently (Box 5.3). In the USA, the average (median) female who reported such orgasms, experienced them at a rate of about once every 3 or 4 months and only about 1% of women experienced more than one per week. Nevertheless, in the USA population as a whole, the frequency of nocturnal orgasms tended to increase with age throughout the fertile years (Box 5.3).

The incidence of spontaneous orgasms in non-human mammals seems to have received little scientific investigation. Spontaneous emission while asleep has been reported in the male cat (Aronson, 1949) and anecdotal accounts of sleeping female dogs in oestrous report apparent sexual activity, including body movements, vocalizations, pelvic thrusts, vaginal swelling and heavy mucus secretion (Kinsey *et al.*, 1953). Male rats (Beach, 1975), hamsters (Beach and Eaton, 1969), cats (Aronson, 1949) and chimpanzees (Rosenblatt and Schneirla, 1962) also seem to ejaculate spontaneously while awake and at least the rodents then consume the ejaculate.

5.3.2 Self-masturbation

The first orgasm experienced by over half of both male (67%, USA) and female (55%, USA) humans is the result of self-masturbation. Moreover, throughout the lives of both sexes, this remains for most people the most common type of non-copulatory orgasm (Box 5.3).

Among males, by far the commonest technique for self-masturbation is manual stimulation of the penis, on occasion preceded, or accompanied, by physical stimulation of other erogenous zones of the body, such as the nipples (Kinsey *et al.*, 1948; Ford and Beach, 1952). Although oral self-stimulation of the penis is common in other mammals, fewer than 1% (0.2%) of the males studied by Kinsey *et al.* (1948) were capable of inducing ejaculation by taking their own penis into their mouth.

Nearly, but not quite, all males experience sexual fantasies during masturbation. Occasionally, males attempt to simulate copulation by moving the penis against or inside an object and even more rarely by intercourse with a non-human animal (though, according to Kinsey *et al.* (1948), 17% of boys raised on farms in the USA experience ejaculation as the result of contact with other species).

Although 2% (USA) of women claim to be able occasionally to reach orgasm through fantasy alone, without any physical stimulation, women most often masturbate by stimulating their clitoris and/or labial region. In the USA, 84% of women who had ever masturbated had done so at some time by stimulating their clitoris. Most often, this was done directly by massaging the clitoris with their fingers. On occasion, however, 10% had massaged their clitoris indirectly by crossing their legs, contracting their thigh muscles, and applying rhythmic pressure. Sitting so that their genitals rested on the heel of a foot also provided a relatively cryptic technique for stimulating orgasm. Rarely (<1%; 25/5793 case histories; USA), a female had experienced orgasm by having a non-human animal (usually a dog; occasionally a cat) lick her genitalia.

On occasion, females use penis substitutes during masturbation (Ford and Beach, 1952) and the classical Greek, Roman, biblical and oriental literature abounds with descriptions of the use of such artefacts (Kinsey *et al.*, 1953). However, only 20% of the Americans surveyed by Kinsey *et al.* (1953) had ever used any form of vaginal insert during masturbation and the Melanesians studied by Davenport (1965) only ever inserted their fingers. Very rarely (3/5940 case histories; USA), females obtain vaginal insertion by allowing or encouraging intercourse by a dog.

Virtually all adolescent males and females of at least some Melanesian populations in the SW Pacific masturbate to orgasm frequently and are openly encouraged to do so by their parents (Davenport, 1965). Nearly all males (92%, USA) and the majority of females (58%, USA; 73%, UK) masturbate to orgasm at least once in their lives. For males, peak activity occurs while 15 years or younger

(Box 5.3). In contrast, female activity increases with age throughout reproductive life, peaking in the early 40s (Box 5.3).

For females, median masturbation rate is lower than mean masturbation rate due to some females at any one time masturbating relatively often. Over 50% of women reported that at some stage in their lives they had masturbated three or more times in a week and over 25% had done so five or more times in a week.

Self-masturbation has been observed in a wide range of both male and female mammals. As a general rule, as in humans (Box 5.3), females seem to masturbate less than males (Ford and Beach, 1952; Kinsey *et al.*, 1953). Moreover, whereas males are frequently observed to ejaculate following self-stimulation, little is known about how often female mammals orgasm as a result of their activities.

A wide range of genital stimulation techniques are used by non-human mammals (Ford and Beach, 1952). Male primates manipulate their genitals with their hand, foot, mouth or prehensile tail or by rubbing themselves against the ground or a branch. Male elephants sometimes manipulate their semi-erect penis with their trunk and male rodents sit upon their hind feet and manipulate their genitals with forepaws and mouth. Captive male dolphins rub their erect penis on the floor of the tank or hold it in jets of water. Male porcupines rub their penis and scrotum vigorously against the ground or protruding objects and male dogs are notorious for first clasping, and then rubbing and stimulating themselves to ejaculation against, the leg of a human.

Female mammals lick or handle their genitalia or rub their vulva against the ground, branches or other suitable objects (Ford and Beach, 1952). Female primates lick and suck their own nipples and female porcupines stimulate their clitoris with a vibrating stick. Similarly, female chimpanzees have been seen to insert the stem of a leaf in their vagina, then sway from side to side so that the leaf flaps and vibrates against other objects.

In neither males nor females of non-human mammals is masturbation limited to individuals without access to a sexual partner, any more than it is in humans. Thus, male Cape ground squirrels orally stimulate themselves (and eat the ejaculate) both during and outside periods of female sexual receptivity (Jane Waterman, personal comunication, 1992). Red deer stags with harems frequently shed whole ejaculates after apparently stimulating themselves by rubbing their antlers against vegetation (Darling, 1963). Dominant male rhesus monkeys manually masturbate to ejaculation even with full access to fertile and receptive females (Carpenter, 1942).

5.3.3 Heterosexually stimulated non-copulatory orgasms

On occasion, a human allows and/or encourages a member of the opposite sex to stimulate them to orgasm without allowing and/or encouraging copulation. Equally, a human on occasion is prepared to stimulate a member of the opposite sex to orgasm without copulation.

The stimuli tolerated or sought are either oral, manual or pseudocopulatory. The latter involves the male rubbing or thrusting his penis anywhere on or in the female's body other than inside the vagina and includes anal intercourse. Methods for stimulating orgasm in a partner have been the subject of explicit description from the earliest art and writings. They are described in the earliest Sanskrit, Chinese and Japanese documents, down through Greek and Roman writings, and on through the earliest Arabic and European literature (see references in Kinsey *et al.*, 1953).

The commonest forms of sexual stimulation (USA), in decreasing order of frequency, were reported to be: (1) kissing; (2) manual stimulation of breasts; (3) manual stimulation of the female's genitalia; (4) oral stimulation of the female's nipples; (5) manual stimulation of the male's genitalia; (6) genital apposition without penetration; (7) oral stimulation of the female's genitalia; (8) oral stimulation of the male's genitalia. All forms of sexual stimulation increased with both age and sexual experience.

Although some cultures avoid kissing, most

of the above stimulation techniques are relatively ubiquitous among different human societies, being particularly common among adolescents (Ford and Beach, 1952). These authors list 33 societies in which, even though sexual experimentation without copulation by children and adolescents occurs, it is prohibited and/or punished. They also list 48 other societies which are semi-permissive, accepting such experimentation as long as it is not too public, and a further 32 which are more or less totally permissive.

In all human societies, whether formally permissive or not, young individuals, before they form their first stable heterosexual relationship, inevitably have a number of near-copulatory 'exploratory' contacts which involve varying levels of sexual stimulation. Nearly a quarter of females (24%, USA) experienced their first orgasm under such circumstances compared with less than 1% (0.4%) of males.

In the USA in the 1940s, nearly 90% of males and virtually 100% of females experienced such exploratory contacts. However, relatively few such encounters ended in orgasm, particularly by the female. Even at its lifetime peak, in a person's late teens and early 20s, heterosexually stimulated orgasm without copulation accounted for only about 3% of male and about 18% of female orgasms. More females (40%) eventually exerienced orgasm in such circumstances than did males (25%).

Kinsey *et al.* (1953) provide no data on the incidence of partner-stimulated non-copulatory orgasms within the normal sexual behaviour of established couples. Our own data (Manchester, UK) provide no suitable measure for male ejaculation. However, they do show that 5% (15/274) of couples' sexual interactions led to female orgasm without copulation (within the period from 1 hour before to 1 hour after the orgasm). Put another way, 7% (15/212) of the females' male-stimulated orgasms did not involve copulation within ±1 hour. In addition, 40% (103/255) of female orgasms associated with copulation (i.e. within ±1 hour of copulation) were also non-copulatory (occurring during foreplay or postplay, without the penis in the vagina).

All of the oral and manual non-coital stimulation techniques shown by humans are also shown by non-human primates and other mammals (see references in Ford and Beach, 1952; Kinsey *et al.*, 1953). These range from the male licking and nuzzling the female's breasts and nipples in dogs and pigs to the nuzzling and licking of the female's genitalia that is almost universal amongst mammals. Female primates touch and manipuate the male's penis with their hands and mouth. Whereas a male mammal often ejaculates during such non-coital 'sex-play' and although it seems likely that such behaviour will often lead to orgasm by the female, there is no clear-cut evidence on the matter.

5.3.4 Homosexually stimulated non-copulatory orgasms

In nearly all human societies, some individuals at some time in their lives experience orgasm through seeking or allowing stimulation by an individual of the same sex.

Male homosexual behaviour in most societies usually involves anal intercourse (Ford and Beach, 1952). Mutual masturbation and oral-genital contacts are more rarely reported. In some societies (e.g. parts of Melanesia; Davenport, 1965) all males seem to engage in anal intercourse with another male at some time in their lives. In Britain, however, although 6% of males reported homosexual experience, only 4% reported intimate genital contact, such as anal intercourse (Johnson *et al.*, 1992). Finally, in a few societies (e.g. the Crow Indians), homosexual anal intercourse apparently is absent though oral–genital contacts are fairly frequent (Ford and Beach, 1952).

Female homosexual behaviour usually involves manual stimulation of the genitals, particularly the clitoris (Ford and Beach, 1952). In the USA in the 1940s this was the commonest form of homosexual stimulation. Other common techniques in decreasing order of frequency were: manual stimulation of the breasts; oral stimulation of breasts; oral stimulation of genitals; and genital apposition.

In some cultures (e.g. parts of Melanesia,

Davenport, 1965) females only ever use their mouths, hands and fingers during homosexual contacts. Elsewhere, however, on occasion artefacts are used, either to stimulate the clitoris or to insert in the vagina. These latter artefacts range from the large calf muscle of the reindeer used by the Chukchee of Siberia to a wooden phallus, banana or sweet potato used by the Azande of Africa (Ford and Beach, 1952). However, stimulation of the inside of the vagina seems to be relatively infrequent during homosexual stimulation (Kinsey *et al.*, 1953). For example, only 3% of homosexually experienced American women surveyed by Henry (cited by Ford and Beach, 1952) stimulated the inside of the vagina (with a finger or artefact) during homosexual contact.

Homosexual stimulation seems to be particularly effective at triggering orgasm in females (Ford and Beach, 1952; Kinsey *et al.*, 1953). Homosexual contacts are reported (USA) to have nearly twice the 'success rate' of heterosexual contacts in producing an orgasm.

The incidence of homosexual behaviour varies considerably from population to population. In about 60% of the societies surveyed by Ford and Beach (1952), homosexual behaviour was common and socially accepted. Thus, among the Siwans of Africa (Ford and Beach, 1952) and the Melanesians on some islands in the SW Pacific (Davenport, 1965) virtually all males engaged in homosexual intercourse at some stage in their lives.

In the remaining 40% (29/76) of societies surveyed by Ford and Beach (1952), homosexual activity was rare and the object of some form of social taboo. Most industrialized western societies also fit into this category. In both Britain (Johnson *et al.*, 1992) and France (ACSF, 1992), only about 4% of males claimed to have had at least one occurrence of intercourse with another male in their lifetime. In the USA in the 1940s, about 37% of males claimed to have had at least one homosexual contact which resulted in ejaculation (Kinsey *et al.*, 1948). A more recent survey, however, concluded that the level in the USA was about the same as in Britain and France (Miller, 1993).

In nearly all human societies, homosexual behaviour seems to be more common among males than females (Ford and Beach, 1952; Kinsey *et al.*, 1948, 1953). On average, about 2–3 times as many males as females seem to show such behaviour. For example, in the USA in the 1940s, 37% of males claimed to have experienced homosexual orgasm in their reproductive lifetime compared with 13% of females (Kinsey *et al.*, 1948, 1953). In the 1990s, comparable figures were 5.1% for males and 2.4% for females in France (ACSF, 1992) and 3.6% for males and 1.7% for females in Britain (Johnson *et al.*, 1994).

For males, homosexual activity is most common amongst the young and prereproductive (Ford and Beach, 1952). In the USA in the 1940s (Box 5.3) and in Britain in the 1990s, most males who were ever to ejaculate via homosexual activity had done so by the age of 20 years. The proportion of males showing homosexual activity then declined with age. In contrast, females who were ever to experience orgasm via homosexual activity first did so at any age up to about 45–50 years (Box 5.3). However, the proportion of females engaged in homosexual activity stayed relatively constant with age (Box 5.3). This pattern arose because the average (median) female who experienced homosexually stimulated orgasm confined her homosexual activity to 1–3 years of her life and to one partner (USA). Only 4% of homosexually experienced females had more than 10 homosexual partners. This is in striking contrast to males with homosexual experience, 22% of whom had had more than 10 homosexual partners. Similar trends were found in Britain in the 1990s (Johnson *et al.*, 1994).

The vast majority of individuals who experience homosexually stimulated orgasm are bisexual, engaging in both homosexual and heterosexual intercourse, either at different or overlapping stages of their lives. Relatively few are exclusively homosexual over an entire lifetime. Estimates of the incidence of such exclusive homosexuality for the USA are about 1% for females (Kinsey *et al.*, 1953) and 1% (Miller, 1993) to 4% (Kinsey *et al.*, 1948, p. 656) for males. Estimates for France and Britain suggest a figure of less than 1% of the population for both sexes (80–90% of people who reported experience of homosexual intercourse also reported experience of heterosexual intercourse; ACSF, 1992; Johnson *et al.*, 1994).

Instances of exclusive homosexuality are cited in the anthropological literature (e.g. individuals among the Mohave Indians of the American southwest) but are also rare. Exclusive homosexuality is virtually unknown even in those Melanesian island populations in which nearly all males engage in homosexual anal intercourse at some stage in their lives (Davenport, 1965).

It seems, therefore, that relatively few people (perhaps <1%) fail to reproduce due to a lifetime of exclusive homosexuality. This is no higher, and is probably lower, than the proportion of people who fail to reproduce during a lifetime of exclusive heterosexuality. In the UK, males who report homosexual activity inseminate as many different female partners over a lifetime as do males who report only heterosexual activity (Blower, 1993).

Homosexual behaviour has been reported from a wide range of non-human animals, including, among vertebrates, reptiles, birds and mammals. In the whiptail lizard, *Cnemidophorus uniparens*, of New Mexico and Arizona, males do not exist and the young develop parthenogenetically (Crews and Young, 1991). Females show pseudosexual behaviour, one mounting another as during copulation in related sexual species. Such mounting facilitates ovarian growth and ovulation in the mounted lizard and pairs reciprocate mounting on different occasions.

At least some males and females of virtually all species of birds and mammals at some stage in their lives engage in homosexual interactions, including mounting (Thorpe, 1974). On some occasions, at least for males, these interactions end in orgasm by one or both partners (Ford and Beach, 1952). How often the same is true for females is unknown.

As far as males are concerned, homosexual mounting and mutual masturbation has been reported for many species of primates, including baboons and chimpanzees, both in captivity and in the wild (Ford and Beach, 1952). Anal intercourse by the males of non-human primates has also been observed and described (Hamilton, 1914; Kempf, 1917). On occasion, both male partners ejaculate during anal intercourse, one of the males masturbating while being penetrated by the other (Kempf, 1917). As in humans, male homosexual behaviour peaks in young, prereproductive individuals (Ford and Beach, 1952).

Female primates both finger and lick the genitalia of other females (Ford and Beach, 1952). In a graphic description of a pair of female chimpanzees, Bingham (1928) also describes rear–rear genital appression after one of the pair had been masturbating. Among most mammals, however, the commonest form of homosexual contact between females is by mounting. Such behaviour has now been observed in 71 mammalian species, including rats, mice, hamsters, guinea pigs, rabbits, porcupines, marten, cattle, antelope, goats, horses, pigs, lions, sheep, monkeys and chimpanzees (Dagg, 1984).

Compared with males, female mammals are more likely to continue their homosexual behaviour into their reproductive life. Often, there appears to be some link between stage of the menstrual cycle and the manifestation of homosexual behaviour (Goy and Roy, 1991). This link is discussed further in section 5.5.

5.4 Interaction between copulation and the different types of non-copulatory orgasm

During phases in their life when humans do not have sexual access to a heterosexual partner, all orgasms are, of course, non-copulatory. However, non-copulatory orgasms also occur during phases when an individual does have such access. During these phases, the occurrence of non-copulatory orgasms relates clearly to the sequence and timing of successive copulations.

Except where otherwise stated, the conclusions in this section are based on our Manchester, UK, study (Box 3.8). The commonest form of non-copulatory orgasm for both sexes in this study was self-masturbation. For ease of description, therefore, in the

Box 5.4 Influence of intercopulation interval on probability of males and females experiencing a non-copulatory orgasm (self-masturbatory or nocturnal) between copulations

(Data from Manchester, UK survey; Boxes 3.8 and 3.9.)

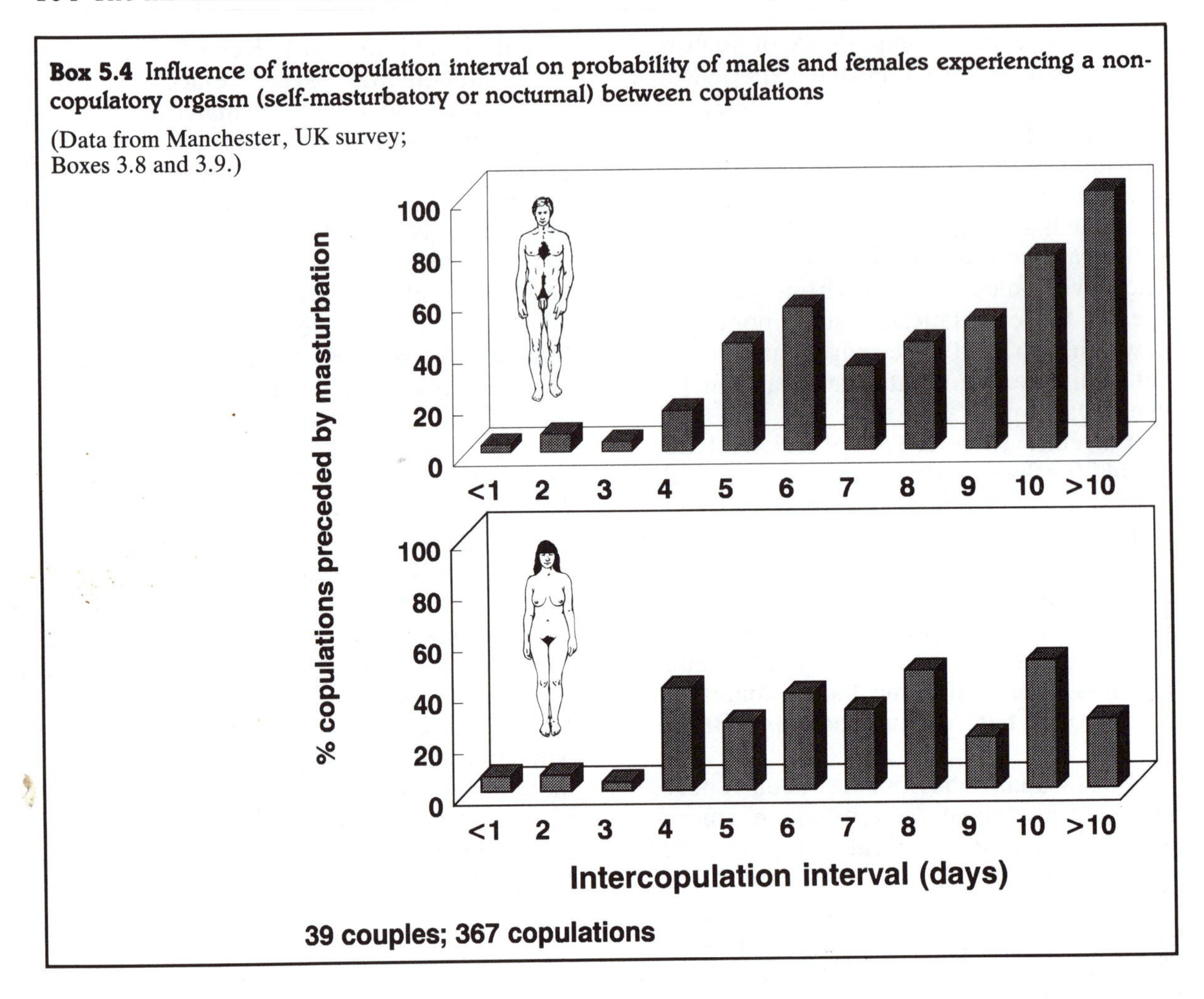

first part of this section we use masturbation as a general term for non-copulatory orgasms. At the end of the section we distinguish between the different types.

In our study, both males and females were much more likely to masturbate when the time interval between copulations was longer (Box 5.4). Such a pattern may seem unsurprising but for neither sex can it be interpreted simply in terms of the time available for masturbation to occur. The reasons are as follows:

For males:

1. Masturbation was rare (c. 5% of occasions) if intercopulation interval was less than 4 days (96 h) (Box 5.4) and only approached 100% of occasions if intercopulation interval was 10 days (240 h) or more (Box 5.4). Mean intercopulation interval on occasions that males masturbated between copulations was 7–8 days (176 h) (Box 5.5).
2. Yet, when a male masturbated between copulations, the mean time interval from copulation to first masturbation was only 2–3 days (62 h) (Box 5.5).
3. It follows that males were not simply masturbating in response to a long time since last copulation. Rather they must have been masturbating (after 2–3 days) in anticipation of a long (>4 days; mean = 7–8 days) intercopulation interval.
4. If copulations had intervened randomly between masturbations, the mean time from last masturbation to next copulation should have been half the mean time between consecutive masturbations. Instead, mean time from last masturbation

Box 5.5 Mean time interval between different sexual events for males and females

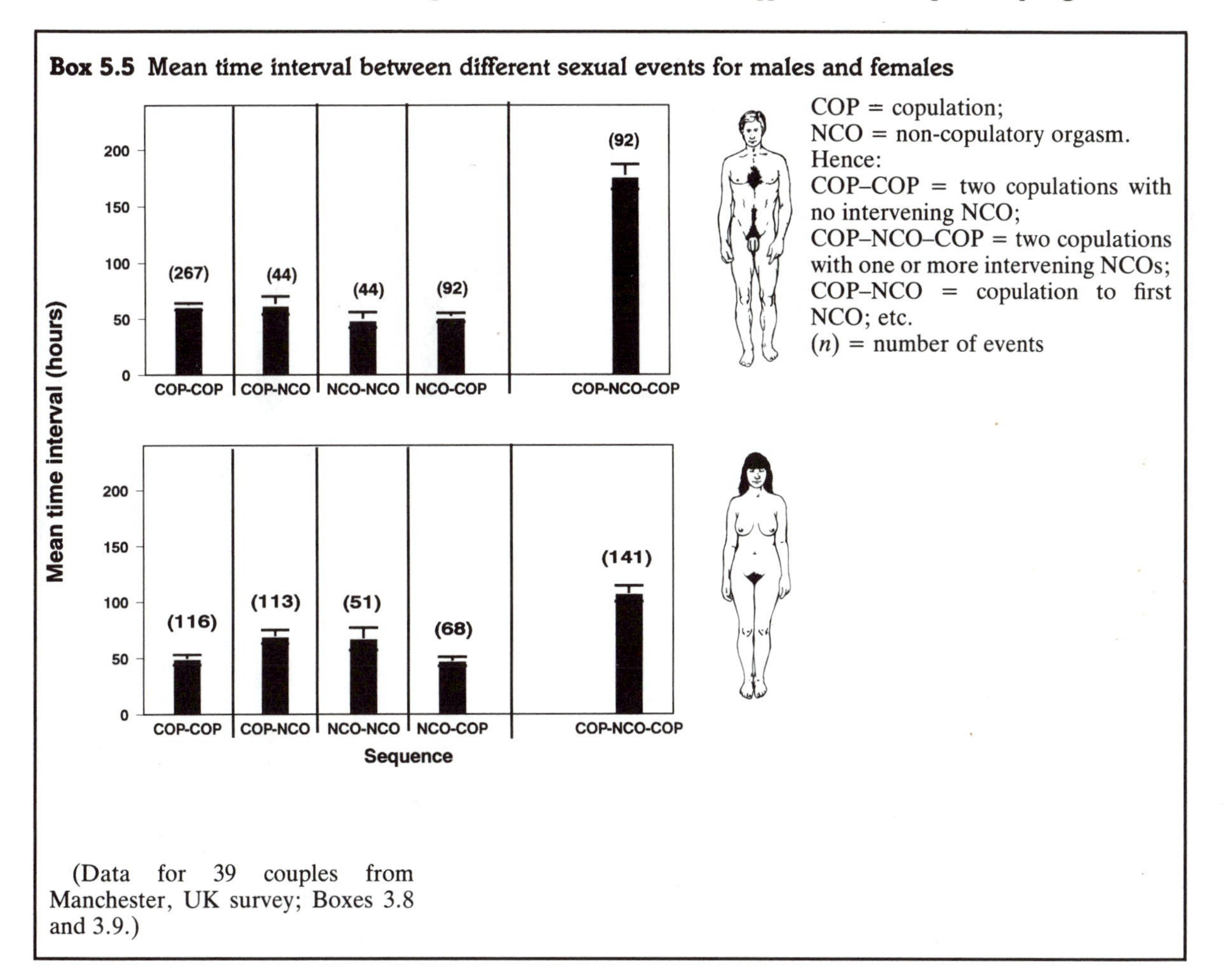

COP = copulation;
NCO = non-copulatory orgasm.
Hence:
COP–COP = two copulations with no intervening NCO;
COP–NCO–COP = two copulations with one or more intervening NCOs;
COP–NCO = copulation to first NCO; etc.
(n) = number of events

(Data for 39 couples from Manchester, UK survey; Boxes 3.8 and 3.9.)

to next copulation (51 h) was just slightly longer than the mean time between consecutive masturbations (49 h) (Box 5.5).

5. Males must therefore have been showing an element of anticipation of the next copulation by masturbating about 2 days (51 h) in advance (or avoiding copulation for 2 days after masturbation).

For females:

1. Masturbation occurred on only one-third or so (30–50%) of occasions even when intercopulatory interval was as long as 4–10 days (96–240 h) (Box 5.4).
2. If masturbation was to occur at all, it did so on average about 3 days (69 h) after the previous copulation (Box 5.5).
3. If females had not masturbated by 4 days (96 h) after their last copulation, they were not increasingly more likely to do so with increasing length of intercopulation interval (Box 5.4).
4. Having masturbated once, females thereafter masturbated about every 3 days (67 h; Box 5.5) until the next copulation intervened.
5. Time from last masturbation to next copulation (47 ± 4 h) was significantly ($P<0.05$) shorter than the time between successive masturbations (67 ± 10 h) as would be expected if copulation occurred randomly with respect to the time of last masturbation.

In summary:

Males most often masturbated in anticipation of an intercopulation interval longer than 4 days (96 h). Whenever they failed to anticipate such a long interval, masturbation became increasingly likely with time.

Box 5.6 Mean time interval from copulation to first non-copulatory orgasm of different types

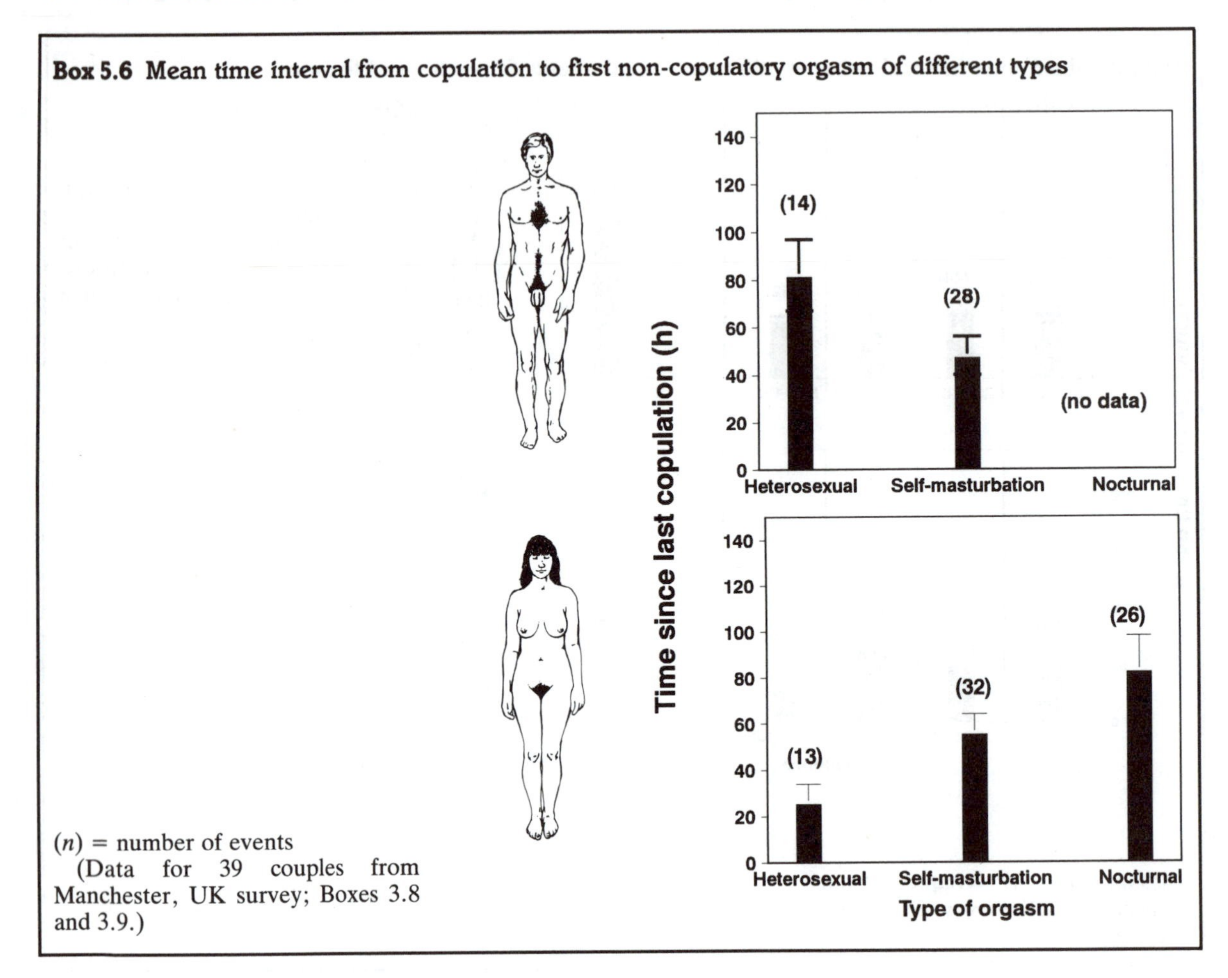

(n) = number of events
(Data for 39 couples from Manchester, UK survey; Boxes 3.8 and 3.9.)

Whether the males successfully anticipated the >4 day interval or not, they contrived in particular to have a gap of about 2 days (51 h) between masturbation and the next copulation.

Females, on one-third or so of occasions, masturbated 3–4 days after copulation and thereafter every three days or so until the next copulation randomly intervened. If they did not masturbate 3–4 days after copulation, they rarely did so until after the next copulation. Deviations from this pattern were primarily linked to the menstrual cycle (section 5.5).

So far in this section we have for convenience discussed intercopulatory orgasms as if they were all due to masturbation. In fact, although masturbation was the commonest source of such orgasms in our Manchester, UK, study, some were spontaneous (nocturnal) and some were due to heterosexual stimulation.

The time interval from copulation to first non-copulatory orgasm differed according to the type of orgasm involved (Box 5.6). For males, the time from copulation to the first non-copulatory orgasm was longest (82 ± 15 h) if the latter was heterosexually stimulated. This suggests that female-stimulated ejaculations were most likely to occur when the male had not anticipated a long (>96 h) intercopulatory interval by self-masturbation.

For females, the interval from copulation to first non-copulatory orgasm was longest (83 ± 15 h) when the first non-copulatory orgasm was spontaneous (nocturnal). This suggests that, for women who experience them, nocturnal orgasms may be some form of 'back-up' event, occurring on occasions when a non-copulatory orgasm is advantageous but when the female has not experienced one from heterosexual- or self-stimulation.

We have no data on the relationship between self-masturbation and nocturnal emissions by males. We expect, however, that the relationship is the same as for females; the probability of nocturnal emission increases as time since last ejaculation increases. In so far as males who masturbate and/or copulate more often show a lower incidence of nocturnal emission (Kinsey *et al.*, 1948), there is some support for this expectation. However, males who are neither copulating nor masturbating rarely experience nocturnal emissions with sufficient frequency for the latter to be considered direct substitutes for other forms of ejaculation.

5.5 The distribution of non-copulatory orgasms through the menstrual cycle in females

In human females, all types of non-copulation orgasm can occur at any stage of the menstrual cycle. Nevertheless, all are more likely to occur at some stages of the cycle than at others.

According to Harvey (1987), a postmenstrual/preovulation peak of occurrence is characteristic of most female-initiated sexual activity. Our own data support this claim (Box 5.7). The preovulation peak is most striking for lesbian and nocturnal orgasms but is also evident for self-mastubatory orgasms. Our study also suggests that the postmen-

Box 5.7 Distribution of different types of non-copulatory orgasm experienced by females during different types of menstrual cycle

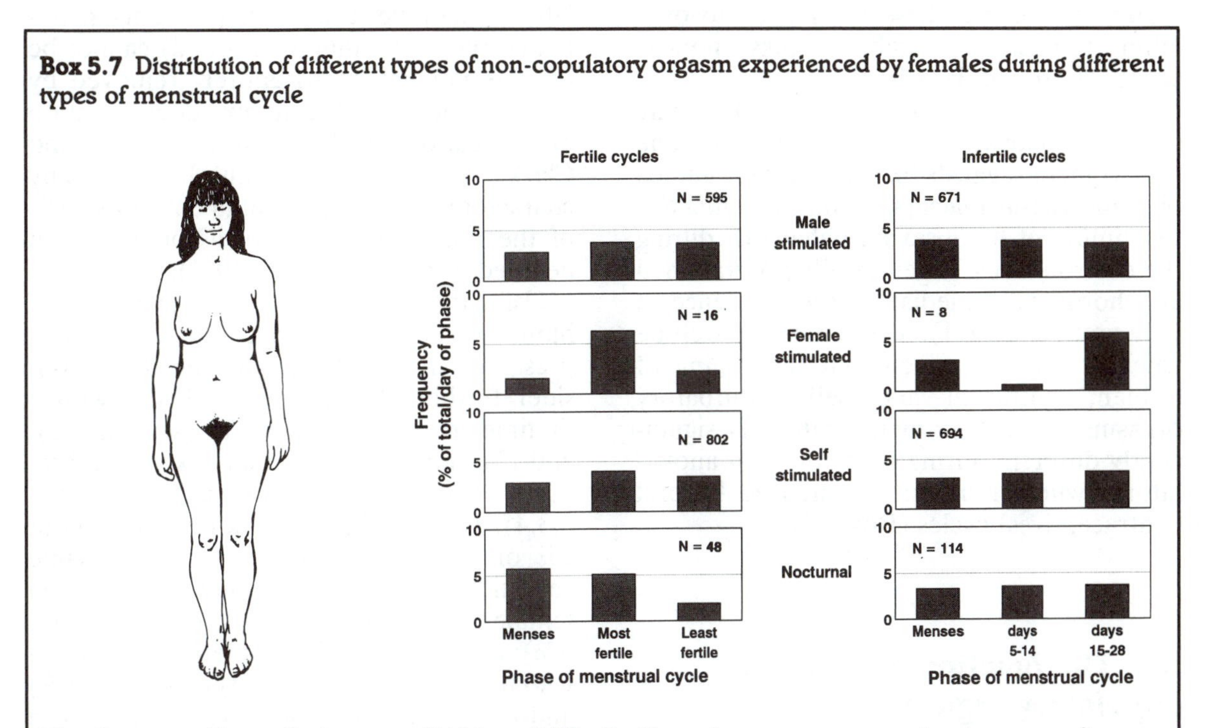

The figure combines all the non-copulatory orgasms recorded during our UK, nationwide and Manchester studies (Boxes 2.6 and 3.8). Frequency is expressed as a percentage of total/day of phase (i.e. 245 of the 694 self-masturbatory orgasms recorded during infertile cycles occurred between days 5 and 14 of the cycle, giving a %/day of phase of $245 \times 100/694/10 = 3.53$). Fertile cycles are cycles shown by females not taking oral contraceptives or with any other induced hormonal disturbance. Infertile cycles are cycles by females taking oral contraceptives plus some naturally anovulatory cycles recorded during our Manchester study, using basal body temperature methods.

During fertile cycles, nocturnal orgasms are least common after ovulation. Female-stimulated (strongly) and self-stimulated (weakly) show peaks in the most fertile phase. Male-stimulated peaks are least common during menses.

All patterns are significantly different during infertile cycles suggesting a hormonal basis to the menstrual timing of non-copulatory orgasms.

strual peak occurs primarily during ovulatory menstrual cycles. The peak is absent when females take oral contraceptives and also seems to be absent during natural anovulatory cycles.

In contrast to lesbian, nocturnal and self-masturbatory orgasms, non-copulatory orgasms in the presence of a male partner do not show a postmenstrual/preovulatory peak, nor are they a significant function of hormonal status (Box 5.7).

Anecdotal accounts for non-human mammals suggest that self-masturbatory, spontaneous and homosexual female orgasms most commonly occur in the period leading up to ovulation (Kinsey *et al.*, 1953), as they do in humans. Most information concerns homosexual behaviour. Often, the female being mounted by another female has, or is soon to, ovulate (rats, dogs, cattle) whereas the mounter is in the phase leading up to ovulation (guinea pigs, rabbits, pigs, horses, cattle) (Ford and Beach, 1952).

Experimental evidence from breeding studies, removal of ovaries, and hormone replacement suggests that, at least for guinea pigs and rhesus macaques (*Macaca mulatta*), the timing of homosexual behaviour during the oestrous cycle is genetically programmed and hormonally mediated (see reference in Goy and Roy, 1991). The same may well be true for humans, given that the timing of lesbian, nocturnal and self-masturbatory orgasms relative to menstruation is significantly different during ovulatory and anovulatory (whether natural or induced by oral contraceptives) cycles (Box 5.7).

5.6 The function of non-copulatory orgasms

If one thing emerges clearly from our consideration of non-copulatory orgasms it is that individuals differ in their orgasm pattern. Some people masturbate, some do not. Some experience spontaneous orgasms, some do not. Some experience homosexually stimulated orgasms, some do not.

Even among those who do experience a particular type of orgasm, rates of experience may vary. As far as non-copulatory orgasms are concerned, humans show what is, in effect, a behavioural polymorphism (i.e. within the population, different people show different behavioural strategies). This conclusion influences strongly the way in which we have to consider the function of the different behaviour patterns.

5.6.1 Behavioural polymorphism

As in all discussion of polymorphism, it is first necessary to determine the cut-off level, that frequency below which a heritable characteristic is so rare that it does not require an adaptive explanation. In all probability, such a genetic characteristic is present in the population either because it is: (1) being maintained purely by mutation; or (2) cannot be (or has not yet been) entirely removed by natural selection. Inevitably, determination of the cut-off level is arbitrary (Hartl and Clark, 1989). Conventionally, however, any characteristic that is present in only about 1% of the population may be discounted from detailed discussion (Cook, 1991).

Using this yardstick, there are two facets of human sexuality which, according to the data presented in this chapter, we need not consider further. These are: (1) males with a lifetime inability to ejaculate (section 5.2); and (2) individuals of either sex with a lifetime of exclusive homosexuality (section 5.3.4). Both facets are shown by only about 1% or less of the population. By the same yardstick, however, we cannot ignore: (1) females with a lifetime lack of orgasm (c. 5%); (2) males (c. 5%) and females (c. 30%) who never masturbate; or (3) males (c. 15%) and females (c. 50%) who never experience nocturnal orgasms. Nor can we ignore bisexuality which, depending on location, may be shown by 2–100% of the populaton.

Polymorphisms, whether morphological or behavioural, are usually considered to evolve and be maintained within the population by one or a combination of two mechanisms (see review by Cook, 1991). One possibility is that the environment is heterogeneous and that different morphs are adapted to different microhabitats (e.g. snails or moths of different colours resting on different coloured substrates). In such cases the relative frequency of the different morphs should, to some extent, reflect the relative frequency of the different microhabitats in the environment (Cook, 1991).

More apposite to sexuality is the possibility that a particular behaviour is only advantageous if it is shown by no more than a certain proportion of the population. If the behaviour becomes too common, the characteristic becomes disadvantageous and the proportion of individuals possessing it declines. If the behaviour becomes less common, the characteristic becomes advantageous and the proportion possessing it increases. As a result, the polymorphism stabilizes when possessors of the different characteristics on average all achieve roughly the same reproductive success. Thus, when considering such a balanced polymorphism, discussion should revolve primarily around the trade-offs that make the different characteristics advantageous when less common but disadvantageous when more common.

5.6.2 Male self-masturbation and nocturnal emission

At first sight, self-masturbation and nocturnal emission by males do not seem candidates for the discussion of a polymorphism. They are shown by most individuals and when they are not shown the cause seems more embedded in differences in circumstance than differences in predisposition. Nevertheless, we suggest that male self-masturbation and nocturnal emission are part of a male polymorphism, the roots of which are firmly implanted in male–male differences in involvement in sperm competition.

The function of male self-masturbation and nocturnal emission seems relatively straightforward, relating in some way to rate of sperm production and how long sperm are stored before being inseminated into a female. As frequency of ejaculation during copulation changes with age, so too does frequency of non-copulatory ejaculation (Box 5.1). The result is that, together, the two outlets maintain a total frequency of ejaculation that, as a male ages, relates closely to his rate of sperm production (Box 5.8). This pattern, combined

Box 5.8 Variation with age in daily sperm production and rate of ejaculation (USA)

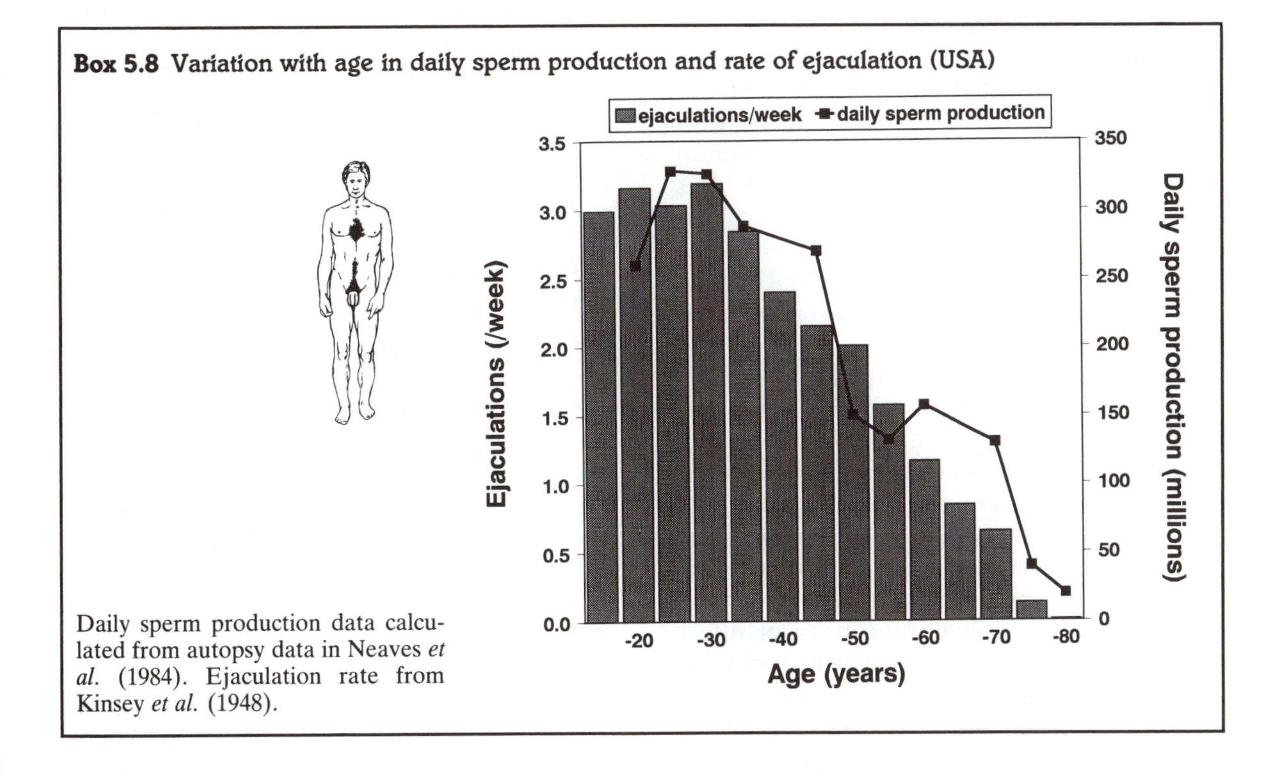

Daily sperm production data calculated from autopsy data in Neaves *et al.* (1984). Ejaculation rate from Kinsey *et al.* (1948).

with the sequence and timing of successive copulations and masturbations (Box 5.5), means that females are rarely inseminated with sperm which have been 'in store' awaiting ejaculation for more than 2–4 days. It is tempting, therefore, to conclude that the function of masturbation, with nocturnal emission as some form of back-up, is in some way to influence the age and number of sperm with which the male next inseminates a female (Chapter 9).

Males seem to differ little in their strategy of masturbation and nocturnal emission. Thus, virtually all males ejaculate during copulation and seem to avoid masturbation if intercopulation interval is relatively short (2–4 days). Whenever intercopulation interval approaches a week or more, males masturbate about 2 days before their next copulation and experience nocturnal emission if time since last ejaculation becomes excessive. Observed differences between males are thus more a reflection of different rates of opportunity, and acceptance of opportunity, for copulation than of different basic physiological strategy. Such relative uniformity of strategy, however, does not mean that there is no behavioural polymorphism in male sexuality.

Ejaculation frequency differs considerably between males. In the USA, Kinsey *et al.*'s (1948) survey included a male who had ejaculated only once in 30 years and another who had ejaculated 30 times each week also for 30 years. Such extremes are exceptional but even within the middle three-quarters of the males surveyed, individual ejaculation rates varied from those averaging one ejaculation per week to those averaging seven ejaculations per week.

Møller (1988c) for nine mammal species and Ginsberg and Rubenstein (1990) for two species of zebra suggested variously that a high risk of sperm competition had selected for larger testes, greater sperm production and higher copulation rates. We suggest that the same selection has acted on humans and has led to the evolution of a balanced polymorphism. This polymorphism is manifest anatomically as variation in testis size, physiologically as variation in rate of sperm production, and behaviourally as variation in rate of ejaculation and involvement in sperm competition. Five factors lead us to this conclusion.

1. Differences in testis size are heritable (as judged from racial differences; section 4.5.1), perhaps due to a Y-linked gene (Short, 1979).
2. Differences in male involvement in sperm competition are a function of differences in testis size (Box 4.17).
3. Differences in rate of sperm production are a function of differences in testis size (Box 4.17).
4. Differences in ejaculation rate are a function of differences in rate of sperm production, both with age (Box 5.8) and between individuals (Box 5.9).
5. Differences in ejaculation rate are a function of testis size (Kim and Lee, 1982). [Kim and Lee (1982) actually concluded for their sample of Korean males that rate of copulation was not a function of testis size. Their conclusion was accepted by Diamond (1986). However, the correlation coefficient reported by Kim and Lee ($r = 1.08$; their Table 6) cannot be correct. Reanalysis of their data (means and ranges) gives a significant positive relationship, not only between testis size and copulation frequency but also between testis size and sperm density (i.e. sperm/ml of seminal fluid during masturbation). In our own study (Manchester, UK; Box 3.8), there was a significant positive association between testis size and ejaculation rate (when controlled for age or copulation rate).]

We suggest, therefore, that males are polymorphic with respect to their involvement in sperm competition. Some males, below average for size of testes, sperm production rate and rate of ejaculation, concentrate on guarding their female partner and spend little time attempting to copulate with other females. Other males, above average for all of these factors, invest less in mate-guarding and spend more of their time and energy in attempting to copulate with a number of females, including those paired to other males.

Box 5.9 Relationship between daily sperm production, mean ejaculation interval and proportion of time spent with partner between copulations (Manchester, UK)

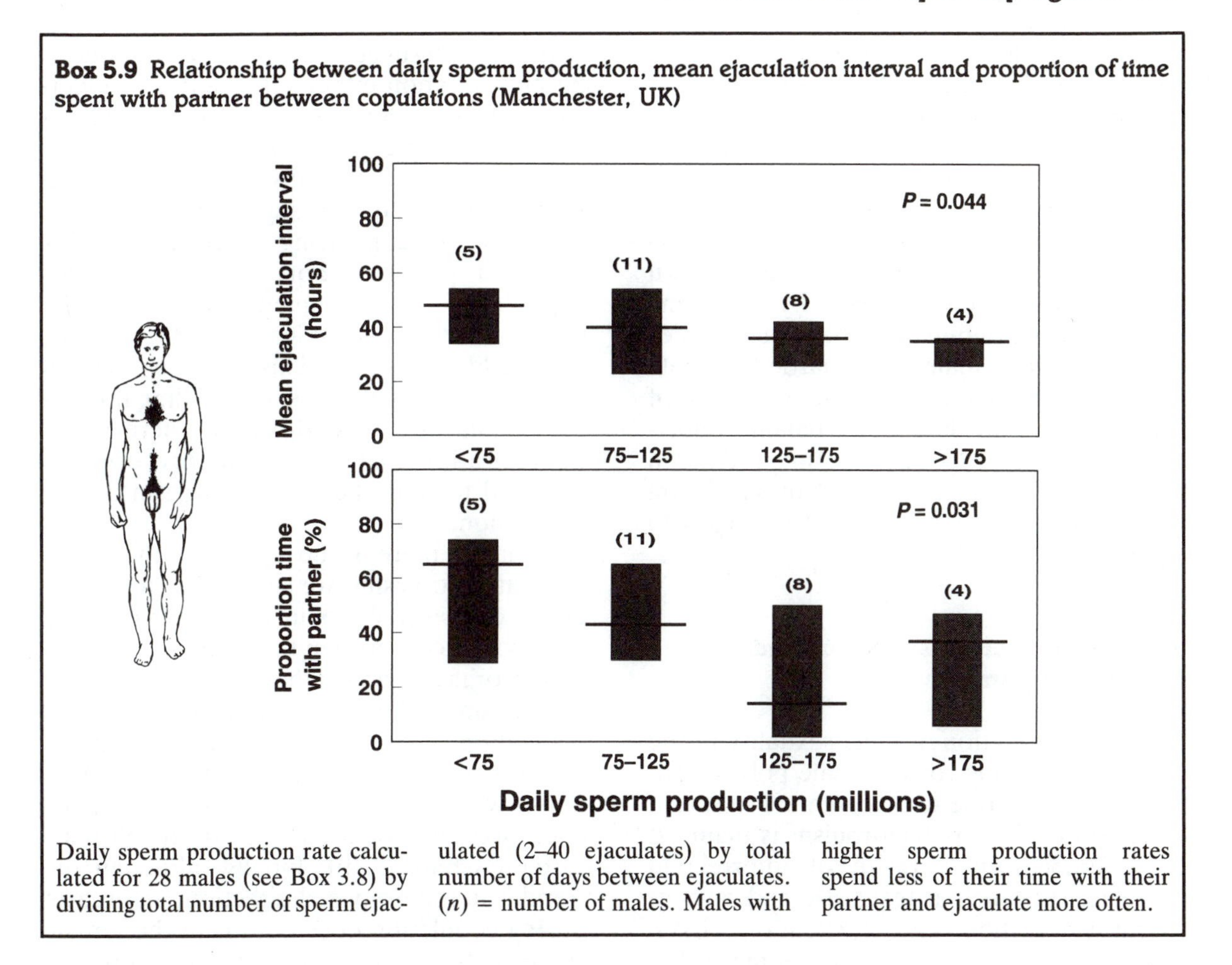

Daily sperm production rate calculated for 28 males (see Box 3.8) by dividing total number of sperm ejaculated (2–40 ejaculates) by total number of days between ejaculates. (n) = number of males. Males with higher sperm production rates spend less of their time with their partner and ejaculate more often.

For the observed variation to be part of a balanced polymorphism, there have to be advantages and disadvantages to both types of behaviour. Moreover, at least one of the types has to suffer a disadvantage whenever it becomes too common in the population.

The disadvantage of small testes should be that their possessors produce fewer sperm per day and thus, all else being equal, must either: (1) ejaculate less often and, on average, inseminate older sperm; or (2) ejaculate as often and, on average, inseminate fewer sperm. Either alternative should impart a disadvantage to the male in sperm competition and thus should favour the male concentrating on other strategies (e.g. mate-guarding) to reduce involvement in such competition.

Our study of males in Manchester, UK (Box 3.8) does indeed suggest that males with lower sperm production rates (and hence smaller testes; Box 4.17) spend more time with female partners (Box 5.9). Such males should also pair primarily with the less polyandrous of females (section 2.4.1). Conversely, of course, males with larger testes and higher sperm production rates (Box 4.17) should pair with the more polyandrous of females, invest less in mate-guarding (Box 5.9), and show behaviour that more often leads them into sperm competition (Box 4.17).

The disadvantage of larger testes seems likely to be threefold: (1) larger testes require more energy to develop and retain; (2) they should be more vulnerable to damage (section 4.5.2); and (3) they may make it more difficult for the male to store the sperm at lower temperatures (section 4.5.2).

As long as males with larger testes (pursuing behaviour with a high risk of sperm competition) are relatively infrequent in the population, their strategy can be advantageous. Such males get the advantage of inseminating more females with more sperm than their competitors while at the same time having a low risk that their own female(s) will be inseminated by another. However, if males with such a strategy become too numerous in the population, each individual's own activities will be counteracted by too many similar males inseminating the same females (including the male's unguarded female partner(s)) with equally large numbers of sperm. The result should be a balanced polymorphism (for testis size, sperm production rate and behavioural strategy) in which, on average, males with the different strategies do equally well.

5.6.3 Female self-masturbation and nocturnal orgasm

If our interpretation of male sexuality in the previous section is correct, male polymorphism is primarily one of involvement in sperm competition. This polymorphism is manifest in differences in a range of interrelated characteristics such as testis size and rates of sperm production and ejaculation. Even though males may differ in rates of masturbation and nocturnal emission, observed variation is the result of polymorphism (and opportunity) not its cause.

The situation seems to be quite different for females. Women show considerable variation in their sexual responses (Kinsey *et al.*, 1953). Some women never experience orgasm, some experience orgasm under a wide range of circumstances. Some never experience orgasm during copulation, some only experience orgasm during copulation. Some experience orgasm both nocturnally and during masturbation and some under only one of these circumstances. Yet others experience a range of orgasm types but more under one circumstance than another and even then the primary outlet varies from woman to woman. The striking impression is that females vary considerably with respect to orgasm pattern and that in its own right this variation within the population is the basis of a balanced polymorphism.

It has been suggested that orgasms induced by masturbation either represent a mechanism by which a female can teach her body to orgasm during copulation or are substitutes for copulatory orgasms (LoPiccolo and Lobitz, 1972). Neither suggestion, however, provides a convincing explanation for observed patterns. For example, females experience more frequent non-copulatory orgasms as they age, at least over their reproductive life (Box 5.1). This increase is shown by all categories of orgasm, including nocturnal and masturbatory (Box 5.3). There is thus no indication that non-copulatory orgasms are used to train the body to orgasm during copulation.

Equally, there is no evidence that non-copulatory orgasms are substitutes for copulatory orgasms. Winokur *et al.* (1959) found no association between the incidence of nocturnal orgasms and either the frequency of copulation or the incidence of orgasm during copulation. In our study, within age-groups, females with more frequent rates of copulation were more, not less, likely to experience orgasms nocturnally or from self-masturbation (Box 5.10). Finally, although the trend is in the proposed direction, an orgasm during copulation does not make the female significantly less likely to orgasm nocturnally or through masturbation in the interval leading up to her next copulation (Box 5.11).

Thus, although masturbation when young may provide an element of self-training and although nocturnal and masturbatory orgasms may to a small extent be influenced by orgasm (or lack of orgasm) during copulation, training and compensation are clearly not their major functions.

The advantages of masturbation, nocturnal orgasm and avoidance of orgasm are discussed in detail in Chapter 10. Our conclusion is that these orgasms (and their lack) are mechanisms by which the female influences events in her reproductive tract. In particular they are mechanisms by which the female influences the retention of sperm from past and future copulations. As such, they have a primary role in male:female conflict over sperm retention and can have a major influence on the outcome of sperm competition.

Box 5.10 **Relationship for females between frequency of copulation and frequency of non-copulatory orgasms (UK)**

Data for self-masturbatory plus nocturnal orgasms from nationwide survey (Box 2.6). Only data for two age-groups are illustrated. In the full data set (sample sizes are shown), and when controlled for age, females who copulate more frequently also have more-frequent self-masturbatory and nocturnal orgasms.

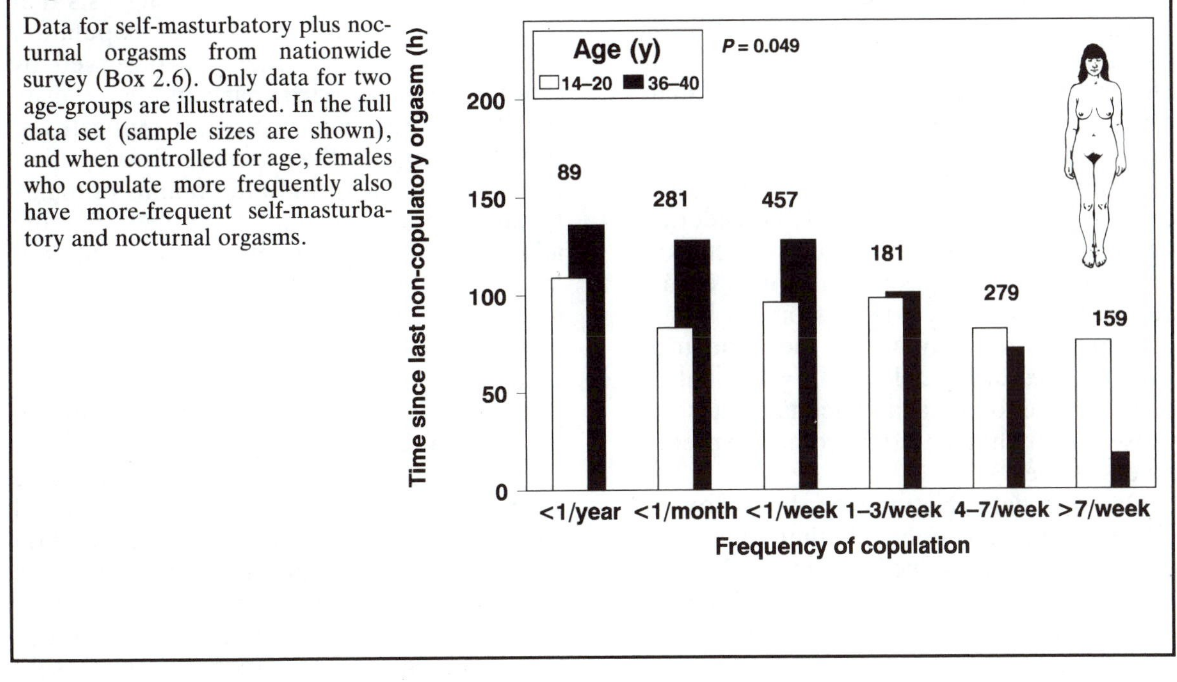

Box 5.11 **Lack of influence of orgasm at last copulation on probability of self-masturbation or nocturnal orgasm before next copulation at three different intercopulation time intervals (females, Manchester, UK)**

(n) = number of occasions. When controlled for interval between copulations, females are not less likely to masturbate or have a nocturnal orgasm if they experienced orgasm at their last copulation.

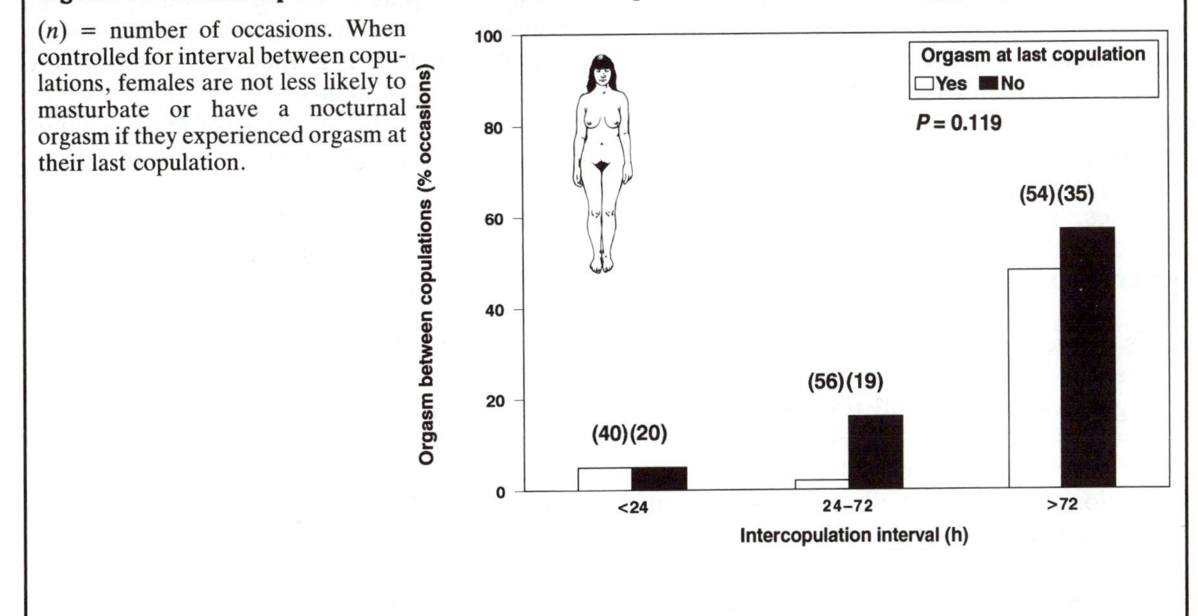

A key feature of female success in male: female conflict over sperm is that the male cannot predict the level of retention that his sperm are going to experience from one copulation to the next (Chapter 10). To this end, a female is aided considerably if males cannot make predictions about sperm retention with one woman on the basis of past experience with another. The situation is exactly that which will promote considerable diversity of female behaviour within the population. All orgasm patterns are most advantageous to their possessors as long as they are not too common in the population. Beyond a certain incidence, a given pattern should cease to be advantageous because the possessors of that pattern become too predictable to males.

Our own data (Nationwide, UK; Box 2.6) allow us to test only whether there is a difference in reproductive success between females who never masturbate and those who do. When controlled for age there is, as expected, no difference in reproductive success between females with these different orgasm strategies.

Female unpredictability is discussed again in more detail in Chapter 6 in relation to infidelity and reproductive crypsis.

5.6.4 Orgasm crypsis

One striking feature of masturbation and nocturnal orgasm in both males and females is their crypsis, occurring while the individual is solitary or at least inconspicuous (Ford and Beach, 1952). Even in the most permissive of societies, orgasms are rarely public events (Davenport, 1965). When they do occur in public, they usually occur discreetly, often at night and/or when in the company of individuals of the same sex (e.g. males masturbating before sleep in a communal hut; females sitting on a heel and gently rocking while with a group of other females; Ford and Beach, 1952; Davenport, 1965). Similar forms of discretion are shown by other primates (Ford and Beach, 1952) though captive males often become more blatant, at least in the presence of humans. The programmed psychology for secrecy probably helps to explain the otherwise irrational level of prejudice that exists concerning such activities by other people.

An apparent urge for crypsis suggests that the advantage of these non-copulatory orgasms is eroded if they are observed by other individuals, particularly of the opposite sex. This in turn has two further implications. First, it suggests that the function of masturbation etc. lies within the realms of male–female conflict. Second, it implies that the advantage of the behaviour is directly physiological, with no benefit, perhaps even a cost, in terms of conveying information to another individual.

In many ways, the advantage of crypsis is probably similar for both sexes and is linked to the advantage of hiding infidelity (Chapter 6). The timing of male masturbation is strongly linked to intercopulation interval and anticipation of the next copulation (section 5.4). If male masturbation were always overt to his female partner, underlying patterns relative to the pair's own copulations should become evident (consciously or subconsciously) to the female. Change in the male's masturbation pattern during periods of unfaithfulness could then alert the female to her partner's infidelity. The easiest way for the male to conceal these changes in pattern is usually to ejaculate cryptically.

In Chapter 10, we show that females change their pattern of orgasm during periods of infidelity in a way that influences sperm competition between their partner and the target of their infidelity. Again, therefore, the easiest way for the female to hide any change in pattern in non-copulatory orgasm is normally to orgasm cryptically.

There may be one further advantage to the male in ejaculating solitarily. The colour and smell of semen is often changed by sexually transmitted diseases. By ejaculating in private, the male can prevent others gaining information on whether he has or has not any such disease.

Although the majority of non-copulatory orgasms are solitary and/or secretive, on occasion both males and females orgasm without copulation in the presence of a heterosexual partner (section 5.3.3). Such orgasms must have a twofold cost. First, they lose the usual advantage of crypsis. Second, they involve an intimancy that brings with it increased costs in terms of time, energy and risk of disease transmission. Thus, to be adaptive, such intimate non-copulatory orgasms must on occasion generate unique benefits not attainable by more solitary activity.

The main point of interest in this context is not why the pair should avoid copulation. Under certain circumstances there are many potential advantages from so doing (Chapters 4 and 6). Instead, it is why, on occasion, the male and/or female should orgasm overtly in their partner's presence rather than cryptically via self-masturbation or at night.

The main feature of overt orgasms is that the climaxing individual is giving their partner information. As far as the ejaculating male is concerned, this information ranges from: (1) demonstration (via manual or oral stimulation) that he is able to ejaculate; and/or (2) (via pseudocopulation) able to copulate; to (3) the timing of this, his most recent, ejaculation and hence the maximum age of his next, perhaps copulatory, ejaculate. Moreover, because the colour and smell of semen changes in the presence of some sexually transmitted diseases (Hargreaves, 1983), males prepared on occasion to ejaculate in a female's presence can demonstrate the health of their ejaculate.

As far as the climaxing female is concerned, the interplay of cryptic and overt orgasms is a major part of her strategy to confuse the male over levels of sperm retention (Chapter 10). Allowing the male to observe a non-copulatory orgasm could be an important element in this strategy.

As, on average, the observer probably gains from witnessing their partner's orgasm, they benefit on occasion from positively encouraging the partner to orgasm in their presence. In addition to the information obtained, a female may gain by influencing the maximum age of sperm in her partner's next copulatory ejaculate. Similarly, a male may gain by influencing the female's retention of sperm at her next copulation, especially if there is a risk that it may be with a different male.

In principle, a female could also gain from stimulating her partner to ejaculate without copulation in order to observe the amount of seminal fluid ejaculated by the male. This could give some information on how long it is since he last ejaculated. However, as seminal fluids recover relatively quickly (Mann and Lutwak-Mann, 1981), the female could probably only tell whether the male had ejaculated in the previous 12 hours or so. Within the context of her partner's infidelity, however, this could still be useful information.

If the transference of this range of information is sufficiently advantageous to both male and female, it could be enough evolutionarily to maintain the observed behaviour. The main topic of theoretical interest then becomes the optimum ratio of cryptic to overt orgasms for male and female performers and observers. This is discussed further for females in Chapter 10.

As most males and females who ever orgasm do so from time to time without copulation in the presence of a heterosexual partner, there may not be any behavioural polymorphism with respect to this particular behaviour. However, in the past the ratio of cryptic to overt non-copulatory orgasms has received relatively little attention in surveys and we know of no suitable data to analyse the situation further.

5.6.5 Homosexuality

In contrast to the lack of past attention to heterosexually stimulated non-copulatory orgasms, homosexually stimulated orgasms have generated a great deal of interest. Most importantly, there is some empirical evidence for the existence of a genetic polymorphism, at least in males.

Genetic studies of male sexual orientation suggest strongly that individual predisposition to homosexual behaviour is genetically programmed (Kallmann, 1952a,b; Heston and Shields, 1968; Eckert *et al.*, 1986; Buhrich *et al.*, 1991; Bailey and Pillard, 1991). Recently, apparently strong evidence has been presented of a linkage between DNA markers on the X chromosome and male sexual orientation (Hamer *et al.*, 1993; Hamer, 1993). Some authors, however, still remain unconvinced by the data (Baron, 1993).

Reported differences in the hypothalamus of the brains of males which may relate to their level of homosexuality could, if real, be part of this genetic programming (LeVay, 1991, 1993) or could in part be induced by differences in life-style (Maddox, 1991; Fausto-Sterling, 1992; Hubbard and Wald, 1993 for critical scrutinies of LeVay's work). Any correlations between physical and personality parameters and level of homosexuality could again in part be genetic and in part be induced.

A lifetime of exclusive homosexuality is rare (1% or less), even in societies where nearly everybody is bisexual (section 5.3.4). In all probability, therefore, exclusive homosexuals are the small and reproductively maladaptive genetic tip of the adaptive iceberg of bisexuality. The level of polymorphism for bisexuality seems to vary between populations from a few per cent in Europe and America to nearly 100% bisexual in some areas of Melanesia. Everywhere, however, as in most mammals, bisexuality is more common among males. In most human populations, there are two to three times as many bisexual males as females.

Understanding this particular polymorphism requires answers to three questions: (1) what is the advantage of bisexual behaviour; (2) why is bisexual behaviour a disadvantage if it is shown by too great a proportion of the population; (3) why is the level at which the polymorphism balances different for males and females and different in different populations?

(a) WHAT IS THE ADVANTAGE OF BISEXUAL BEHAVIOUR?

Most evolutionary biologists have been tempted to conclude that homosexual behaviour is either maladaptive or a non-adaptive corollary of other adaptive behaviour. Alcock (1989), for example, concluded that homosexuality was a non-adaptive reflection of selection for males to be relatively urgent and indiscriminate. Such an hypothesis does not, of course, explain female homosexuality. Wilson (1978) suggested that the advantage of homosexuality was that the individuals gave up their own reproduction in order to aid the reproductive effort of close relatives. Even if Wilson's hypothesis explained the giving up of reproduction, however, it does not explain the adoption of homosexual behaviour.

We suggest that homosexual behaviour, in the context of bisexuality, has a reproductive advantage in its own right.

Two examples of bisexual behaviour in invertebrates have well-documented reproductive advantages, neither of which seems likely to apply directly to humans. Both examples concern homosexual contacts between males.

Males of the acanthocephalan worm, *Monoliformes dubius*, a parasite in the gut of rats, engage in homosexual copulation during which they attempt to use their seminal fluid to cement over the genitalia of other males (Abele and Gilchrist, 1977). This reduces the risk of sperm competition by reducing the number of males in the vicinity who are capable of efficient insemination. Obviously there is no direct parallel here with human homosexuality (unless perhaps males who contract a sexually transmitted disease can reduce the competitiveness of other males by infecting them with the same disease).

In the cave bat bug, *Xylocoris maculipennis*, males inject sperm and seminal fluid directly into the body cavity of other males, using a sharp and pointed penis. The recipient benefits by 'digesting' the seminal fluid, thus gaining a nutrient rich meal. Many of the sperm are also killed and 'digested' by the recipient but a few survive, migrate to, and infiltrate, the recipient's testes (Carayon, 1974). When the recipient next inseminates a female, he also inseminates some of the donor's sperm. By inseminating other males, therefore, a male increases his chances of having sperm inseminated into a female, albeit by proxy, even if in the meantime he himself dies of starvation (a major risk to these insects). In humans, however, except perhaps in bisexual orgies, it is unlikely that sperm from one male will end up in a female as a result of having been ejaculated into, or onto, an intermediate male.

We know of no direct evidence for vertebrates that individuals who show homosexual behaviour ever positively benefit in terms of

Box 5.12 Reproductive success of bisexual and heterosexual women at different ages (UK)

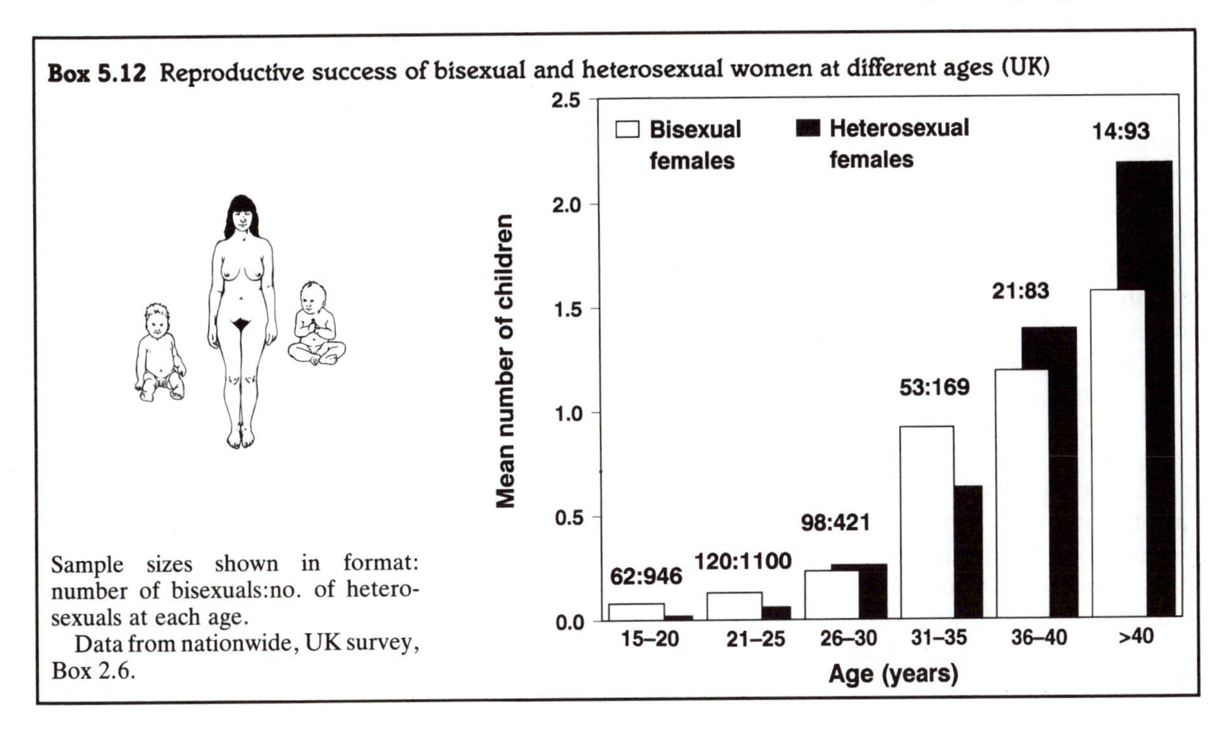

Sample sizes shown in format: number of bisexuals:no. of heterosexuals at each age.

Data from nationwide, UK survey, Box 2.6.

reproductive success. An exception, in a sense, is the whiptail lizard, *Cnemidophorus uniparens*, of New Mexico and Arizona, in which homosexual mounting by females seems to be necessary to stimulate egg development (Crews and Young, 1991). As males do not occur in this species, however, the situation is unusual.

Equally, we know of no direct evidence that male and female vertebrates who engage in homosexual behaviour suffer a reduction in reproductive success. As in humans, most individual vertebrates who show homosexual behaviour are bisexual. Thus, in most species of primate observed, both males and females who show homosexual behaviour also copulate with the opposite sex (Ford and Beach, 1952). More specifically, a male macaque who was observed to have regular anal intercourse with a younger male partner was no less active than other males in copulating with females (Hamilton, 1914). Similarly, female ungulates and other mammals who mount other females often copulate and ovulate soon afterwards (Goy and Roy, 1991).

We know of no data for male humans on the number of children born to bisexuals compared to heterosexuals. However, our own nationwide UK survey (Box 2.6) has allowed us to carry out such an analysis for females. In the absence of any evidence to the contrary, we shall assume that the same pattern is shown by males. If, eventually, this proves not to be the case, then the arguments that follow will need to be revised for males.

In their study of nearly 300 women in Los Angeles, Essock-Vitale and McGuire (1985) found that by age 39 years, exclusive heterosexuals (n = 287) tended to have had more children than women with some lesbian experience (n = 11). Our much larger UK study of nearly 4000 women produced the same result for this age group (Box 5.12). However, in our study, females with homosexual experience began to produce children earlier (Box 5.12). Thus, up to age 25 years, bisexuals had very significantly more children than heterosexuals. On average, bisexuals were four times more likely to have children than heterosexuals by age 20 years (8% as opposed to 2%) and were still twice as likely by age 25 years (13% as opposed to 6%). The difference in mean number of children disappeared and then reversed with increasing age, though in our study, even after a reproductive lifetime, exclusive heterosexuals did not have significantly more children than those with some homosexual experience.

Reproductive success can be increased by reproducing earlier as well as by producing more children (Box 4.9). The pattern shown in Box 5.12 in which bisexuals reproduce earlier but may have fewer children over a lifetime could well indicate a balanced polymorphism, bisexuals and heterosexuals achieving equal reproductive success over time but through different strategies.

The implication, therefore, is that homosexual experience allows an individual to make an earlier or more efficient start to reproduction. It is tempting to link this observation with that other universal feature of bisexuality: that homosexual contacts are most common in young and/or prereproductive individuals. Among the different species of birds and mammals, homosexual behaviour is practically universal in the prereproductive stages (Thorpe, 1974) but becomes less common as reproduction proceeds.

As in other mammals, human males are much more likely to engage in homosexual behaviour when young and inexperienced with females (e.g. Kinsey *et al.*, 1948; Davenport, 1965; ACSF, 1992). In Kinsey's survey in the USA, most males (74%) who were ever going to have homosexual contacts had done so by age 15 years and virtually all (98%) had done so by age 20 years (Box 5.3). The same was true in the Melanesian population studied by Davenport (1965) and in the most recent study in Britain (Johnson *et al.*, 1994).

Young males of many vertebrates are notoriously inefficient at copulation, initially even having difficulty in choosing which end to mount. Male chimpanzees, for example, require several months or even years of practice and experience in sexual performance to develop maximum coital efficiency (Ford and Beach, 1952). Inexperienced male monkeys and chimpanzees, when encountering a receptive female, become strongly sexually aroused but are often so awkward at attempting intromission that the mating is never completed. Adults who have been denied the opportunity to gain sexual experience when younger are often unable to copulate (Ford and Beach, 1952). A level of experience and practice with other males would be of obvious advantage in increasing the success of an individual's first mating opportunities with a female.

There may well be some advantage in using other males as targets for practice rather than females. Females, because of the risk of conception, may less often than males be prepared to allow males the opportunity to experiment. Both males in a homosexual partnership, however, could gain from their alliance if practice is mutual (= reciprocal altruism; Trivers, 1971) as seems to be the case. In the southwest Pacific, for example, anal intercourse between young male humans is fully reciprocal, the males taking it in turns to take active and passive roles (Davenport, 1965).

Such alliances between males will naturally be most effective for experiment and practice if the males are able to some extent to mimic female behaviour, not only during copulation but also during courtship and stimulation. A level of effeminacy in the behaviour of bisexually inclined males will therefore be an advantage as long as this does not make them less able to acquire female partners for copulation.

Naturally, anal intercourse is only a reasonable model for vaginal intercourse in those mammals, such as primates, in which the male rectum is a passable analogue of the vagina. When inseminating a female, the male then has only to learn to penetrate the vagina rather than the rectum. In most birds, however, homosexual mounting should generally be good practice for heterosexual copulation, the single cloacal opening raising fewer problems than the dual rectal and vaginal openings of mammals.

The proposed advantage of homosexual contact to males, therefore, is that such males gain more and earlier experience at stimulation of, and copulation with, a partner. As a result, they are better able to court, stimulate and inseminate females when their first lifetime opportunities arise. Such males may thus be less likely to miss the first opportunities for reproduction through inexperience. In addition, better stimulatory and copulatory technique could give the homosexually experienced male an early edge in terms of sperm competition over less experienced rivals.

At first sight, the fact that some males continue to show homosexual behaviour well into their heterosexual life-phase may seem inconsistent with this 'practice' hypothesis. Such continuation of homosexual behaviour has been reported for a great many mammals and a few birds, such as geese (Thorpe, 1974), as well, of course, as for humans. In the southwest Pacific, for example, many men overtly continue homosexual behaviour even once they have begun to reproduce (Davenport, 1965). As males age, however, they increasingly interact with younger rather than peer or older males. They also increasingly take the active role in anal intercourse, in return providing their younger male partner with money and/or other resources rather than the opportunity to reciprocate (Davenport, 1965).

One possible function for continuing homosexual behaviour beyond first heterosexual activity is to experience differences between individuals. In particular, it could be advantageous to experience differences in their response to different techniques of courtship, stimulation and copulation.

As we have shown in this chapter and discuss further in Chapter 10, females differ markedly in their courtship, copulation and orgasm strategies. Consequently, experience with one female does not necessarily fully prepare a male for copulation with another. This means that, even once heterosexual activity has begun, experience with different partners may continue to be advantageous. Experience of different males may well provide benefits for later experiences with different females. Again, a level of ability to mimic aspects of female behaviour during courtship and copulation should be advantageous to such males. Those who are better mimics should attract more partners through offering better practice. As a result, such males themselves gain more experience of a wider range of partners. In addition, of course, once such behaviour is established, the opportunity presents itself for such males to exploit other males, trading sexual access for resources in the same way as do female prostitutes.

Therefore, in so far as courtship and copulation with different males can provide a male with information useful for courtship and copulation with different females, the continuation of homosexual behaviour beyond first reproduction may well be advantageous. The fact that men who report a high number of homosexual partners also report a high number of female partners (Blower, 1993) is consistent with this suggestion. By being better able to adjust patterns of courtship, stimulation and copulation to suit the individual partner, a male may not only increase the number of females he inseminates, he may also gain an advantage via sperm competition over less experienced rivals.

Females, because of the risk of disadvantageous conception (see the discussion of family planning; Chapter 4), are also likely to gain from homosexual rather than heterosexual practice at copulation. However, whereas inexperienced male primates are relatively inept at copulation, inexperienced females appear to be much less disadvantaged (Ford and Beach, 1952). In any case, a level of ineptness may be a positive benefit to young females, perhaps offering a mechanism for mate selection. Only the more competent males may be able to copulate successfully with a relatively inexperienced female.

Thus, in contrast to males, any benefit that young females might gain from practice seems likely to be concerned with facets of copulation other than simple intromission. As a result, there is less selective pressure on females to gain their practice and experience when very young. This seems to be reflected in the much slower rate at which females come to experience homosexual interactions compared with males (Box 5.3). Even so, about half (54% USA; 40% UK) of females who were ever going to have homosexual contacts had done so by age 25 years and over three-quarters (77% USA; 79% UK) had done so by age 30 years (USA from Kinsey *et al.*, 1953; UK from nationwide survey, Box 2.6). As for males, such contacts were primarily prereproductive with a second peak for females who were between heterosexual partners.

Our nationwide UK study has allowed us to identify a number of similarities and differences between females who had and had not had lesbian experience (Box 5.13). When

Box 5.13 Index of differences between women with and without lesbian experience when controlled for age (UK)

Data for 368 women with lesbian experience, 2812 without (nationwide UK survey, Box 2.6). Index is the *z*-value from Meddis' specific rank sum test (Meddis, 1984) and tests whether women with lesbian experience are significantly more active in the areas indicated.

From left to right, women with lesbian experience:

DO NOT copulate more or less often;
ARE NOT more or less likely to orgasm associated with copulation.

However, they:

ARE MORE LIKELY to have orgasms which promote high sperm retention (Chapter 10);
HAVE MORE FREQUENT non-copulatory orgasms;
ARE MORE LIKELY to masturbate;
ARE MORE LIKELY to have more than one concurrent male partner; and
ARE MORE LIKELY to double-mate (Chapter 8), thus promoting sperm competition, their last copulation being less likely to be with their main male partner.

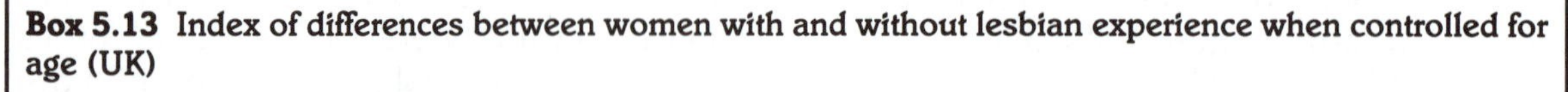

compared to females with no lesbian experience and controlled for age, females with lesbian experience were no different ($P>0.05$) in their:

1. copulation rates;
2. probability of experiencing an orgasm during copulation; or
3. rate of nocturnal orgasm.

However, they were significantly ($P<0.05$; Box 5.13) more likely to:

4. time their orgasm during copulation in a way that maximized sperm retention (see Chapter 10);
5. masturbate (86% v 71%) and to masturbate more often;
6. have a higher rate of non-copulatory orgasms (50–400% higher, depending on age);
7. have more than one concurrent male sexual partner; and
8. double-mate.

The picture that emerges for bisexual females is of individuals more likely to become involved in sperm competition (Box 5.13) and more likely to reproduce earlier (Box 5.12). As part of their reproductive strategy, they possess (or develop) abilities at masturbation and timing of copulatory orgasm that are different from the abilities possessed by women without lesbian experience.

For females, the interplay of orgasm and

Box 5.14 **Distribution during the menstrual cycle of male-associated events experienced by females with and without lesbian experience**

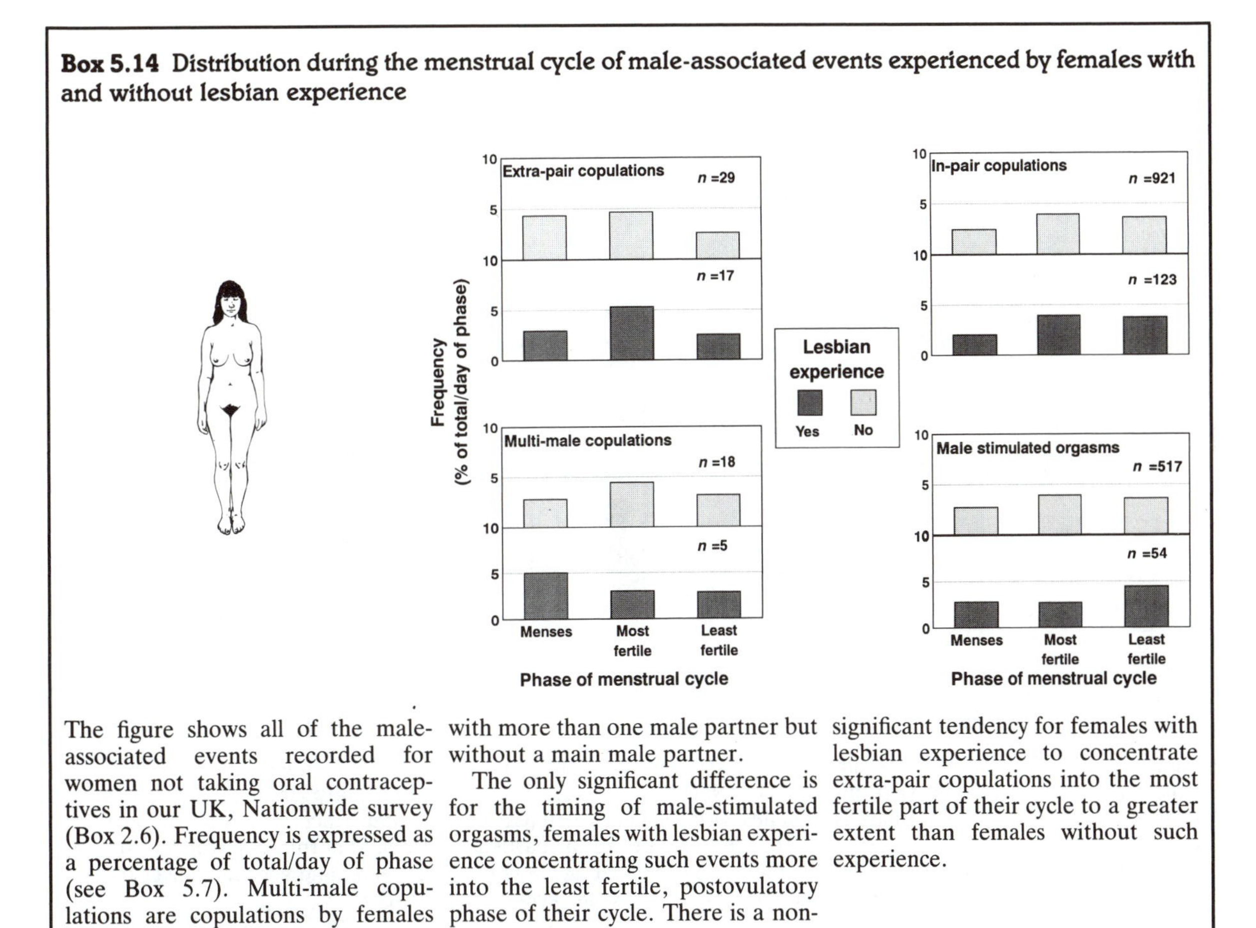

The figure shows all of the male-associated events recorded for women not taking oral contraceptives in our UK, Nationwide survey (Box 2.6). Frequency is expressed as a percentage of total/day of phase (see Box 5.7). Multi-male copulations are copulations by females with more than one male partner but without a main male partner.

The only significant difference is for the timing of male-stimulated orgasms, females with lesbian experience concentrating such events more into the least fertile, postovulatory phase of their cycle. There is a non-significant tendency for females with lesbian experience to concentrate extra-pair copulations into the most fertile part of their cycle to a greater extent than females without such experience.

copulation in their heterosexual relationships represents a complex and sophisticated strategy that has an important influence on sperm retention (Chapter 10). This strategy involves not only rates and timings of orgasm but also has elements of deception and subterfuge. If a female can experiment with these strategies with another female, she should be better armed, and armed better when younger, for the male:female conflict that will be the major determinant of her reproductive success.

As males vary far less than females in their orgasm responses and strategies (sections 5.6.2 and 5.6.3), females have less to gain than males from experience with different partners. Rather, females have more to gain from practising their own responses in the context of a close relationship. These differences in selective pressure may account for the fact that female homosexuality most often takes the form of a relatively long-term relationship with one or few partners during a relatively short period (1–3 years) of their lives (section 5.3.4). In contrast, male homosexuality is much more likely to involve multiple partners.

There is an indication in our data (Box 5.14) that women with lesbian experience (not taking oral contraceptives) concentrate their extra-pair copulations more tightly in their most fertile phase and male partner-stimulated orgasms into their least fertile phase. It is tempting to interpret these differences as reflecting greater expertise at manipulating males as a result of their lesbian experience. Without more data, however, other less interesting interpretations are possible.

In summary, we suggest that, in the context of bisexuality, homosexual behaviour by both men and women is a form of speeded-up

Box 5.15 Variation with age in differences in risk of genital tract problems for women with and without lesbian experience (UK)

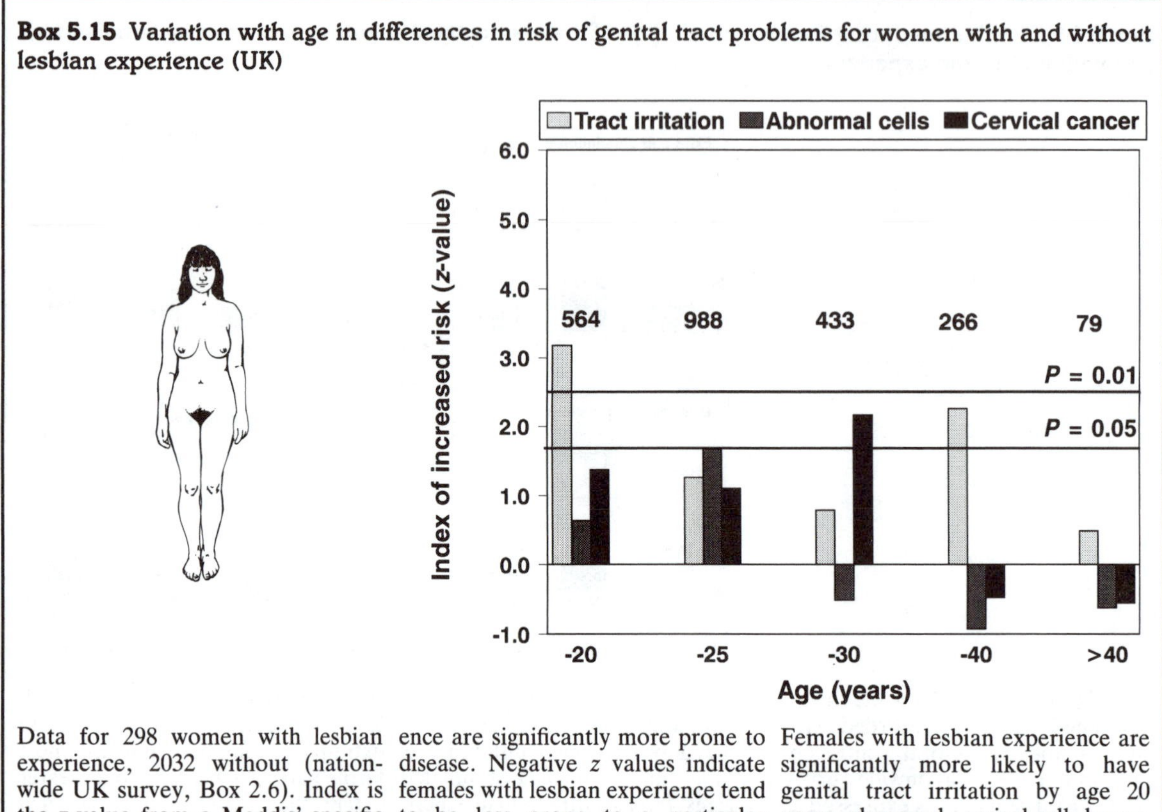

Data for 298 women with lesbian experience, 2032 without (nationwide UK survey, Box 2.6). Index is the z-value from a Meddis' specific rank sum test (Meddis, 1984) for whether women with lesbian experience are significantly more prone to disease. Negative *z* values indicate females with lesbian experience tend to be less prone to a particular problem at a particular age. n = total number of females in age-group. Females with lesbian experience are significantly more likely to have genital tract irritation by age 20 years, abnormal cervical cells by age 25 years, and cervical cancer by age 30 years.

preparation for the heterosexual interactions that will be the basis of their lifetime reproductive success. In both sexes, homosexual outlets provide more opportunity for practice than could be achieved early in life via purely heterosexual outlets. The result is that bisexuals are better prepared than homosexuals for early reproduction and for success in and through the promotion of sperm competition.

(b) WHY IS BISEXUAL BEHAVIOUR A DISADVANTAGE IF IT IS SHOWN BY TOO GREAT A PROPORTION OF THE POPULATION?

The disadvantage of homosexual behaviour is perhaps much more obvious than the advantage.

Since the advent of AIDS, the life expectancy of human males who show homosexual behaviour is almost certainly lower than that of males who do not. Even before AIDS, homosexual males may always have been exposed to a higher risk of contracting other sexually transmitted diseases (Johnson *et al.*, 1992) and thus to have had a shorter life expectancy.

In our own study of females (Nationwide, UK; Box 2.6), women with past lesbian experience were significantly more likely to be suffering from genital tract infection by age 20 years. By age 25 years, they were more likely to have cells identified as 'abnormal' in cervical smear tests and by age 30 years they were more likely to have cervical cancer (Box 5.15).

In part, this greater risk of disease could be a direct result of the lesbian behaviour itself. There are at least two documented cases of HIV transmission from female to female (Chu *et al.*, 1992). The herpes virus and those strains of the HPV (Human Papilloma Virus) associated with, and maybe the cause of, cervical cancer (Adler, 1991) could also be spread from female to female. In part, however, the increased risk of disease may be indirect, a result of the increased polyandry and involvement in sperm competition that bisexual women also show and for which, we argue, their lesbian behaviour is preparation. Whatever the cause, the consequence is the same: the behavioural syndrome of which female bisexuality is a part carries a higher risk of disease than exclusive heterosexuality.

According to our hypothesis, bisexuals have the advantage over exclusive heterosexuals in being better prepared for early reproduction, particularly with respect to sperm competition. The size of the advantage depends at least in part on what proportion of the population is not bisexual and thus less well prepared. The size of the disadvantage of bisexuality probably also depends on what proportion of the population are bisexual.

During the early, and often more virulent, stages of the spread of a given infectious organism through a population, the more individuals who show homosexual behaviour and the more partners they each have, the greater the risk to each individual of contracting the disease.

The nett advantage of bisexuality in any given population is thus a function of the proportion of bisexuals in the population and the relative incidence and virulence of sexually transmitted diseases in the bisexual and heterosexual communities. The most likely outcome of such selection is a balanced polymorphism, the proportion of bisexuals stabilizing at the level at which the reproductive success of bisexuals and exclusive heterosexuals are about the same. The pattern of reproductive success shown by the two groups in Box 5.12 could be interpreted as showing just such a balance.

(c) WHY IS THE LEVEL AT WHICH THE POLYMORPHISM BALANCES DIFFERENT FOR MALES AND FEMALES AND DIFFERENT IN DIFFERENT POPULATIONS?

If the above model is correct, we should expect the bisexual–heterosexual polymorphism to balance at the point at which the advantage of faster, and perhaps better, preparation for heterosexual behaviour is just negated by increased risk of disease. On average, males probably have more to gain from early homosexual exerience, largely because inexperienced males are so inept at even basic heterosexual activities, such as mounting and penetration. In contrast, female benefits from homosexuality are probably more subtle, being perhaps less concerned with basic copulation and more with developing an ability to exploit the promotion of sperm competition.

If the increased risk of disease were equal for the two sexes, these different benefits of homosexual behaviour would be enough to generate more male bisexuals than females. In the absence of information on whether male bisexuality carries with it a greater risk of disease than female bisexuality, we cannot judge the extent to which disease risk changes this simple expectation. However, almost universally in mammals, including humans, males are more likely to be bisexual than females (Ford and Beach, 1952). This suggests that the difference in advantage of bisexuality experienced by the two sexes is greater than any converse difference in disease risk.

In large, mobile human populations, such as Europe and the United States, bisexuals are relatively uncommon. This suggests that in such societies there is either only a small advantage to bisexuality (Box 5.12) and/or a large difference in the incidence and virility of disease in the bisexual and heterosexual populations (Box 5.15). In at least some small island communities, however, bisexuals can often greatly outnumber exclusive heterosexuals (Davenport, 1965). The implication is that in such small communities, the increased risk of disease associated with bisexuality is rather small.

The more isolated a population, the less likely it is to be invaded by new diseases. When a disease does invade, its spread is rapid and inevitable. Only the offspring of those immune to past diseases have survived to be the current population. It is quite possible, therefore, that, at least until the last few decades, bisexuality in such populations was not associated with an increased risk of disease. The advantage of bisexuality would then drive it through virtually the whole population.

The balance between the manifestation of bisexuality and the risk of disease could influence factors other than the proportion of the population programmed for bisexuality. For example, at the individual level bisexuals should ideally be programmed to adjust their number of homosexual partners according to the observed risk or virulence of disease in their current environment. In the past decade, since the advent of AIDS, the mean number of homosexual partners per male has decreased (UK; Blower, 1993).

The balance may also be manifest through variation in the form of homosexual behaviour. Different forms of homosexual expression must have different levels of advantage and different risks of contracting disease. For example, anal (or, in birds, cloacal) intercourse may well be the form of homosexual behaviour that offers the best preparation for heterosexual activity but it probably also carries the greatest risk of disease.

The observed pattern for many birds and mammals is for various forms of homosexual behaviour, including mutual masturbation and mounting, to be common whereas penetration and ejaculation is less common (Thorpe, 1974). In humans, about 94% of European (British and French) males avoid homosexual contacts altogether, about 2% engage in homosexual contact but avoid anal intercourse, and about 4% engage in anal intercourse (ACSF, 1992; Johnson *et al.*, 1992). In contrast, in some small island populations in the southwest Pacific, virtually 100% of young males claim to engage in anal intercourse (Davenport, 1965).

(d) HOMOSEXUALITY: SUMMARY OF A HYPOTHESIS

We suggest that, in the context of bisexuality, the advantage of homosexual behaviour for both sexes is that it is a form of preparation for heterosexual copulation and sperm competition.

In young males homosexual behaviour is a mechanism for learning the crude basics of courtship, stimulation, mounting and intromission. In older males it is a mechanism for refining and 'individualizing' courtship and copulation tactics by experiencing a range of differences of response by different partners.

In females, homosexual behaviour is in small part a mechanism for learning the basic responses of courtship and copulation but is primarily a mechanism for developing an efficient yet deceptive orgasm pattern that will empower her with cryptic control over sperm retention.

In both sexes, the advantages of bisexuality have to be traded against the disadvantage of increased risk of disease and perhaps lower life expectancy. Bisexuals, therefore, are programmed for a shorter generation interval through earlier reproduction, more involvement in sperm competition, and probably earlier death (the James Dean syndrome: 'live fast, die young'). At the population level, bisexuals increase in number until advantages and disadvantages just balance. The result is the bisexual–heterosexual polymorphism characteristic of each sex in each population.

5.7 'Push-buttons': the clitoris, nipples and masturbation

5.7.1 Push-button power

Humans, like other mammals, if constrained, will eventually void the contents of their bladder spontanteously and uncontrollably. From time to time, they may do the same with the contents of their nose, stomach, and intestines, particularly if suffering from appropriate infections. Normally, however, the need for such evacuations first triggers preliminary behaviour by which the individual seeks out a location appropriate both ecologically and socially for the particular act. Most social mammals, for example, have a specific latrine area within their home range.

The advantage of this directed preliminary behaviour is that the individual can select a location that minimizes the risk of: (1) self reinfection; (2) infecting other individuals important to its reproductive success (e.g. mates, offspring); (3) attracting enemies or predators while vulnerable; and/or (4) giving information damaging to itself (e.g. that it is infectious) to other individuals. For the last two considerations, speed is often important and, at least in the case of vomiting, 'push-button' mechanisms are sometimes adopted (e.g. dogs and cats eating emetic substances; humans putting fingers down their own throats).

Push-button techniques may be more powerful than purely psychological techniques which may be insufficient or inefficient as well as being too slow. Perhaps the best illustration of the power of a push-button over conscious control is that of trying to stay awake while drowsy. Even on occasions that falling asleep may be positively dangerous (e.g. when driving a car), conscious control may not always be enough and a push-button technique (e.g. self-pinching) may help.

Masturbation shares many similarities with these other phenomena:

1. The urge to masturbate is generated by internal events. [Males seem to be triggered to masturbate by the increasing age of stored sperm in (presumably usually subconscious) anticipation of the time of the next copulation. Females seem to masturbate in response to stage of menstrual cycle and time since last copulation.]
2. The final act of masturbation, however, involves conscious selection of a suitably cryptic time and location.
3. Masturbation uses push-buttons to speed the process.
4. In the absence of masturbation, ejaculation or orgasm often eventually occurs spontaneously (nocturnally).

The techniques used in masturbation by humans and other mammals were described in section 5.3.2. To a large extent, the masturbation process mimics the sequences and uses the same push-buttons as those involved in copulation. However, there are some differences.

Males most often masturbate by stimulation of the penis in a way that, to varying degrees, mimics the stimulation normal for copulation. Females most often masturbate by direct stimulation of the bulb of the clitoris (henceforth, simply the clitoris) rather than the inside of the vagina. Of course, the human clitoris may receive some stimulation during copulation but rarely is it as direct or intense as during self masturbation. More so than any other anatomical feature, the human clitoris seems now to function primarily as a push-button for masturbation. Again, this technique allows stimulation while reducing the risk of infection that could result from the insertion of fingers or artefacts into the vagina.

Copulation, because of the advantage, before attempting penetration, of an erect penis and vaginal lubrication often involves a 'primer' stage during which non-genital 'erogenous' parts of the body (such as the nipples) are stimulated. There would seem to be less advantage to a primer stage for masturbation than for copulation and in the former it is more often omitted. The fact that such a stage does sometimes occur during masturbation could yet again reflect an advantage in reducing the risk of infection by reducing, however slightly, direct contact with the genitals.

5.7.2 The evolutionary history of male nipples

Most male placentals have nipples (though they have been lost in rats and mice; Daly, 1979) which are obvious homologues of the nipples of females. As far as is known, however, no species has males who use their nipples for lactation and feeding the young (Daly, 1979). Even fruit bats in which the male produces small amounts of milk show no evidence of having suckled (Francis *et al.*, 1994). Indeed, male nipples are unlikely ever to have been functional for lactation for they first evolved in animals which were most unlikely to have shown any form of paternal care (Chapter 6).

Whether male nipples have any direct function, other than perhaps a minor role in sexual stimulation, is unclear. One possibility is that nipples have been retained by males because mammary tissue has been retained. Only in rats has mammary tissue virtually disappeared in males and only in this group are there no nipples.

Any secretory tissue needs an outlet to relieve pressure both in the course of normal secretory activity and sometimes in the case of infection. In the case of mammary tissue, the evolved outlet is the nipple. Perhaps, therefore, the question should be not 'why have males retained nipples?' but 'why have males retained mammary tissue?' There is no clear answer but perhaps the most profitable approach would be to seek a secondary, perhaps endocrine, secretory function for mammary tissue that is applicable to both sexes (but which is reduced in rats).

If such a function exists, then the retention of mammary tissue and hence nipples by both sexes during mammalian evolution would be unsurprising. The advantage of nipple sensitivity in females for suckling and lactation would have led to the evolution of a relatively high concentration of nerves (the density of nerve endings in female nipples exceeds that of most other areas of the skin surface; Wakerley *et al.*, 1988). Such sensitivity would in turn predispose the female nipple for a role in the primer stage of sexual stimulation and hence for adoption as a primer push-button in masturbation.

The male homologue would similarly inherit such innervation through selection on the female. In the absence of selection on males against sensitive nipples, any slight advantage of a sensory function during copulation would retain this innervation. In most mammals in which the male mounts the female from behind, the male nipples probably do receive some stimulation during copulation. Debatably, the evolution of an upright stance and a reduction in body hair that has taken place in the human lineage over the past few million years may have led to a recent increase in the still minor role of male nipples as primers. Men are less sensitive than women to nipple stimulation (Short, 1976) but do show some measurable physiological response (Kolodny *et al.*, 1972). This, albeit limited, sensitivity in turn could pre-adapt male nipples for adoption as minor primers during human masturbation.

The origin of nipples in the human lineage was presumably simultaneous for males and females (say, 100 million years ago). Their minor sensory role in sexual stimulation during copulation has probably been relatively unchanged ever since. However, in the past three million years or so, the evolution of an upright stance and loss of hair in the human lineage may have increased their role sufficiently for them to be adopted as minor and occasional push-buttons during masturbation.

5.7.3 The evolutionary history of the human clitoris

The evolutionary history of the human clitoris has been the subject of considerable discussion (e.g. Symons, 1979; Alcock, 1987; Gould, 1987a,b; Sherman 1988, 1989 a,b; Jamieson, 1989). Perhaps some of the most relevant facts are as follows:

1. All female mammals have a clitoris which, developmentally and presumably evolutionarily, is the homologue of the penis (Box 5.16).
2. In many mammals, including many primates, the clitoris, like the penis, has a clear role in directing urine away from the body (Box 5.16). This role of the clitoris seems to have disappeared in most old world monkeys, apes, and humans.
3. Across mammals, even across primates (Box 5.16), the size, shape and position of the clitoris is highly variable. It ranges from an organ as large as, or even larger, than the penis, to a small inconspicuous organ as in most old world monkeys, apes and humans.
4. The sensory innervation of the bulb of the human clitoris is three times more concentrated than the innervation of the end of the penis (Goldenson and Anderson, 1986).

With this perspective, the evolution of the clitoris in the human lineage seems relatively clear. Initially evolving in early mammals in response to selection on males for a penis via pressure of copulation (Chapter 6), the female homologue was not exposed to selection to be minimized or to disappear. Instead, this penis homologue was exposed to selection for separate adaptive shaping.

The earliest selective pressures seem to have been concerned with carrying the urine away from the body. Inheriting the sensory innervation of the penis, however, the clitoris probably also had an early sensory role during copulation, giving a direct association with copulatory orgasms (section 3.5.3). This preadapted the clitoris for a push-button role in masturbation, giving the advantage of triggering non-copulatory orgasms without the increased risk of infection that would be associated with vaginal penetration by fingers or penis substitutes.

Different lineages will have been exposed to different selection for the form of the clitoris depending on the relative advantages of separating copulatory stimulation, non-copulatory stimulation by other individuals, and self-stimulation. The human clitoris, like that of most old-world monkeys and apes, is small and cryptic (indeed, prior to the 17th century there was no English word for the clitoris; Goldenson and Anderson, 1986).

Such a small, external, and relatively inaccessible clitoris near, but not too near, the vagina as in most old world monkeys, apes and humans seems adaptive to: (1) self-stimulation; (2) some, but not excessive, stimulation during copulation; and (3) access, but only with relative difficulty, for non-copulatory stimulation by a partner. Larger, more obvious, clitorises like those of lemurs and new world monkeys seem adaptive to stimulation by a partner, both during copulation and at other times.

The clitoris is thus likely first to have appeared in the human lineage simultaneously with the penis, probably about 100 million years ago during the early evolution of placental mammals. Its small size may have evolved during the early radiation of old-world monkeys (say, 35 million years ago; Martin, 1993). Its modern human form, however, could be much more recent and may not have stabilized until about 3 million years ago during hominid evolution.

Box 5.16 The clitoris

The human clitoris

Up to about 7 weeks after conception, the genitalia of male and female human fetuses are virtually identical. At 10 weeks, the penis is slightly larger than the clitoris and labia minor, which form from the same primordial cells in the female. At 12 weeks the differences are more pronounced and the male scrotum has formed from the tissue that becomes the labia major in the female. At 34 weeks the distinctive features of the genitalia of the two sexes are clear.

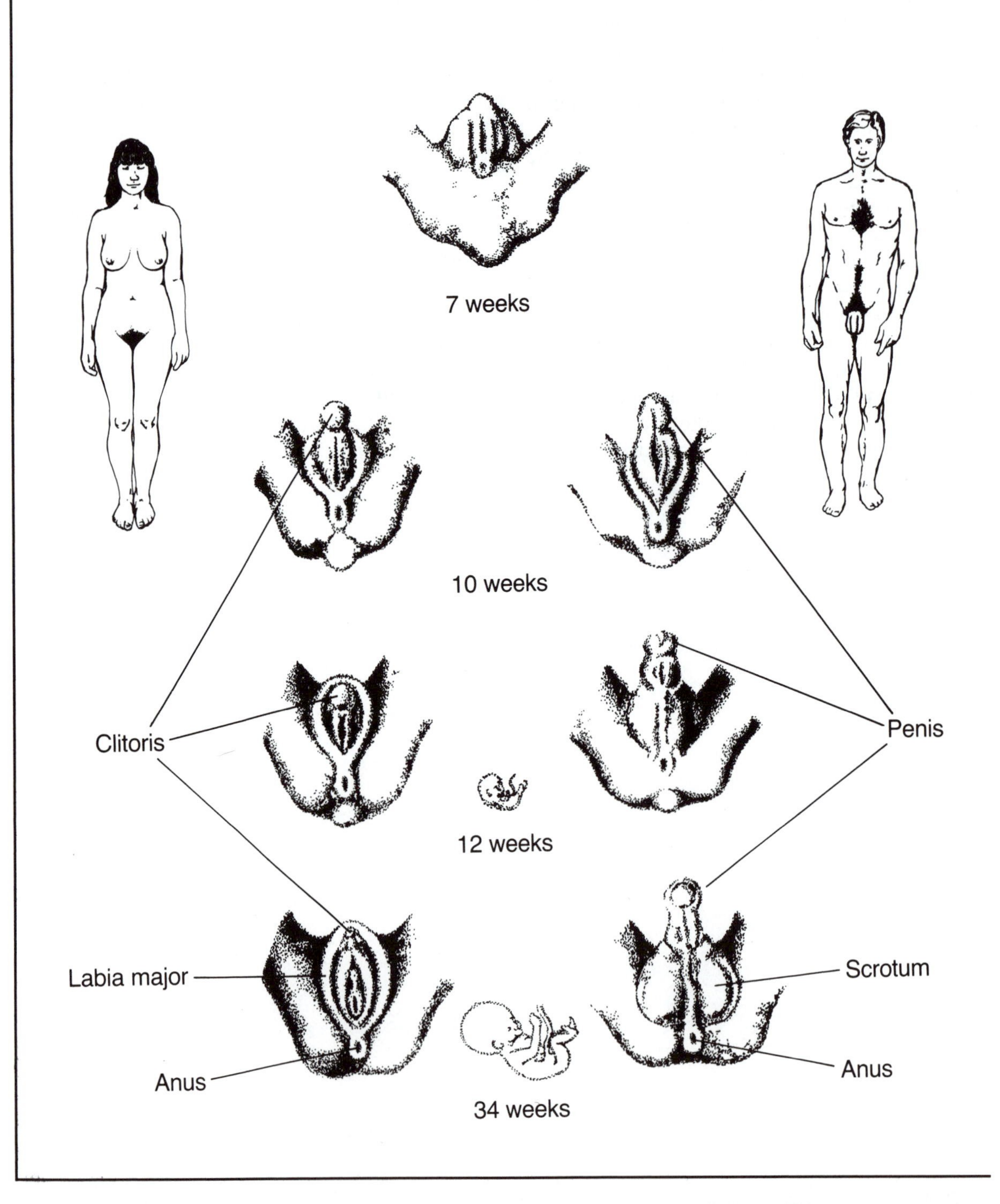

In adult females, the bulb of the clitoris is relatively small, separate from the urethral opening, and covered by a hood which can be drawn back/up to expose the bulb of the clitoris. At most, the stream of urine from the urethra can flush out only those sperm and seminal fluids that have already flowed back down the vagina into the vestibule.

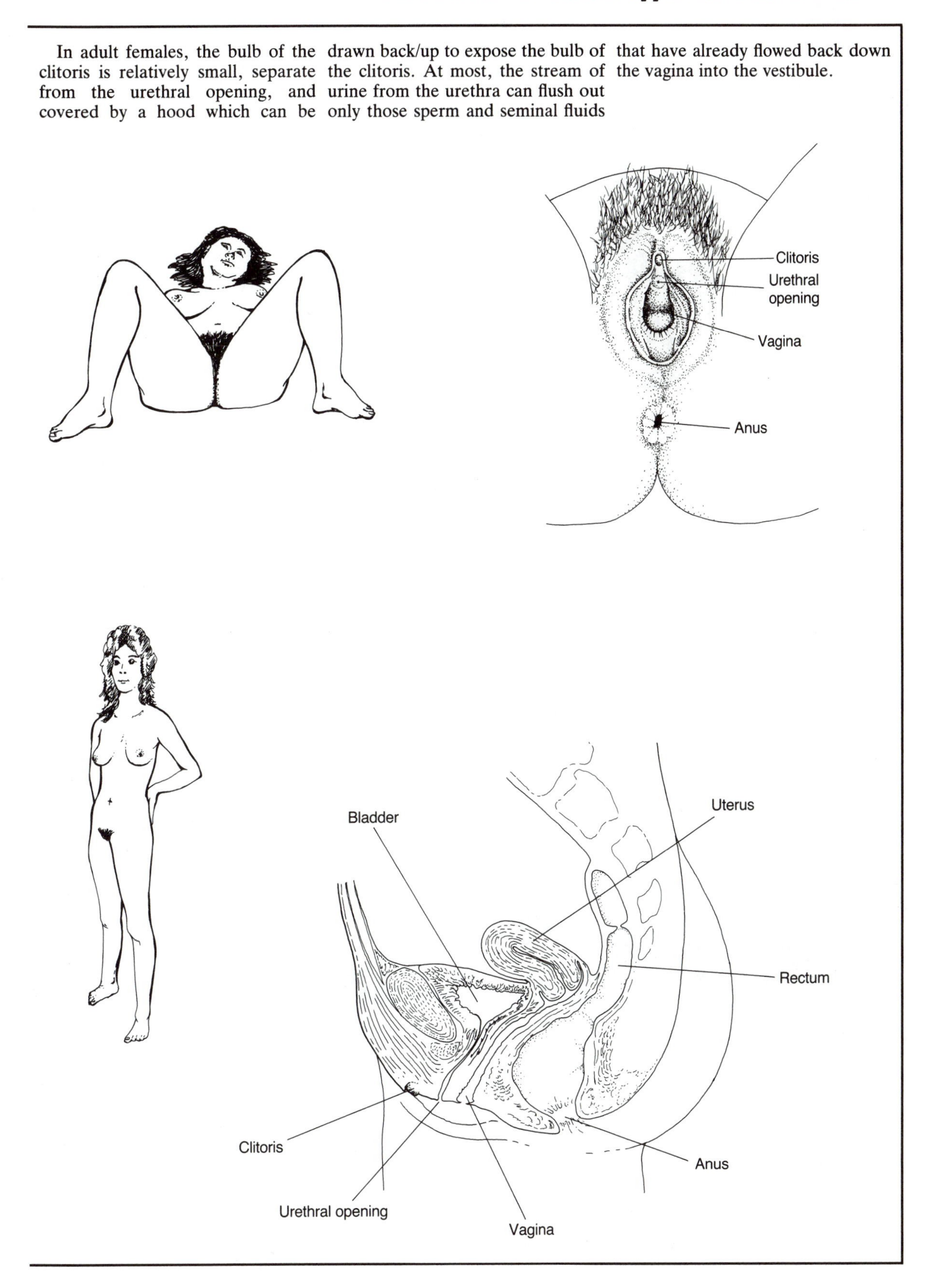

Box 5.16 (*continued*)

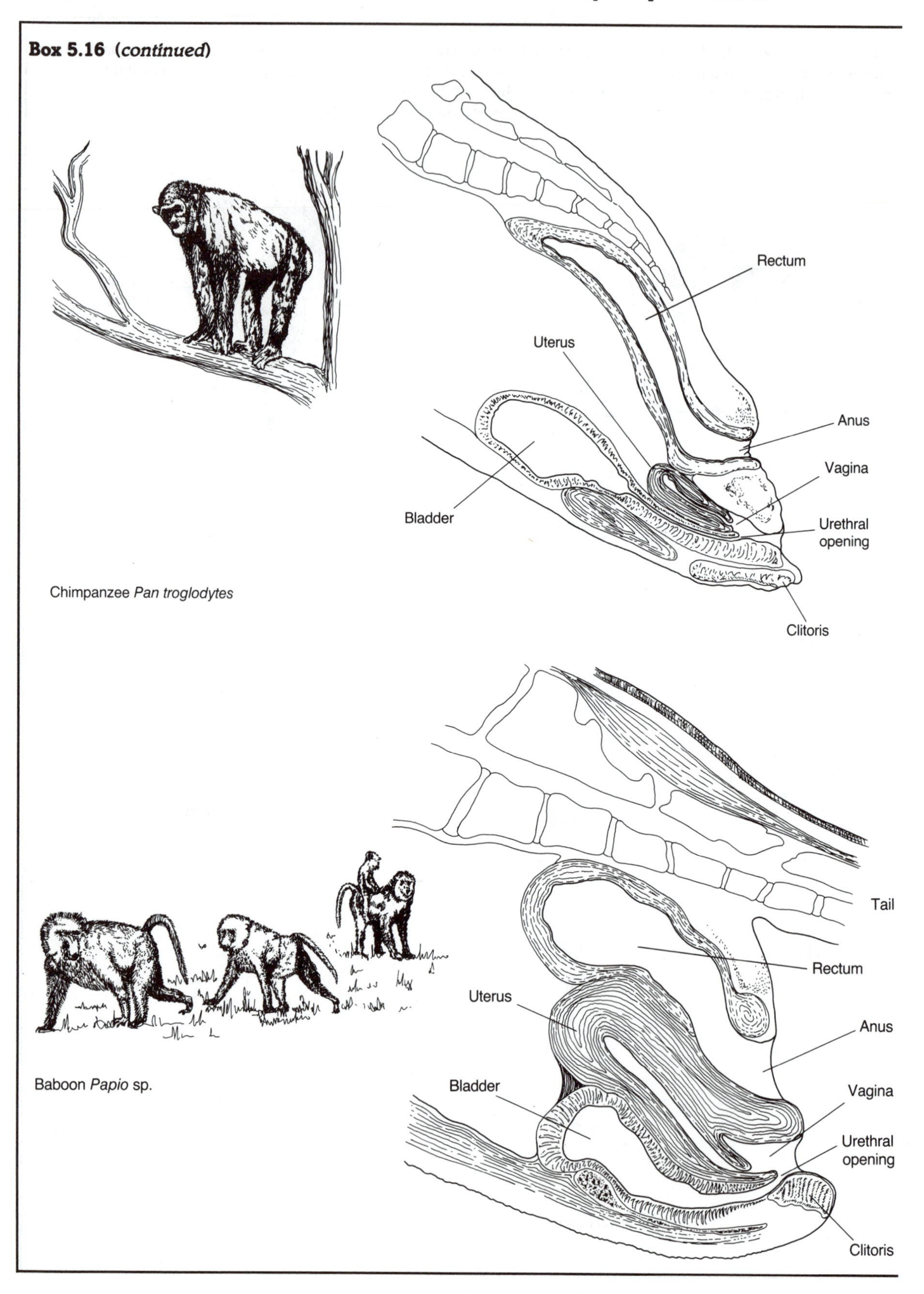

Chimpanzee *Pan troglodytes*

Baboon *Papio* sp.

The clitoris of apes and old world monkeys

A small, relatively inconspicuous, clitoris as in humans is the rule rather than the exception in old world monkeys and apes. Some old world monkeys have a somewhat large and conspicuous clitoris but in no species is it ever so elongate as to be pendulous. Also as in humans, the clitoris and the urethral opening are invariably relatively separate and in some, such as the chimpanzee, the urethral opening is inside the vagina. In most species, therefore, the clitoris has no role in the directing of urine away from the body. However, at least in species such as the chimpanzee, in which the urethra opens directly into the vagina, urine could actually be used to flush sperm and seminal fluids away from the cervix giving the female an influence over sperm retention that is perhaps even greater than through normal flowback. In other species, however, as in humans, any similar role must be restricted to flushing away sperm and seminal fluids that have already flowed down the vagina and collected around the vulva.

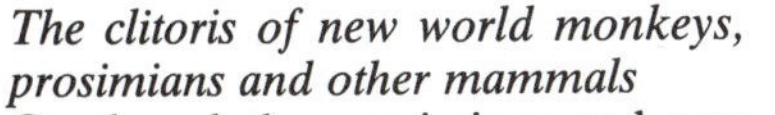

The clitoris of new world monkeys, prosimians and other mammals

On the whole, prosimians and new world monkeys have a large, conspicuous and often pendulous clitoris. Smaller clitorises do occur but they are the exception rather than the rule. Thus, in lorises the clitoris can be as large as the penis and in the new world spider monkey, *Ateles belzebuth*, the clitoris, at 47 mm, is actually longer than the penis. Moreover, the clitoris is grooved and contains a cartilaginous os clitoridis. Across other mammals, as across primates, the size, shape and position of the clitoris is highly variable.

Spider monkey *Ateles* sp.

The position of the opening of the urethra relative to the clitoris and the involvement of the clitoris in urination also varies considerably across mammals. In insectivores and many rodents the urethra perforates the clitoris (Eckstein and Zuckerman, 1956). In many primates (e.g. lorises) the urethra even opens at the tip of the clitoris which can have a prepuce like the penis. In such species, the clitoris carries and ejects urine away from the body in the same way as does the penis. In lemurs, the urethra opens near the base of the clitoris. However, whether perforated by the urethra or not, the clitoris, like any projection near the urethra, can direct urine away from the body.

Dwarf bush baby
Galagoides demidovii

The penis: copulation and urination

Unlike the clitoris, the penis of most mammals contains a common urinogenital tube which opens at the tip. The penis thus serves the dual function of carrying both sperm and urine along its length to a common opening at the penis tip. In monotremes, however, the penis carries sperm but not urine, the urethra opening near the penis base. Unless this represents a secondary adaptation, it seems likely that the initial evolution of the penis was governed primarily by selection on copulation with the demands of urination playing a secondary role. Even so as a projection near to the urethral opening, even the earliest penis would probably have directed urine away from the body.

Compiled from Keeton (1980) and Hill (1953, 1962, 1970b).

6 The primate inheritance: paternal care, sexual crypsis and the façade of monandry

6.1 Introduction

By the time the human lineage entered its primate phase about 100–60 million years ago (Box 6.1), the genetic programme that orchestrated sexual anatomy, physiology and behaviour was recognizably similar to that of modern humans. Relatively few elements were to be added during the remaining years of prehominid primate evolution and even fewer during the final 15 million years or so of hominid evolution. Nevertheless, those few elements that were to evolve, particularly during the last 2–4 million years, during the *Australopithecus* and/or early *Homo* phases, added a veneer of strategic subtlety that endows the study of human sperm competition with a pleasing level of complexity.

The key evolutionary step was the evolution of parental care by males and the façade of monandry that it induced in the females of a number of primate lineages, including humans. The result was the evolution of a powerful female psychophysiology; an attribute that now allows females, subtly and cryptically, to promote and exploit sperm competition while seeking to reassure at least one of the competing males that he really is the father of her offspring and should thus provide her and them with paternal care.

6.2 The primate inheritance: parental and mating systems

6.2.1 Paternal care

Male humans are programmed to interact with their offspring in such a way as to enhance their offsprings' survival and development. They share this behaviour with the males of many fish, a few anuran amphibians, a few reptiles, some ungulates, carnivores and cetaceans, and the vast majority of birds and primates (Ridley, 1978; Clutton-Brock, 1991). As in most birds and a few mammals, human males share their parental care with a female partner or partners, the mother(s) of their offspring. The evolutionary pressures and pathways that have led to biparental care in all of these animals, including humans, are relatively clear (Box 6.2).

Although humans are generally considered to be unusual in their level of paternal care (Alexander and Noonan, 1979; Lovejoy, 1981), they are, in the wider primate perspectives, far from being top, being classified by Whitten (1987) as affiliaters rather than intensive caretakers. Rates of interaction with infants are generally low. Of the 80 cultures included in a world survey of paternal relationships, fathers were rarely or never near their infants in 20% (West and Konner, 1976)

Box 6.1 A suggested course for the human lineage during its primate and hominid phases of evolution

Primate phylogeny
Redrawn and modified from Martin (1993).

Hominid phylogeny
Compiled from Wood (1978, 1992).

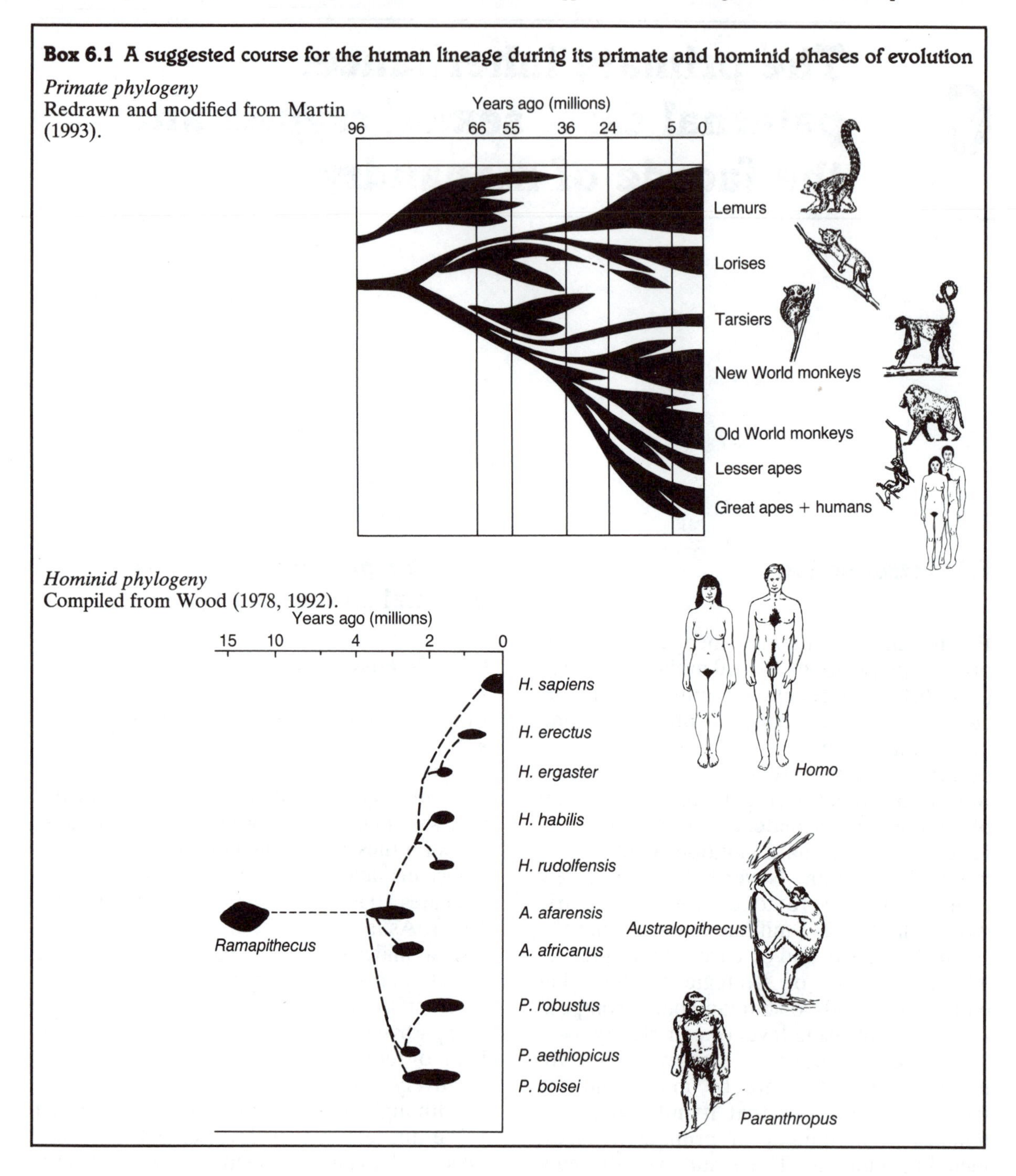

and in only 4% was there a close father–infant relationship. Even in these latter societies, such as the !Kung San of Africa, fathers spent only 14% of their time interacting with infants (Konner, 1976). This is about the same as the most active fathers in industrial societies (3 h/day; Lamb, 1984). However, some fathers in modern industrial societies spend as little as 45 minutes each week in interaction with infants (Lamb, 1984).

Human fathers rarely assume major responsibility for child care and actual care-taking is rare (Whitten, 1987). !Kung fathers contributed no more than 6% of infant care-

Box 6.2 The evolution of paternal care

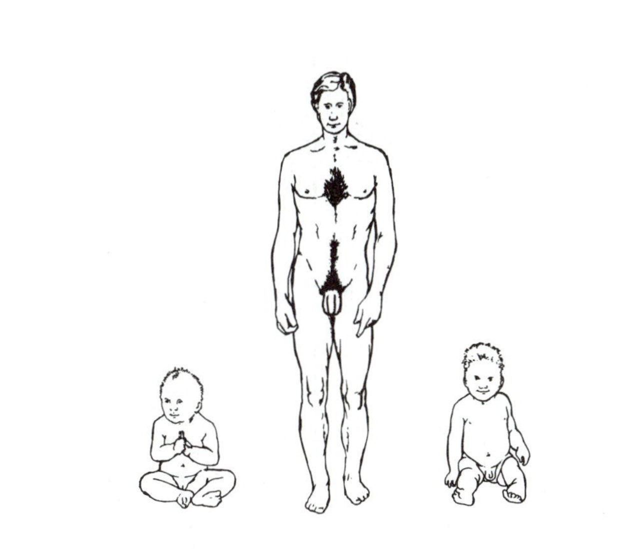

Paternal care is relatively common in external fertilizers, such as fish. This is partly because females have the first opportunity to desert, leaving the male in the 'cruel bind' of either abandoning the young or attempting to raise them single-handedly (Box 4.2). The converse is true in internal fertilizers, such as birds and mammals. Yet paternal care does occur in both of these latter groups, usually in the context of parental care by both parents. The most likely evolutionary pathway for such behaviour is as an extension of male mate-guarding (Box 2.11).

In some ecological situations, particularly when competition for good reproduction sites is high, males have a relatively low chance of gaining sexual access to a female in a good site for reproduction. In such cases, once a male has obtained sexual access to one or more females, there is a premium on retaining that access beyond one female fertility cycle. Those males that succeed in monopolizing a particular female or group of females thus overlap with and influence their offspring primarily because of their (the male's) long-term proximity to the mother.

Once males overlap in space and time with their offspring, selection then begins to operate on males to invest the amount of time and energy in their offspring that maximizes the male's reproductive success. Optimum paternal care in any particular lineage will depend on a number of factors, such as: (1) the probability that the female's offspring are the male's genetic offspring (i.e. certainty of paternity); (2) the extent to which the male is physically capable of enhancing the survival and development of his genetic offspring; (3) the extent to which the species' ecology allows the male to enhance the survival and development of his genetic offspring; and (4) the extent to which paternal care for one female's offspring reduces the male's chances of siring offspring via another female.

We assume that, wherever biparental care occurs, both males and females have a higher reproductive success than they could achieve if only one, or neither, sex showed parental care. This situation will occasionally arise as a product of the physical and other attributes of the species and the species' ecology. We further assume that the major part of the reproductive success of both males and females will be realized within the reproductive unit.

taking, even though they were free of subsistence activities for at least half of their time. Instead, in both industrial and non-industrial societies, the major form of male–infant interaction is play (Lamb, 1984). Only in their provision of food do human males resemble the intensive caretakers among primates (Whitten, 1987). Even here, however, subsistence is typically provided to the mother rather than directly to the offspring.

Despite the relatively small paternal investment made by human males, there is clear evidence that it plays a significant part in the survival and success of the female's offspring (Borgerhoff Mulder, 1991). Thus, among the Ache, a child's chances of mortality are significantly increased if its father dies before it reaches 15 years (43% versus 19%; Hill and Kaplan, 1988). The father's contribution becomes increasingly important after the age of 2 years. Loss of a father, although less damaging than loss of a mother, also influenced mortality risk of children in 18th century Ostfriesland, Germany (Voland, 1988). In contrast, loss of a father did not influence the survival of Kipsigi (Kenya) children, but the resources made available by the father did influence the number of offspring had by both sons and daughters (Borgerhoff Mulder, 1991).

Human males are selective over the infants towards whom they direct their paternal care.

Box 6.3 Monogamy and polygamy

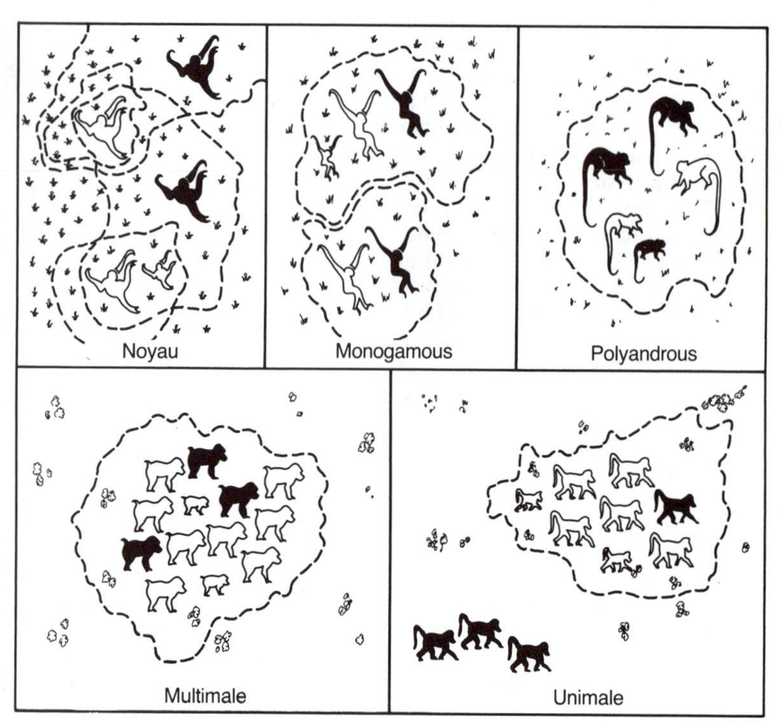

When both males and females show parental care, the parental system is either monogamous (1 male + 1 female) or polygamous (1 or more males + one or more females). Which occurs depends on the species' ecology. When ecological conditions vary from location to location within a species' range, the species may vary from mainly monogamous to mainly polygamous from place to place.

The social structures associated with the different mating systems are typically described as: polyandrous (monogynous males, polyandrous females; e.g. tamarins, Goldizen, 1987); monogamous (monogynous males, monandrous females; e.g. gibbons, Leighton, 1987); unimale (polygynous males, monandrous females; e.g. gorilla, Stewart and Harcourt, 1987); and multimale (polygynous males, polyandrous females; e.g. chimpanzee, Nishida and Hiraiwa-Hasegawa, 1987). In addition, some species (e.g. orangutan; Rodman and Mitani, 1987) have a much more solitary existence. Individuals have their own home ranges but those of some males overlap the home ranges of several females with whom the male mates. Some authors (e.g. Fleagle, 1988) term this social organization noyau.

(Figure re-drawn from Fleagle, 1988.)

Factors when male parental care is minimal (i.e. extended mate-guarding)

Monogyny tends to occur when ecological conditions (e.g. a relatively uniform availability of space, food and shelter for females and their offspring) necessitate females spacing themselves evenly through the environment. There is a degree of female separation that makes it difficult for a male to guard more than one female at a time.

Polygyny tends to occur when resources are spatially clumped. Females then aggregate in areas of most suitable habitat. Males who can defend these areas against intrusion by other males, thus have access to several females.

Factors when male parental care is significant

In most such cases, both males and females contribute to the feeding, protection and/or education of the young.

Monogyny tends to occur in species exposed to physical and/or ecological constraints such that males are physically and energetically unable successfully to show full parental care of the offspring of more than one female. For example, most passerine birds physically feed their young. Females lay the number of eggs that approach the optimum for the production of grandchildren, given feeding of the young by both parents. Males are unable to feed and protect the number of offspring produced by more than one female.

Polygyny tends to occur by the same route as with extended mate-guarding but, instead of just guarding the females against other males, the male also contributes to the care and protection of the young (e.g. gorillas). Groups of males share defence of any given group of females.

Primarily, they are paternal towards infants who have a high probability of being their genetic offspring though, as with some other primates (Whitten, 1987) they may at least initially show some (step-) paternal care towards unrelated offspring as a ploy to gain mating access to the mothers.

In a succession of papers on a variety of human societies, Daly and Wilson (e.g. 1987) have shown that child abuse and neglect occur

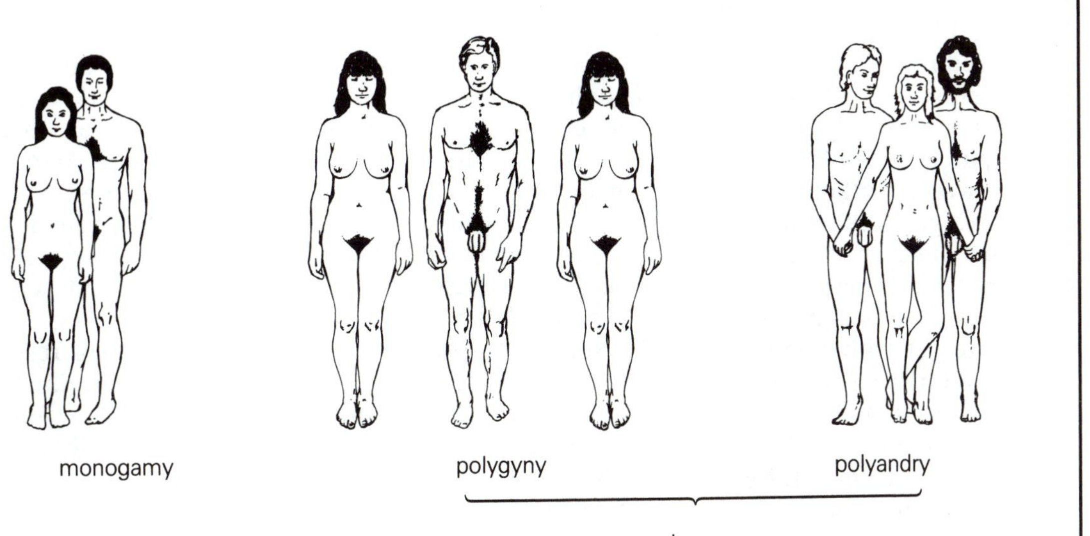

Evolution of the conditional response
Whether monogyny or polygyny is the optimum system for any particular lineage will, according to the above considerations, vary with time and location. Species are unlikely, therefore, to evolve to be exclusively monogynous or polygynous. Rather, species have evolved such that individuals adopt, or attempt to adopt the mating and parental system that is optimum for the circumstances in which they find themselves. Selection thus acts to favour individuals who interpret the optimum system for their circumstances most accurately.

This does not mean, of course, that lineages do not differ in the proportion of time they are, and presumably have been, exposed to selection in polygynous or monogynous situations. Thus, at the present time gibbons are predominantly monogynous but occasionally polygynous. Gorillas are occasionally monogamous but predominantly polygamous, females being largely monandrous, males being largely polygynous (Schaller, 1963). Chimpanzees are occasionally monogamous but largely polygamous, males being polygynous, females polyandrous (Tutin, 1979), and so on.

Mate choice
We assume that in both polygamous and monogamous situations, both males and females achieve the major part of their reproductive success via the offspring they produce and raise within the reproductive unit. Mate choice is therefore based on an attempt to select a partner with whom the individual will produce most grandchildren. This does not just involve selecting the partner with 'best genes'. As, if not more, important is selecting a partner with the genotype that is most complementary to the genotype of the chooser.

Even in polygamous situations, if there is any bi-parental care, there is some pressure to choose a partner or partners who will be the most complementary in the raising of offspring. However, numerical and ecological factors are also important. Some attention should be given to: (1) how many members of the same sex will be competing for parental care and resources from the partner or partners available for selection; (2) the availability of resources offered by the environment over which the partner or partners have control and to which they offer access.

In monogamous situations, the most grandchildren are produced by individuals who accurately base their choice of partner on whether the individual will be complementary in the formation of a reproductive unit (e.g. kittiwakes, Coulson, 1966). In most cases studied, the longer pairs are together, the more successful they become at raising offspring. Pairs are most likely to 'divorce' following failed attempts at reproduction (e.g. kittiwakes, Coulson, 1966).

more often in households with a stepfather than in households with the genetic father. When the father lives in a household containing both genetic and stepchildren, he is more likely to abuse the latter than the former (Lightcap *et al.*, 1982).

A further indication that males are programmed to direct paternal care towards their genetic offspring is provided by the tendency of males to provide less support for their sons in societies in which the females are more polyandrous. This trend has been noted for over a century in anthropology (reviewed by Hartung, 1985). In such societies, the males usually redirect their care and resources towards their sister's offspring, rather than their partner's. Conversely, males may direct care and resources to (potentially their) offspring born to women other than their own main partner (e.g. Roman men who supported particular children of slaves or other men's wives; Betzig, 1992a).

The extent to which male non-human primates interact with infants varies considerably between species ranging from intensive caretaking to simple tolerance or even use and abuse (Whitten, 1987). The most intensive caretaking is shown, for example, by the New World titis, night monkeys, tamarins and marmosets and by the Old World siamang and Barbary macaque (the scientific names of primates are listed in Boxes 6.7 and 6.8). In these species, males look after infants for a large portion of the day, carry them and even share food with them.

Males in many Old World (e.g. stump-tailed macaques; savannah baboons; mountain gorillas) and a few New World (e.g. black howler monkey) species have less intensive affiliations. They carry the infants infrequently but sometimes look after them for brief periods. Primarily, however, their relationships are characterized by frequent proximity, close friendly contact, and protection from other group members.

In all primates with parental care, as in humans, male–infant relationships are selective, particular males within the group associating with particular infants. The primary factor, as with humans, is probably genetic paternity. Thus, the primates with the most intensive male–infant affiliations are those who normally live in monogamous groups and have a high probability of paternity (Whitten, 1987). Moreover, in primates as a whole, infants are at risk to males who cannot be their fathers. In almost all cases of infanticide, males killed infants who were unlikely or impossible offspring (Struhsaker and Leland, 1987; Whitten, 1987). However, factors other than probability of paternity also seem to be involved. These are discussed in section 6.3.

The level and nature of human paternal care probably evolved hand in hand with the evolution of hunting and gathering. Three species of the larger primates have evolved hunter–gatherer behaviour (baboons, Strum, 1981; chimpanzees, Goodall, 1986; humans, Lee and DeVore, 1968). In all three, the hunting of large animals (e.g. hares, small ungulates, other primates) is a primarily male domain. In this respect, these primates differ from other hunting mammals (e.g. felids, canids) in which the females hunt as much as, if not more than, the males (e.g. Schaller, 1972). The difference may well result from female primates having physically to carry their young while foraging. The females of other hunting mammals do not have such an encumbrance.

The human lineage was probably well on the way to being bipedal by 4 million years ago, even if still living in forest (Wood, 1993). The first differentiation of the genus *Homo* and the first appearance of stone tools seems to have taken place around 2.4 million years ago (Hill *et al.*, 1992). By this stage, the lineage may well have had a hunter–gatherer ecology not very different from some recent societies (Lee and DeVore, 1968), males concentrating on long-distance hunting, females on local gathering. At this stage, paternal care had probably reached its current level (Lovejoy, 1981).

Through the provision of protein-rich meat, the male human acquired an increased ability to enhance his offspring's survival, albeit indirectly through the mother. However, a sexually partitioned hunter–gatherer ecology seems to favour both monogamy (Box 6.3) and periods of separation of male and female partners. The repercussions on sperm competition are considerable.

6.2.2 Systems of biparental care: monogamy and polygamy

When males and females cooperate over parental care, the reproductive unit may be either one male plus one female (**monogamy**), or one or more males plus one or more females (**polygamy**). Males who are exclusively paired to just one female for the purposes of mating and parental care are **monogynous**; those paired to more than one

Box 6.4 Human mate choice

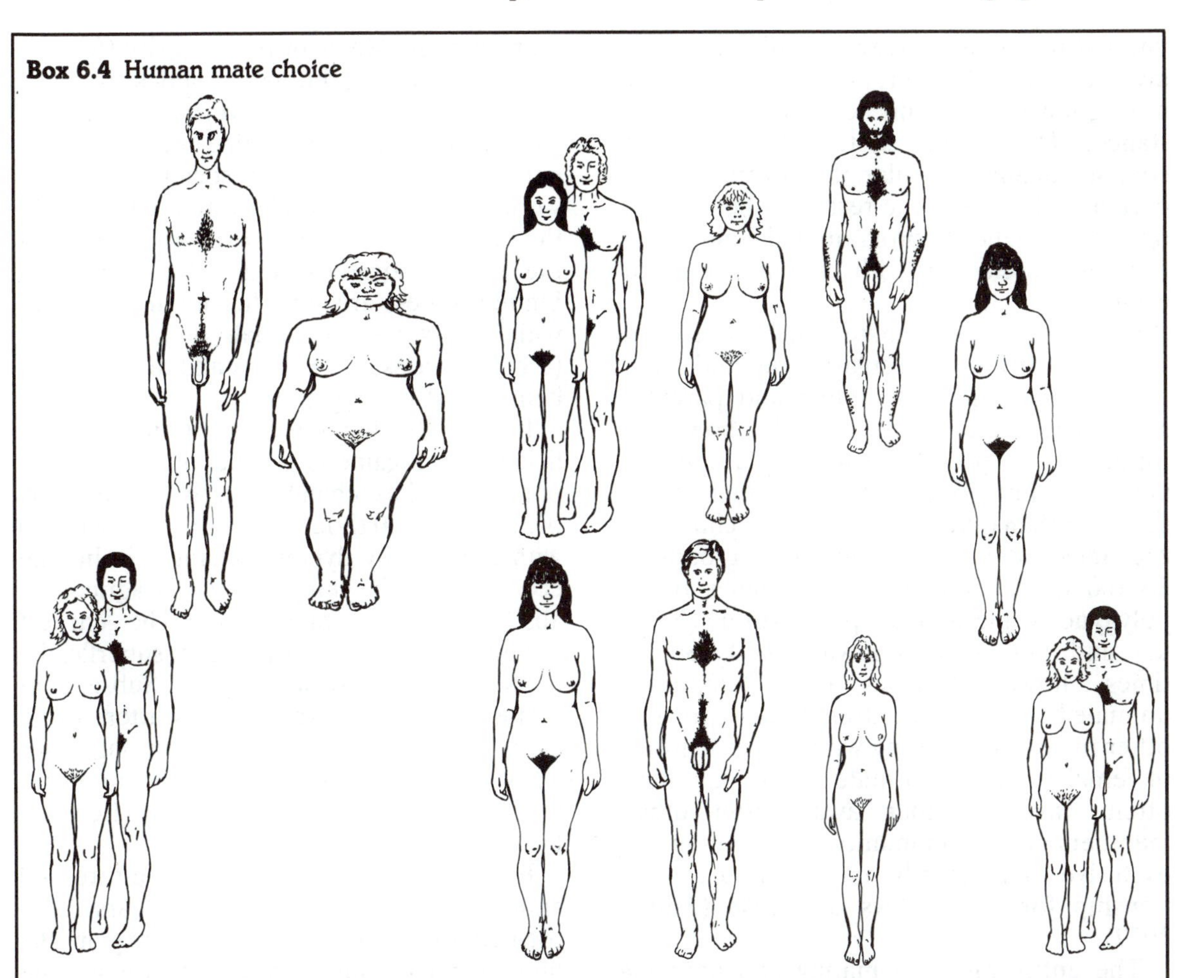

Several authors have now considered mate choice and retention by humans in an evolutionary perspective (e.g. Symons, 1979; Daly and Wilson, 1983; Betzig, 1988; Borgerhoff-Mulder, 1991; Demarest and Schoch-Ciuffreda, 1992; Mills, 1992; Singh, 1993), including a major cross-cultural study (Buss, 1989). On the whole, the same factors emerge as important for humans as for other animals.

In choosing a long-term partner, within the framework of complementariness, males are most attracted to healthy females of maximum potential fertility; females are attracted to high status, resource-wealthy males. One result of this dichotomy is that males tend to prefer younger females and females tend to prefer older males. This leads crossculturally to a characteristic age-difference between partners; males tending to be the elder of the two. Males use a variety of physical cues (e.g. waist:hip ratio; Box 4.7) in their attempt to assess female potential fertility.

The partner with whom any individual finally begins to reproduce is then the compromise between the most suitable available and the one that can be 'obtained' given one's own attributes and competition from other individuals. Once partnerships have been established, infertility of the partner is one of the major factors leading to separation.

female are **polygynous**. Females who are exclusively paired to just one male are **monandrous**; those to more than one male **polyandrous** (see also section 2.4.1). Females are still monandrous even if they share their single male partner with other females. Males are still monogynous even if they share their single female partner with other males. The social structures associated with the different mating systems are illustrated in Box 6.3.

The selective pressures which act on parental systems and on mate choice within each system are outlined in Box 6.3. Human mate choice is discussed briefly in Box 6.4. Of particular importance is the conclusion that selection acts primarily to produce a conditional response strategy by individuals. Individuals are thus programmed to adopt monogamy or polygamy according to circumstance. The mating system that is most common at any particular place or time in any particular lineage is the result of the interaction of a genetically programmed conditional response interacting with circumstance and ecology.

Most species of primate which have been studied in any detail display some, and often an extreme, range of mating and parental systems (Fleagle, 1988). Usually, however, for any given species in any one place and at any one time, one system is much more common than others. For example, chimpanzees most often live in multi-male groups of 20–100 or so individuals, the males being polygynous, the females polyandrous (Nishida and Hiraiwa-Hasegawa, 1987). Sometimes, however, males form temporary unimale harem groups, attempting to prevent other males from mating with one or a group of females. Finally, a male and female may often travel together away from other members of the community and maintain an exclusive, temporarily monogamous, relationship for a few days or weeks (Tutin, 1979).

The entire range of mating and parental systems is also shown by modern humans (Betzig *et al.*, 1988). Nevertheless, some forms are more common than others. In fewer than 1% (4/849; Murdock, 1967) of human societies is the female overtly polyandrous. Thus, the typical adult human female, like the females of 50% (34/68) of anthropoid primate taxa (Sillén-Tullberg and Møller, 1993) is superficially monandrous with one principal male partner from whom she and her offspring receive care, protection and resources (Borgerhoff Mulder, 1991). Males protect their sexual access to each female partner via various forms of mate-guarding (section 2.4.4) and, as in many other primates, other males and females recognize the relationship (Day and Wilson, 1978).

Although the typical female human is superficially monandrous, the typical male (like the males of 84% (57/68) of anthropoid primate taxa; Sillén-Tullberg and Møller, 1993) is not necessarily even superficially monogynous. Males in the majority (83.4%, i.e. 708 of 849) of human societies surveyed by Murdock (1967) were overtly polygynous, two or more females sharing the same male partner. In relatively few societies (16.1%) were the males superficially monogynous. At the present time, however, males in some of the largest societies (e.g. China, Europe, United States) are superficially monogynous with only some individuals or local groups (e.g. Mormons, USA; Faux and Miller, 1984) being overtly polygynous.

It is likely that the proportion of superficially monogamous couples in the human population has changed fairly dynamically over the past 20 thousand years in association with the equally dynamic changes in human ecology (i.e. the hunter–gatherer/agriculturalist/industrialist transitions). The earliest hominids were probably mainly superficially monogamous (Lovejoy, 1981). Subsequent hunter–gatherers, exposed to scattered and limiting resources spread evenly among males (Box 6.3), probably retained the monogamous mating and parental system (Lee and DeVore, 1968).

In contrast, the advent of agriculture and animal husbandry (c. 15 000 years ago) seemed to herald a universal swing in the human population towards polygyny and extreme reproductive inequality between males (Betzig, 1988). Most of the 849 predominantly polygynous societies surveyed by Murdock (1967) were agriculturalists. The critical factor in this swing seemed to be the clumping of resources associated with agriculture and husbandry and the inevitable increase in differences between males in the resources they could accumulate, defend and offer.

A number of studies over a wide range of human societies have now shown that males of higher status and/or who can provide more resources (wealth or food): (1) are more polygynous (e.g. have larger harems or more wives); (2) achieve more EPCs; and (3) are putatively cuckolded less often (see references in Betzig, 1988). Moreover, there is direct evidence that, as a result of these advantages, males with greater wealth and resources produce more grandchildren (e.g. Kipsigis; Borgerhoff Mulder, 1991).

Historical studies of Italy (Betzig, 1992a) and elsewhere in Medieval Europe (Betzig, 1992b) suggest that gross reproductive inequality between males persisted in reality until the industrial revolution. However, with industrialization superficial monogamy again became common. The critical factor may have been that industrialization once more brought about a relatively more even spread of resources among males. The inequalities that do still exist in industrial societies, however, still result in males of higher status and wealth having more access to more females (Betzig, 1988), despite the superficial appearance of monogamy.

Humans, perhaps better than any other primate species, illustrate the conditional and opportunistic nature of sexual programming for parental and mating systems.

6.3 The primate inheritance: infidelity

Monandrous females and monogynous males are superficially faithful to their partners. By definition, their copulations are in-pair copulations (= IPCs). If either sex copulates with an individual other than their overt partner, the copulation is an extra-pair copulation (= EPC) and the behaviour is described as infidelity.

Monogamy is rare among mammals (3–5% of species compared with 90% of birds; Mock and Fujioka, 1990) and is characteristic of only 17% of anthropoid primate taxa (Sillén-Tullberg and Møller, 1993). The most recent analysis suggests that in the human lineage, monogamy is part of the hominid inheritance, evolving from a unimale (harem) system (Sillén-Tullberg and Møller, 1993) among emerging hunter–gatherers (*Australopithecus* or *Homo*) some 2–4 million years ago. Infidelity, however, pre-dates monogamy and has its roots much earlier in primate evolution. For example, the harem females of a unimale group are being unfaithful to the harem male if they mate with any other male.

Even the female members of multimale groups have males to whom they can be unfaithful. There are a number of levels at which the males of multimale groups can relate to individual females (Hrdy and Whitten, 1987). Sometimes, the males are simply possessive, consorting with fertile females and attempting to prevent them from mating with other males. More subtly, however, males of multimale groups may form very strong affiliations with particular infants (Whitten, 1987).

Often these infants are likely to be the male's own but equally often they are not. In fact, one of the most intense examples of paternal caretaking is found in a multimale species (Barbary macaque) in which the males have one of the lowest certainties of paternity of any particular infant (Whitten, 1987). However, there is considerable evidence for a number of multimale species (see references in Whitten, 1987) that by showing care towards an infant, whether genetically his or not, the male gains an increased chance of mating with the mother when she is next fertile. He may then continue to show true paternal care to her next offspring.

Whatever the basis of the male–female relationship, females who, for whatever reason (section 2.4.1), then mate with a male other than the consort male are, in effect, being unfaithful to that male.

Even though both males and females of species which show biparental care achieve the major part of their reproductive success within their reproductive unit (Box 6.3), both may increase their reproductive success via infidelity (Box 6.5). The interplay of selective pressures is such, however, that infidelity will inevitably be a subtle and sophisticated form of behaviour (Box 6.6).

One of our favourite anecdotes, told by Nick Davies, is of a monogamous pair of a bird, the dunnock (*Prunella modularis*). Davies described the pair feeding together, hopping in parallel lines towards a bush. On reaching the bush, the male passed one side, the female the other. Immediately out of the male's sight, the female flew into nearby undergrowth and copulated with a waiting male. She then flew back to the bush and hopped round to rejoin her mate on the far side as if nothing had happened.

In recent years, a number of excellent

Box 6.5 The behavioural ecology of infidelity

We assume that in both polygamous and monogamous situations, both males and females achieve the major part of their reproductive success via the offspring they produce and raise within the reproductive unit. Even though both sexes gain most of their reproductive success through the offspring raised within the unit, both sexes may also increase their reproductive success through being unfaithful to their partner(s) within the reproductive unit.

Males may actually increase the number of offspring they sire by mating with females other than their partner(s). Obviously, mating with a female who then has to raise the offspring single-handedly is less advantageous than cuckolding another male (i.e. mating with a female who is, or will be, paired to some other male). This is because the former female is less likely either to raise the offspring or to raise it/them to the same quality as the latter.

Females cannot substantially increase the number of offspring they produce simply by being unfaithful to their male partner(s) within the reproductive unit. However, their infidelity can generate all of the advantages listed in association with polyandry (section 2.4.1) as well as the advantage of the promotion of sperm competition (section 2.5.6).

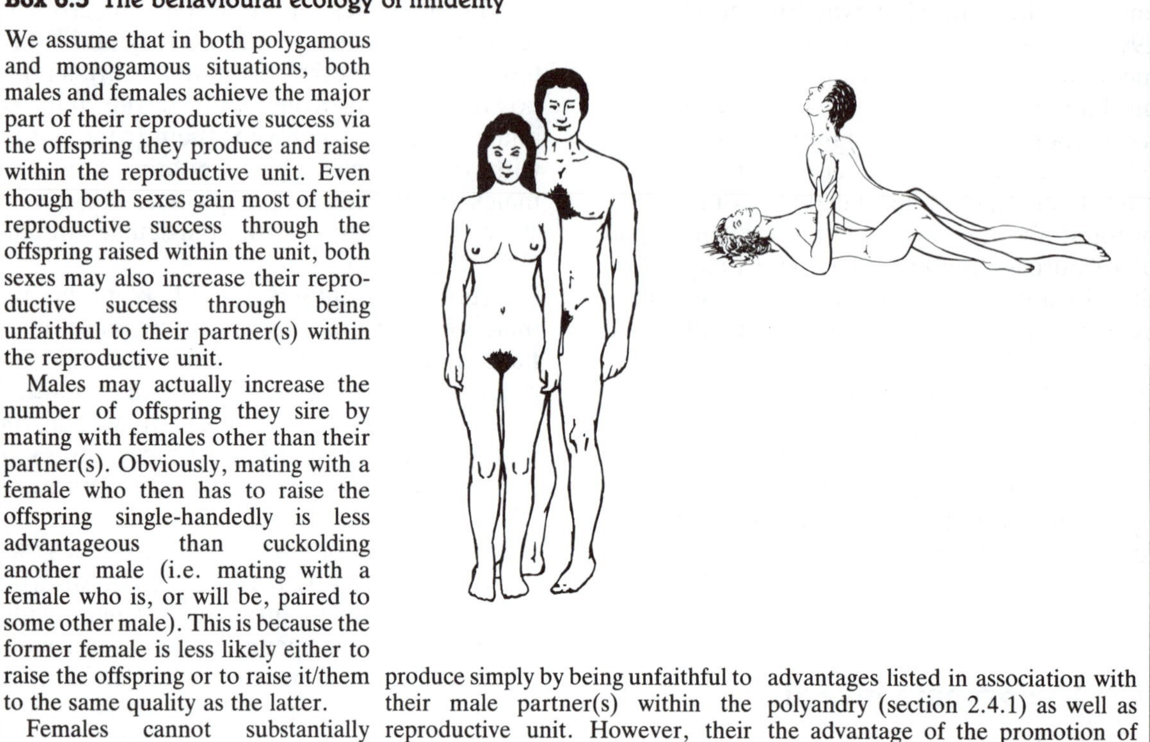

sequences have been filmed of female primates, particularly macaques, achieving EPCs despite the close attention of a dominant or consort male. Most such sequences involve rapid copulation with a nearby male, perhaps behind a tree or rock, while the consort male's attention is temporarily distracted. The most striking aspects of the behaviour, however, are the way that the extra-pair male may hide his erection from the consort male with his hand and the 'nothing happened' pose adopted by the female within seconds of her infidelity.

The reader can probably provide his/her own examples of equivalent behaviour by some of their human acquaintances.

6.3.1 Female infidelity

Female infidelity is probably by far the most important generator of sperm competition in humans and other superficially monandrous animals (Smith, 1984b). As already described, however, the behaviour has its origins further back in primate evolution than the emergences of monogamy.

In all such superficially monandrous species in which the male contributes significant paternal care, a female discovered to be unfaithful risks retribution from her partner. In part, such retribution could function to minimize the consequences (to the male) of his partner's infidelity. In part, it could function to dissuade the partner from being unfaithful again. The main options open to the male are: physical aggression; forced copulation; a reduction in paternal care; and desertion.

In detailed studies of monogamous birds (see references in Birkhead and Møller, 1992), there is little evidence of aggression but relatively good evidence of forced copulation and a reduction in parental care. There is no evidence of the male deserting his female partner totally. There is, however, an example of a male magpie responding to his female partner having an extra-pair copu-

Box 6.6 Why is infidelity so subtle and sophisticated?

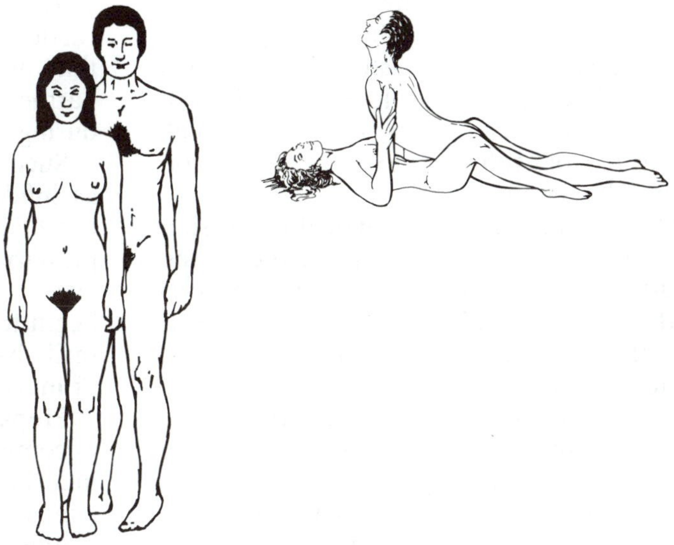

We assume that both sexes may increase their number of grandchildren through their own infidelity (Box 6.5). Equally, however, both may suffer a reduction in their number of grandchildren through their partner's infidelity. Males risk the heavy penalty of being cuckolded and expending a substantial part of their lifetime's reproductive effort in raising the offspring of another male. Females risk their male partner(s) contributing part of their parental effort to another female's offspring. Both risk their partner deserting them for the object of their infidelity, leaving them in the 'cruel bind' (Box 6.2) of having either to abandon their current offspring or to attempt to raise them single-handedly. Both options may reduce the number and/or quality of their offspring and thus of the multiplication of their descendants. Both sexes also risk an increase in their chances of contracting sexually transmitted diseases.

As both sexes may suffer from the infidelity of their partner(s), both should gain from making some effort to prevent their partner(s) from being unfaithful. All monogamous and polygamous animals show some level of mate-guarding. Males suffer more than females from their partner's infidelity because being cuckolded carries a heavier evolutionary penalty than partial loss of parental effort. Consequently, in most cases males are the more active in guarding their mates. Even females (e.g. birds, Birkhead and Møller, 1992; primates, Smuts, 1987), however, behave aggressively toward any other female in which their male partner shows sexual interest or which show an interest in their male partner.

If an individual obtains clear evidence that its attempts at mate-guarding have failed, the partner having succeeded in being unfaithful, the individual has two options. Either: (1) it can remain with its partner and tolerate any disadvantage that its partner's infidelity has produced; or (2) it can desert, seek another partner, and leave its former partner to raise any mutual offspring single-handedly along with any offspring that result from their infidelity. Which option offers the individual the greatest multiplication of descendants depends on the actual outcome of its partner's infidelity and the speed and probability with which the individual can find another partner. Any heritable characteristic that results in a response to the partner's infidelity that is nearest to the optimum will spread through the population.

So, individuals gain from their own infidelity and from preventing or detecting their partner's infidelity but suffer from their partner's infidelity and from being prevented from infidelity by their partner. The result should be a continuing arms race between males and females for the possession of subtle and sophisticated mechanisms to succeed in undetected infidelity on the one hand and to prevent or at least detect any infidelity by their partner on the other.

lation in the middle of her fertile period by abandoning their current clutch but then initiating a new breeding attempt with the same female.

In non-human primates, there is evidence only that males rapidly remate a female partner who copulates with another male (Busse and Estep, 1984; Estep *et al.*, 1986). In her review of male aggression towards females, Smuts (1987) only describes male baboons and vervets aggressively trying to prevent female contact with other males. There is no specific reference to aggression in response to females succeeding in mating, nor of desertion. In contrast, the whole range of such retaliatory behaviour is clearly shown by male humans.

The majority of male desertions in humans follow female infidelity. Daly *et al.* (1982) found that, crossculturally, men were more likely than women to cite their partner's adultery as a cause for divorce. Essock-Vitale and McGuire (1985) found that, among 300 Los Angeles, USA, women, those who claimed to have been unfaithful were divorced on average about twice as often as those who did not. Finally, Betzig (1986) found a wife's infidelity listed by males as a cause for conjugal dissolution in 48 of a sample of 104 societies. In others, the female was at risk of even more extreme retribution to end the union.

Much of the recorded physical aggression of males towards their female partner is the result of the male's knowledge (or suspicion) of the female's infidelity (Daly *et al.*, 1982). The killing of the female partner by the male is also often attributable to the female's infidelity: up to 33% of such killings in the United States (Daly *et al.*, 1982); 85.3% in Canada (Chimbos, 1978); 45% in British colonial Africa (Bohannan, 1960); and similarly large proportions in the Sudan (Lobban, 1972), Uganda (Tanner, 1970); and India (Daly *et al.*, 1982).

Only 13 (11.2%) of 116 societies indexed by Broude and Green (1976) were reported to allow female infidelity to occur without retribution. This does not mean, of course, that even in these societies all individual males were tolerant. In the remainder (88.8%) of societies, female infidelity, if discovered, was reported to provoke retribution.

Despite the potential risks if discovered, female infidelity is universal among human societies. Of 56 societies reviewed by Broude and Green (1976), virtually all women engaged in EPCs in 12.5% of the societies, a moderate proportion in 44.6%, and relatively few, but some, in 42.9%.

Kinsey *et al.* (1953) found that about 26% of adult females in the USA had engaged in EPCs; rather less than the figure of over 50% for the female readers of *Cosmopolitan* magazine surveyed by Wolfe (1981). In a 1974 USA survey (Diagram Group, 1981), 12% of females had been unfaithful in their first year of marriage and 38% had been so after 10 years or more of marriage. In the UK, a computer-aided telephone interview study of randomly selected households found in 1990 that of women aged 18–35 with a main male sexual partner, 8.2% reported another sexual partner during the previous year (10.6% in London; 5.9% Edinburgh; 3.1% Glasgow) (D. McQueen in Macintyre and Sooman, 1992). The much larger direct interview study carried out by Johnson *et al.* (1994) suggests that around 10% of women with a main partner have at least one phase of infidelity during a 5-year period.

Such cryptic polyandry only really qualifies as infidelity via EPCs once the female has reached the stage in her life when she has a primary male partner. Many females, however, pass through an early phase when they begin to copulate with males without necessarily having any particular male behave as a primary partner.

Prepairing copulations by young females is uncommon in only 20.2% of 114 societies indexed by Broude and Green (1976) and is almost universal in 49.1%. For example, young and unpaired !Kung females leave their village with a male to copulate in the surrounding bush, often eventually pairing with one of the males concerned (Howell, 1979). Trobriand Island females were polyandrous when adolescent but, once paired to a particular male, a female's infidelity often provoked her being killed (Malinowski, 1929).

According to Chagnon (in Smith, 1984), some adolescent US females use cryptic polyandry to acquire a primary partner. They copulate repeatedly with a number of males until fertilized, then name the favoured phenotype (or best provider) as father.

The proposed benefits of infidelity (Box 6.5) and the costs of discovery impose selection on females to avoid detection (Box 6.6), despite similar selection on males to be vigilant (Box 6.6). On the whole, females seem to be moderately successful at escaping detection. For example, in the USA in the 1940s, over half (51%) of 470 women had (so far) succeeded in EPCs without detection by their partner (Kinsey *et al.*, 1953). In a further 9%, their partner only suspected the infidelity.

6.3.2 Male infidelity

As with female infidelity, male infidelity is universal among human societies. Of 56 societies reviewed by Broude and Green (1976), virtually all men engaged in EPCs in 12.7% of the societies, a moderate proportion in 56.4%, and relatively few but some in 30.9%. Kinsey *et al.* (1948) found that about 50% of adult males in the USA in the 1940s had engaged in EPCs. In Britain in the 1990s, in a five-year period around 20% of males had engaged in EPCs (Johnson *et al.*, 1994).

In both humans and other superficially monandrous animals, it is much less common for male infidelity to trigger retribution from their female partner(s) or from the larger social group than it is for female infidelity. Only 23% of 116 human societies have any legal prohibitions of male infidelity (Broude and Green, 1976) but these laws seem intended to create the illusion of, rather than to enforce, equity between the sexes (Smith, 1984b). Even in these societies, unlike females, males are not actually punished for their indiscretions or, if so, less severely.

6.4 The primate inheritance: male contraception

Contraceptive behaviour allows females to continue to reap the non-reproductive advantages of copulation while avoiding the untimely production of offspring (section 4.4). Females are favoured who optimize, among other things, the timing and spacing of attempts at reproduction (= family planning) without abandoning copulation. Selection for contraceptive behaviour acts with greatest force on females who show extended parental care. The same is likely to be true for males.

In species with biparental care, such as many primates, males should gain an advantage from 'family planning' in the same way as do their female partners. Less obvious for males, however, are the non-reproductive benefits of copulation which are so diverse and clear for females (section 2.4.1). We suggest two possibilities: spite and assessment, the latter in essence being pseudocopulation or withdrawal (i.e. copulation without insemination).

6.4.1 Spiteful copulation

Spiteful copulations have a long evolutionary history and are primarily the result of selection via sperm competition. Males of many species with internal fertilization respond to a copulating pair by attempting to oust the incumbent male in order immediately to take over insemination of the female themselves. Such behaviour is found in insects (Parker, 1970b), birds (Birkhead and Møller, 1992) and mammals (Adler and Zoloth, 1970).

In all of these species, females seem to be exceptionally receptive to rapid re-mating by a different male, perhaps as part of the advantage of promoting sperm competition (section 2.5.6). Although female humans do at times engage in very rapid double-matings (Box 2.9), we know of no evidence that women do or do not have a heightened receptivity to insemination by one male during, or just after, insemination by a different male.

In all of the non-human species studied (e.g. rats, *Rattus norvegicus*; Matthews and Adler, 1977), males gain most by mating with the female as quickly as possible after the previous male; maximally (e.g. dung flies, *Scatophaga stercoraria*; Parker, 1970b) by actually interrupting the other male and resuming where he left off. For this to occur, perception of a copulating couple needs to generate sexual arousal in males and rapid preparation of copulation (as it does, for example, in pig-tailed macaques, *Macaca nemestrina*; Busse and Estep, 1984). Male humans seem to show just such a response, the reaction being the basis of voyeurism and the appeal of hard-core pornography.

A number of studies of non-human animals have shown that when a male, having just (<15 min) inseminated a female, observes that female copulating with another male, the time interval to his next insemination of that female is significantly reduced (e.g. roof rats, *Rattus rattus*, Estep, 1988; pig-tailed macaques, Busse and Estep, 1984; stump-tailed macaques, *M. arctoides*, and rhesus macaques, *M. mulatta*, Estep *et al.*, 1986).

At worst, a male who rapidly copulates with a female just inseminated by another male reduces the success of the first male at fertilizing the female's egg(s). At best, the second male gains the lion's share of fertilization chances (Schwagmeyer and Foltz, 1990). Between these two extremes, the copulation is spiteful. The male does not fertilize any of the female's eggs himself but successfully prevents the other male's ejaculate from doing so. Such an outcome has been clearly demonstrated for rats, *Rattus norvegicus* (Matthews and Adler, 1977).

We know of no clear evidence of a spiteful outcome to copulation in humans. In artificial insemination by donor (AID) treatments for infertility, however, the couple may be advised not to copulate for the two days prior to AID. This is because cases have been recorded in which the husband's inseminate, although not itself fertile, has prevented fertilization by the donor's artificial inseminate (Quinlivan and Sullivan, 1977).

Discussion of the possible mechanics of spiteful copulation is delayed until Chapter 11. For present purposes, however, it is sufficient that such a copulation strategy has evolved.

In species with biparental care, optimum family planning may not be the same for male and female, particularly in the context of infidelity. There may often be circumstances in which a child with the partner may not be timely for either male or female but that a child via EPC and cuckoldry may be advantageous for the female (but not, of course, for the in-pair male).

In such circumstances, the in-pair male has two options: no copulation or spiteful copulation. Avoidance of copulation gives the male more certain information concerning paternity (the offspring is not his own) but increases the risk of his partner conceiving to another male. Spiteful copulation risks denying him information concerning paternity (the offspring may or may not be his) but increases his chances of delaying his partner's conception until the time is optimum for him. Which option is best at any one time will depend on the male's precise circumstances.

6.4.2 Withdrawal

Pseudocopulation by the male probably has an evolutionary history equally as long as spiteful copulation. Primates have long been known on occasion to penetrate the female without ejaculating (e.g. orangutan and gorilla; Nadler and Collins, 1991). In some instances male primates may follow such pseudocopulations by self-masturbation to ejaculation (Kempf, 1917).

Many rodents show intromission, the male inserting his penis, thrusting, and then withdrawing without ejaculation (Adler, 1969). After several intromissions, the male usually then ejaculates. Intromission probably has several functions (see section 6.8.2) but in so far as the female rat does not transport sperm from the vagina to the uterus without at least one intromission (Adler, 1969), an element of female assessment may be involved.

The phenomenon of female coyness has in many species led to prolonged courtship behaviour while the female assesses whether the male is a suitable partner to whom to commit her gametes (section 2.3.2). Ideally, the female should seek information concerning all aspects of the male's sexual performance, including efficiency at copulation, in order to assess the potential quality of any sons he may give her (Box 2.12).

The dilemma for the female is how to assess the male's copulatory performance without risking conception. Although a female of any species may gain from such assessment, the opportunity is particularly important for females of species in which the male may be a potential long-term partner, as in species with biparental care. It is also in such species that males may benefit most from cooperating with the female during assessment, exchanging the opportunity for a single insemination at the present with longer term, perhaps multiple, opportunities in the future. In part, pseudocopulation (i.e. the male inserting his penis but not ejaculating), is likely to be an adaptation to these selective pressures. In humans, pseudocopulation is colloquially known as 'withdrawal'.

In such situations, the male strategy is to reassure the female that insemination will not occur in order to be allowed to copulate. The male may then either cooperate with the female, exchanging short-term disadvantage for possible long-term advantage, or deceive the female, inseminating sperm anyway. Cooperation is clearly contraceptive but deceit can also qualify.

Optimum male strategy may sometimes be to give the appearance of cooperating, by avoiding an obviously full ejaculation, but actually to release a few, perhaps very fertile, sperm with little fluid. If the probability of fertilization from such behaviour is reduced relative to a full ejaculation, the behaviour still qualifies as contraceptive. The male may gain two possible advantages from his apparent, but not actual, cooperation with the female's family planning. He gains if: (1) more copulations at reduced fertility summate to a greater total fertility than fewer copulations with full insemination; and/or (2) he achieves future access to the female for a series of full inseminations.

Human withdrawal behaviour (the male removing his penis from the vagina before ejaculating) has many of the characteristics of such a strategy. The male appears to cooperate with the female in so far as ejaculation occurs outside of her body. Nevertheless, the probability of conception is only a little reduced (12–40% conceptions in 100 women years) compared to full ejaculation (60–75%) (Johnson and Everitt, 1988). In part, the failure rate is probably due to the male occasionally not withdrawing before ejaculation. In part, however, the failure could indicate that any sperm which 'leak' into the female (without proper ejaculation) before the male withdraws (Johnson and Everitt, 1988) may be very fertile (Chapter 11).

The low number of sperm leaked before withdrawal is not, in itself, a total bar to conception. There are a number of cases on record of vasectomized males apparently fathering offspring, despite inseminating perhaps only tens of sperm. Of course, there is always the possibility that such cases owe more to female infidelity than sperm fertility. Recently, however, a case has been reported in which not only was the vasectomized male known to ejaculate very few and apparently only immotile sperm but also his paternity was confirmed by DNA fingerprinting (Thomson *et al.*, 1993).

6.5 The hominid inheritance: female sexual crypsis

The females of many primates, like most mammals, show oestrus (i.e. a period of heightened interest in copulation; Heape, 1900; Rowell, 1972). Oestrus is often associated with a variety of hormonal and histological changes (Beach, 1976). In primates, various bodily, particularly vaginal, secretions change their nature during oestrus, and the area around the anus, vulva and sometimes chest may swell and become more brightly coloured (Hrdy and Whitten, 1987).

The ovarian cycle of species with conspicuous peaks to female receptivity, however these are manifest, is often termed an oestrous cycle. In many primates, whether they have an oestrous cycle or not, the ovarian cycle is also manifest as a menstrual cycle (i.e. from one period of menstrual bleeding to the next; Rowell, 1972). However, not all primates have menstrual cycles, the menstrual blood being either retained in the uterus (e.g. stump-tailed macaque, *Macaca arctoides*; Hafez, 1971) or being too slight to be seen at the vulva (e.g. blue monkey, *Cercopithecus mitis*; Rowell, 1972).

Nor do all primates have oestrous cycles. Some (e.g. marmosets, *Callithrix* spp.; black-and-white colobus monkeys, *Colobus guereza*; humans, Sillén-Tullberg and Møller, 1993), are **sexually cryptic** and appear not to advertize their level of fertility. As part of this apparent crypsis, females may initiate or allow copulation throughout their menstrual cycle as well as during pregnancy (Hrdy and Whitten, 1987).

Thus, some primates have neither a visible oestrous nor menstrual cycle (e.g. blue monkey, *Ceropithecus mitis*; orangutan, *Pongo pygmaeus*), some have only an oestrous cycle (e.g. stump-tailed macaque, *Macaca arctoides*), some have only a menstrual cycle (e.g. humans) and some have both (e.g. chimpanzee, *Pan troglodytes*) (references in Hrdy and Whitten, 1987).

6.5.1 The behavioural ecology of sexual swellings

Primate species with females with sexual swellings (Box 6.7) most often live in multi-male groups (Clutton-Brock and Harvey, 1976; Sillén-Tullberg and Møller, 1993). Three main hypotheses for the evolution of sexual swellings have been proposed.

1. The 'best male' hypothesis (Clutton-Brock and Harvey, 1976) in which sexual swellings are postulated to function to incite competition between troop males for sexual access to the signalling female. This increases the chances that the female will be fertilized by the male best able to defend mating access, thereby increasing the chances that her male offspring will have the same attribute (Box 2.12).
2. The 'confused paternity' hypothesis (Hrdy, 1981). This suggests that sexual swellings increase the female's opportunities to mate with several different males. In so far as males may offer more parental care and/or less aggression to the offspring of females with whom they have mated, a female who mates with many males could increase the total amount of paternal care (Taub, 1980) and decrease the total aggression (Hrdy, 1979) her future infant will receive.
3. The 'obvious ovulation' hypothesis (Hamilton, 1984). Sexual swellings are suggested to serve to pinpoint the timing of ovulation, thereby increasing paternity confidence and hence paternal investment by one or two males.

In general, there is undoubtedly some relationship between the timings of sexual swellings and fertility. Thus, in the savanna baboon, *Papio cynocephalus*, the skin around the vulva begins to swell during or just after menstruation (Hrdy and Whitten, 1987). The size of the swelling gradually increases for 2–3 weeks, reaching peak size a few days before ovulation. Immediately after ovulation, the swelling rapidly decreases in size and subsides into the 'flat' phase of the cycle.

Female receptivity to copulation is not limited to the time of maximum swelling (Rowell, 1972). A common basic pattern is that shown by savanna baboons (Hall and DeVore, 1965; Hausfater, 1975). Early in the cycle, as the skin begins to swell, juvenile and subadult males mate with the female. Later in the cycle, as the size of the swelling increases and the female nears ovulation, fully adult males actively compete to monopolize mating opportunities by forming exclusive 'consort' relationships with females. In a typical cycle, consortships with adult males begin about a week before ovulation and continue until the swelling begins to detumesce.

Superimposed on this basic pattern, however, are many other influences. Males show strong partner preferences (Smuts, 1985) and there is no simple relationship between the maximum size of a female's sexual swelling and her attractiveness (Scott, 1984). Nor is there any clear correlation between a male's dominance rank and his success at gaining access to, guarding, and eventually fertilizing females (Ginsberg and Huck, 1989).

In savannah baboons, dominant males consistently herded females about to ovulate (Cheney and Seyfarth, 1977) but Smuts (1985) found that females still managed to consort with a mean of 4.8 males during each oestrus. A male's success in consortships did not correlate with independent studies of his rank but did correlate with his age (Smuts, 1985). However, consortships themselves did not fully predict paternity unless they occurred just before ovulation (Bercovitch, 1987).

In Chimpanzees, 70% of copulations occurred openly and promiscuously and only 2% occurred when the male and female were isolated in consortships (Tutin, 1979). Nevertheless, the majority of conceptions occurred when a female was in consortship with a single male. Consortships occurred around the time of maximum tumescence and hence around the time of peak fertility. Neither age nor dominance rank was a good predictor of a male's success in forming consortships.

DNA fingerprinting studies of four macaque species (Japanese, crab-eating, Barbary

Box 6.7 Some taxa of anthropoid primates with sexual swellings

Taxa		Visual signs of ovulation	Mating system
New World monkeys			
Capuchin	*Cebus*	S	MM
Squirrel	*Saimiri*	S	MM
Black banded saki	*Chiropetes*	S	MM
Red Howler	*Alouatta seniculus*	S	U/MM
Mantled Howler	*A. palliata*	S	MM
Old World monkeys			
Allen's swamp	*Allenopithecus*	C	MM
Talapoin	*Miopithecus*	C	MM
Patas	*Erythrocebus*	S	U
Spot-nosed	*Cercopithecus nictitans*	S	U
Barbary macaque	*Macaca sylvanus*	C	MM
Lion-tailed macaque	*M. silenus*	C	MM
Pig-tailed macaque	*M. nemestrina*	C	MM
Black macaque	*M. nigra*	C	MM
Bonnet macaque	*M. radiata*	S	MM
Stump-tailed macaque	*M. arctoides*	S	MM
Crab-eating macaque	*M. fascicularis*	S	MM
Taiwan macaque	*M. cyclopis*	C	MM
Rhesus macaque	*M. mulatta*	S	MM
Japanese macaque	*M. fuscata*	S	MM
Gelada	*Theropithecus*	C	U
Mangabey	*Cercocebus*	C	MM
Hamadryas	*Papio hamadryas*	C	U
Yellow baboon	*P. cynocephalus*	C	MM
Mandrill	*Mandrillus*	C	U
Banded leaf monkey	*Presbytis melalopha*	S	M/U
Dusky leaf monkey	*P. obscura*	S	U
Proboscis monkey	*Nasalis larvatus*	C	U/MM
Simakotu	*Simias concolor*	C	U
Dove langur	*Pygathrix nemaeus*	S	U/MM
Snub-nosed	*Rhinopithecus*	S	MM
Black colobus	*Colobus satanus*	C	MM
Olive colobus	*Procolobus*	C	MM
Red colobus	*Piliocolobus*	C	MM
Apes			
Gibbons	*Hylobates*	S	M
Gorilla	*Gorilla*	S	U
Chimpanzees	*Pan*	C	MM

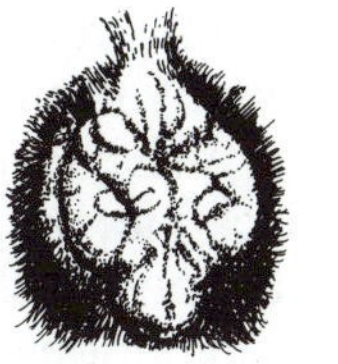

Talapoin

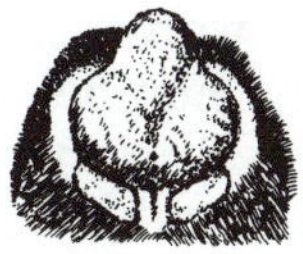

Black macaque

Crab-eating macaque

Rhesus macaque

Common chimpanzee

Signs of ovulation: S = slight; C = conspicuous.
Mating system: M = monogamous; U = unimale; MM = multimale.

(Table constructed from figures in Sillén-Tullberg and Møller, 1993; swelling illustrations redrawn from Wickler (1967).

and rhesus) have all shown that copulatory patterns are not accurate indicators of paternity (Inoue *et al.*, 1992; de Ruiter *et al.*, 1992; Mènard *et al.*, 1992; Berard in Bercovitch, 1992). Moreover, in Barbary macaques, *M. sylvanus*, male dominance rank was significantly correlated with actual paternity only when subadults were included in the analysis (Paul and Kuester in Bercovitch, 1992). Between adult males, however, there was no correlation between paternity success and rank. In a captive chimpanzee population, males of low rank sired relatively few offspring but high rank by itself conferred no reproductive advantage (Ely *et al.*, 1991).

In the Japanese macaque, *M. fuscata*, male dominance rank was positively correlated with the number of copulations accompanied by ejaculation (Inoue *et al.*, 1991). However, the number of copulations with ejaculation was not correlated with the number of offspring. Primiparous females in particular were often fertilized by males other than the male or males with whom they copulated most often. Although low-ranking males generally had only limited opportunities for copulation, they were nevertheless able to sire offspring.

One factor that may interfere with rank/mating relationships is a widespread female interest in mating with new troop members (Smuts, 1987). Such males usually begin low in rank but are high in interest to females. In one troop of Japanese macaques, a newly immigrant male and thus the lowest ranking of all fully adult males, mounted more different females than did any other male (Takahata, 1982).

The general conclusion is that although males do compete for females showing conspicuous signs of fertility, females also exercise considerable influence over who gains access to them at different stages of their cycle (Hrdy and Whitten, 1987; Ginsberg and Huck, 1989). Usually, the female contrives to ensure that she receives sperm from more than one male. Sometimes the female's polyandry is overt (e.g. chimpanzees; Tutin, 1979) but often, when a dominant male consorts with the female and attempts to guard her from other males, polyandry is devious and cryptic as already described. Even the females of species with sexual swellings, therefore, have and use an ability to manipulate males in their (the females') own best interests.

In part, a polyandrous female may be promoting sperm competition (section 2.5.6); in part she may be giving some opportunity for fertilization to younger males who she assesses may have the genetic potential to be successful competitors in the future. In all cases, however, she will be confusing paternity and thus eliciting for her future offspring the support of more, and/or the antagonism of fewer, males than if she had mated with just one (Hrdy, 1979, 1981).

6.5.2 The evolution of sexual crypsis and monandry

The majority of primates (53% of 68 anthropoid taxa) have females with slight or conspicuous sexual swellings and phylogenetic analysis strongly suggests that at least slightly overt cycles were characteristic of the earliest primates (Sillén-Tullberg and Møller, 1993). However, as primates radiated into different ecological niches, sexual swellings were lost on a number of independent occasions, one of which was on the line leading to modern humans.

In a careful phylogenetic analysis of the anthropoid primates, Sillén-Tullberg and Møller (1993) have concluded that sexual crypsis has appeared 0–1 times under monogamy but 8–11 times in a non-monogamous context. On the other hand, monogamy was inferred to have evolved independently 4–6 times in the presence of sexual crypsis but only 1–3 times in its absence.

Thus, it seems that sexual crypsis is most likely to evolve in lineages with females who are overtly polyandrous. Having evolved, however, sexual crypsis seems to promote females who adopt a superficially monandrous strategy.

6.5.3 Infidelity's accomplice: the behavioural ecology of sexual crypsis

Sexual crypsis changes the emphasis in male: female interactions. Males still compete for sexual access to females and mate-guarding still occurs (Hrdy and Whitten, 1987; Smuts, 1987). Inevitably, however, both competition and guarding become more evenly spread through time because the male has less, or no, indication of the time of peak female fertility. Mate-guarding in particular is forced to be of a longer-term nature and thus cannot be carried out at the short-term expense of other activities, such as feeding and sleeping. As a result, females have more time and opportunity to engineer polyandry whenever polyandry is advantageous.

Sexual crypsis seems to have evolved most often under multimale social conditions (Sillén-Tullberg and Møller, 1993). Potential advantages driving such evolution are that crypsis is: (a) an aid to confuse paternity (Hrdy, 1979, 1981); and/or (b) a mechanism to allow the female more influence over which male succeeds in fertilization. The probability that, having evolved under multimale circumstances, sexual crypsis then promotes monandry (Sillén-Tullberg and Møller, 1993) is a clear indication of the nature of that monandry.

Females could well often be disadvantaged by actual monandry, losing all of the benefits of polyandry (section 2.4.1) in exchange for paternal care from just one male. Whereas all the males in the vicinity are potential genetic fathers to a female's offspring, not all are available to show paternal care, being already paired to another female or females.

Potential paternal carers will always, therefore, be less numerous than potential genetic fathers. Moreover, simply because a particular male is available to show parental care to a female's offspring does not mean that he is necessarily the male in the vicinity with the most competitive sperm and/or the male who would give the female genetically the most successful sons (Box 2.12). On the contrary, if at any one time the 'better' males are more likely already to be paired, the males available for paternal care may on average be suboptimal (MacKinnon, 1978). However, a female who could elicit high levels of parental care from an available male partner by creating the illusion of monandry, while still cryptically gaining the benefits of polyandry, could gain a number of advantages.

The most successful females will be those who succeed in retaining intensive parental care from one male (their primary partner) while engineering either sperm competition and/or fertilization by the 'best' male in the vicinity. In this, sexual crypsis could be a considerable advantage. A female, by soliciting copulation with her partner at infertile stages of her cycle but with the most favoured male at the most fertile stages, could contrive to retain her partner's services as a parent while having offspring by the more favoured male.

Of course, the penalties of discovery (Box 6.5) still apply but sexual crypsis gives the female much more opportunity to separate paternity from paternal care according to her own best interests. Having evolved in multimale groups (Sillén-Tullberg and Møller, 1993), therefore, sexual crypsis may in turn endow females with a weapon by which they can create the façade of monandry without losing the benefits of polyandry via infidelity. In many lineages, therefore, such as the human, sexual crypsis may have promoted the evolution of females who appear to be monandrous but who, in reality, are often polyandrous.

We know of no direct evidence that female humans or other primates actually benefit from such a strategy of superficial monandry. However, the females of a monogamous bird, the blue tit, *Parus caeruleus*, show just such a strategy and do seem to gain a reproductive benefit as a result.

Male blue tits in the population studied varied in that some males (= high quality males) had longer tarsi, were more likely to survive to the following year, and produced young who were also more likely to recruit into the next year's breeding population (Kempenaers *et al.*, 1992). Females paired to high quality males were more faithful than those paired to lower quality males. The latter females were more likely to visit the territories of, and mate with, high quality males. Genetic fingerprinting showed that these EPCs did produce young sired by the high quality males, thus cuckolding the females' lower quality partners. As a result of being unfaithful to their partners, yet retaining their paternal care, the females produced young with a higher reproductive potential.

Monandrous, sexually cryptic females of non-human primates (Box 6.8) have most often been observed being unfaithful to their partner when they encounter unfamiliar, or at

Box 6.8 Some taxa of anthropoid primates with sexual crypsis

Taxa		Mating system
New World monkeys		
Owl monkey	*Aotus*	M
Titi monkey	*Callicebus*	M
Gould's marmoset	*Callimico*	M
Tamarin	*Saquinus*	M
Lion tamarin	*Leontopithecus*	M
Common marmoset	*Callithrix jacchus*	M
Pygmy marmoset	*Cebuella*	M
Sakis	*Pithecia*	M
Uakari	*Cacajao*	MM
Black howler	*Alouatta caraya*	U/MM
Woolly monkey	*Lagothrix*	MM
Spider monkey	*Ateles*	MM
Woolly spider	*Brachyteles*	MM
Old World monkeys		
Diana monkey	*Cercopithecus diana*	U
Vervet monkey	*C. aethiops*	MM
L'Hoest's monkey	*C. lhoesti*	U
De Brazza's monkey	*C. neglectus*	M
Hamlyn's monkey	*C. hamlyni*	U
Moustached monkey	*C. cephus*	U
Mona monkey	*C. mona*	U
Blue monkey	*C. mitis*	U
Toque monkey	*Macaca sinica*	MM
Assamese macaque	*M. assamensis*	MM
Silvered leaf	*Presbytis cristata*	U
Gray langur	*P. entellus*	U/MM
Guereza	*Colobus quereza*	U
Apes and humans		
Orangutan	*Pongo pygmaeus*	U
Human	*Homo sapiens*	M/U

Mating system: M = monogamous; U = unimale; MM = multimale

(Table constructed from figures in Sillén-Tullberg and Møller, 1993.)

least relatively unfamiliar males (Hrdy and Whitten, 1987). Females not previously receptive may become so within days of new males contacting their troops. In an extreme case, a female red howler monkey, *Alouatta seniculus*, was receptive whenever she encountered males from nearby troops (Sekulic, 1982). Female infidelity with extra-group males has also been observed in patas monkeys, *Erythrocebus patas*, redtails, *Cercopithecus ascanius*, and blue monkeys, *C. mitis* (Cords, 1987).

There are reports, also, of the deceptive nature of infidelity by sexually cryptic primate females. For example, from time to time, male grey langurs, *Presbytis entellus*, have to fight off rivals from nomadic all-male coalitions who attempt to oust them from their harem (Hrdy, 1977). The harem females often take advantage of their male's distraction during such fights by being unfaithful, either with males from the coalition or with other extra-group males (Mohnot, 1984).

According to Sillén-Tullberg and Møller (1993), sexual crypsis has evolved 8–11 times in the anthropoid primates. In the human lineage, however, sexual crypsis probably did not evolve until after the lineage's separation from the other great apes. Sexual crypsis, and the facade of monandry that it facilitated, are thus both part of our hominid inheritance. As such, these facets of human sexuality may not have appeared until c. 5 million years ago.

6.6 How cryptic is female fertility in humans?

The success of sexual crypsis as a female strategy to separate paternity from paternal care depends on the female being able to hide times of peak fertility from males, particularly her primary partner. Benshoof and Thornhill (1979) go further and suggest that the success of the strategy will also depend on the female concealing these times from her own conscious detection. Their grounds are that some degree of self-deception would facilitate the deceit of males. Women are notoriously inaccurate at recalling menstrual and other sexual events, even from their immediate past (Bean *et al.*, 1979).

6.6.1 Ancient beliefs, modern knowledge

In the discussions that follow, it should be remembered that any conscious mechanisms by which the male or female may attempt to identify ovulation and mid-cycle are very much a modern phenomenon. Any such attempt is based on the knowledge that: (1) conception involves the meeting of a sperm and egg; (2) females produce an egg at roughly mid-cycle; (3) sperm and egg live for relatively short periods; and (4) elements in the female menstrual cycle are associated with fertility. Such knowledge is very recent.

Until the 1920s, humans had a wide range of beliefs concerning fertility and conception, none of which were accurate. This fact by itself must bear some testimony to the effectiveness of the sexual crypsis of human females.

According to Dunham *et al.* (1991), many societies could see no link between copulation and conception. Some thought that babies entered the mother from the environment, usually water (e.g. Aborigines, Australia; Tapirape Indians, Brazil; various tribes, South Africa); others thought that babies were formed entirely from menstrual blood accumulating in the female (e.g. Ashanti, West Africa).

Some societies considered that the male primed the female's body for conception but then played no further part (e.g. Trobriand Islanders, South Pacific) (Dunham *et al.*, 1991). Sometimes, the priming was considered to occur just once, far in advance of conception (e.g. at the girl's first ever copulation; Imerina and Betsileo, Madagascar).

In contrast, yet other societies considered that only the man was involved (Dunham *et al.*, 1991). They thought that whole babies originated in the male (e.g. such as in his brain, from where, after 40 days' incubation, they passed to his penis; the Malay) and were then 'seeded' into the female in the seminal fluid. The seeding hypothesis, supported by Aristotle, received a boost about 300 years ago when sperm were first seen down a microscope by Van Leeuwenhoek. Scientists immediately thought they could see tiny whole humans in human sperm, donkeys in donkey sperm, etc. The entities were hence named 'seed animals' (spermatozoa).

Hippocrates may have been the first European to suggest that the female also played a role in conception (Dunham *et al.*, 1991). He suggested that the menstrual blood, once it stopped flowing to the outside, accumulated in the mother to form the baby's flesh, the seminal fluid then forming the brains and bones. However, few preferred Hippocrates' suggestion to the seeding hypothesis.

It was not until the late 1800s, when sperm were first seen fertilizing eggs (in starfish and sea urchins), that a role for the female in conception was acknowledged. Finally, it was not until the 1920s that it was accepted that mid-cycle, not the end of menstruation, was the most fertile time of the menstrual cycle (Johnson and Everitt, 1988).

Clearly, therefore, any threat to sexual crypsis due to conscious attempts by male or female to detect mid-cycle events is very recent. Throughout most of the hominid phase of human evolution, any threat to sexual crypsis could come only from cycles of detectable signals. These would need to emanate from the female and to trigger subconscious, cyclical responses by the male at the appropriate time.

6.6.2 Menstruation and sexual crypsis

Some facets of the physiological changes that underlie the human menstrual cycle are difficult to hide. Menstruation itself is one example for, in many primate species, including humans, the blood and other products flow out of the vagina to the outside where they may be detected both by sight and smell (Table 30–1 in Hrdy and Witten, 1987).

We often encounter the view that menstruation is a rare, almost pathological, event for wild mammals, including non-industrial humans. The suggestion is that the 'natural' state for a female is either to be pregnant or lactating. However, with: (a) conception occurring in at most one in three cycles (even when a fully fed, healthy, unstressed woman copulates on the optimum day of the cycle); and (b) the relatively high incidence of anovulatory cycles (particularly when stressed, at certain seasons, and/or during lactation); even the most fecund woman probably averages at least five cycles per conception. Field observations of monogamous white-handed gibbons *Hylobates lar* also report fertile females passing through several cycles despite frequent in-pair and occasional extra-pair copulations (U. Reichard, personal communication). Cycles are, therefore, an important part of sexuality and, inevitably, are most likely to occur in the months leading up to conception.

It is probably to the mutual benefit of both male and female to copulate less during menstruation: females (and males) may be at greatest risk of contracting disease; males have least to gain due to reduced female fertility. It has been argued (Knight, 1991) that such conspicuous menstrual bleeding, inevitably detected by the female and fairly obvious to a male partner, therefore evolved to advertise a phase of reduced mutual benefit and thus to reduce the incidence of copulation at this time.

Many primates will copulate during menstruation but even those that do copulate show a considerable reduction in frequency at this time (e.g. rhesus macques, *Macaca mulatta*, Rowell, 1972). Certainly, all studies of industrial societies (e.g. USA; Udry and Morris, 1977), including our own study (UK; Bellis and Baker, 1990) have shown a reduced incidence of copulation by humans during menstruation (section 6.6.8). In some other human societies, this reduced interest has generated, and is now reinforced by, social taboos (Knight, 1991).

Despite the reduced sexual interest shown by primates during menstruation, there is no clear evidence that visible menstrual bleeding has evolved as a specific signal. Even primates that have no visible bleeding show a reduction in copulation rate during shedding of the uterine lining (e.g. marmosets, *Callithrix jacchus*).

Whether menstruation evolved to inhibit copulation or not, it could clearly mark, both for the male and the female, a period of reduced fertility. It does little, however, to signpost periods of maximum fertility, either in the future or in the past.

Even when the ensuing cycle is ovulatory and fertile (section 4.4), the time of maximum fertility bears little temporal relationship to the previous menstruation. This is because the time interval from menstruation to the surge of oestrogen (oestradiol, E_2) that heralds ovulation is programmed in humans to be a highly variable 5–28 days (Box 4.6). However, a variable follicular phase is not confined to sexually cryptic species, being found also in baboons, chimpanzees and rhesus macaques (Rowell, 1972), all three of which have visible menstruation (Hrdy and Whitten, 1987) and sexual swellings (Box 6.7). Associated with this variability in such species is considerable variability in the length of maximum tumescence. In the pygmy chimpanzee, *Pan paniscus*, for example, some individuals may reach maximum tumescence up to four times in a single follicular phase, with slight detumescence in between, before eventual ovulation and real detumescence (Dahl, 1986).

It seems unlikely, therefore, that a variable follicular phase evolved specifically as a part of sexual crypsis. Perhaps the most reasonable explanation for the variability of the follicular phase is that it is a period during which the female is primed and 'waiting' for ovulation (Jochle, 1973, 1975). Jochle's suggestion is that follicular development occurs after menstruation up to the point at which an oestrogen surge would trigger ovulation in about 2–3 days. The female then 'waits' for events to unfold in her environment, hence the variability in length. Depending on events, she may then either: (a) abort ovulation to produce an anovulatory cycle; (b) ovulate to enhance the chances of conception (e.g. an oestrogen surge in response to copulation); or, we suggest, (c) ovulate to decrease the chances of conception (e.g. an oestrogen surge in the absence of a recent copulation).

Such a waiting strategy, which inevitably promotes variability in cycle length, would be advantageous both to species with and without sexual swellings. In all probability, therefore, the evolution of sexual crypsis began with females already having the advantage of a variable follicular phase.

In contrast to the variability of the follicular phase, once the female human experiences an oestrogen surge, the sequence of events seems to be much more tightly programmed (Collins *et al.*, 1981; Garcia *et al.*, 1981; Vermesh *et al.*, 1987). A day after the oestrogen surge, there is a surge of urinary LH (luteinizing hormone). A day (20–44 h; mean = 30 h) after that ovulation occurs, and 13 (12–15) days after that the next menstruation begins.

Peak fertility more or less coincides with the oestrogen peak and all fertility effectively ends a few hours after ovulation (Barrett and Marshall, 1969). As such, therefore, menstruation is a more accurate but purely retrospective indicator of the time (i.e. 14–18 days previously) of the female's last fertile period. However, what the information really signifies is the lack of conception during the preceding cycle (but see section 6.6.7) and the start of a whole new cycle of a male's trying to gain paternity.

The lack of value to the male of retrospective assessment from menstruation means that a relatively invariable luteal phase is unlikely to prejudice sexual crypsis. Such an invariable phase is also found in baboons and chimpanzees, both of which have conspicuous sexual swellings (Rowell, 1972). In contrast, the rhesus macaque, *Macaca mulatta*, which has slight sexual swellings (Box 6.7), has a luteal phase that is reported to be just as variable in length as the follicular phase (Rossman and Bartelmez, 1946).

The conspicuous nature of menstruation in humans and many other primates, therefore, does not in principle seem to be a threat to female crypsis, given an ancestral female with an already variable follicular phase. Phylogenetic analysis seems to confirm this conclusion.

Employing the same technique for phylogenetic analysis as Sillén-Tullberg and Møller (1993), and using their taxonomic, mating system and mating sign classifications, we conclude that the ancestral condition for the anthropoid primates was an absence of visible menstruation. We further estimate that visible menstruation has since evolved independently most often in lineages with visible signs of ovulation (3–5 times under conspicuous signs; 5–8 times under slight signs, and 0–3 times under sexual crypsis). Similarly, we estimate that it has evolved most often under multimale mating systems (6–10, multimale; 0–5 unimale; 0–2, monogamy).

In contrast, a reduction in signs of ovulation and monogamy have evolved more or less irrespective of the conspicuousness of menstruation. (Reduced signs of ovulation: 2–4 times under visible menstruation; 3–5, non-visible menstruation; monogamy: 1–3, visible; 3–5, non-visible.)

It seems, therefore, that both a multimale mating system and visible signs of ovulation facilitate the evolution of visible menstruation but that, having evolved, a visible menstruation does not hinder the evolution of sexual crypsis and monogamy. Although the sequence in the human lineage is somewhat equivocal, it seems most likely that visible menstruation evolved first followed some time later by sexual crypsis and then superficial monandry.

6.6.3 Cervical mucus and sexual crypsis

Another characteristic of the menstrual cycle that could threaten crypsis is the cyclical change in the flow rate and chemical nature of the cervical mucus (Boxes 3.5 and 4.6). During the most fertile phase, the mucus is produced in greater quantity, has lowered viscosity, is more elastic (= Spinnerbarkeit), and 'ferns' more as it dries (Hilgers *et al.*, 1978; Blackwell, 1984).

These changes might seem to be a particular threat to the female strategy of crypsis because, as adaptations to sperm transport, they show their greatest and most conspicuous changes at precisely that time of peak fertility it is most important for the female to hide. However, there can be as much as a 9-day variation in the temporal relationship between peak mucus symptoms and ovulation, peak symptoms occurring up to a day after ovulation (Box 6.9). On any one occasion, therefore, mucus characteristics do not provide an unequivocal indication of the time of maximum fertility, even if detected.

Females can detect, at least with training, the physical changes in their cervical mucus and such detection is one mechanism used by women consciously attempting to conceive or to avoid conception (Hilger *et al.*, 1978). In addition to these physical changes, the chemical changes in the female's mucus impose changes in the intensity and 'pleasantness' of its smell.

Doty *et al.* (1975) asked both male and female subjects to rank the intensity and pleasantness of the smell of vaginal secretions collected overnight on sterile tampons worn by four female volunteers. Over 23 menstrual cycles, the most intense and most unpleasant secretions were those collected during menstruation. The least intense and least unpleasant odours were those from samples collected during the fertile phase.

Some elements of male courtship behaviour (e.g. nuzzling and licking the vulva, inserting fingers into the vagina) are common to most primates, including humans (Chapter 5). They may well have evolved as a mechanism to aid assessment of the stage of the female's menstrual cycle via the nature of her cervical/vaginal secretions. However, another and perhaps more potent function may be to assess whether, and how recently, the female has been inseminated by a previous male (section 3.3.2).

To collect such intimate information concerning the female's cervical mucus, the male must already be in a near-copulatory situation with the female. Such information does not, therefore, help a male to decide in advance which females and when are the most profitable to inseminate. For this, some form of more distant (say at least a metre) pheromone is necessary.

6.6.4 Pheromones and sexual crypsis

There has been much debate over whether humans and other anthropoid primates have distant pheromones that influence sexual behaviour (see reviews by Comfort, 1971; Doty, 1976; Rogel, 1978; Stoddart, 1990). It has been claimed, for example, that female rhesus macaques produce an oestrogen-stimulated vaginal secretion that males can detect by smell and that raises the male's sexual interest in the female at oestrus (Michael and Keverne, 1970; Keverne 1976).

Dogs can discriminate between the sweat of women at different stages of their menstrual cycle (Kloek, 1961) and claims have been made that humans can consciously detect the odours of androgens, oestrogens and progestogens. Moreover, it has been claimed that males and females, and females at different stages of their cycle, have different sensitivities to these substances (references in Rogel, 1978). Past reviewers, however (e.g. Doty, 1976; Rogel, 1978) have consistently considered the evidence to be equivocal, even for rhesus macaques.

More recently, reviewing the same evidence, both Hrdy and Whitten (1987) and Stoddart (1990) seem prepared to accept the evidence that the hormonal status of female rhesus macaques may be transmitted to males olfactorily. Stoddart (1990), however, concludes that there is no unequivocal evidence of anything similar occurring in humans.

Box 6.9 The relationship between cervical mucus symptoms and ovulation

Ferning refers to the pattern made by cervical mucus as it dries which is different before ovulation (top pattern) from other stages of the menstrual cycle (bottom pattern). Spinnbarkeit refers to the elasticity of the mucus which, at its peak, can be stretched more than 10 cm before it breaks.

Data show the days on which peak cervical mucus symptoms were observed by 32 'normal' women over 71 cycles in relation to the observed peak of luteinizing hormone in blood serum. On average, this peak occurred 10 ± 5 h before ovulation, the latter being determined by observing the follicle or corpus luteum from the ovary by laparoscopy.

Ferning patterns re-drawn and modified from Speroff *et al.*(1989). Graph drawn from data in Garcia *et al.* (1981).

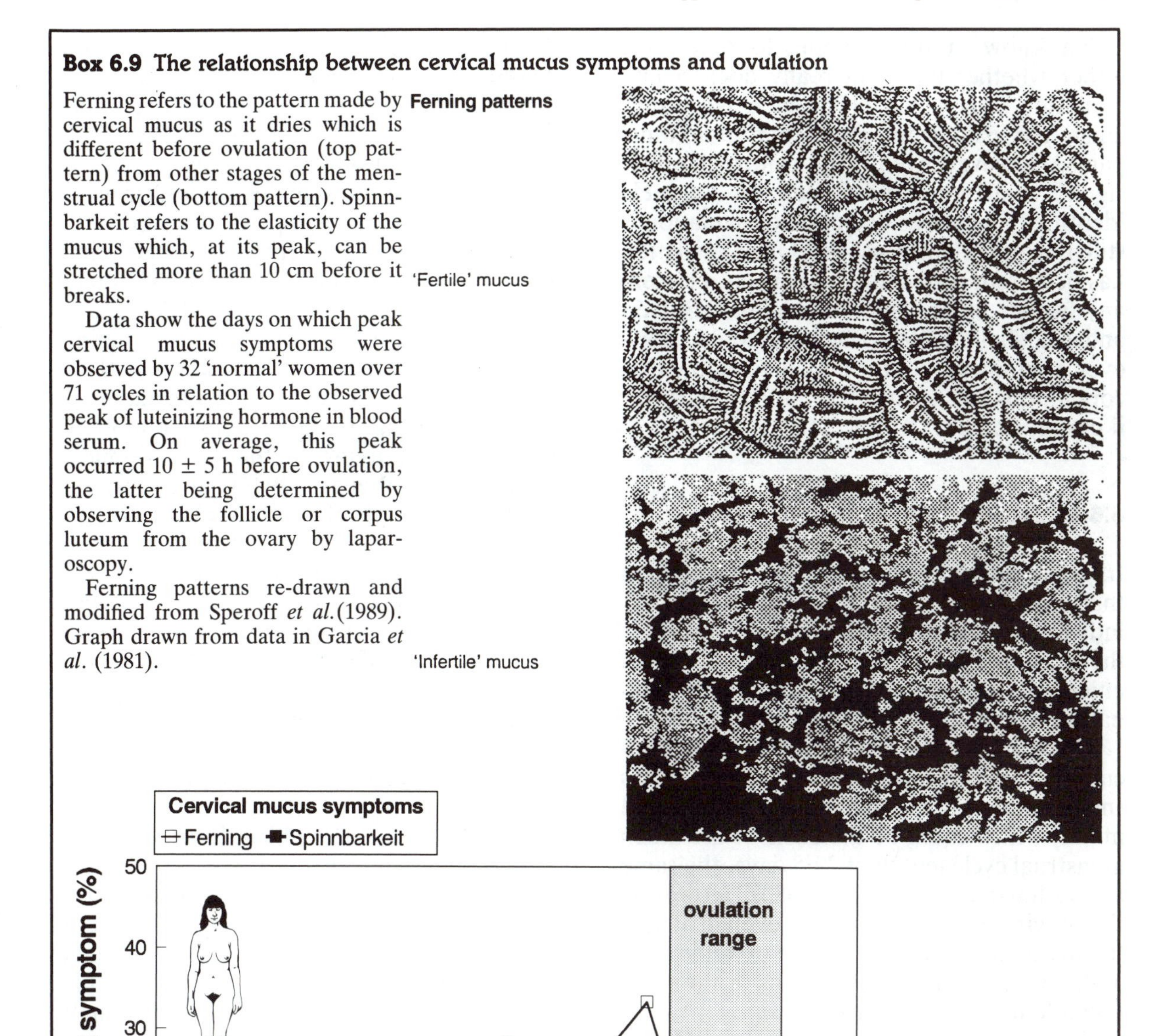

6.6.5 The pain of ovulation?

Some women claim to be able consciously to recognize that ovulation has occurred as a result of mittelschmertz, a pain in the lower abdomen occurring midway through the female's intermenstrual interval. This pain may be caused by irritation of the pelvic peritoneum by blood or other fluid escaping from the ovary at ovulation (Osol, 1972).

We know of no attempts to determine either whether this pain really does signify ovulation or how reliably women can distinguish the pain from any other abdominal pains. However, even if the pain really does signify ovulation, and it can be uniquely identified, conscious detection of ovulation is still not particularly useful. Once ovulation has occurred, probability of conception from any further insemination is minimal (Barrett and Marshall, 1969). The pain could, however, usefully trigger the female to embark on copulations that are most advantageous to her if conception does not occur.

6.6.6 Menstrual synchrony: no threat

In apparent contradiction of the conclusion that menstrual cycles are cryptic to other individuals, and even to the female herself, is the existence of one of the currently most enigmatic features of human female sexuality: menstrual synchrony.

There is some indication of a weak association between menstrual and lunar cycles (menstruation being more likely around full moon) which could be the basis of the mean menstrual cycle length of 29.5 days, the same as the lunar cycle (Cutler, 1980). However, this environmental influence, even if it is a factor, cannot expain either the extreme variability in menstrual cycle lengths or the main social features of synchrony.

Over the decades, nuns, female students, women prisoners, sisters, and mothers and daughters have often remarked on the unusual menstrual synchrony that seemed to develop within their communities. Menstrual synchrony has also been documented anthropologically (e.g. Yurok Amerindians; Buckley, 1982). In 1971, scientific investigation seemed to show that the phenomenon of menstrual synchrony was real and not simply due to chance or to a communal response to the wider environment (McClintock, 1971). Since then, there have been a number of investigations the main conclusions of which have been briefly as follows.

A move towards synchrony is very rapid, being greatest during the first cycle and more or less complete after 3–4 cycles (McClintock, 1971). A common environment, such as a shared house or room, may promote synchrony (McClintock, 1971; Quadagno *et al.*, 1981; Jarett, 1984; Weller and Weller, 1993) but often is not by itself enough. Pairs of women who are room-mates are less prone to synchronous cycles than pairs of women who spend a large proportion of their time together outside of their living place (Graham and McGrew, 1980; Goldman and Schneider, 1987). Synchrony occurs both in a mainly female (McClintock, 1971) and a mixed-sex (Graham and McGrew, 1980; Quadagno *et al.*, 1981) setting. It occurs between mothers and daughters living in the same domicile (Weller and Weller, 1993) and between lesbian couples (Weller and Weller, 1992).

Secretions have been collected on small absorbent pads placed under the armpits of women with regular periods, extracted with ethanol, and then placed three times a week under the noses of other women. Such experiments seemed to show that the recipients synchronized their menstruations with the donors, indicating the action of a pheromone (Russell *et al.*, 1980; Preti *et al.*, 1986). However, these experiments have been strongly criticized (that of Russell *et al.* for not being double-blind; that of Preti *et al.*, by Wilson (1987), for errors in tabulation and analysis).

Similar experiments using chemicals collected from the armpits of males also seemed to show an influence on menstruation, primarily reducing the variability in cycle length of the recipient females (Cutler *et al.*, 1986). These results complement the observation that women with high involvement with males have shorter and less variable menstrual cycle lengths (McClintock, 1971; Goldman and Schneider, 1987) and are more likely to ovulate (Veith *et al.*, 1983). The critical factor was the physical presence of a man (Cutler *et al.*, 1986), not copulation or genital stimulation (Cutler *et al.*, 1985). The implication, therefore, is that when close contact with males is reduced, females have longer, more variable cycles, presumably because of a greater incidence of anovulatory cycles. Contact with males, however, does not seem directly to influence menstrual synchrony between females (Goldman and Schneider, 1987).

Evolutionary interpretations of human menstrual synchrony range from the whimsical to the mathematical. Knight (1991) suggests that, combined with a taboo on copulation during menstruation, hunter–gatherer women used to go on a communal monthly strike, forcing the men to go hunting with the offer of sex on their return. The proposed association between menstruation and the full moon (Cutler, 1980) is seen as being consistent with this story. In contrast, Knowlton (1979) models the possibility that menstrual, and hence ovulation, synchrony decreases the likelihood of a male being unfaithful to his partner and thus increases his level of paternal investment in her offspring. Finally, Kiltie (1982) suggests that menstrual synchrony in modern women is either non-adaptive or a vestige from some adaptive past.

Interpretations of menstrual synchrony invariably make two mistakes. First, they overestimate the level of synchronization which is never so precise in any community as significantly to influence, say, the opportunities for male infidelity. Second, they assume that menstrual synchrony reflects ovulation synchrony. Yet it is precisely those circumstances (communal female living) that promote synchrony (McClintock, 1971) that also promote the highest incidence of anovulatory cycles (Metcalf and Mackenzie, 1980).

Wilson (1992) has pointed out that McClintock's protocol for investigating menstrual synchrony includes three errors which bias towards the discovery of synchrony. He argues that only studies which perpetuated those errors have been able to demonstrate synchrony and suggests that synchrony is not therefore a real phenomenon. Studies which do not perpetuate these errors find no evidence of synchronization (Wilson *et al.*, 1991).

Our own pilot study of synchrony in Manchester (Bellis and Baker, unpublished), which investigated 130 females in 30 living groups, did not include these errors and failed to show synchrony. Nearly as many groups showed a tendency for desynchrony (menstrual dates became further separated) as showed a tendency for synchrony. However, the distribution of synchronous and desynchronous groups was not random. Synchrony was most likely among groups of women with the lowest incidence of ovulation. Moreover, our results suggested that when a new group of women started to live together, a high level of anovulation promotes the development of synchrony, not vice versa. Conversely, females who are ovulating actually desynchronize with time, eventually menstruating further apart than would be expected by chance.

If the replications that are in progress confirm these findings, they will suggest that, whatever the advantage of synchronization, it is not a function of synchronized ovulation. It is instead a response of anovulatory women, ovulatory women responding to communal living by avoiding synchrony.

Ovarian cycle synchrony is also found in other mammals, but its characteristics vary between species. In some species, the synchrony is clearly seasonal and presumably based on environmental influences (e.g. squirrel monkeys, *Saimiri* spp.; Baldwin, 1970). In others, it is based on social interactions. Apparent synchrony is reported in harems of hamadryas baboons, *Papio hamadryas* (Kummer, 1986) and gelada baboons, *Theropithecus gelada* (Dunbar, 1980). Synchrony seems to occur within but not between harems in a troop, implying some form of social rather than environmental influence. Similar conclusions have been reached for golden lion tamarins *Leontopithecus rosalia* and chimpanzees, *Pan troglodytes* Wallis, 1985).

Female rats, which do not menstruate, show synchrony of ovulation after being housed together for 3–4 cycles. By passing air from one cage to another, it has been shown that airborne chemicals (pheromones) from one female can extend or shorten the cycle of another. Preovulatory rats can speed up (and postovulatory rats can slow down) ovulation in another group. Similarly, preovulatory cervical mucus mixed with water and sprayed into the noses of cows advances the time of oestrous.

When groups of female mice live together their oestrous cycles are mutually disrupted, the females showing an increase in pseudopregnancy (Lee and Boot, 1955) and anoestrus (Whitten, 1959). In golden

Box 6.10 The timing of in-pair (IPCs) and extra-pair copulations (EPCs) in relation to fertility during different phases of the human menstrual cycle

When females take oral contraceptives there is, of course, no peak of fertility during their 'menstrual' cycle. They copulate more with their partners (IPCs) later in their cycle and EPCs are not significantly more frequent before day 16 than after ($P = 0.233$).

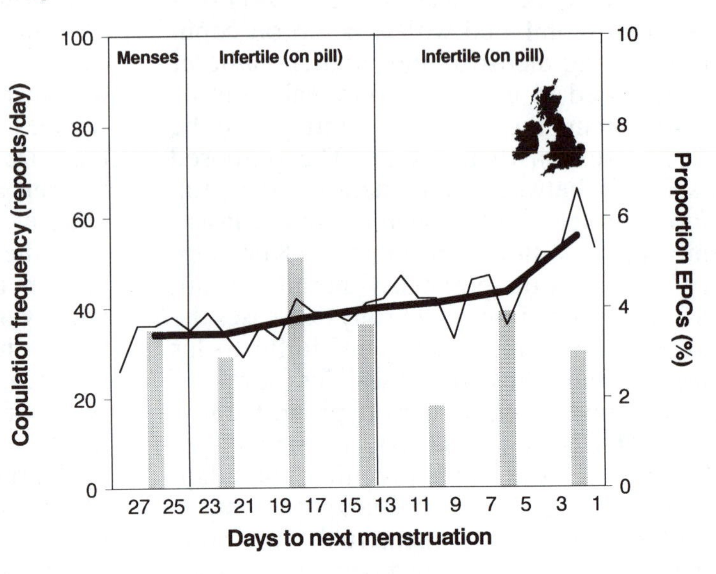

When females do not take oral contraceptives, probability of conception shows a clear peak from about day 6 of the human menstrual cycle until about the time of ovulation (vertical line at about day 16). Peak fertility occurs before, rather than at ovulation.

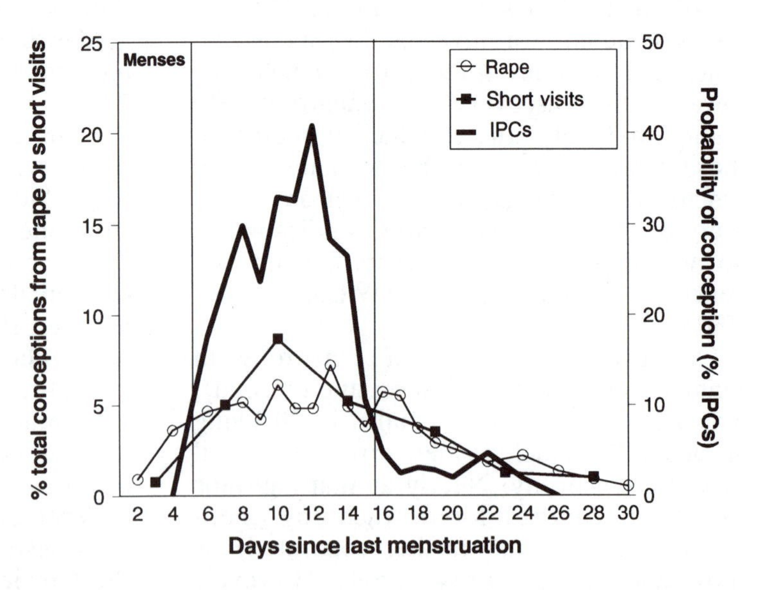

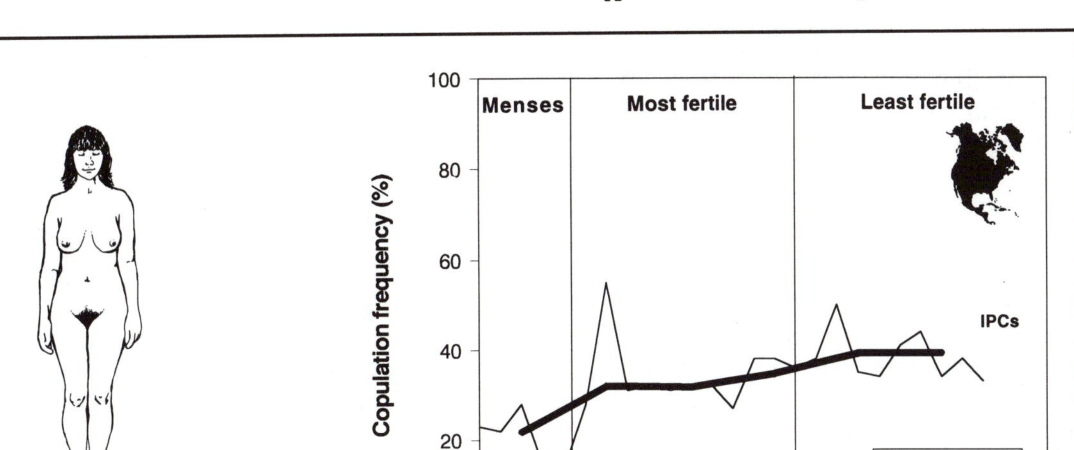

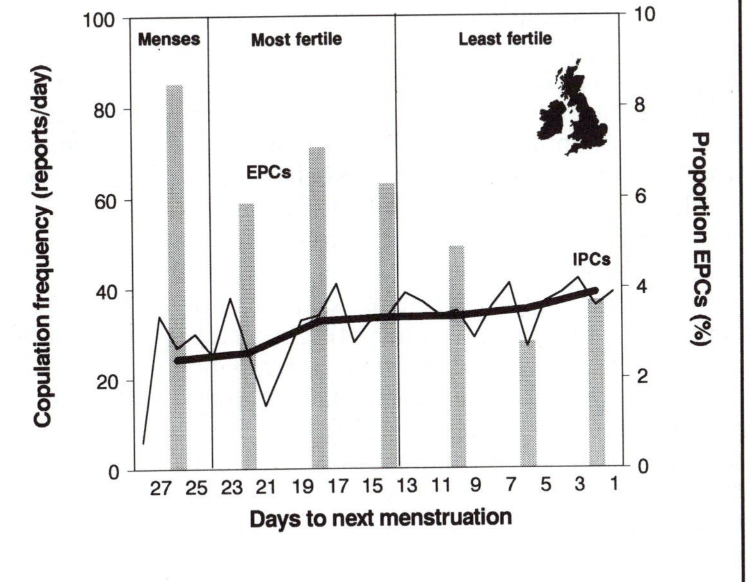

Females not taking oral contraceptives copulate more with their partners (IPCs) during the infertile luteal phase but more with extra-pair males during the fertile pre-ovulatory phase. There is a significant positive association between EPC incidence and probability of conception ($P = 0.018$).

Conception data from Jochle (1973; Table 2). USA copulation data from Udry and Morris (1977); UK data expanded from Bellis and Baker (1990).

hamsters, females housed together continue to cycle but lose their receptivity to males (Lisk *et al.*, 1983). In mice the disruption is in part mediated by a non-volatile 'contact' pheromone, probably present in the excreta (Archunan and Dominic, 1991). Release from the female pheromone (Marsden and Bronson, 1965) or exposure to a male (Whitten, 1959) or male urine (Marsden and Bronson, 1965) reinstates receptivity and fertility (Whitten, 1959). When release is synchronized, so too is the next, but not subsequent, cycles (Whitten, 1959).

Thus, in humans and other mammals, females who live in groups with reduced male contact respond by an increase in anovulation. Renewed contact with males, mediated by male pheromone, promotes ovulation. In the continuing absence of males, cycle synchronization occurs. In animals with an oestrous phase, this synchronization is of ovulation. In at least humans among sexually cryptic animals, this synchronization is of menstruation rather than ovulation. As yet, therefore, there is no indication that either menstrual synchrony or any associated pheromone are a threat to sexual crypsis.

6.6.7 Copulation during pregnancy: the final touch

Women often continue externally to show a menstrual cycle for the first 2–3 months of pregnancy and female sexual libido does not decrease and may even increase during the first few months (Masters and Johnson, 1966). !Kung women do not reveal or acknowledge their pregnancy until it begins to show unambiguously after the third or fourth months (Howell, 1979).

Many other female primates continue to copulate well into pregnancy (e.g. vervet monkey, *Cercopithecus aethiops*; chimpanzee, *Pan troglodytes*; see Table 30.1 in Hrdy and Whitten, 1987). Hrdy (1979) interprets the behaviour as yet another way that the female can recruit paternal care for (and/or reduce antagonism towards) her future offspring from a wider range of males.

Hrdy's suggestion does not explain why a monandrous female should continue to mate with her long-term partner. We suggest that hiding the onset of pregnancy is the cryptic female's final touch in rendering it extremely difficult for even a very attentive male partner to make any retrospective assessment of paternity. Polyandrous females in multimale groups may similarly confuse a previous consort, just as if they were monandrous.

Phylogenetic analysis (as in section 6.6.2) suggests that the ancestral anthropoid primate occasionally copulated during pregnancy and that since then in some lineages the behaviour has increased in importance and in some it has decreased. There is no clear link with mating system, copulation during pregnancy having both increased and decreased in the more polyandrous situations. On the other hand, the behaviour does seem to have increased in importance more often in lineages that show female sexual crypsis (increased: twice under crypsis; 2–3 times under slight signs of ovulation; 0–1 under swellings; decreased: 0, crypsis; 1, slight signs; 1, swellings).

In contrast, lineages with frequent copulations during pregnancy do not seem then to be more likely to decrease visual signs of ovulation though there is an indication that they are more likely to evolve monogamy (three times under frequent copulation during pregnancy, once under occasional copulation, and never in lineages in which copulation during pregnancy is rare).

The implication, therefore, is that the loss of visible signs of ovulation is followed by an increase in copulation during pregnancy which in turn facilitates the evolution of superficial monandry. Such a sequence could well have occurred in the human lineage.

6.6.8 How successful is sexual crypsis by human females?

The overall impression is that peak fertility in the human menstrual cycle really is cryptic to the male and is probably also largely cryptic to the female, at least consciously. However, whatever one concludes from discussions of potential threats to female crypsis (sections 6.6.1–6.6.7), the final test of crypsis is whether males nevertheless still manage to inseminate the female more often during her fertile than infertile phases.

Various authors (e.g. McCance *et al.*, 1937; Udry and Morris, 1968; James, 1971) have claimed that human in-pair copulations peak during the fertile phase and are depressed during the infertile luteal phase. These claims were refuted by Bancroft (1987) who argued instead that peak sexual interest occurred when the chances of conception were low in the week before or the week after menstruation.

The peaks and trends to which the earliest authors refer are unimpressive to the point of being invisible (Fig. 1 in Udry and Morris, 1977) and depend to a considerable extent on the way in which the menstrual cycle is expressed (e.g. standardized, forward, reversed). The most reliable study (Udry and Morris, 1977) monitored luteinizing hormone, the surge of which should precede ovulation by about 1 day. There is no indication of a peak in copulation during the fertile phase. Rather, their data show the same postovulatory increase that we found for in-pair copulations in the UK (Bellis and Baker, 1990) (Box 6.10). The pattern is the same as found by both Udry and Morris (1970) and ourselves (Box 6.10) for females taking oral contraceptives. The pattern is not, therefore, an artefact of couples consciously trying to avoid conception.

The overwhelming impression from our own data and the data in Udry and Morris (1977) is that, apart from the menstrual trough, in-pair copulations are erratically but relatively evenly spread through the menstrual cycle. Only statistical smoothing reveals that in-pair copulations tend to be most common during the luteal phase of reduced fertility. Males give no indication of being able to concentrate copulations during their partner's fertile phase. On the contrary, by some means they seem to be shunted into concentrating on the infertile phase.

Concern that the human pattern is a cultural artefact (Stoddart, 1990) can be countered by data from other primates with sexual crypsis. In free access tests in which the male tended (often forcefully) to initiate copulation, orangutans, *Pongo pygmaeus*, mated throughout the cycle (Nadler and Collins, 1991). Blue monkeys, *Cercopithecus mitis*, and vervet monkeys, *C. aethiops*, show no clear change (excluding a menstrual trough) in copulatory activity during the female's cycle and have, if anything, like humans, a greater frequency during the luteal phase (Rowell, 1972).

In contrast, species with conspicuous sexual swellings (e.g. anubis baboon, *Papio cynocephalus*) show a clear, and often abrupt, decrease in sexual activity once detumescence signals the beginning of the luteal phase (Rowell, 1972). Species with slight, but relatively inconspicuous sexual swellings, show a more varied pattern. Thus, gorillas, *Gorilla gorilla*, show a mid-cycle peak in free-access tests (Nadler and Collins, 1991). In contrast, rhesus macaques, *Macaca mulatta*, have no clear peak of sexual activity but, like humans, have a tendency to copulate more during the luteal (postfertile) phase (Rowell, 1972).

The circumstances of all of these data were such that females were essentially being monitored for either in-pair copulations or at least copulations with males with whom the female had lived for some time. When the circumstances change to those more nearly representing extra-pair copulations, the pattern changes. Thus restricted access tests have been carried out on orangutans and gorillas in which the female has the choice of whether or not to enter the cage of a male with whom she has not been housed (Nadler and Collins, 1991). Furthermore, the females were tested over a period with more than one male. Under such circumstances, females show a clear peak of copulation during the fertile phase of her cycle.

We have shown previously (Bellis and Baker, 1990) for humans that the incidence of extra-pair copulations during the menstrual cycle is significantly different from the incidence of in-pair copulations. Extra-pair copulations peak during the female's fertile phase (Box 6.10). This implies a cyclical change in female behaviour and preferences, rather than an ability of males to detect that the female is fertile (Bellis and Baker, 1990).

Measurements of the distances females walk on different days of the menstrual cycle using a pedometer (Morris and Udry, 1970) show a mid-cycle peak (Box 6.11). The

Box 6.11 Female activity in relation to fertility during the human menstrual cycle

All diagrams show data for women not on the pill and currently with a main male sexual partner. Two diagrams (b,d) also include data for women without a current partner. The USA data (a) measured activity by pedometer.

The UK study (b–d) was of 45 students who were asked to keep daily records of their activity over a 2-week period.

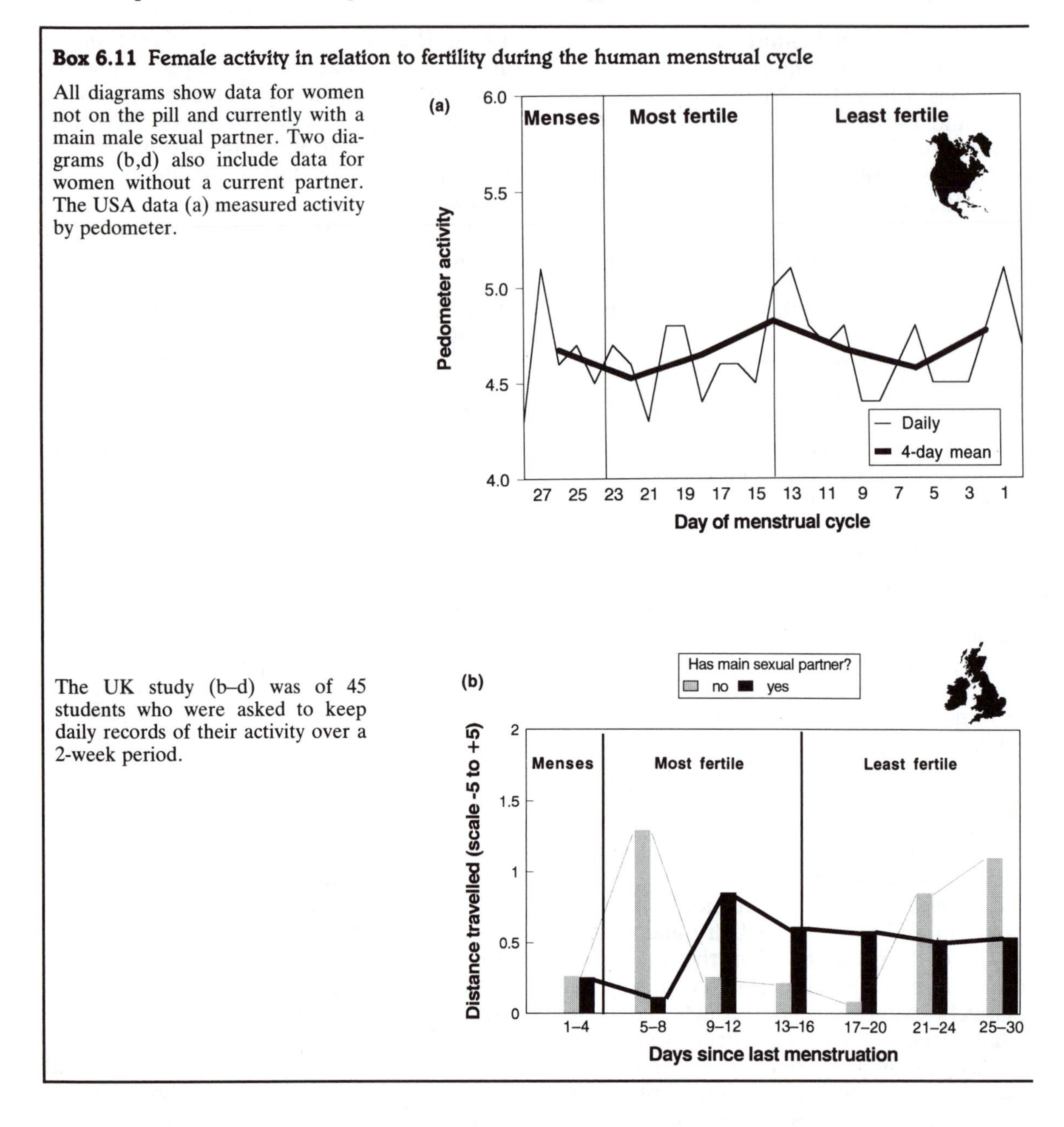

females in this study all had sexual partners and their pattern of activity was similar to the incidence of EPCs in our UK study (Box 6.10).

Our own, less direct study, of the distance females travel at different stages of the menstrual cycle, also shows a peak during the fertile phase for females with a main sexual partner (but not taking oral contraceptives) (Box 6.11). Such females also spend less time with their partner in mid-cycle and more time on their own. In contrast, females without a male partner show a reduction in distance travelled in mid-cycle and a decrease in time spent on their own (Box 6.11) The data are just as predicted if females without a partner try to avoid mid-cycle contacts with new males whereas those with a partner seek such contacts.

In a recent study of female visitors to an Austrian discotheque, there was a significant correlation between the oestradiol level in

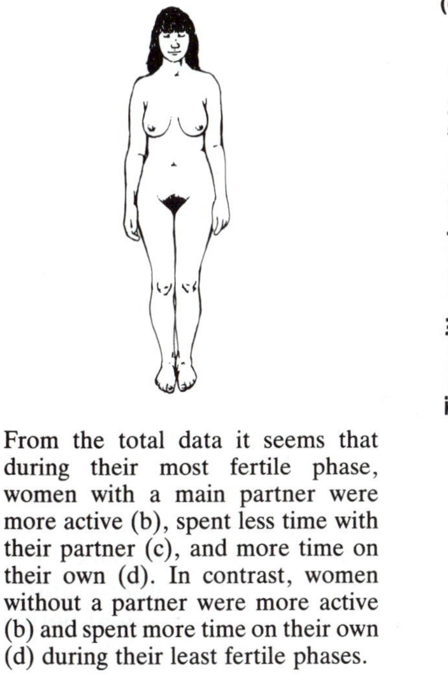

From the total data it seems that during their most fertile phase, women with a main partner were more active (b), spent less time with their partner (c), and more time on their own (d). In contrast, women without a partner were more active (b) and spent more time on their own (d) during their least fertile phases.

USA data from Morris and Udry (1970). UK data from Baker and Bellis (unpublished).

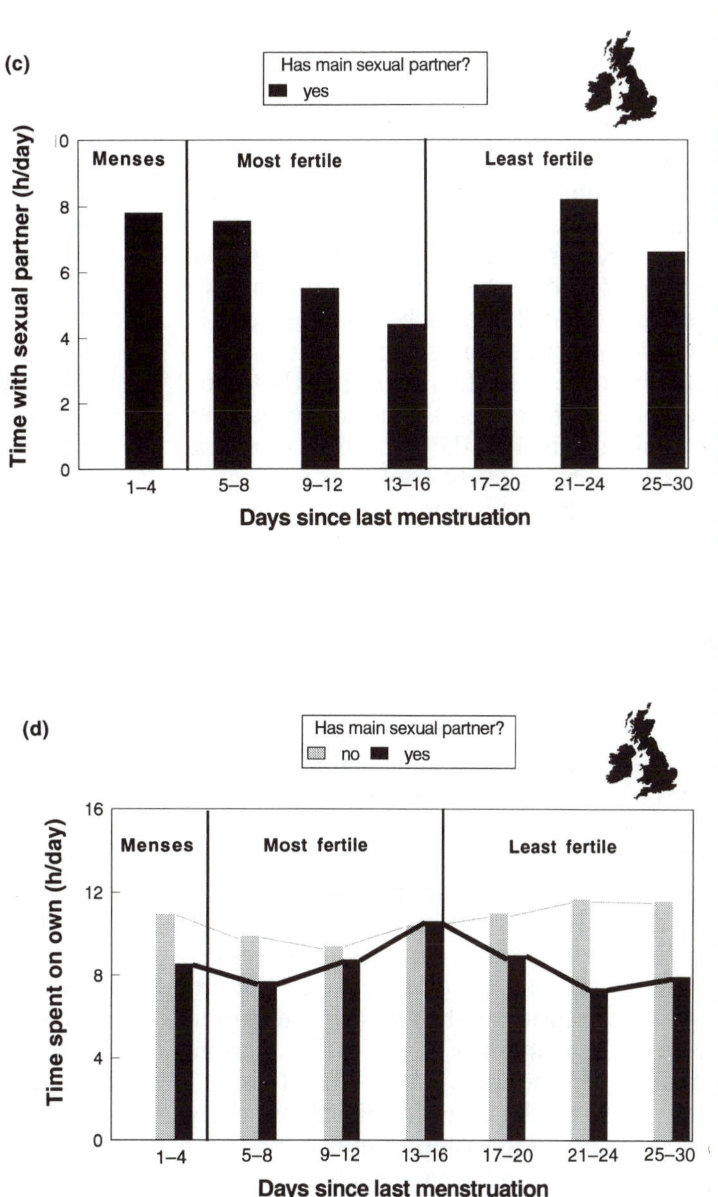

saliva (peak level c. 2 days before ovulation) and the form of the female's clothing for females not taking oral contraceptives (Grammer *et al.*, 1993). Females in their most fertile phase wore tighter clothing and exposed more bare skin than females in non-fertile phases. All of these females arrived unaccompanied by males. Unfortunately, no information is available on what proportion of these females had a current male partner.

Some cyclical variation in mood and behaviour is of course inevitable. Examples are: greater readiness to copulate and greater activity during the fertile phase (Box 6.11); perhaps an antisocial phase premenstrually (e.g. rhesus macaques, Rowell, 1972; humans, Dalton, 1964, 1966). However, such essential variations are relatively easy to mask simply by the female showing the same behaviour patterns at other times with relatively random mood and behaviour swings in between. Such a pattern is strikingly obvious in all published data (e.g. Morris and Udry, 1970; Udry and Morris, 1977).

The essential behavioural cycle underlying the menstrual cycle may be detected statistically (as in Boxes 6.10 and 6.11) but peaks and troughs are probably too erratic to be perceived unequivocally by the male. Scientifically, no unambiguous behavioural marker that may be perceived at the time of peak fertility (4–1 days before ovulation) has yet been identified. If no such behaviour can be identified by rigorous scientific study, it is unlikely to be identified by either the casual male observer or even the attentive partner. It would seem that the evolution of sexual crypsis has been fairly complete in humans and some other primates.

6.7 The hominid inheritance: permanently pendulous breasts

Human females are unique among the primates in that they develop pendulous breasts at puberty and retain them throughout life (Short, 1979). All other primates, including the great apes, develop pendulous breasts only at the onset of lactation and their breasts remain pendulous only until lactation ceases.

Human breast development at puberty involves expansion of the adipose and stromal tissue of the mammary gland, rather than the glandular epithelium itself. Breast size is determined by the quantity of adipose tissue and bears no relationship to milk production ability (Tortora and Anagnostakos, 1984).

The males of most societies (Ford and Beach, 1952), including western society (Smith, 1984) take a sexual interest in female breasts, though preferred shapes may differ from society to society (e.g. long and pendulous; small and upturned; Ford and Beach, 1952) and perhaps between individual men. Many authors have been fascinated by the permanently pendulous breasts of female humans and several unlikely evolutionary scenarios have been elaborated.

One of the most sociobiological discussions was by Smith (1984) who suggests that the permanently pendulous breasts of female humans evolved as part of the process of female sexual crypsis (section 6.6). A permanently pendulous condition would make it more difficult for a male to detect when a female ceased lactation and thus would contribute to the female's ability to separate paternal care from paternity. On Smith's interpretation, therefore, there is an evolutionary and functional link between human breast anatomy and sperm competition.

In fact, breasts do vary in size during the menstrual cycle (Milligan *et al.*, 1975) but maximum size occurs during the luteal (post-fertile) stage, just before menstruation. Changes in breast size, therefore, are no more of a threat to sexual crypsis than menstruation itself (section 6.6.2). Some selection on breast size and constancy during the evolution of sexual crypsis as suggested by Smith (1984) is thus a possibility. It is still necessary, however, to explain why only humans among sexually cryptic species have evolved permanently enlarged breasts which first develop at puberty, not at first pregnancy.

First, in a species such as humans in which the majority of, but not all, fertile females would be either pregnant or lactating (Rowell, 1972) and thus have enlarged breasts, males could well use such breasts as an indication of female fertility. Females who first began to develop breasts at puberty might then be at an advantage in attracting male interest.

The advantage could be especially marked after the beginning of the move towards monogamy. Under such circumstances, a male prepared to offer paternal care is a limiting factor (section 6.5). There is clear evidence for humans (Betzig, 1988) that female fertility is an important, albeit subconscious, factor in a male's choice and retention of a female partner. Any female who advertizes fertility, therefore, may well obtain a prospective partner faster than one who does not. Thus, there could well be a link in the human lineage between the evolution of sexual crypsis, the move towards monogamy, and the evolution of permanently enlarged breasts after puberty.

Still unexplained by this scenario is why other sexually cryptic monogamous primates did not also evolve permanently enlarged breasts. It is tempting to suggest that hominids were finally tipped toward such breasts by that other adaptation more or less unique to their lineage, bipedalism. At the very least, a vertical posture made breasts more conspicuous. It is possible, however, that there was a further factor.

From about 4 million years ago (Wood, 1993), hominids have been both incipiently bipedal and hunter–gatherers. Such a lifestyle involves both males and females, with their young, going on relatively long daily journeys, either together or apart. One of the major factors influencing the rate of production of offspring was the time for one offspring to be large enough to walk these distances without being carried, thus freeing the female to have her next offspring which could then be carried (Blurton Jones, 1989).

Human babies have relatively large heads and it is some time before their neck muscles are strong enough to support their weight. Even older infants, when asleep, have 'wobbly' heads. The typical carrying posture of children in naked humans is on the hip with the head of the sleeping child resting on the woman's breast. A swollen breast thus acts as a cushion to the sleeping baby's or infant's head. As infants need to be carried beyond the end of their mother's lactation, as females share carrying of each other's offspring on a reciprocal basis, and as preparous females take part in this reciprocal arrangement, selection may well have favoured swollen breasts as soon as females are large enough to play a major role in the carrying of young.

We suggest, therefore, that sperm competition, sexual crypsis, monogamy, bipedalism and a large head all interacted during the last 2–4 million years of human evolution to select for females with permanently enlarged breasts.

6.8 The hominid inheritance: penis size, shape and function

The evolution of internal fertilization, intromittent organs, and copulation was discussed in Chapter 3. The mammalian penis is a new organ, not derived from the intromittent organs of reptiles (section 3.3). As the organ that delivers sperm into the female, it would be suprising if it played no part in sperm competition.

6.8.1 Penis size and shape

The variation in shape and structure of the penis of primates, including humans, is described in Box 6.12. The human penis is unusual in four respects: (1) it is large, both in length and width; (2) it has no bone; the baculum (Box 6.12); (3) the prepuce or foreskin attaches to the shaft very close to the glans; and (4) in contrast to apes, in which the preputial opening is flush to the body wall, the shaft permanently protrudes (Short, 1979).

Most of the time, the human penis is flaccid and hangs downwards, resting against the testes. When erect and ready for copulation, the human penis is on average about 13 cm long (Short, 1979). About 80% of penises are between 11 and 15 cm long (Short, 1979) and the range, excluding extreme cases, is c. 8–24 cm; Hessell, 1992). In part, the variation is racial (Box 6.12).

The erect human penis is straight and piston shaped with a smooth terminal protuberance (= glans, Box 6.12). The rear part of the glans has a greater diameter than the penis shaft and has a back edge that is more-or-less vertical.

Many authors have been fascinated by the size of the human penis which, for a primate, is both relatively and absolutely large. It is, for example, nearly twice as long and over twice as wide as that of the chimpanzee. Short (1980, 1981) and Halliday (1980) have reviewed speculation on why the human penis has evolved to be so large. Suggestions have included aggressive display, attractive display, and facilitation of a variety of copulatory positions to enhance female sexual stimulation during coitus. Smith (1984) has argued convincingly that none of these suggestions seem realistic.

Aggressive display with the penis seems unlikely. Penis size is inherited independently of body size (Schonfield, 1943) and it seems unlikely that a large man with a small penis

Box 6.12 Penis size, shape and structure in primates

The penis of primates, including humans, and other mammals is composed chiefly of cavernous (erectile) tissue and is traversed by the urethra (Kinzey, 1974). The skin is thin and freely movable. The penis usually terminates in a usually swollen 'cap', the glans penis. Towards the base of the glans the skin forms a free fold, the prepuce or foreskin, which overlaps the glans to a variable extent. Humans are unusual in having the prepuce attached very close to the glans.

The primate penis, like that of several other orders of mammals, contains a bone (the os penis or baculum) (Kinzey, 1974). The bone is completely absent, however, from the tupaiids, *Tarsius*, the ateline monkeys, *Cebuella*, *Callimico*, and humans. In *Aotus* it is cartilaginous and in the other primates it varies considerably in length, from less than 3 mm in some gibbons to 55 mm in the stump-tailed macaque. In all primates except the lemurs, in which the baculum is forked, it is straight or slightly curved. Positioned within the glans, it parallels the long axis of the penis and is dorsal to the urethra.

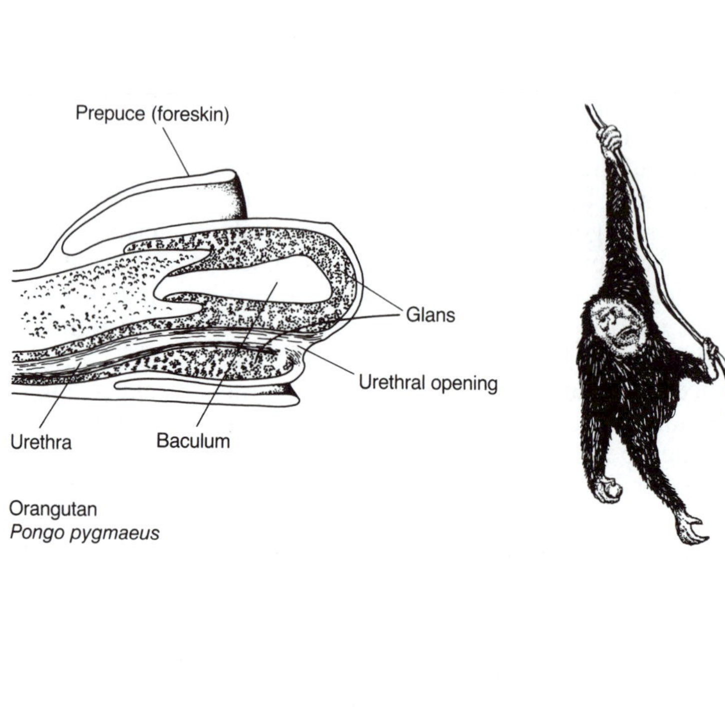

All of the penis covered by the prepuce (i.e. the glans in humans; the glans and various lengths of penis shaft in other primates) is covered by skin intermediate between ordinary epidermis and mucous membrane. This is the area of skin with most contact with the vagina walls when the male penetrates the female. In humans, the skin of the glans is exceptionally smooth; in some other primates the glans epithelium is specialized for local keratinization giving rise to horny papillae (e.g. langurs), spicules (e.g. chimpanzee), or even

would be physically intimidated by a small man simply because he has a large penis. There are no anthropological data to suggest the use of the penis in aggressive display or to suggest that males have erections during aggressive encounters (Smith, 1984).

Any argument based on female attraction or greater female stimulation seems equally unlikely. In surveys, at least, females are singularly indifferent to the penis, paying much more attention to the male's buttocks (Diagram Group, 1981) (perhaps for the same reasons as males pay attention to the female's waist:hip ratio; i.e. as an indication of hormonal status and fertility; Singh, 1993). Only 2% of women indicated an interest in the penis compared with 39% in the buttocks (Diagram Group, 1981).

This indifference of females towards at least the flaccid penis is reflected in the fate of US magazines featuring photographs of nude males. These have consistently failed to secure a substantial female readership compared with the male market for photographs of nude females which supports a dozen or so periodicals in the United States (Smith, 1984). Human females are singularly unimpressed and decidedly unaroused by 'flashers' (Smith, 1984). Finally, penis size has little to do with female stimulation during

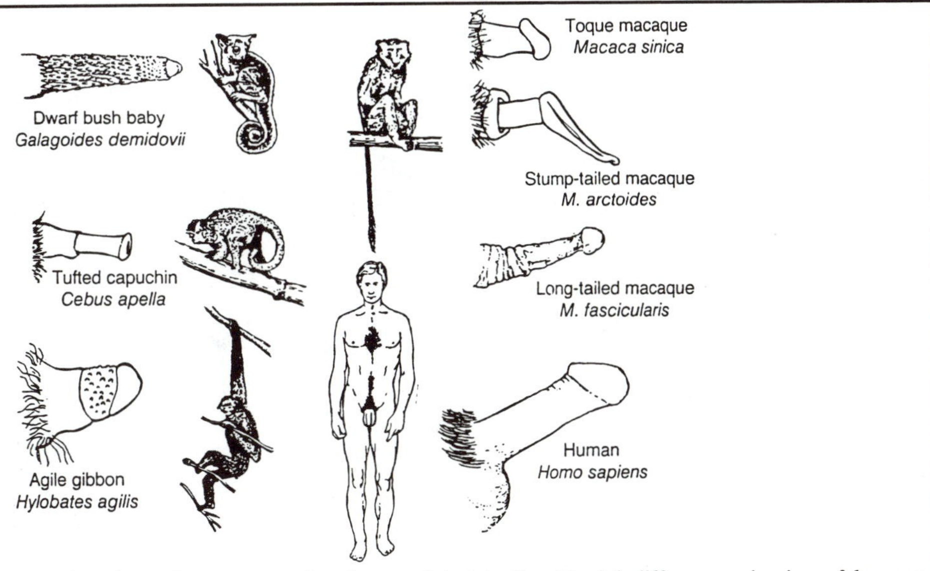

quite large recurved hooks (e.g. Galagidae and Indriidae) (Hill, 1953).

A few species have no differentiated glans (e.g. howler monkey, *Alouatta*; chimpanzee, *Pan*). The shape of the glans is highly variable in prosimians and New World monkeys. However, in all the cercopithecid monkeys (except the stump-tailed macaque, *Macaca arctoides*), apes (except the chimpanzee) and humans the glans is more or less acorn shaped (Kinzey, 1974). Of all the primates, the glans penis of the gorilla is most like that of humans (Hill & Matthews, 1949). The shape of the glans in macaques has been related to the shape of the female genital canal (Fooden, 1967).

Except in the Tupaiidae, the primate penis is invariably pendulous to some degree. In this respect the primates differ from most other mammals, except bats and bears. The penis is particularly long in macaques and baboons (Kinzey, 1974).

The human penis is the longest and widest of all of the hominoid primates. There are, however, racial variations which are greater than might be expected simply from variation in body size.

Racial differences in size of human penis

Race	Length (cm)	Width (cm)
Mongoloid	10–14	3.2
Caucasoid	14–15	3.3–4.1
Negroid	16–20	5.0

Data for erect penises from Rushton and Bogaert (1987).

Data on baculum length from Kinzey (1974) from whom many of the penis shapes are redrawn. Other penis shapes from Kanagawa and Hafez (1975). Drawing of human penis original.

copulation or with the probability of female orgasm (Masters and Johnson, 1966; Hite, 1976; Wolfe, 1981). Nor does small size prevent adoption of any standard copulatory position (Smith, 1984).

Smith (1984) appears to have been the first to suggest that the size of the human penis may be related to sperm competition. He proposed that penis length has been selected to deliver sperm further up the female tract than competing males. On this basis, however, the chimpanzee, with more intensive sperm competition, should have been exposed to even greater selection to increase penis size. Moreover, Smith's hypothesis does not explain the relatively large width and shape of the human penis. While still invoking sperm competition, we think there is a more likely interpretation of the size, shape and function of the human penis.

6.8.2 Plugs and penises

Males of mammalian species which produce hard copulatory plugs (section 3.3.3) which bind to the lining of the female tract (e.g. rats; Voss, 1979) usually have a thin, pointed (= flagellar) penis often with backward pointing spines or barbs on the penis. These penises

Box 6.13 The human penis as a suction piston for sperm removal

The shape of the human penis, its fit in the vagina, and the pumping/thrusting action during copulation, all suggest a function in the removal of a soft/liquid plug from the upper vagina. As the penis pulls back, sperm, seminal fluid and cervical mucus will be sucked down the vagina. On the next forward thrust, the glans should push through this material, then on the next backward pull, the back edge of the glans should pull the material back down the vagina. The greater the suction, the greater the chance of removing cervical mucus with perhaps older sperm from the cervix itself.

Inspired by drawings in Hessell (1992).

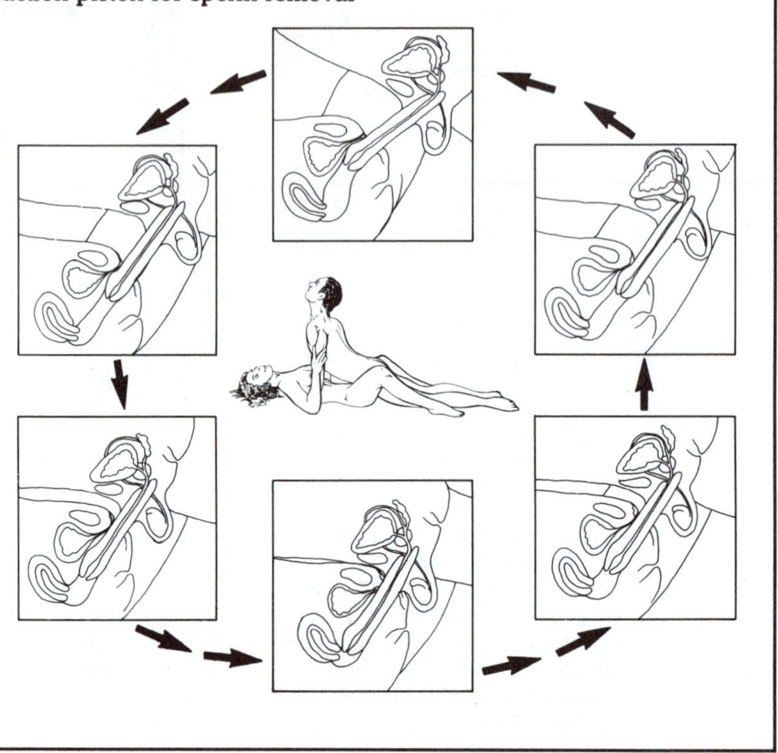

are adapted to remove hard plugs. We suggest (Box 6.13) that straight, piston-shaped penises with a smooth, acorn-shaped glans as in humans are also adapted to remove plugs and sperm but of the soft or liquid coagulum type, not the hard.

Several studies have demonstrated the role of flagellar penises in the removal of hard copulatory plugs. Rats, for example, usually precede insemination by a series of intromissions in which the male penetrates the female, performs numerous rapid thrusts of the penis, and then withdraws without ejaculation (though a few sperm may be inseminated). Usually the male then examines and grooms the penis tip before initiating another intromission. Once the male has: (a) successfully removed the plug; or (b) determined no plug was present; the female is inseminated. In rats, 69% of plugs are successfully removed, most (82%) after only three or fewer intromissions (Mosig and Dewsbury, 1970). The remaining 31% of plugs, however, successfully resist removal. In deer mice, *Peromyscus maniculatus*, males may inseminate the female even if they fail to remove the previous plug (Dewsbury, 1985). This may lead to an accumulation of plugs in the female tract.

If a rat plug is disturbed by just one intromission by a second male within a minute or two after it has been deposited, far fewer sperm from the first male reach the uterus than if the plug is not disturbed until 6 min after ejaculation (Matthews and Adler, 1977). It seems inevitable, therefore, that plug tenacity and the ability to remove plugs with the penis are important factors in the outcome of sperm competition in rats. The same may not necessarily be true for deer mice, *Peromyscus maniculatus* (Dewsbury, 1988).

Of primary importance to rats in their ability to remove copulatory plugs with their penis is the efficient functioning of their striated penile muscles (Wallach and Hart, 1983). The flagellar penis works around, dislodges, and then removes the sometimes broken plug from the female tract.

Many insects use barbs or 'combs' on the

penis (or pseudopenis) to remove sperm from the female tract (e.g. dragonflies, Waage, 1979; beetles, Gage, 1992) and it seems likely that the various hooks and barbs on the penises of various rodents and primates (Box 6.12) may also have some role to play in plug and/or sperm removal.

6.8.3 The piston penis

Humans and other species with piston-shaped penises and smooth, acorn-shaped glans would probably be unable to remove a hard plug. However, a soft coagulum or decoagulated seminal pool could be sucked and then pulled out by an appropriately shaped piston penis. Backward and forward thrusting of the penis during copulation, combined with the shape of the penis in a distended vagina (Box 6.13) should successfully remove a major part of any soft copulatory plug or liquid seminal pool. The shape of the glans seems perfectly adapted to such a push–pull–suck–push–pull–scrape function. Depending on the suction pressure that the thrusting penis can generate, the penis may even successfully remove the older part of the cervical mucus column along with any contained sperm and leucocytes (Chapter 11).

Ultrasound scanning (Hessel, 1992) shows that, during human copulation, the penis distends the vagina to a greater or lesser degree. The vagina is not a cavity but a slit, with the walls pressed against each other. Consequently, the penis touches all parts of the vagina (Box 6.13). As copulation proceeds, the upper part of the vagina dilates like a balloon (Box 6.13), presumably filling with some combination of air and cervical mucus, as if creating a reservoir to receive the seminal pool (Box 3.5). Otherwise, the walls of the vagina are completely passive during intercourse. In contrast, the entire uterus moves vigorously up and down in the female pelvis as the male thrusts.

In order to distend the vagina sufficiently to act as a suction piston, the penis needs to be a suitable size. The relatively large size (12–16 cm deep; Hessell, 1992) and distendability of the human vagina (especially after giving birth) thus imposes selection, via sperm competition, for a relatively large penis. In support of this suggestion, penis and vagina size covary racially in parts of the French West Indies (Rushton and Bogaert, 1987).

The dimensions and elasticity of the vagina in mammals are dictated to a large extent by the dimensions of the baby at birth. The large head of the neonatal human baby (384 g brain weight compared with only 227 g for the gorilla; Harvey *et al.*, 1987) has led to the human vagina when fully distended being large, both absolutely and relative to the female body size. This is particularly so once the vagina and vestibule have been stretched during the process of giving birth, the vagina never really returning to its nulliparous dimensions. Racial differences in size of penis (mongoloid < caucasoid < negroid; Box 6.12) reflects racial differences in birth weight (Hardy and Mellits, 1977) and hence, presumably, racial differences in size of vagina.

The large size of the human vagina may thus be relatively recent. If vaginal size is a function of the size of the baby's head at birth, the rapid increase in brain size (australopithecines, 500 cm^3; *Homo erectus*, 1000 cm^3; *Homo sapiens*, 1400 cm^3; Jerison, 1973; Passingham, 1982) over the last 3 million years should have been associated with a relatively large increase in vaginal dimensions. The relatively large vagina, and hence penis, in humans may therefore be as recent as 1–2 million years.

6.8.4 The evolution of plugs in primates

Some species of mammal have evolved hard copulatory plugs whereas others have not, or have lost them. Phylogenetic analysis (as described in section 6.6.2) suggests that the ancestral anthropoid primate had a soft/liquid plug and a piston penis. This character state may even have been inherited from far back, perhaps preprimate, in its mammalian history. The subsequent evolutionary sequences in two main types of lineage in the anthropoid primates are summarized in Box 6.14. The sequences shown represent the most parsimonious solutions to produce the current distribution of characters, as far as this distribution is known.

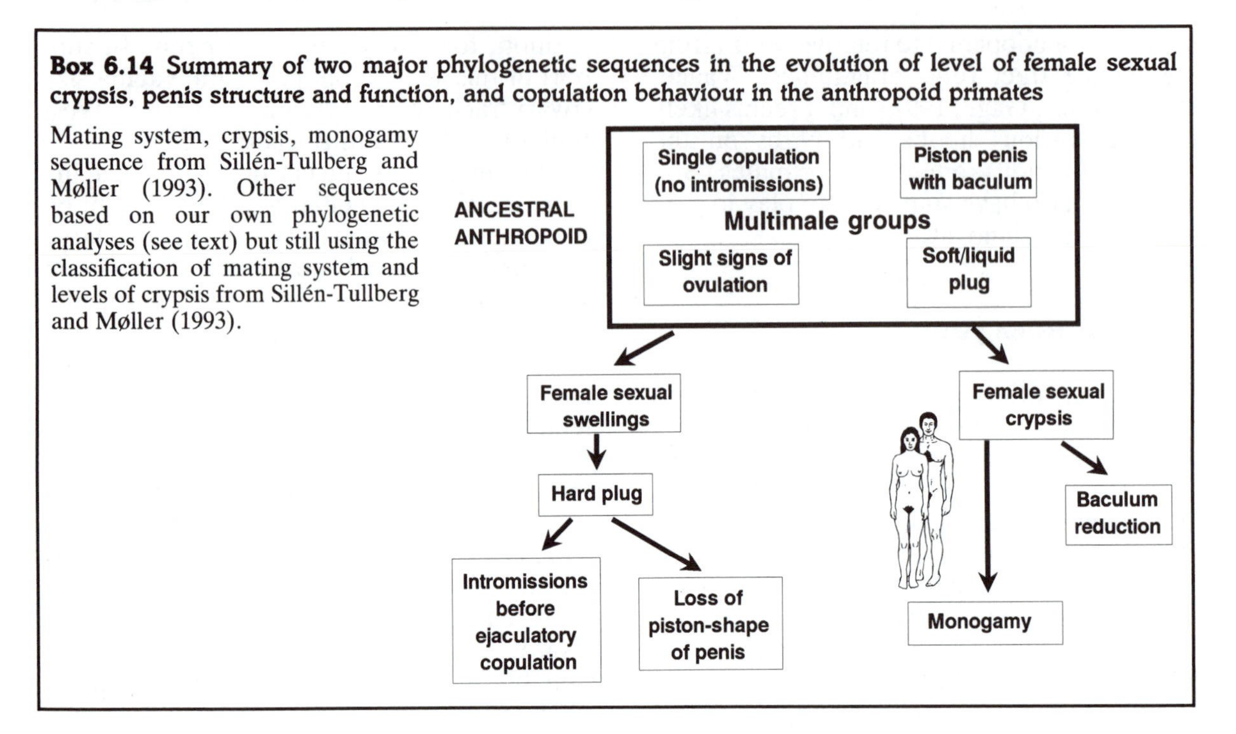

Box 6.14 Summary of two major phylogenetic sequences in the evolution of level of female sexual crypsis, penis structure and function, and copulation behaviour in the anthropoid primates

Mating system, crypsis, monogamy sequence from Sillén-Tullberg and Møller (1993). Other sequences based on our own phylogenetic analyses (see text) but still using the classification of mating system and levels of crypsis from Sillén-Tullberg and Møller (1993).

Hard plugs seem mainly, if not only, to have evolved in lineages with visible signs of ovulation and a multimale mating system (e.g. *Cebus*; the *Macaca–Cercocebus* lineage; *Pan*; see Roussel and Austin, 1967). Intromission behaviour (in which the copulation leading to ejaculation is preceded by a series of thrusting mounts without ejaculation, e.g. various independent platyrrhines; *Cercopithecus nictitans-mitis* lineage; *Macaca silenus-nigra* lineage; *M. mulatta-fuscata* lineage; see Hrdy and Whitten 1987) and loss of a piston-shaped penis (e.g. ancestral platyrrhine; *Macaca sinica-arctoides* lineage; *Pan*; see drawings in Dixson (1987) and in the various volumes by Hill) are also more likely to evolve in such lineages, particularly after the evolution of hard plugs. The set of characters and hence, perhaps, the evolutionary sequence and pressures seem to parallel the situation described for rodents in section 6.8.2.

Lineages which evolve crypsis, such as the human lineage, retain the soft/liquid plug and piston penis characteristic of their ancestors. Such lineages, however, are more likely to show an evolutionary reduction in the size, or even loss of the baculum than are those which evolve a hard plug, intromission behaviour, and/or a non-piston type of penis (Dixson, 1987).

6.8.5 Thrusting and copulation duration

The average duration of copulation in humans is said to be around 2–10 minutes (Hrdy and Whitten, 1987) and the average number of thrusts to be 100–500 with a thrust rate of up to 2/second.

If the human penis functions as a suction piston during copulation, the selective pressures on copulatory duration become clear. On the one hand, males will be favoured who shorten the time from penetration to ejaculation due to the risk of being disturbed or of the female suddenly showing rejection responses. On the other hand, the longer the duration (particularly the greater the number of thrusts) the more material will be removed from the female tract, thus improving the competitiveness of the ejaculate.

How long copulation should last on any given occasion will depend on the optimum trade-off between these two factors. This depends in part on the risk of disturbance or rejection and in part on whether the female tract contains any hindrance to sperm retention and how quickly and easily it can be reduced or removed. The probability and influence of the female experiencing copulatory orgasm (Chapter 10) is an additional factor.

It is a moot point whether a primate with a piston-shaped penis trying to remove a soft or liquid plug requires a longer or shorter time or more or fewer thrusts than a male with a more flagellar penis trying to remove a hard plug. We could find no phylogenetic association between copulation duration or number of thrusts and either mating system, sexual crypsis or type of plug.

6.8.6 Copulatory position

Copulatory position would seem to make little difference to the efficacy of the penis as a suction piston (Box 6.15). All primates,

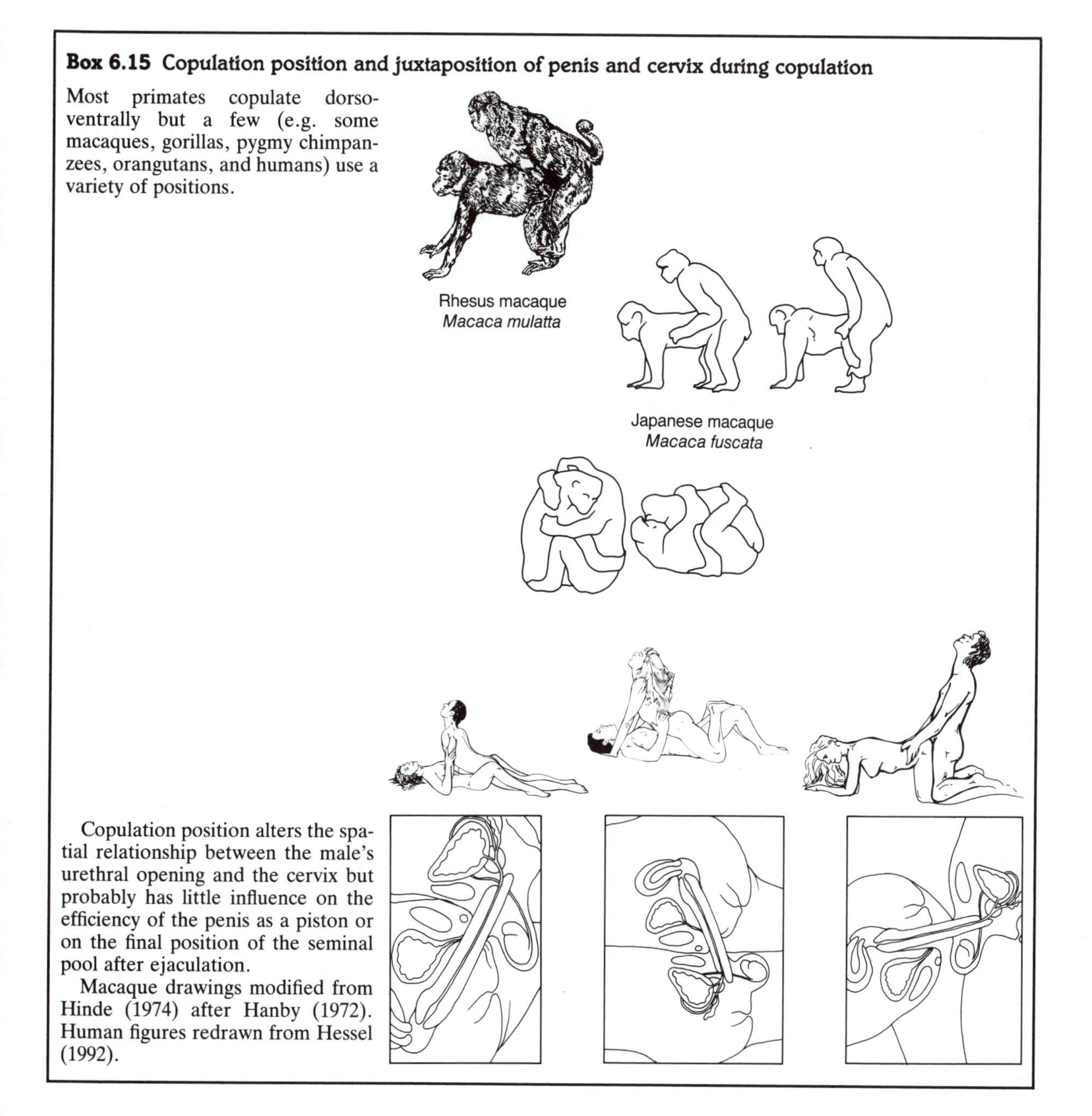

Box 6.15 Copulation position and juxtaposition of penis and cervix during copulation

Most primates copulate dorsoventrally but a few (e.g. some macaques, gorillas, pygmy chimpanzees, orangutans, and humans) use a variety of positions.

Copulation position alters the spatial relationship between the male's urethral opening and the cervix but probably has little influence on the efficiency of the penis as a piston or on the final position of the seminal pool after ejaculation.

Macaque drawings modified from Hinde (1974) after Hanby (1972). Human figures redrawn from Hessel (1992).

including humans, copulate dorsoventrally and in the majority of species this is the only mating position used. A few primates, however, also sometimes (e.g. gorilla) or usually (e.g. orangutan, human) adopt a ventral–ventral position (Short, 1979). There is some hint of a link between the use of ventral–ventral positions and bipedalism.

Anthropologically, there is considerable variation in the copulatory position adopted (Ford and Beach, 1952), much probably depending on the habitat and substrate in which copulation normally occurs. The need for speed, secrecy and vigilance during copulation may also influence position. The dorsal–ventral position is probably faster and allows greater vigilance.

The precise influence of copulatory position on the efficiency of copulation is still a matter of debate. Some authors (e.g. Masters and Johnson, 1966; Smith, 1984b) consider that the dorsal–ventral position has most advantages. They suggest that penile penetration is deeper, gravity may assist semen uptake (Box 6.15) and even that the male may ejaculate through the cervix into the uterus. The latter suggestion, however, is clearly misinformed, the plug of cervical mucus securely preventing such an event.

6.8.7 The human penis: an evolutionary sequence

We suggest, therefore, that the sequence of events in the final shaping of the human penis were as follows. During the anthropoid and perhaps hominoid phases, the human lineage lived in multimale groups (Sillén-Tullberg and Møller, 1993), females had at least slight sexual swellings (Sillén-Tullberg and Møller, 1993), the males produced soft/liquid plugs, the relatively small but piston penis contained a baculum and had an acorn-shaped glans. Copulation involved thrusting.

Some time in the hominoid to hominid transition, the human lineage began to evolve females who first showed sexual crypsis, then adopted superficial monandry (section 6.5). Finally, when the lineage leading to the genus *Homo* began to evolve large brains and hence large vaginas, selection was imposed, via sperm competition, on males with larger penises.

Throughout this sequence, the most successful males were those who could most successfully remove sperm, fluids and coagula from the female tract. In so far as males with larger penises may have been better able to remove sperm, selection would favour an increase in penis size to parallel the increase in size of vagina.

Our analysis leads us to make a further prediction. If penis size is an important factor in sperm competition, it would be surprising if males and females did not have some reaction to penis size. First, males should perceive males with a penis larger than themselves as more of a threat if they ever show a sexual interest in the same woman. Second, females should prefer to mate with males who will give them male descendants with a penis most efficient at removing a rival's sperm. The sexual selection that we propose is similar to that proposed by Eberhard (1985) except that we do not invoke sexual stimulation.

It has long been known that males are often pre-occupied by the size of their penis relative to that of other males (Diagram Group, 1981). However, our prediction seems to run counter to the apparent female indifference to the penis already described (Smith, 1984b). However, most such studies have been outside of a sexual situation and have often concentrated on female reaction to a flaccid penis or to the penis as a source of stimulation to orgasm.

Our prediction relates specifically to the piston-like qualities of the erect penis in relation to the size of the female's own vagina. Up to a point, the female preference should be for a larger penis. Beyond a certain penis size, however, a female may risk internal damage and optimally her preference should be curtailed accordingly.

7 The modern scenario: contraception, fecundity and the illusion of conscious control

7.1 Introduction

In Chapters 2–6 we discussed in detail the evolution of human sexuality, that genetic programme which orchestrates not only the development of human sexual anatomy and physiology but also the subtle and sophisticated interplay of male and female strategies. Hopefully, these chapters have provided some, perhaps occasionally novel, insights into why modern humans may be programmed to behave sexually the way that they do. Millenia of natural selection on our pre-human and human ancestors have moulded a species that, like all other species, should consist of individuals each striving, consciously/subconsciously, to leave more descendants than their contemporaries.

Thus, in modern humans, sperm competition takes place within the context of the sexual anatomy and behaviour that has evolved throughout the history of the human lineage. However, it also takes place against a backdrop of efficient modern methods of male and female contraception and a rhetoric extolling the supremacy of conscious choice over subconscious, physiological programming as a mechanism to limit fecundity.

We know, from past reactions to our work, that many people find the concept of modern humans striving to leave more descendants (say, grandchildren) than their contemporaries difficult to accept. They point out that much of their own sexual activity involves a conscious decision to avoid conception by using contraceptives. How then, they ask, can we suggest they are employing strategies to aid their reproductive success, particularly in so intangible a process as sperm competition? To such people, our interpretation of their behaviour and their body's physiology is totally alien to their own conscious rationalization of their motivations and emotions. Modern embellishments to human sexual behaviour seem at first sight to make nonsense of our whole approach.

7.2 Modern embellishments

Modern humans have inherited a programme for sexual development and behaviour that has been shaped by 4000 million years of evolution, perhaps particularly by the pressures generated by sperm competition. Relatively few facets of current behaviour are modern innovations. Those few that are, however, could well help to shape the sexual programme of the generations of the human lineage still to come.

Box 7.1 Modern contraceptives

A brief history

A female in ancient Egypt had almost as many contraceptive options as the modern woman. She could choose from: (1) fumigating her genitals with emmer grains; (2) inserting a lint tampon soaked in honey and a spermicidal herbal liquid made from acacia tips; or, the most popular, (3) inserting a mixture of crocodile dung, sour milk and honey. About 2400 years ago, in Greece, Aristotle recommended smearing cedar oil, white lead or frankincense on the female genitals whereas Pliny suggested rubbing sticky cedar gum over the penis.

Most modern contraceptives have ancient prototypes. The spermicidal sponge was foreshadowed in ancient Persia by natural sea sponges (soaked with alcohol, iodine, quinine or carbolic acid) inserted in the vagina before copulation. The cervical cap had prototypes in the scooped-out halves of pomegranites provided by Greek doctors, the squeezed lemon halves given to his lovers by Casanova, the vegetable seed-pods inserted by Djuka (South Africa) women, and the cervical plug of grass used by the women of the Kasai Basin (central Africa).

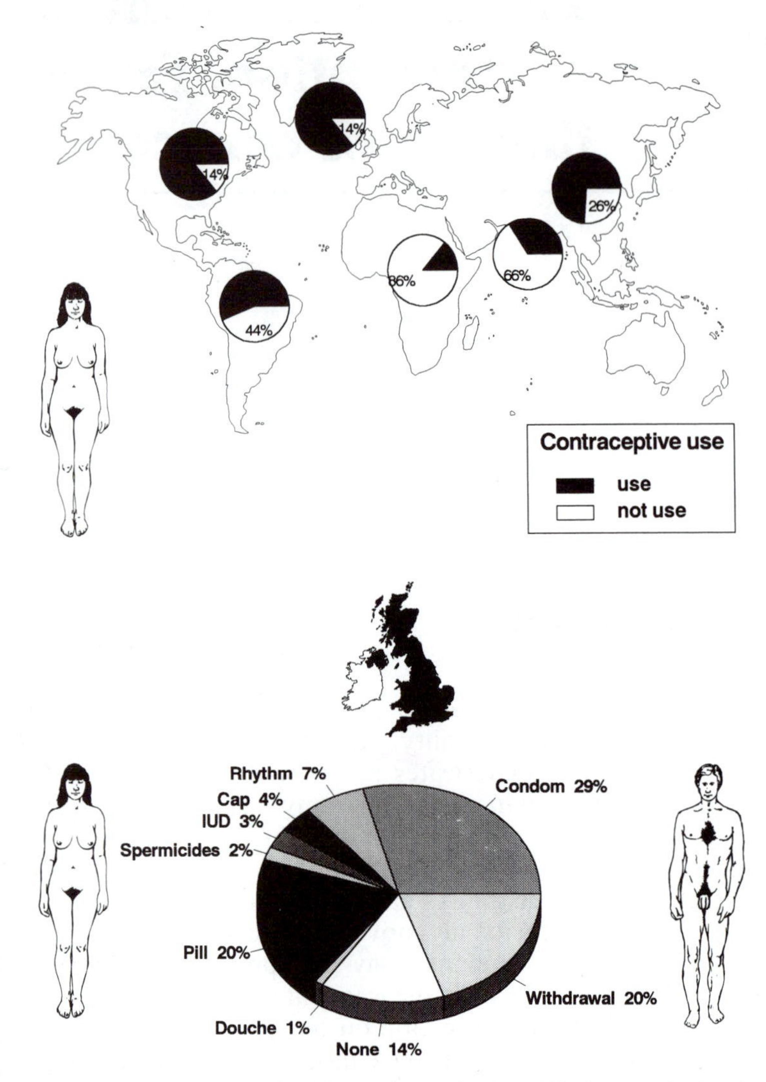

The condom has been known since Roman times and was widely used in Europe by 1700. Fallopio designed the first medicated linen sheath in the 1500s but the item gained its name from the personal physician to King Charles II, the Earl of Condom, who recommended its use to the king as an aid to prevent the contraction of syphilis.

By the 1890s, all of the barrier methods of contraception in use today were on sale openly in the UK and intrauterine devices had been developed. However, the successful use of contraceptives was unlikely to have been widespread until well into the twentieth century (Box 7.3). Research on sterilization techniques began in the 1850s but surgical techniques were not adequate for their general use for nearly 100 years. Finally, oral contraceptives began to become generally available around 1961–65.

7.2.1 Modern contraception

(a) FEMALE CONTRACEPTION

Family planning and contraception by females has a long evolutionary history in the human lineage, stretching back probably some 100 million years or so to the evolution of female parental care and placental mammals (section 4.4). Even oral contraception may not be a human invention, female chimpanzees perhaps chewing *Aspilia* leaves as an aid to contraception (Page, 1993).

Modern contraceptive techniques (Box 7.1) available to women merely render the ancient mammalian practices more effective. In most cases, the modern techniques mimic or extend naturally evolved methods. Oral

Contraceptive usage by women (c. 1980)
The level of contraception varies widely in different parts of the world from about 14% of women in Africa to an average of about 70% of women in the 'developed' countries. In Britain, nearly half of couples rely on male-implemented contraception methods.

History: general from Dunham *et al.* (1991); UK from Macfarlane and Mugford (1984). Failure rates from MacLeod (1953), Ryder (1973), Johnson and Everitt (1988), Vessey *et al.* (1982), Bounds *et al.* (1992). Usage from Johnson and Everitt (1988) and WHO (1990).

Failure rates

Failure rates of different forms of modern contraceptives (expressed as number of pregnancies per 100 woman years (= 10 women for 10 fertile years, or 100 women for 1 fertile year, etc.))

	Pregnancy rate (%)		Pregnancy rate (%)
Female methods		**Male methods**	
'Natural'			
No protection	60–75		
Age <25 years	94[a]		
>35 years	16[a]		
Lactation for 12 months	25–40		
		Withdrawal	7–40
Abstinence	0	Abstinence	0
Modern			
Douche alone	20–40	Condom	3–30
Foams, jellies, creams	5–40	Vasectomy	0.02–0.15
Rhythm	13–40		
Diaphragm + spermicide	2–30		
Female condom	4–26		
Intrauterine device	2–4		
Oral contraceptives			
Combined pill	<1–10		
Sequential	1–2		
Progestagens	5–10		
Morning-after (oestrogen)	0–2		
Morning-after (progestagen)	5–30		
Tubal ligation	0.04–0.13		
Hysterectomy	0.0001		

[a] % of women becoming pregnant in 6 months of unprotected intercourse.

contraceptives either unfalteringly induce anovulatory cycles (combined pill) or indefinitely extend the infertile luteal phase (minipill) (Johnson and Everitt, 1988), thus enhancing sexual crypsis. In the absence of oral contraceptives, the monitoring of basal body temperature (Roetzer, 1977) or hormone levels (Vermesh *et al.*, 1987) enhances the efficiency of restricting copulation to the infertile phase of the menstrual cycle. Barrier methods (e.g. cervical cap) and spermicides enhance the selective or blocking function of the cervical mucus and leucocytes (Johnson and Everitt, 1988). Intrauterine devices enhance the prevention of implantation and induced abortions enhance the efficiency of spontaneous abortions (Johnson and Everitt, 1988). Sterilization procedures (e.g. removal of the uterus; removal or ligation of the oviducts; Johnson and Everitt, 1988) in essence hasten menopause. Finally, formalized adoption and fostering procedures allow some females to avoid maternal care without infanticide.

The majority of copulations by female primates with sexual crypsis are contraceptive. They are adapted to provide the female with the non-reproductive benefits of copulation, not to lead to fertilization (section 4.4). Except for the greater involvement of conscious decisions, neither the behaviour nor the psychology associated with modern female contraception thus differs in any qualitative way from that associated with pre-existing methods. Only the efficiency of contraception has changed (Box 7.1). The modern woman, therefore, has even greater opportunity to reap the benefits of enhanced sexual crypsis (sections 6.5 and 6.6), multiple mating and polyandry (section 2.4.1), including promoting and influencing sperm competition (section 2.5.6), while still attempting to optimize family planning (section 4.4).

(b) MALE CONTRACEPTION

To some extent, the same perspective holds for the modern male. Modern contraceptive methods involve the same elements of spite, pseudocopulation and deception as do evolved methods (section 6.4).

Condoms are an extension of the evolved method of pseudocopulation. They may allow a male to persuade a female to extend exploratory petting as far as to allow copulation, using the double reassurance that insemination will not occur and that the risk of sexually transmitted diseases is reduced. Yet condoms retain a high chance of fertilization (Box 7.1). Although, when used properly, the risk of conception with a condom may be as low as three pregnancies per 100 woman years, in normal usage the risk varies from 5 to 30 pregnancies per 100 woman years. This is up to about half the risk experienced by a fertile couple with no protection.

To what extent these figures reflect condom failure, male incompetence or male strategy is a moot point. As with natural pseudocopulation/withdrawal, males may actually attempt to inseminate the female while appearing to offer copulation with low conception risk. Prostitutes report males regularly trying secretly to remove condoms. If deception is part of the male strategy, it raises intriguing possibilities about the proposed male contraceptive pill, depending as it does on the male's sympathetic cooperation with the female.

An interesting question, which we cannot answer, is whether a male's reproductive success is lower or actually higher if he uses a condom. The critical factors in the trade-off are: (1) the increased number of copulations the male achieves through using a condom; (2) the reduced (but not zero) chances of fertilization per copulation due to the contraceptive effect of the condom; and (3) the reduced risk of contracting sexually transmitted diseases.

Male sterilization through vasectomy (Johnson and Everitt, 1988) also has elements of pseudocopulation. Some vasectomized males advertise their condition via badges such as ties and medals in an attempt to increase their number of copulations. The efficiency of vasectomy as a contraceptive method is 0% during the first 2–3 months while preoperative sperm are passing through the male tract but should be 100% thereafter, as long as the operation was carried out competently (Johnson and Everitt, 1988). In actual practice, it is a little less than 100% (Box 7.1). Even so, it is sufficiently low that after 2–3 months the vasectomized male is unlikely to achieve an increase in reproductive success through deceit as a result of the operation. However, vasectomy probably does allow the male to retain an element of spite through his ejaculation of seminal fluids.

A vasectomized male who has reached optimum reproductive level with his partner gains the advantages of family planning with his partner, having (fewer) children who then grow to be more successful and thus perhaps give the male more grandchildren. The insemination of sperm-free semen allows the male at least some opportunity to reduce the chances of his partner conceiving via some other male (Quinlivan and Sullivan, 1977). More importantly, vasectomy allows the male greater certainty that any later children conceived by his partner are not his own. He is then in a better position to decide on optimum strategy (i.e. continued paternal care or desertion) for response to his partner's infidelity (section 6.3.1). The irreversibly vasectomized male does, of course, lose the opportunity of furthering his reproductive success via EPCs or through switching to a new long-term partner.

Depending on the age and reproductive status of males who opt for vasectomy, any given individual may or may not enhance or reduce his reproductive success as a result. In England and Wales, however, most men are sterilized when still only in their 30s (Macfarlane and Mugford, 1984) and it is possible that, on average, irreversible vasectomy is maladaptive for males. The same may not be true for reversible vasectomy.

In contrast, a woman may well gain reproductively through persuading her partner to undergo vasectomy. By so doing, she reduces the chances both that he will father children via some other female and thus perhaps also that he will desert. The result is more certain parental care for her own offspring while retaining for herself the option of further offspring fathered by other males.

7.2.2 Reduced fecundity

It is a common and in some ways optimistic view that the fecundity of modern humans is in some way 'unnaturally' low and that this reduced fecundity is attributable to the efficiency of modern methods of contraception (Robey *et al.*, 1993). We suggest, however, that the situation is not that clear.

At the turn of the century, a hunter–gatherer female using nothing but physio-behavioural contraception would on average give birth to about four children in her lifetime (Carr-Saunders, 1922). Once born, survival prospects were relatively high and two or three would survive to maturity. Hunter–gatherer children had a good diet and lived in small, mobile and relatively isolated groups with a low incidence and transmission of infectious diseases (Lee and DeVore, 1968; Barnett, 1971). This scenario could well have been typical of the last million years or so of human evolution.

Females who lived in agricultural situations characteristic of the last 10 thousand or so years of human evolution had a much higher birth rate (Barnett, 1971). Families into double-figures were commonplace. However, once born, survival prospects were relatively low. Agriculturalists' children had a high carbohydrate diet and lived in large, sedentary groups with a high incidence and transmission of disease. Whole families were sometimes wiped out by virulent infections. In order to have a reasonable chance of two or three children surviving to maturity, many more had to be produced.

Finally, the modern female in an industrial society using modern contraceptive techniques has a lifetime reproductive prospect of two to three children (Box 7.2). Survival prospects are high and the majority of children survive to maturity.

Over the past few centuries, the birth rate in different societies has always followed changes in the death rate with a time lag of a few decades. This is the so-called 'demographic transition' from a pre- to post-industrial type of population growth which has characterized many western nations between 1870 and 1940 and parts of the developing world over the last 40 years. The main stages were summarized by Bourgeois-Pichat (1951).

Before the demographic transition in any given society, a rough equilibrium existed between a very high birth rate (= number of live births per female lifetime) and a high death rate. The first stage of the transition is an initial decline in the death rate without a corresponding change in birth rate, with the result that population size and density increase rapidly. Scandinavia and France entered this stage as early as the 1750s or so. The second stage is marked by a decline in the birth rate. In most of western Europe this stage was reached about 1880 but France entered this stage before 1800. The third stage consists of an even more pronounced decline in the birth rate until, as in some western European countries at present, the birth rate is less than the death rate.

During the demographic transition,

Box 7.2 Female fecundity in various countries (c. 1980)

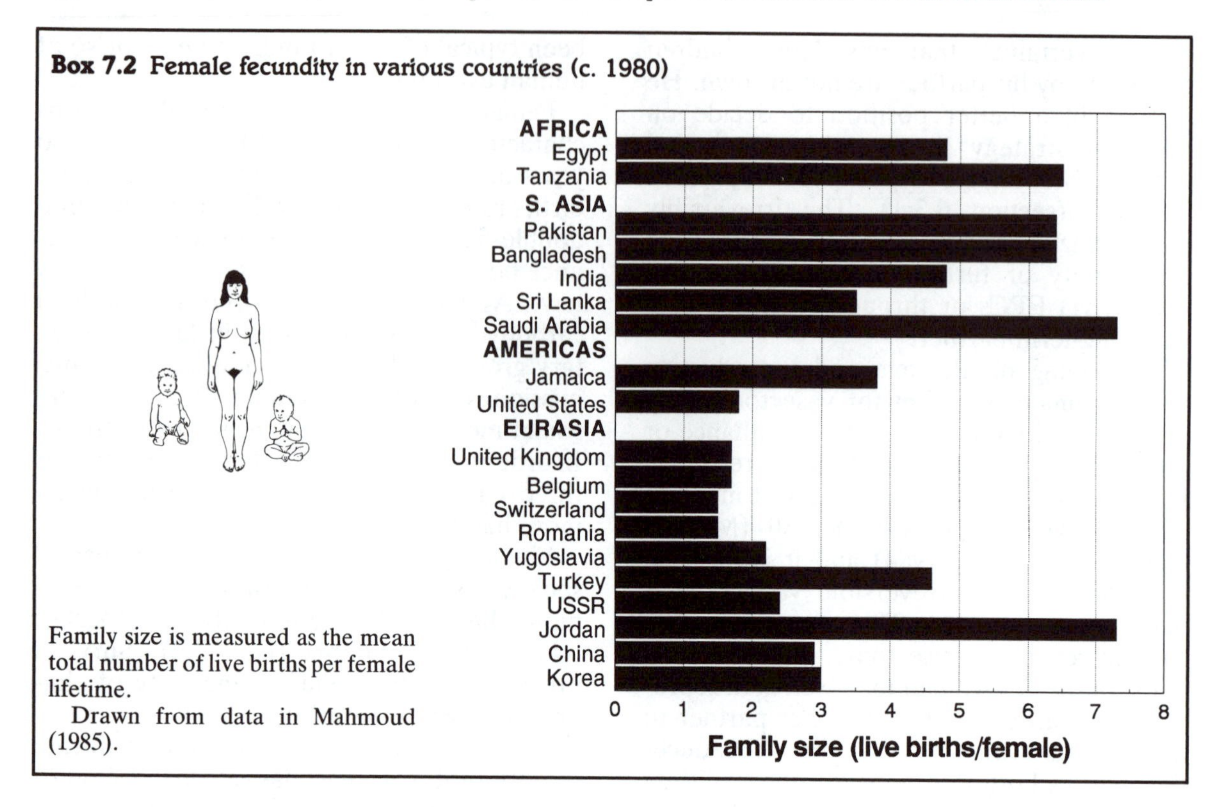

Family size is measured as the mean total number of live births per female lifetime.

Drawn from data in Mahmoud (1985).

females seem to adjust the number of children they attempt to raise according to the prospects of their offspring surviving to maturity (Box 7.3). In most of the countries which were the first to show the transition, the decline in birth rate clearly preceded the widespread use of modern contraceptives (e.g. Box 7.3). Thus, for most of the relevant historical period, the contraceptive techniques on which the declining birth rate depended were largely physiological and behavioural.

Modern contraceptive techniques require more conscious preparation than their physiological mammalian counterparts and involve artefacts. They are more effective (Box 7.1) but do not lead to a birth rate much lower than during hunter–gatherer times. Indeed, when the completed family size (live births/female) is plotted against life expectancy (Box 7.4), it is difficult, if not impossible, to detect any influence of the use of modern contraceptives. Even social and economic differentials in family size within a society (e.g. UK; Benjamin, 1965) reflect life expectancy within each stratum.

Viewed in this perspective, modern forms of contraception appear as part of an evolutionary continuum rather than as a qualitatively new phenomenon. They have become part of the mechanism by which females continue to do what they seem always to have done: adjust their family size according to their offsprings' survival prospects. Females seem to be adapted to give birth to the number of offspring that, given extant mortality risks, will give them a particular number to raise through to reproductive maturity. This in turn implies that (subconsciously?) to make the necessary adjustment, the female must monitor the survival prospects of infants and children in her environment, then formulate her family plan accordingly.

It appears from the above figures that, since hunter–gatherer times, human females have given birth to the number of offspring that should allow them, on average, to raise two or three children through to maturity. We assume that this number is the one that, in combination with offspring 'quality' through the food, care and attention that they receive, maximizes the female's multiplication rate

Box 7.3 The demographic transition and use of contraceptives in the UK

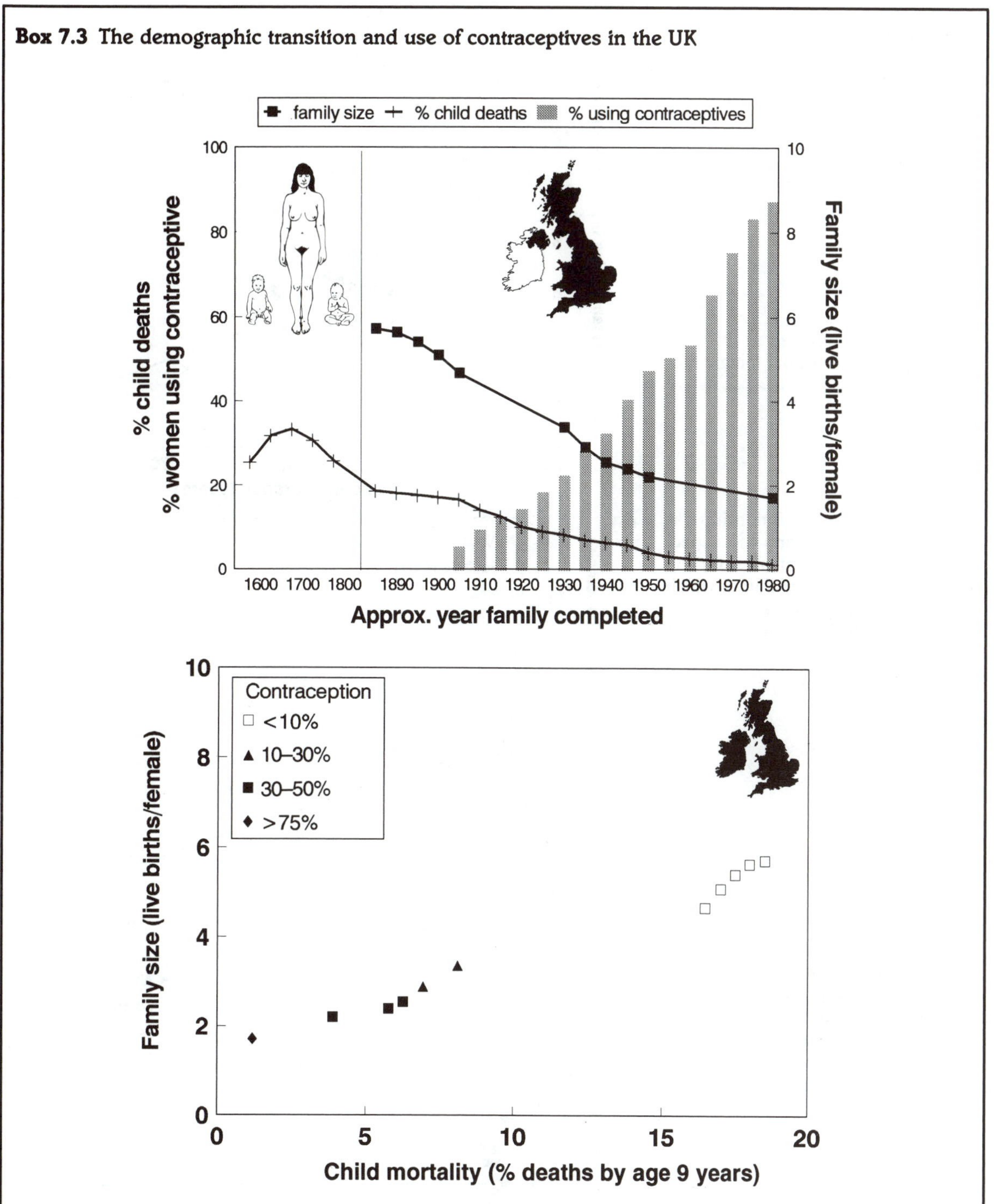

Family size is expressed as the total number of live births per female lifetime. Child death rate is expressed as the proportion of children who die by age 9 years (by 14 years, pre-1800). Pre-1800, the death rate is for Colyton, Devon, and is probably low relative to the rest of the UK. Post-1800 the death rate is for England and Wales.

Childhood death rate begins to fall some time between 1700 and 1800 and is followed by a fall in family size sometime after 1800, well before modern contraceptives are in general use. The close relationship between family size and children's survival prospects does not seem to be dependent on the availability and use of efficient contraceptives.

Drawn from data in Macfarlane and Mugford (1984) and Mahmoud (1985).

Box 7.4 Family size in relation to life expectancy and use of contraceptives in various countries

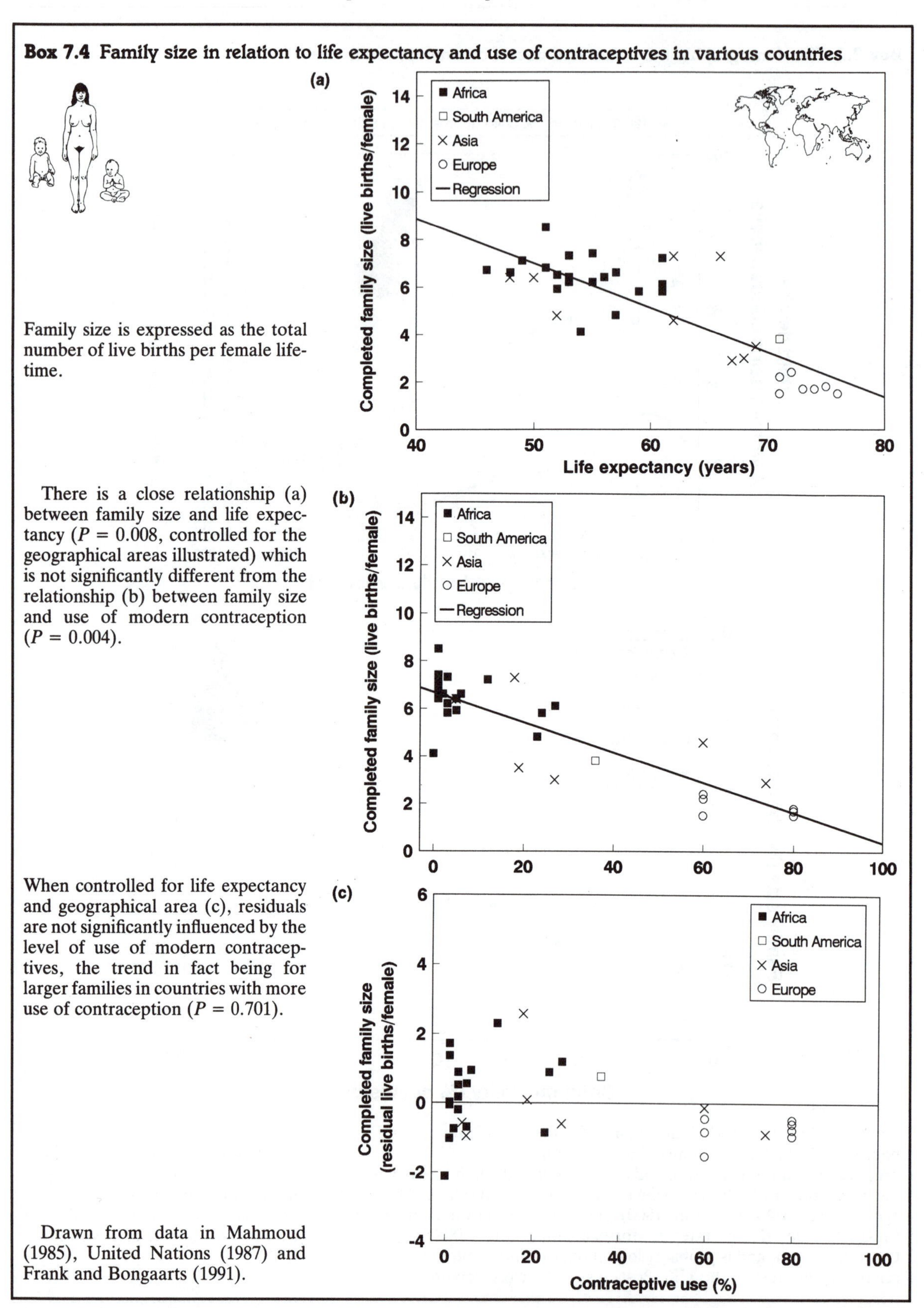

Family size is expressed as the total number of live births per female lifetime.

There is a close relationship (a) between family size and life expectancy ($P = 0.008$, controlled for the geographical areas illustrated) which is not significantly different from the relationship (b) between family size and use of modern contraception ($P = 0.004$).

When controlled for life expectancy and geographical area (c), residuals are not significantly influenced by the level of use of modern contraceptives, the trend in fact being for larger families in countries with more use of contraception ($P = 0.701$).

Drawn from data in Mahmoud (1985), United Nations (1987) and Frank and Bongaarts (1991).

(i.e. is her optimum clutch size; Lack, 1954). Producing any fewer children, or attempting to raise any more, could each lead to reduced reproductive success when measured over the ensuing generations.

More specifically, the suggestion is that by limiting the number of first-generation offspring in a highly competitive environment parents might, through focused investment of wealth and resources, be able to enhance the reproductive value of those offspring (Hartung, 1985). Parental wealth is known to influence the number of grandchildren produced via both sons and daughters (e.g. Kipsigis, Kenya; Borgerhoff Mulder, 1989). The final question, therefore, is under what circumstances, if any, does limiting the number of first generation offspring actually result in more grandchildren than having more first generation offspring.

A first attempt to model the situation was made by Rogers (1990). The conclusions were that at the lower wealth ranges of a population, long-term fitness is maximized by using the currently available wealth to maximize family size. The more wealthy ranges, however, gain relatively little from increasing family size and thus may benefit, in long-term reproductive success, from limiting family size so that those few offspring raised are reproductively more successful. As Rogers recognizes, the conclusions are sensitive to a number of assumptions. At the very least, however, the model shows that the reduction in family size during the demographic transition could well have been a response that actually increased individual reproductive success.

In summary, therefore, we suggest that modern contraceptive techniques have the following characteristics.

1. They exploit and/or enhance psychological predispositions and strategies evolved much earlier in mammalian history.
2. Far from reducing individual reproductive success, they enable females more efficiently to have the size of family that, on average, they would by different methods probably have had anyway, given extant survival prospects.
3. They become part of the mechanism by which some women actually increase their reproductive success, not via an increase in number of children but via an increase in number of grandchildren etc. This increase is achieved by more efficient family planning and more efficient strategic manipulation of males.
4. For males, in contrast, modern contraceptive methods (particularly condoms) may enhance the reproductive success of some individuals. However, there is a much greater chance that for other males (particularly vasectomized) modern methods may be maladaptive.

7.2.3 Elevated infertility: clinical and behavioural

Infertility affects up to 15% of the potentially fertile human population (WHO, 1990).

In view of the overwhelming evolutionary importance of reproduction (Box 2.2), the one facet of the genetic programming of any animal that would be expected to work at peak efficiency would be reproduction. Thus, the failure of some people to reproduce raises an important question: why have these people's genetic programmes, shaped and honed so ruthlessly over past millenia, failed to translate survival and actions into reproduction?

Two main causes of infertility may be recognized: clinical and behavioural. When the cause is clinical, the reproductive behaviour of the individual is not unusual yet reproduction does not occur. When the cause is behavioural, it is the individual's behaviour that induces a failure to reproduce.

Currently, clinical infertility seems high, particularly in western societies. A study coordinated by the World Health Organization (WHO) in 25 countries and covering more than 10 000 infertile couples showed that tubal blockage for females and testicular malfunction for males were the main causes in the majority of definitive diagnoses (WHO, 1990). Most of these causes are probably the result of past or current infections, leading to pelvic inflammatory disease (PID) (Khatamee, 1988).

At least half the cases of female subfertility in the USA are probably the result of sexually transmitted diseases (Khatamee, 1988). So too are many of those not uncommon cases of blocked epididymides in males, which may result from gonorrhoea (Johnson and Everitt, 1988). This is not surprising as the STD organisms, themselves, may well have been selected for an ability to induce infertility (section 3.2). That they often succeed merely reflects the ability of parasites often to outmanoeuvre their hosts in evolutionary arms races. Other causes of infertility are suggested to be exposure to a wide range of modern environmental pollutants (e.g. cigarette smoke, alcohol, carbon monoxide, X-rays) or particular responses to oral contraceptives and intrauterine devices (Dunham *et al.*, 1991).

There are various forms of behaviour directly associated with failure to reproduce. Some individuals simply never find a mate despite actively searching. Others (e.g. chastate nuns and monks, exclusive homosexuals) shun reproductive opportunities. Yet others, particularly males, by virtue of below average potency in any of the strategies discussed in Chapters 2–6, simply fail in either male:male or male:female competition. By far the most potent behavioural contribution to reduced reproduction in modern western society, however, is self-inflicted via contraception (i.e. using contraceptives so assiduously throughout life that reproduction never occurs). Why, then, does the modern genetic programme often fail?

In many parts of the world, the modern human environment is very different from previous environments. The explosion of environmental poisons over the past century presents a novel chemical environment. The huge increase in mobility over the same period generates a rate of exposure to different infectious organisms that must be orders of magnitude greater than ever before. The efficiency of at least some of the modern artefacts for contraception is potentially (i.e. if used continuously) far greater than was ever the case using the physiological and behavioural techniques of other mammals (Box 7.1).

The overall result is that not all combinations of the sexual programme, both physiological and behavioural, that are the human inheritance, are now as well adapted as they would have been in past environments. Physiological and behavioural responses that would have led to reproduction in past environments now falter or fail, some individuals never encountering in their lifetime the particular circumstances that in previous generations would have triggered reproduction.

To anybody other than an evolutionary biologist, this apparent maladaptiveness raises no philosophical problem. They would simply point to the fact that many of the apparently maladaptive behaviour patterns (e.g. lifelong contraception; monasticism; exclusive homosexuality) are simply the result of individuals making a conscious choice over what is important in their lives. Such conscious choice then overrides any evolved predispositions. This view demands consideration for it strikes at the heart of the discussions in the chapters to follow.

7.3 Conscious and subconscious sexual strategy

As behavioural ecologists interested in human behaviour, we are much more interested in what people and their bodies actually do than in what they say they want to do. Thus, in this book, we restrict ourselves only to what people say they have done. We pay little or no attention either to their opinions as to why they think they behaved as they did or to their opinions as to what they think they might do in the future. Thoughts and wishes are not by themselves evolutionary fodder unless they translate into actions and reproduction (Box 2.2).

Already in this book we have encountered examples of deception, some of which (e.g. sexual crypsis) may even work best if there is a strong element of self-deception. The conscious mind is by no means always an accurate barometer of adaptive, subconscious mechanisms. A woman who assiduously insists on her partner using contraception at every copulation, then engages in EPC, without contraception, at the most fertile phase of her menstrual cycle, may genuinely believe that she only wanted children by her partner. A man who 'accidentally' slips out of a condom, leaving it in the vagina full of sperm, may genuinely believe that he did not wish to inseminate the female. Natural selection does not care. The behaviour and the child that results are evolutionary fodder. The beliefs, no matter how genuine, are not.

The remainder of this book is concerned with the mechanics and consequences of sperm competition in humans. As such, we are concerned with subconscious, physiological decisions by the body rather than with conscious, behavioural decisions. Of course, we are well aware that these decisions by the body are taking place against a backdrop of conscious rhetoric. Often, conscious and subconscious strategies will concur but often they will not. When they do not, we concern ourselves with what the body does, not with what the person might think it is doing. As in many aspects of behaviour, it seems likely that physiological programming has a more powerful influence than any conscious rationalization. Bearing in mind the benefits of self-deception and the deception of others, we suspect that subconscious programming is actually likely to be a more accurate indicator of adaptive function than is conscious rationalization.

We stressed at the very beginning of our discussion of human sexuality (Chapter 2) that all heritable aspects, even conscious thought, must be a manifestation of an evolved programme. The predisposition to make this or that conscious decision will ultimately relate to some heritable feature, even if that heritable feature is a conditional feature (i.e. when in this circumstance, do this, when in that circumstance, do that). Predispositions to react to the environment in a maladaptive way will be selected against.

Conscious decisions not to reproduce are as maladaptive as any other feature of behaviour. The individual may not care that after 4000 million years of continuity his/her personal lineage will end if he/she fails to reproduce. But then evolution will not care either and extinct his/her personal lineage will go! Natural selection continues.

7.4 Natural selection on modern humans

There is a widespread misconception that modern humans are immune to natural selection. Even in principle, the view is clearly fallacious. As long as people differ in the number of children they produce and as long as the facets of anatomy, physiology or behaviour associated with these differences are at all heritable, natural selection continues to be a force.

As we have seen in this chapter, physiological and behavioural responses that would have led to reproduction in past environments now sometimes falter or fail. More people than ever before are failing to reproduce. In the USA, even clinical infertility doubled between 1980 and 1990 (Khatamee, 1988; Dunham *et al.*, 1991). Natural selection is thus now more intense, not less. It is acting on people whose reproductive success is below average just as ruthlessly as it has ever done in the evolutionary past. Whatever cause or conscious motivation for low reproductive performance any given individual may have, any associated heritable characteristics will either not be present in future generations or will be underrepresented.

At the same time, of course, individuals who would have failed to reproduce in past environments may now have some opportunity to take advantage of assisted conception techniques (e.g. *in vitro* fertilization; Matson *et al.*, 1989; Morroll *et al.*, 1992, 1993) and to reproduce. Any heritable characteristic that influences the individual's decision whether or not to take advantage of this assistance, and their physiological ability to respond to assistance, will also be a target for natural selection. The result may be the persistence and expansion of lineages which are only adapted to reproduce in the modern environment.

As in the past, it is the heritable characteristics of people who have the higher reproductive successes in the current generation which will numerically dominate the generations to come. The only reward for failing to reproduce, no matter how high-minded the motivation, is the extinction of individuals with that motivation (always allowing for the manifestation of sterility imposed by parental manipulation, Box 4.15). Not the least maladaptive trait is the predisposition, when confronted with an infallible means of contraception, to use that means so efficiently as to fail to reproduce altogether. Such predispositions will die out, the population becoming dominated by individuals who are predisposed to use contraception to maximize, not decrease, their reproductive success.

Eventually, if the recent ecological, social and technological environment stays stable long enough, the human population will come to be dominated numerically by individuals who achieve the greatest long-term reproductive success in that environment. These will be the individuals who reproduce most despite the levels of environmental toxins, sexually transmitted diseases, social pressures, and efficiency of contraception. Natural selection will have caused the sexual programme of the human lineage to have changed, slightly but significantly, yet again.

8 Levels of human sperm competition

8.1 Introduction

In Chapters 2–6 we argued that sperm competition has been one of the most potent selective forces to shape all aspects of the sexual programme of modern humans. In Chapter 7 we pointed out that selection continues to act as ruthlessly as ever to shape human sexual behaviour albeit in many countries in a relatively novel environment. Our grounds for this assertion were that individuals still differ in their reproductive success and that inevitably these differences are largely a reflection of heritable characteristics.

Although selection on the sexual programming of modern humans is still intense, it does not automatically follow that sperm competition has a high profile in that selection. Whether it has or not will depend on the frequency of sperm competition events in modern populations and, in particular, what proportion of children are being conceived via sperm competition. That is, what proportion of conceptions occur while the female has sperm from more than one male in her reproductive tract or, more generally, what proportion of conceptions involve the sperm from one male being influenced by the sperm from another male. Depending on the way in which sperm competition proceeds (section 2.5), there is an important difference between these two scenarios.

To consider the frequency of sperm competition in humans, three important factors need to be known: (1) how long a time can elapse between inseminations from different males for sperm competition still to occur; (2) how often does this happen; and, particularly, (3) how often does it lead to conception.

8.2 Sperm competition: where and for how long do sperm live in the human female?

The following description is our final interpretation of the events in the human female tract following copulation. It is based on the discussions and analyses presented in Chapters 2–6. Not everybody interested in mammalian sperm competition will agree with our final picture (e.g. Gomendio and Roldan, 1993) and the main differences of opinion are discussed in section 8.2.3.

Copulation and insemination in humans results in the male depositing a pool of semen containing millions of sperm in the upper part of the vagina adjacent to the cervical opening (Box 3.5). There, semen may mix to some extent with cervical mucus dripped, ejected or sucked (by penis action; section 6.8.3) into the vagina earlier on during copulation. Within less than a minute of insemination into the upper vagina, the human ejaculate coagulates to form a soft, spongy structure (section 3.3.3). After about 10–20 minutes, the structure decoagulates and, at least in part, is eventually ejected in the flowback (section 3.5.2).

At insemination, the cervix dips into the seminal pool and establishes an interface between the semen and the column of mucus which fills the cervical channel (Box 3.5). This interface consists of a series of finger-like projections of semen into the cervical mucus. Swimming sperm are then directed along diagonal mucus channels, many entering sperm storage organs, the cervical crypts (section 3.5.1). Some other sperm seem to pass quickly through the mucus and may enter the uterus directly. Once vanguard sperm have passed through the mucus the macromolecular structure of the mucus seems to be disrupted and any following sperm make much slower progress. Such sperm may never leave the cervical mucus, eventually being phagocytosed or shunted trapped in mucus back into the vagina. Alternatively, they may eventually reach either the uterus or the cervical crypts. (Sperm distribution 1 h after insemination is shown in Box 3.5).

The number and proportion of sperm which leave the seminal pool is very variable and depends at least in part on the age of the sperm (Chapter 9) and the female orgasm pattern (Chapter 10). The remainder are ejected in the flowback.

8.2.1 Sperm competition and the flowback

The flowback is a mixture of seminal fluid, female secretions, sperm and other cells (originating in both the male and female) that emerges from the female tract and is lost to the outside after insemination (section 3.5.2).

Most of the seminal fluid and about 130 million (= about 35%) of the sperm inseminated by a human male remain in the female for on average only 30 mins (section 3.5.2). These fluids and sperm can thus play a role in sperm competition only on occasions that the female is inseminated by a second male within about 30 min of a previous male (or, more precisely, before flowback of the inseminate from the first male). Their arena for competition is in the upper vagina and, presumably, the contest is for passage to the cervix and perhaps the cervical crypts.

Even these vaginal sperm, however, may be reduced in number before they can compete. An unknown proportion may be removed from the vagina before the second male ejaculates by the suction generated via the copulatory thrusting of the second male's piston-shaped penis (section 6.8.3).

After flowback, sperm competition can involve only those 65% or so of sperm (Chapter 10) which remain behind in the female tract.

8.2.2 The fate of sperm which remain in the female tract

Depending on the stage of the female's menstrual cycle, vanguard sperm from the seminal pool may swim directly through channels of cervical mucus. They may then travel on, with or without assistance from the female tract, into the uterus and perhaps up to the oviduct. With assistance (from the uterus and/or oviduct), the first sperm arrive in the oviduct 5 min after insemination (Settlage *et al.*, 1973). Without such assistance, the first sperm may still arrive after only 30 min (Harvey, 1960).

Once in the oviduct, the sperm may attach to oviductal epithelial cells in the lower isthmus (section 3.5.1) and effectively enter storage. After a period of residence in the isthmus, they may detach from the oviductal epithelial cells in response to a 'signal' released by the female tract around the time of ovulation, and start to swim up the oviduct towards the egg (Hunter, 1988; Smith and Yanagimachi, 1991). The sperm which reach the egg first are the ones most likely to succeed in fertilization (Cummins and Yanagimachi, 1982).

Most of the sperm which reach the cervix after insemination are probably destined either to remain in the cervical mucus or to enter the cervical crypts (section 3.5.1). While in the cervical crypts, sperm become immotile or slow, are relatively safe from phagocytosis and may even be nourished (section 3.5.1). There are up to ten thousand crypts in the walls of the human cervix, potentially capable of accommodating millions of sperm. The subsequent expulsion of sperm from each crypt is a sudden phenomenon occurring at different times from each crypt. This staggered ejection results in a steady but eventually decreasing traffic of sperm out of the crypts over a period that may be as long as 8–10 days after insemination.

As the sperm leave the crypts, they re-enter the cervical mucus (Harper, 1988). Peak numbers are found in the mucus 2 hours and again 24 hours after (artificial) insemination (Settlage *et al.*, 1973) perhaps reflecting two main phases of release from the crypts. The more upwardly mobile sperm (Bellis *et al.*, 1990b) may pass through the mucus into the uterus where peak numbers have been seen about 24 hours after copulation (Settlage *et al.*, 1973). Some of these sperm may reach the oviducts and may join the earlier sperm in storage in the lower isthmus. Others may pass through the oviducts, some even exiting into the female's body cavity (Ahlgren, 1975).

At any one time, peak numbers of sperm in the whole oviduct reach 2000–5000 (Croxatto *et al.*, 1973). Most of these, however, are probably in the isthmus; the maximum number seen at the site of fertilization, the ampulla of the oviduct, is 200 (Ahlgren, 1975). Yet, if the oviduct is ligated, up to 23 000 may collect in the ampulla (Ahlgren, 1975). The impression is one of a steady passage of sperm through the oviduct, pausing longest in the secondary storage area of the isthmus.

The less upwardly mobile sperm (Bellis *et al.*, 1990b) from the cervical mucus and crypts probably remain in the cervical mucus. Here they are gradually phagocytosed by leucocytes (section 4.4.8) and/or are eventually carried by the glacier-like flow of mucus back into the vagina.

How long human sperm stay alive in different parts of the female tract is uncertain. In the hostile conditions of the vagina (Box 3.4), live sperm are found for a maximum of 12 hours after copulation (Vander Vliet and Hafez, 1974). In contrast, live sperm have been found in the cervix eight (Austin, 1975) or even ten (Baker, Bellis, Creighton and Penny, unpublished data) days after copulation. In the uterus, live sperm have been found up to two days after copulation (Rubenstein *et al.*, 1951) and in the oviduct up to nearly four days (85 h) after copulation (Ahlgren *et al.*, 1975; Croxatto *et al.*, 1975). Fertilization is most likely two days after copulation (Barrett and Marshall, 1969), but continues to be possible for at least five days after copulation (Barrett and Marshall, 1969; Ferin *et al.*, 1973; Thomas *et al.*, 1973; Baker and Bellis, 1993b).

We conclude, therefore, that for sperm from two different males to interact in the vagina, the female must double-mate within about 30 minutes (to involve the flowback sperm) or 12 hours (to involve the sperm which remain after flowback). For sperm to interact in the uterus the female must double-mate within about two days, in the oviduct, within 4–5 days, and specifically around the egg, within about five days. However, for sperm to interact in the cervix, either in the cervical mucus, or the cervical crypts, the female can double-mate any time within 8–10 days.

Smith (1984) opted for 7–9 days as the time interval within which females need to double-mate for sperm competition to occur. In order to avoid overestimating the frequency of double-mating and sperm competition in humans, we have opted previously (Bellis and Baker, 1990; Baker and Bellis, 1993a,b) for five days as the critical time interval. In the context of our '*Kamikaze Sperm Hypothesis*' (section 2.5.3 and Chapter 11), however, 8–10 days might be more realistic.

8.2.3 Differences of opinion

The above picture of human (and mammalian) sperm competition seems to us to be the most reasonable interpretation of the existing literature. The most recent review of mammalian sperm competition, however, paints a quite different picture. According to Gomendio and Roldan (1993): (1) eutherian female mammals do not store sperm; (2) only the fertile, not the active, life span of sperm is relevant to sperm competition; and (3) mammalian sperm are generally fertile for up to 48 hours after ejaculation, human sperm being fertile for 48–72 hours. In this section, we discuss the main differences between the two interpretations.

(a) SPERM STORAGE

According to Gomendio and Roldan (1993) one of the most crucial differences between birds and mammals is that female birds have sperm storage organs whereas eutherian mammals do not. They base this on the observation that sperm survive in birds longer than in mammals and on their argument that female birds need to store sperm (because they produce eggs sequentially over a number of days) whereas female mammals do not (because their egg(s) in any given cycle are usually produced simultaneously). Neither of these reasons seems compelling.

On average, it is undoubtedly true that sperm survive longer in birds than mammals (Birkead and Møller, 1992) but the difference is quantitative not qualitative. The range of sperm survival times in birds is 6–42 days (Birkhead and Møller, 1992); in mammals it is 0.5–42 days (or 0.5–198 days if we include bats). In order to promote their argument, Gomendio and Roldan exclude bats, disbelieve the claim of 42 days for hares (following Stavy and Terkel, 1992), and dismiss dogs (11 days; Doak *et al.*, 1967) on the grounds that only sperm mobility, not fertility, was established. A figure of 7 days for humans was dismissed on similar grounds in favour of a figure of 48–72 hours (from France, 1981).

The argument that female birds benefit from storing sperm whereas female mammals do not is at best weak. Equally powerful arguments for an advantage in sperm storage in female mammals can be presented (section 3.5.1 and sections 4.4.4–4.4.6). These same arguments could also be claimed to provide a more powerful explanation for sperm storage in birds than the sequential ovulation argument suggested by Gomendio and Roldan (1993).

In fact, having stated that there is no sperm storage in mammals (their p. 95), Gomendio and Roldan do later (pp. 96–97) seem to accept some degree of sperm storage by female mammals at least in the isthmus of the oviduct. However, they specifically dismiss sperm storage in the cervical crypts, at least in humans (their Box 2). Their only reasons are that sperm remain alive in the cervix (7 days) for longer than they have been recovered from the oviduct (3.54 days) and that the invasions of leucocytes after copulation would prevent further sperm migration from the cervix to the oviducts. Thus they conclude that 'The cervical crypts . . . are unlikely to act as a store to re-stock oviductal "reserves".'

Their reasoning is weak in two ways. First, the longest that sperm have been found in the oviducts (3.5 days) should be regarded as a minimum, not a maximum. The possibility of sperm being present in low numbers for longer always being impossible to rule out. Second, their argument that further sperm migration from cervix to oviduct would be prevented by leucocytes beyond the 3.5 days that sperm have been found in the oviducts ignores at least three factors.

1. If it were true, a male inseminating his partner more frequently than once every 3.5 days would only ever have sperm reach the oviduct on the first occasion.
2. Some sperm, particularly those that migrate to the oviduct, are immune to leucocyte attack (section 4.4.8).
3. In any case, leucocytosis has declined virtually to base levels by 24 h postinsemination (Pandya and Cohen, 1985) which is just when the peak of sperm numbers is seen in the uterus and the second peak is seen in the cervical mucus (Settlage *et al.*, 1973).

About 20 years ago, Morton and Glover (1974a) presented a compelling case for the cervical crypts of female mammals being sperm storage organs. Gomendio and Roldan ignore both that and more recent evidence (see section 3.5.1 except where stated).

1. The sperm storage tubules of birds and the cervical crypts of mammals are found in analogous positions in the female tract and are at least analogous to, if not homolo-

gous to, each other and similar structures in some reptiles.

2. The proportion of inseminated sperm which enter the storage tubules of birds and of humans is about the same (<1%).
3. In rabbits, ruminants and humans, the channels of the cervical mucus help to direct sperm to the cervical crypts.
4. While in the crypts, sperm are quiescent, safe from phagocytosis, and may receive nutrition. Moreover, they both gain and retain fertility and viability (Morton and Glover, 1974a).
5. Staggered over days, sperm are released and/or expelled from the crypts and re-enter the female tract. In the rabbit, sperm leave the cervix at the rate necessary to maintain the uterine and oviductal sperm counts at their observed levels (Morton and Glover, 1978a).

We agree with Gomendio and Roldan that the first sperm to appear in the oviducts after insemination have probably by-passed the cervical crypts (Box 3.5) and that these sperm enter storage in the isthmus (section 3.5.1). We also agree that it is possible, but consider it unlikely, that sperm subsequently released from the cervical crypts never reach secondary storage areas in the oviductal isthmus. We do not agree, however, that sperm from the cervical crypts play no part in sperm competition.

(b) FERTILE LIFE VERSUS ACTIVE LIFE

Gomendio and Roldan (1993) state categorically that 'In the context of sperm competition only sperm fertile life matters' not 'the duration of sperm motility' (their Box 1).

They hypothesize that mammalian sperm competition occurs only in the oviduct, between the lower isthmus and the ampulla (the site of fertilization). They envisage sperm from different males in residence in the isthmus, responding to a signal of impending ovulation by then racing to reach the descending ova, the first to arrive having the greatest chance of fertilization. Such a scenario, of course, would indicate that only fertile life, not active life, is important.

Any form of interaction between sperm from different males that influences the number or speed of sperm which enter this final race would change the situation totally. If sperm from one male, lower down the track (say in the cervix), can influence the number and speed of sperm from another male which enter the oviductal race, then active life can be just as important as fertile life.

For example, suppose a human male (A) inseminates his partner on day 1. By day 6, none of his sperm are fertile but at least some will continue to live (in the cervix) until day 9. On day 8 his partner is inseminated by another male (B) but the sperm from male A in the cervix interact with the sperm from B. The result is that only half as many sperm from B reach the oviduct as would otherwise have been the case. On day 9, male A inseminates his partner, his sperm interact with the live sperm from male B, and a certain number of A's sperm reach the oviduct. On day 11 the female ovulates and the oviductal sperm from A and B race to the egg.

Whatever the outcome, the result will have been influenced by the presence of live but no longer fertile sperm in the female tract. Unless Gomendio and Roldan can rule out all possibility that still active (but infertile) sperm influence the passage of sperm from other males through the female tract, which on present evidence (Baker and Bellis, 1993b; see Chapter 10) they cannot, their argument that the motile life of sperm is irrelevant in the context of sperm competition cannot be sustained. Indeed, we should argue that the very fact that sperm motility in most species is considerably longer than sperm fertility is by itself a strong indication that motile but infertile sperm have an important part to play in sperm competition.

(c) THE FERTILE LIFE OF HUMAN SPERM

As far as the active life of human sperm is concerned, Gomendio and Roldan and ourselves are more or less in accord, accepting a time of about 7 days (Gomendio and Roldan, 1993), 8 days (Austin, 1975) or 10 days

(Baker, Bellis, Creighton and Penny, unpublished). As far as the fertile life is concerned, we have assumed (Bellis and Baker, 1990) a period of at least 5 days, based primarily on the huge study by Barrett and Marshall (1969), supported by the observations by Ferin *et al.* (1973) and Thomas *et al.* (1973) and re-assured by a conception in our own study (Baker and Bellis, 1993b; Box 8.1). Gomendio and Roldan (1993), in contrast, opt for a fertile life of 48–72 hours, citing France (1981) and Wood (1989). They reject the evidence for a longer fertile life span on the grounds that such studies (presumably Barrett and Marshall's, though they cite no reference) estimated fertile life span using basal body temperature to indicate ovulation. They comment that 'regrettably, this method does not detect ovulation accurately'.

A number of investigations have been carried out on the accuracy with which basal body temperature curves (Boxes 4.6 and

Box 8.1 The relationship between Basal Body Temperature (BBT) curves and ovulation

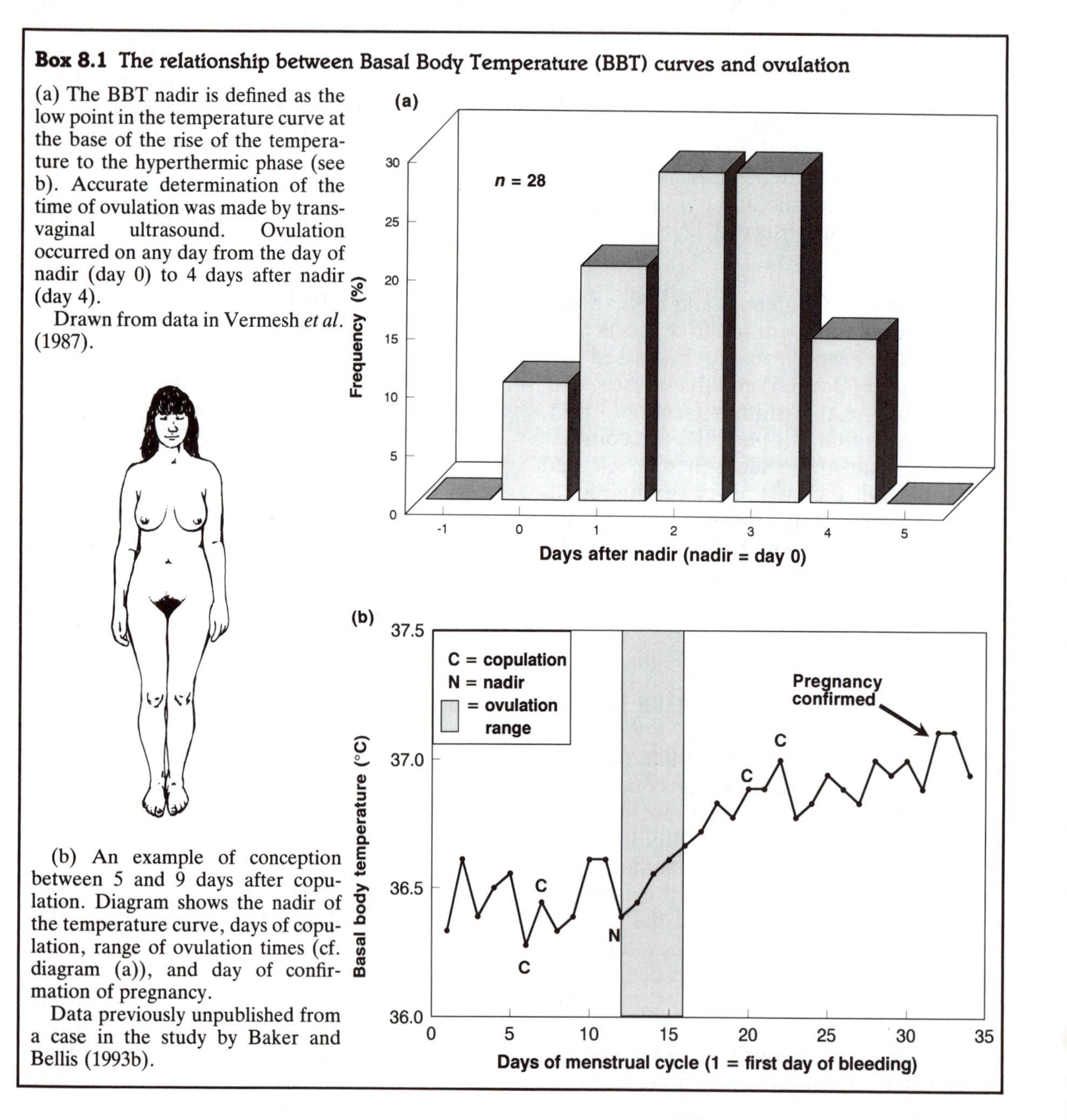

(a) The BBT nadir is defined as the low point in the temperature curve at the base of the rise of the temperature to the hyperthermic phase (see b). Accurate determination of the time of ovulation was made by transvaginal ultrasound. Ovulation occurred on any day from the day of nadir (day 0) to 4 days after nadir (day 4).

Drawn from data in Vermesh *et al.* (1987).

(b) An example of conception between 5 and 9 days after copulation. Diagram shows the nadir of the temperature curve, days of copulation, range of ovulation times (cf. diagram (a)), and day of confirmation of pregnancy.

Data previously unpublished from a case in the study by Baker and Bellis (1993b).

4.10) can be used to identify the day of ovulation (e.g. Morris *et al.*, 1976; Hilgers and Bailey, 1980; Quagliarello and Arny, 1986). All of the studies depended to some extent on comparing temperature curves with other means of detecting ovulation (e.g. LH surge) which themselves are now known to have their own variability (Lloyd and Coulam, 1989).

Recent studies have been able accurately to monitor ovulation using transvaginal ultrasound to observe the disappearance of the dominant follicles (e.g. Vermesh *et al.*, 1987; Lloyd and Coulam, 1989). Several studies have now accurately identified the relationship between the nadir of the basal body temperature curve (Box 8.1) and ovulation (Garcia *et al.*, 1981; Vermesh *et al.*, 1987; Martinez *et al.*, 1992). The conclusions are consistent. Ovulation can occur on any day from the day of nadir to up to 4 days afterwards, ovulation being most likely 1–3 days after nadir. Ovulation does not occur before the nadir (Box 8.1).

These results of course confirm all previous workers' conclusions that the BBT curve is an inaccurate predictor of the time of ovulation, an error margin of 5 days being unacceptable for many purposes. They do not, however, justify Gomendio and Roldan's (1993) decision to ignore Barrett and Marshall's (1969) study of the fertile life of sperm. Barrett and Marshall used the BBT nadir to mark the day before ovulation and thus to calculate the time from last copulation to ovulation. The recent studies now show that Barrett and Marshall will rarely have overestimated the time from copulation to conception. On the contrary, they will on average have underestimated that time interval by a little over one day.

Far from discrediting Barrett and Marshall's study, therefore, recent investigations of BBT curves suggest that the calculation of a sperm fertile half-life of 2 days (Barrett and Marshall, 1969) should in fact have been about 3 days, giving a maximum sperm fertile life of 6 days. The single example of conception from our own study (Box 8.1) is also attributable to a sperm fertile life of about 5–7 days. Moreover, the study by Thomas *et al.* (1973) contains a 5-day copulation–ovulation conception in which the timing of ovulation was identified by a combination of methods, including LH surge.

We see no reason, therefore, to accept Gomendio and Roldan's (1993) view that sperm competition only occurs when female humans mate with two different males within 2–3 days. Whether we are taking sperm fertile life (5–7 days) or active life (7–10 days) the time window for double-mating is considerably longer than they suggest. Smith (1984) was probably correct, therefore, in taking 7–9 days to be the appropriate period. Nevertheless, to be conservative, we shall continue to use 5 days as our figure.

8.3 How often, how many, how long: the statistics of double-mating

In a review of the circumstances leading to double-mating in humans, Smith (1984) identified five primary categories: (1) communal sex; (2) forced copulation; (3) prostitution; (4) courtship; and (5) facultative polyandry.

We prefer to recognize only three categories: (1) forced copulation; (2) overt polyandry; and (3) cryptic polyandry. We thus emphasize that the critical factors are: (a) whether the female's polyandry is driven by the male or the female; and (b) the copulating male's level of certainty that his sperm will find themselves in competition with sperm from another male.

Forced copulation can lead to double-mating in two ways (section 3.4). First, a sexually active female who is forced to copulate within five days of the last insemination by her most recent partner will contain sperm from her partner and the rapist. So, too, will a female who is inseminated by her partner within five days of being forced to copulate. Second, when a female is forced to copulate, not by a single male, but by a group (= 'gang rape'; perhaps up to 70% of all forced copulations; Steen and Price, 1977), each member of the group inseminates the female in turn. Such 'gang rapes' generate a high level of sperm competition, involving flowback sperm as well as those sperm retained by the female.

Overt polyandry takes a number of forms, the best known being 'prostitution' and various form of communal sex such as 'wife sharing', 'wife swapping' and 'orgies' (section 2.4.1). 'Prostitutes' are inseminated by a succession of different males, often with short time intervals between each male. Their reproductive tracts are therefore sites of intensive sperm competition (unless, as is now increasingly the case, they use condoms). All forms of communal sex similarly lead to intensive sperm competition, again often with short time intervals between successive inseminations.

Cryptic polyandry is usually the result of either EPCs (= infidelity; Box 6.5) or of females experimenting with a number of partners before making a decision over which to use as a long-term partner (section 6.3.1). By their very nature, EPCs, whether cryptic or overt, are often likely to be double-matings. Successive in-pair copulations maintain an almost continuous presence of the primary male's sperm in the female's tract (Chapter 9). Any EPC, then, will place the sperm from the primary male partner in competition with the sperm from the extra-pair male(s). Even if sperm from the in-pair male are not present when the female is unfaithful, the pressure for crypsis and the advantage of promoting sperm competition may well favour her rapidly (double-) mating with her partner.

The exploratory polyandry of young girls and women 'between partners' may or may not lead to sperm competition, depending on the rate at which copulation occurs with new partners.

8.3.1 What proportion of males place their sperm in competition?

There is no survey to our knowledge of the proportion of males who at some time in their lives force a female to copulate, either on their own or as a member of a gang. Nor do we have any figures of what proportion of raped females are likely to have been inseminated by a different male (e.g. their partner) in the five days before or after the forced copulation. We cannot, therefore, estimate what proportion of males are likely, at least once in their lifetime, to experience sperm competition via forced copulation (perpetrated either by themselves on another male's partner or by another male on their partner). Whatever the proportion, however, it is likely to be relatively small but to vary as a function of location and circumstance.

About 600 000 forced copulations are estimated to occur in the USA each year (Green, 1980). Let us assume that one-third of these are gang rapes (cf. Steen and Price, 1977) with a mean gang size of 2.5 but that only half of the women mate with their partners in the five days before or after the forced copulation. This means that 300 000 partners and 700 000 rapists (i.e. about 1% of reproductive-age males) find their sperm entering into sperm competition each year as a result of forced copulation.

The increase in incidence of forced copulation during warfare is due primarily to more males raping rather than simply some males raping more (Brownmiller, 1975; Shields and Shields, 1983). One estimate (section 3.4) suggests that the proportion of males who find their sperm entering competition as a result of forced copulation more or less doubles during times of war.

Males who inseminate prostitutes invariably expose their sperm to sperm competition. Among the male population of ancient Greece and Rome the insemination of female prostitutes was almost universal (Bullough and Bullough, 1978). In the USA in the 1940s, 69% of caucasoid males had inseminated a prostitute at least once and 15% did so on a regular basis (Kinsey *et al.*, 1948). In the UK in the 1990s, 10% of men aged 45–59 years had paid for sex at least once in their lives (Johnson *et al.*, 1994). There is no indication that men who pay for sex are individuals who cannot obtain copulations in any other way. In the UK, men who had paid for sex were also more likely to have had large numbers of unpaid partners (Johnson *et al.*, 1994).

In the USA in the 1970s, about 4.5% of couples surveyed had engaged in overt mate swapping with other couples but the majority of these (about 3.5% of couples) had done so only once or twice (Athanasiou, 1973).

In the 1940s a male in the USA had a

roughly 25% chance that his female partner would be unfaithful at some time during their relationship (Kinsey *et al.*, 1953). In addition, he had a 50% chance of inseminating another female during the same period. These figures will now be much higher (Wolfe, 1981). The probability that the female targets of the male's infidelity are at the same time sexually active with another male is unknown. Nevertheless, the chances that a male will experience sperm competition via his own or his partner's infidelity seem relatively high.

The probability that the sperm of a male in western society will experience competition with flowback sperm seems relatively low and is mainly likely to occur via the male's involvement with gang rape, communal sex or prostitutes. In contrast, the probability that his sperm will experience competition with sperm retained in the cervix seems relatively high, primarily due to female infidelity but also via prostitutes and less often, directly or indirectly, via forced copulation.

It would seem that even in peacetime, the majority of males in westernized societies will place their sperm in competition with those from another male at some time in their lives. In other societies (e.g. those in which overt polyandry is the rule; see section 2.4.1) and at other times (e.g. wartime) the frequency with which males place their sperm in competition will be correspondingly greater.

8.3.2 What proportion of females generate sperm competition?

In contrast to the situation for males, we are in a position to make much more specific estimates of the proportion of females who might generate sperm competition for at least one population (Britain, 1989; Box 2.9).

According to our Nationwide survey of nearly 4000 women (Box 2.6), 17% double-mate within their first 50 copulations, 50% within their first 500, and over 80% after 3000 copulations (all circumstances combined, e.g. forced copulation, communal sex, prostitution, infidelity). Increasing the critical time-interval from 5 days to 10 days makes little difference to these figures. After 3000 copulations, 1% of women had double-mated in less than 30 minutes and 13% in less than 1 hour, most such occasions probably generating sperm competition involving flowback sperm.

The majority of British women, therefore, could generate sperm competition at least once in their lives. However, in order to judge the selective impact of double-mating and sperm competition, it is important to know not only the proportion of women who ever double-mate but also when and how often they do so. Box 8.2 shows variation in double-mating rate with stage of reproductive life. For example, EPCs and double-mating were least common after having just one child. In families with more than one child, EPCs and double-mating were most common before the first and last. These are the children who blood-group studies suggest are also the least likely to have been sired by their mother's long-term partner (Michigan, USA; Schacht and Gershowitz, 1963).

8.3.3 Double-mating, contraception and sperm competition

One problem with using the incidence of double-mating to measure the incidence of sperm competition is that, except when they fail (Box 7.1), some modern contraceptives prevent sperm competition (e.g. condom; vasectomy). Female barrier and other methods (e.g. the cap, intrauterine devices) allow sperm to interact but prevent their access to the egg. When used in conjunction with spermicides, they also drastically reduce sperm interaction. In contrast, female oral contraceptives allow sperm to compete freely but rob them of the prize (fertilization). However, in this respect they produce circumstances little different from anovulatory cycles or the infertile stage of a normal, ovulatory menstrual cycle. Even in the absence of modern contraceptives, fewer than 20% of inseminations (USA, 18%; UK, 18%; see Box 6.11) take place in the five days preceding ovulation and thus have any chance of leading to fertilization. The influence of contraception is simply to reduce this proportion.

In the UK, about 20% of copulations deny

Box 8.2 Variation in human double-mating rate with age and sexual experience (UK)

Variation in EPC and double mating rate with age.

Variation in EPC and double-mating rate with number of children.

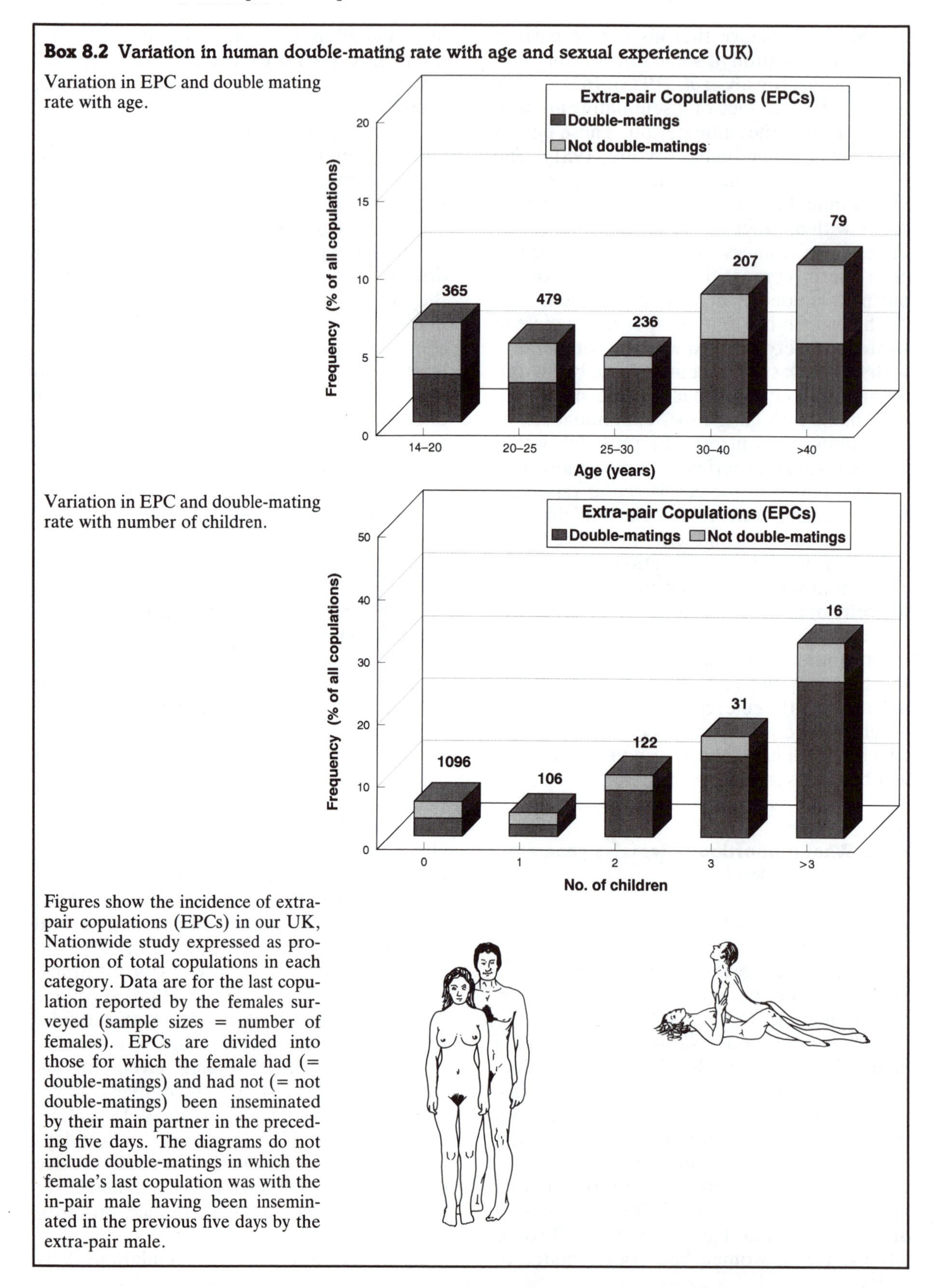

Figures show the incidence of extra-pair copulations (EPCs) in our UK, Nationwide study expressed as proportion of total copulations in each category. Data are for the last copulation reported by the females surveyed (sample sizes = number of females). EPCs are divided into those for which the female had (= double-matings) and had not (= not double-matings) been inseminated by their main partner in the preceding five days. The diagrams do not include double-matings in which the female's last copulation was with the in-pair male having been inseminated in the previous five days by the extra-pair male.

sperm competition completely through use of a condom (Box 7.1). About 10% would allow sperm to interact but would prevent access to the egg and nearly 30% (20% pill; 7% rhythm) would allow sperm to interact freely but with no prize. An additional 20% (withdrawal) might allow a few sperm real opportunity for sperm competition should double-mating occur. About 14% of copulations allow full opportunity for sperm competition. Our own UK Nationwide survey gave a figure of 11% (316/2871; Box 8.3). All things being equal, therefore, contraception in the UK seems at first sight to reduce the proportion of copulations which could be potential targets for sperm competition from 18% to, say, 2%. However, it seems that not all things are equal.

Just because some contraceptive methods prevent sperm competition, it does not follow necessarily that women use them to such an end. As was pointed out in section 7.2.1,

Box 8.3 Association between double-matings, contraceptive use, and fertility (UK)

Figure shows the incidence of extra-pair copulations (EPCs) in our UK, Nationwide study expressed as proportion of total copulations in each category. Data are for the last copulation reported by the females surveyed (sample sizes = number of females). EPCs are divided into those for which the female had (= double-matings) and had not (= not double-matings) been inseminated by their main partner in the preceding five days. The diagrams do not include double-matings in which the female's last copulation was with the in-pair male having been inseminated in the previous five days by the extra-pair male.

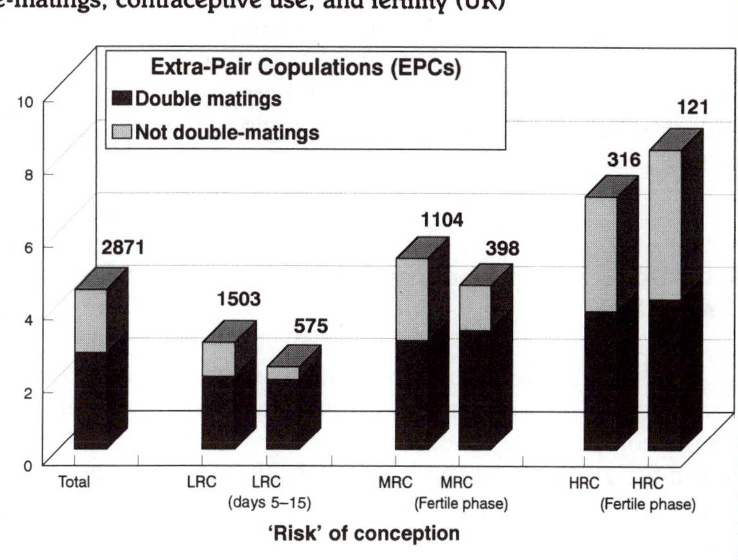

Risk of conception is categorized as follows (see also Box 7.1):

LRC: low risk of conception (oral contraceptive; sterilization; menopause).
MRC: medium risk of conception (barrier methods; intra-uterine devices).
HRC: high risk of conception (no 'modern' contraceptive).

In each category, copulations during the fertile phase (or days 5–15 in cycles with oral contraceptives) are shown separately.

There is a clear increase in the incidence of EPCs, including double-matings, when the risk of conception is greater.

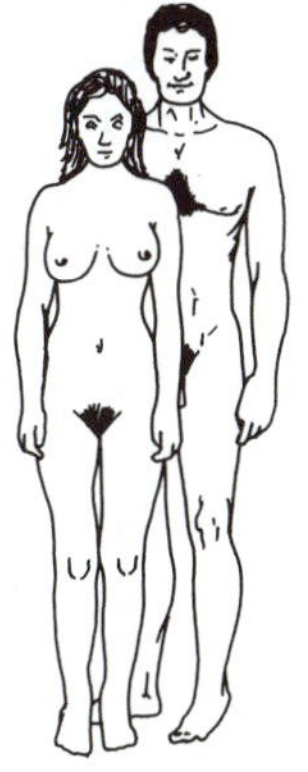

modern contraceptives may be used by women (and men) to enhance their sexual strategies for reproduction and our data provide some evidence of this. The incidence of EPCs and double-matings as a proportion of total copulations for our total UK sample is very much a function of potential fertility (Box 8.3). The more fertile the female (i.e. in terms of stage of menstrual cycle and type of contraceptive), the higher the proportion of copulations that are double matings (cf. Bellis and Baker, 1990).

Women are significantly more likely to use contraception during IPCs than during EPCs, particularly double-matings (Box 8.3). There is thus more than a hint that women are using contraceptives to favour the success of extra-pair males (see also Chapter 10) and to enhance the efficiency of their sexual crypsis with respect to their partner (see discussion in section 6.3.1).

8.3.4 Do females promote sperm competition?

In section 2.5.6, we argued that natural selection should favour females who promote sperm competition. There is a strong indication for women that this occurs (Bellis and Baker, 1990).

IPCs are relatively evenly spread through the menstrual cycle (Box 6.10) but with an increased incidence during the infertile post-ovulation phase. EPCs, on the other hand, are most common during the fertile preovulation phase. During the few days of the fertile phase, there is no correlation between the incidence of IPCs and probability of conception (Bellis and Baker, 1990). There is, however, an overall trend for EPCs, and particularly double-matings, to peak two days before ovulation at the time of maximum probability of conception. By spreading copulations with her partner throughout the menstrual cycle but concentrating EPCs into her most fertile period, a female gains the maximum advantage from her sexual crypsis while reducing the chances of her infidelity being discovered by her partner.

8.4 The pay-off

Interesting and important though it is to be able to assess the frequency with which male and female behaviour leads to sperm competition, we should stress that there is only one real way to estimate the selective impact of sperm competition. The critical factor is the proportion of children in each generation who are conceived via sperm competition. This is the only true measure of the level of natural selection on factors that influence the outcome of sperm competition.

The above discussion suggests that at some time in their lives the majority of males in western societies place their sperm in competition with sperm from another male and the majority of females contain live sperm from two or more different males. The prize, a fertilizable egg, will not be present every time sperm find themselves in competition but then that has been the case for at least the last five million years, since the evolution of female sexual crypsis.

Our behavioural data, from the UK nationwide survey, suggests that on between 4% and 12% of the occasions that a prize is present (i.e. no contraception during the fertile phase of the menstrual cycle), females generate sperm competition via double-mating. The lower figure (4%) is for the occasions on which the female's last copulation was with an extra-pair male, during her fertile period, and was within five days of previously being inseminated by her partner (Box 8.3). The upper figure (12%) allows for two further possibilities: (1) an equivalent (4%) number of the fertile phase inseminations by the partner will have been preceded within five days by insemination by an extra-pair male; and (2) before the female ovulates, a proportion of the 4% fertile phase EPCs not preceded by mating with her partner (Box 8.3) may be followed by insemination by her partner.

On this basis, we should expect that 4–12% of children in Britain are conceived to a sperm that has prevailed in competition with sperm from another male. This is such an important figure, it would be reassuring if it were not based on a single data source, such as our Nationwide survey.

8.4.1 What proportion of children are the result of sperm competition?

Another way of approaching the question of the pay-off to sperm competition via live births is to examine the frequency with which children are not the genetic offspring of their putative father (= cuckoldry; = paternal discrepancy).

Medical students are usually taught that the level of paternal discrepancy is 10–15% (Macintyre and Sooman, 1992). The figure of 10% is most widely used in DNA studies and is quoted in standard genetics textbooks. As Macintyre and Sooman (1992) point out, however, rather few of these studies have been published in any detail and there is a real need for proper investigation using modern techniques. A number of estimates of paternal discrepancy are given in Box 8.4 together with a brief discussion of some of the problems associated with such investigations.

Estimated levels range from a low of 1.4% for caucasians in Michigan, USA, and 2% for the !Kung to a high of around 30%. In Britain, values range from about 6% in a large urban community in London to around 30% for the inhabitants of tower blocks of flats in both northwest and southern England.

Of course, paternal discrepancy should equal EPC rate, not double-mating rate. Not all EPCs are double-matings (i.e. within 5 days of the last or next insemination by the female's primary partner). According to our survey, only about 50% (Box 8.3) of EPCs not involving contraception during the fertile phase of the cycle are double-matings leading to sperm competition. This suggests that approximately 50% of all cases of paternal discrepancy in Britain are children conceived to the sperm of an extra-pair male after competition with sperm from the primary male partner.

If we take the most conservative measure of paternal discrepancy for Britain (6%; Box 8.4), which is also the one based on the largest sample, we obtain a figure for children conceived to the extra-pair male following sperm competition of about 3%. However, this should be an underestimation of the incidence of sperm competition for in addition to occasions on which the EPC male wins, there should be other occasions on which the IPC male wins. It is tempting simply to double the figure of 3% (to 6%) on the grounds that EPC and IPC males have equal chances of winning. This, however, is not necessarily the case.

8.4.2 Which male does best: first or last, extra-pair or in-pair?

One of the few clear conclusions so far to emerge from studies of sperm competition in other animals is that which male wins is not random but depends very much on mating sequence and time interval between matings. In insects, with the exception of for example bees and silkmoths, the last male to mate with the female before her eggs begin to travel down her oviduct invariably has by far the greatest probability of fertilization (Parker, 1970a). That the same is true for birds was first noticed by Aristotle about 2300 years ago (Payne and Kahrs, 1961) and has since been well established (Birkhead and Møller, 1992). For example, the proportion of eggs fertilized by the last male to mate with the female (= P_2) is about 80% for the dung fly, *Scatophaga stercoraria* (Parker, 1970b) and 70–80% for the zebra finch, *Taeniopygia guttata* (Birkhead *et al.*, 1989). Across species in birds, the figure is relatively constant at about 80% (Birkhead and Møller, 1992). The precise figure, however, is influenced by the time interval between matings.

Generally, studies of the chicken, turkey, mallard and zebra finch show that last male precedence is the rule if copulations are separated by 4 h or more. If the copulation interval is less than this, paternity is proportional to the number of sperm inseminated by each male (see review by Birkhead and Møller, 1992).

Mammals show a much less clear-cut influence of mating order. The first male to mate has the advantage in house mice, the last male in prairie voles, with no clear order

Box 8.4 Studies of paternal discrepancy

The data

Area and sample	Sample size	Paternal discrepancy (%)	Authors
North America			
Michigan			
Caucasoid	1417	1.4	Schacht and Gershowitz (1963)
Negroid	523	10.1	Schacht and Gershowitz (1963)
South America			
Yanomamo	132	9.0	Neel and Weiss (1975)
Europe			
West Middlesex	2596	5.9	Edwards (1957)
NW England	?	20–30	McLaren in Cohen (1977)
SE England	c. 250	30.0	Philipp (1973)
France	>171	<14.6	Salmon *et al.* (1980)
Africa			
!Kung bushmen		c. 2	Harpending in Trivers (1972)
Oceania			
Hawaii	2839	2.3	Ashton (1980)
Median		9.0	

The problems

1. *Measurement*

All of the above data on paternal discrepancy were derived from blood group studies. Such data are not ideal in a number of ways but primarily because only certain inheritance patterns can exclude paternity with certainty. Other inheritance patterns, or extra-pair males with the same blood group as the putative father, can lead to missed examples of cuckoldry. There is obviously no control for the latter problem but the former problems, at least, can be controlled to some extent. The procedure is either (when sample size is large) to use only those mother–father blood group combinations that do allow a decision to be made or to make some statistical allowance for the proportion of combinations that cannot be detected. The advent of DNA genetic fingerprinting has the potential to raise the study of paternal discrepancy to a new level of accuracy and the technique has already been used for this purpose for birds (e.g. Burke, 1989) and primates (Chapter 6). No formal study has yet been carried out on humans, for obvious reasons.

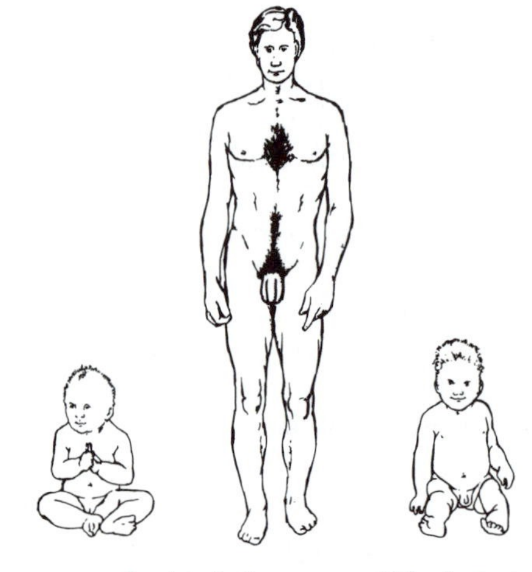

2. *Interpretation*

There is one major interpretational problem inherent in all studies which use discrepancy as a method to judge levels of infidelity and double-mating: the female may change partners after the event.

First, a female may conceive with one male partner, but then obtain paternal care from another (with or without the latter's knowledge) with no temporal overlap in the two relationship. In such cases, the paternal discrepancy involves neither infidelity nor sperm competition. This possibility was used to explain the supposedly high level of paternal discrepancy in Michigan (Smith, 1984).

Second, a female may be paired to one male, conceive by another (via infidelity and perhaps sperm competition), but then move on to the genetic father before being tested for paternal discrepancy. In which case, infidelity and even sperm competition was involved in conception but no paternal discrepancy was evident. This factor is the probable explanation for the apparently low level of paternal discrepancy (2%) found in the !Kung (discussed in Smith, 1984), well known for their high incidence of infidelity (Howell, 1979). These two factors influence the relationship between paternal discrepancy and infidelity in opposite ways and the hope is that they nullify each other.

effects in laboratory rats or deer mice (Dewsbury, 1984). Equally, there were no order effects with swine. In most cases, however, much depended on the time interval between first and second males as it does for birds.

In mice, if the time interval is very short (minutes), the second copulation seems to be spiteful and the female may not conceive at all (section 6.4.1). Thereafter, as time interval increases, the proportion of eggs fertilized by the first male increases. Longer time intervals also favour the first male in hamsters, *Mesocricetus auratus* (Huck *et al.*, 1985). In 13-lined ground squirrels, *Spermophila tridecemlineatus*, short time intervals swing paternity strongly in favour of one or other male, medium time intervals in favour of mixed paternity, and longer time intervals in favour of the first male. Success of the second male depended critically on the length of copulation (Schwagmeyer and Foltz, 1990).

Poultry breeders have long interpreted the second male advantage shown by birds in terms of stratified sperm storage (see review by Birkhead and Møller, 1992). Females are receptive well in advance of ovulation and the sperm are stored in blind-ended storage organs. Sperm from later males could thus be nearer the entrance (and thus nearer the exit) than sperm from earlier males. On a 'last in, first out' principle, storage organs could promote a second male advantage. However, recent but limited empirical evidence provides no support for the stratification model and Birkhead and Møller (1992) suggest that limited sperm storage coupled with some form of sperm displacement may be a better explanation. This is similar to the explanation preferred for insects (Ridley, 1989; Parker *et al.*, 1990; Rothschild, 1991).

Whatever the eventual explanation for the fairly consistent pattern of mating precedence in birds and insects, it seems clear that there can be no simple story for the much more variable outcome in mammals. This outcome probably depends on some variable interaction between sperm number (Chapter 9), sperm removal (Chapter 6), and sperm fertilizing capacity (Chapter 12) as well as on a variety of female influences ranging from sperm retention (Chapter 10) to selective sperm passage through the female tract (Chapter 4).

Female humans are continually receptive and sperm are fertile (5 days) and alive (8–10 days) a lot longer than the optimum time from insemination to fertilization (2 days). The scope for sperm–sperm and female–male interactions in influencing the outcome is considerable. Essentially, however, we have as yet no idea for humans or other primates whether there is a first or last male advantage, if either. Our prediction would be that in humans, all else being equal, it is the last male to inseminate a female near to 2 days before ovulation who has the advantage.

Even given an asymmetry between males depending on mating sequence, the relative probability of fertilization by EPC and IPC males may depend on the number of sperm they each ejaculate (Chapter 9) and the number of sperm from each that the female retains (Chapter 10). We have already given some indication that the female may favour the EP male through her differential use of contraception (sections 8.3.3 and 8.3.4). She may also favour the extra-pair male in her timing of EPCs and IPCs relative to the timing of ovulation (Box 6.10). In Chapter 10, we present evidence that the female may even favour the EP male via the relative number of sperm she retains from each.

Given the possibility that EP males may have a higher probability of fertilization during sperm competition, we may not be justified in calculating the proportion of children conceived via sperm competition by simply doubling 3% to obtain 6%. The figure may be lower (but more than 3%). For the moment, however, we shall be conservative and assume a figure of only 4%.

Both our Nationwide survey, therefore, and blood group studies, suggest that over the last few decades in Britain at least 4% of children (but perhaps as many as 6–12%) have been conceived to a sperm that has prevailed in sperm competition. In other words, in any group of 25 people, at least one owes their existence to the fact that their father's ejaculate had the characteristics necessary to outcompete that of another male. Such a level of reproductive pay-off is relatively hefty in evolutionary terms. In Britain, at the present time, therefore, sperm competition is still a major source of selection on the human sexual programme. Whether the figure is higher or lower in other communities elsewhere remains to be determined.

9 Optimizing inseminates: ejaculate adjustment by males and the function of masturbation

9.1 Introduction

In the previous chapters, we have argued that sperm competition has been the main force to shape the genetic programme that drives human sexuality. Modern humans have not escaped this force. In fact, far from sperm competition being negated by modern techniques of contraception, these techniques seem actually to have been incorporated into the behavioural strategies underlying sperm competition. In evolutionary terms, a substantial proportion of children are still being conceived while their mother has sperm from two or more men in her reproductive tract.

In this chapter, we look at sperm competition from the viewpoint of the ejaculatory behaviour of the modern male. In particular, we are concerned with the way the total number of sperm he inseminates varies from ejaculation to ejaculation. Our premise is that this variation is an adaptation evolved in response to past selective pressures generated by sperm competition.

Essentially, the chapter is divided into five main sections:

1. variation in the number of sperm inseminated during in-pair copulation;
2. variation in the number inseminated during extra-pair copulation;
3. the features and function of masturbation;
4. the possibility of seasonal and long-term changes in the number of sperm ejaculated during copulation and masturbation; and
5. the question of whether males show restraint over the number of sperm inseminated for short-term benefit or long-term economy.

Our approach is always to begin with the predictions of sperm competition theory (section 2.5), and then to consider what is known about variation in ejaculate size as it relates to these predictions. Unless stated otherwise, the data we use are taken from the studies introduced in Box 3.8.

9.2 Variation in number of sperm inseminated during in-pair copulation (IPC)

Whatever the mechanism of sperm competition (i.e. lottery, race or warfare; section 2.5), theory predicts that sperm competition will favour males who inseminate larger ejaculates (i.e. more sperm/ejaculate). There is some direct evidence (for chickens) in support of this prediction (section 2.5.1). When sperm from different males are in

Box 9.1 Predicting the number of sperm ejaculated during IPC and masturbation

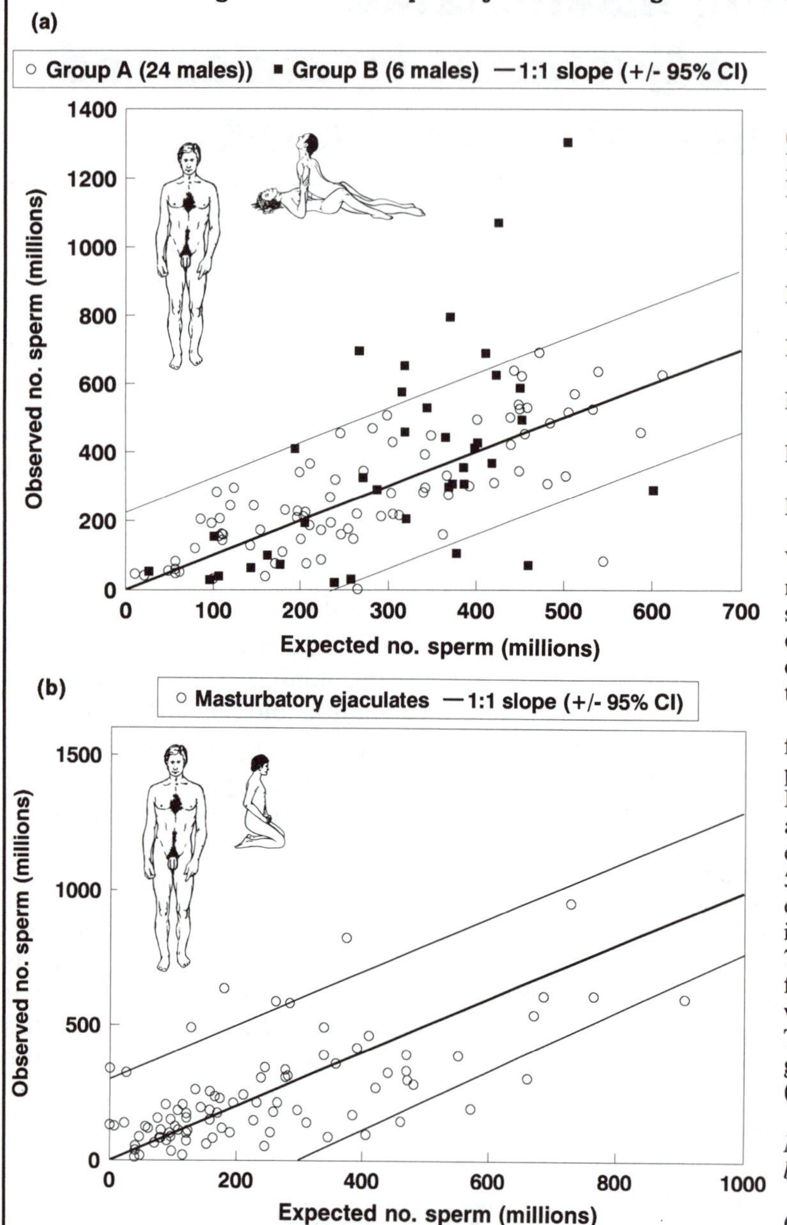

Number inseminated during IPC

(1) NS = PVequ2 − (((PVequ2-mid)/(limC-mid))^2 × (limC-limO))

NS = number of sperm (millions) predicted to be inseminated during IPC;
PVequ2 = the predicted value for NS from equation (2);
mid = mid-point of number of sperm in observed ejaculates (350 million);
limC = the minimum (−359 million) or maximum (919 million) calculable limit to number of sperm from equation (2); and
limO = the minimum (2 million) or maximum (692 million) observed number of sperm in IPC ejaculates in the study concerned.

When PVequ2 > mid, maximum limits should be used; when PVequ2 < mid, minimum values should be used.

(2) PVequ2 = 1.94 × HIPC − 3.40 × PCT + ((2.41 × HMAS − 228) × MC) − 1008 + 23.37 × FW

PVequ2 = number of sperm inseminated (millions);
HIPC = hours (up to 192) since last IPC;
PCT = per cent time the pair have spent together since their last IPC;
HMAS = hours (up to 72) since last masturbation;
MC = 0 for IPC-IPC ejaculates and 1 for MAS-IPC ejaculates;
FW = female weight (kg).

When a parameter exceeds the stated maximum (e.g. 192 for HIPC), the stated maximum should be used. The equation should be used only with caution for female weights outside the range of 52–66 kg.

(a) This equation was calculated for IPC ejaculates from Group A (24 pairs; 84 ejaculates; Baker and Bellis, 1993a). It was then applied to a further 38 ejaculates from a different six pairs (Group B). It explained 58% of the variance in Group A ejaculates; 32% in Group B and 43% in the two groups combined. Nearly 75% of the ejaculates from Group B fall within the 95% confidence intervals about a 1:1 slope for Group A. The regression coefficient for the two groups combined was 1.05 ± 0.05 (forced through origin).

Number ejaculated during masturbation

(3) NS = 42 + 4.63 × HEJ − 6.37 × ABS (MAGE-25)

where NS = number of sperm ejaculated in millions; HEJ = hours since last ejaculation; MAGE = male age in years.

(b) This equation was calculated for masturbation ejaculates from 26 males; 88 ejaculates. When forced through the origin, the regression of observed number of sperm on expected had a slope of 0.85 ± 0.05 SE and explained 33% of the variance.

competition, all else being equal the male who provided most sperm fertilized most of the eggs.

This does not mean, however, that a male should always inseminate as many sperm into the female as possible. Sperm competition theory further states (section 2.5.5) that the number of sperm a male should actually inseminate into a female should be a trade-off between two opposing pressures. On the one hand, the risk of sperm competition favours an increase in the number of sperm inseminated; on the other, potential disadvantages of inseminating too many sperm select for restraint. The predicted result is that on each occasion a male should inseminate a female with the number of sperm that is the optimum trade-off between the risk of sperm competition and the need for restraint (section 2.5.5).

This theory has generated at least three predictions: (1) the greater the risk of sperm competition, the more sperm should be inseminated (Parker, 1982); (2) the greater the reproductive value of a female, the more a male should invest in sperm competition and hence the more sperm should be inseminated (Dewsbury, 1982); and (3) under some circumstances males do better in sperm competition to partition sperm between a succession of in-pair copulations (IPCs) rather than inseminate all available sperm in a single IPC (Parker, 1984).

One of the consequences of our study (Baker and Bellis, 1993a) of the factors influencing the number of sperm ejaculated by humans is that we have produced equations that predict the number ejaculated during IPC and masturbation. These equations are presented in Box 9.1 and reference is made to them at intervals throughout this chapter.

9.2.1 Male response to risk of sperm competition

The first of the predictions of sperm competition theory (that males should inseminate more sperm when the risk of sperm competition is greater) has now been tested at four levels: (1) interspecific (butterflies, Svärd and Wiklund, 1989; birds, Møller, 1988a; primates, Harvey and Harcourt, 1984; Møller 1988b; and ungulates, Ginsberg and Rubenstein, 1990); (2) intraspecific (beetles, Gage and Baker, 1991; flies, Gage, 1992; rats, Bellis *et al.*, 1990a; and humans, Baker and Bellis, 1989b); (3) intermale (humans, Box 4.17); and (4) intramale (beetles, Gage and Baker, 1991; humans, Baker and Bellis, 1993a). So far, all tests are consistent with the predictions of sperm competition theory: males inseminate more sperm when the risk of sperm competition is higher.

As far as humans are concerned, the less time a male spends with his female partner between IPCs, the greater the probability that the female will engage in extra-pair copulation (including double-mating) (Box 2.11). Even when all else is equal (e.g. time since last copulation), males respond to this increased risk by inseminating more sperm into their partner at their next IPC (Box 9.2).

The probability of a female double-mating varies with age (Box 8.2). However, we do not yet know whether the critical factor is female age or age of the male partner. Cross-culturally, on average, the age of male and female partners are similar but with the male being just a few years older (Buss, 1989; see also Box 6.4). Thus, rather than females aged about 22 years being least likely to double-mate, it is possible that females are least likely to be unfaithful to males aged around 25 years but are more likely to be unfaithful when paired to younger, and particularly older, males. In which case, the number of sperm inseminated by a male should vary accordingly with age. Our study of whole ejaculates (Box 3.8) has found just such a positive relationship (Box 9.2).

Perhaps surprisingly, the relationship is still very significant even when sperm number is controlled for variation in interejaculation interval, percentage time together, and other factors (using residuals from the equation in Box 9.1). This is so despite the fact that peak daily sperm production occurs at around age 25 years and then declines with age (Box 5.8). The explanation seems to be that males younger and particularly older than 25 years ejaculate proportionally fewer of their stored

Box 9.2 Number of sperm inseminated during IPC varies according to the risk of sperm competition and testis size.

The risk of sperm competition varies erratically from copulation to copulation as a negative function of the level of mate-guarding and in a more predictable way according to male age.

Level of mate-guarding

The less time a male spends with his partner between copulations, the greater the chance of sperm competition (Box 2.11). Also, the less time he spends with his partner, the more sperm are inseminated. This is evident both for males who produced only one sample (top figure) and for males who produced more than one sample (middle figure). In the middle figure, the darker lines (with dots) are for the four males who produced samples in more than two categories of percentage time together. There are conspicuous differences between males in the number of sperm inseminated. Even so, 77% (10/13) of these males showed a decline in sperm numbers with an increase in time with their partner. Two of the three who did not show a decline inseminated few (<60 million) sperm.

The association between percentage time together and the number of sperm inseminated is not an artefact of time since the male's last ejaculation. The association is still evident within any given range of times since last ejaculation (bottom figure) (Sample sizes are number of ejaculates.)

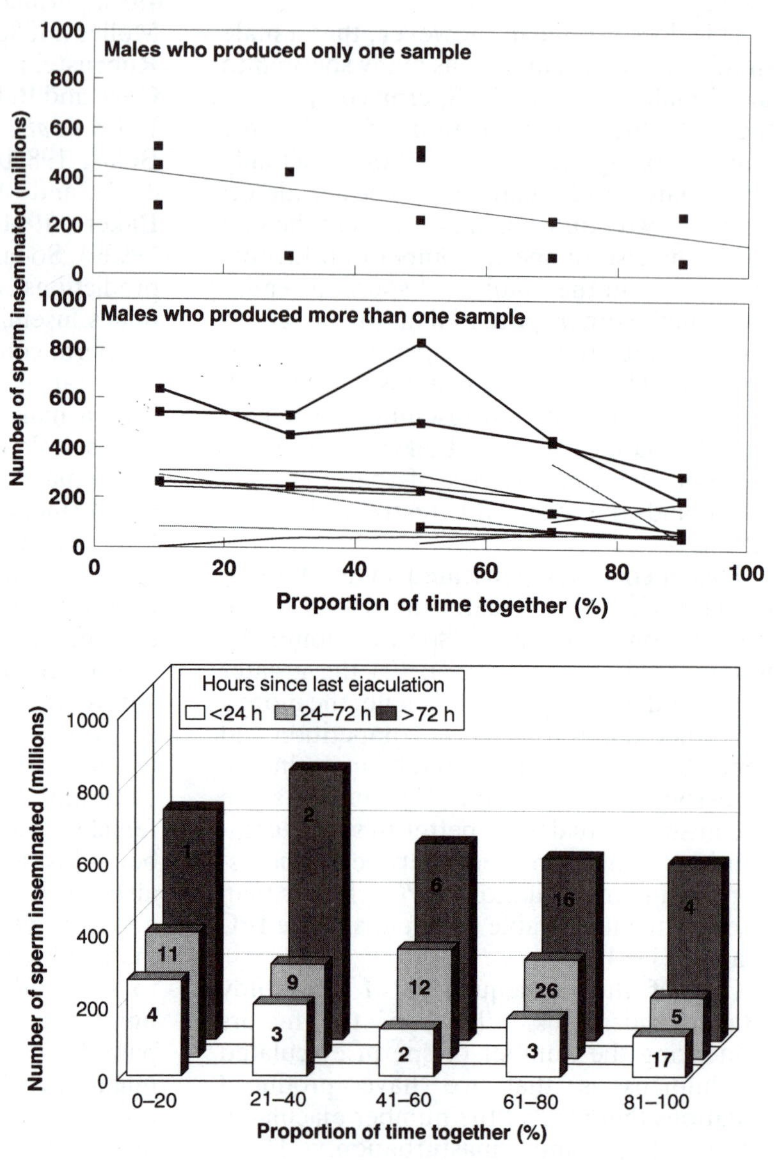

sperm during masturbation (Box 9.6). They thus conserve more sperm for insemination during copulation. The average age of sperm inseminated into their partners by older males will thus be older than those inseminated by younger males. This pattern itself may be adaptive (section 11.3.1).

9.2.2 'Topping-up' the partner

On average, human pairs engage in IPC at median intervals of about every three days (72 h). However, frequency varies with age, length of relationship and phase of menstrual cycle (Boxes 4.8 and 6.11) as well, of course,

Male age

Variation in the number of sperm inseminated during copulation is a significant function of the probability of double-mating by females of comparable age. This remains true even when factors such as interejaculation interval are controlled (by calculating residuals from the equation in Box 9.1). Sample sizes are $n(N)$ where n is the number of males; N the number of ejaculates.

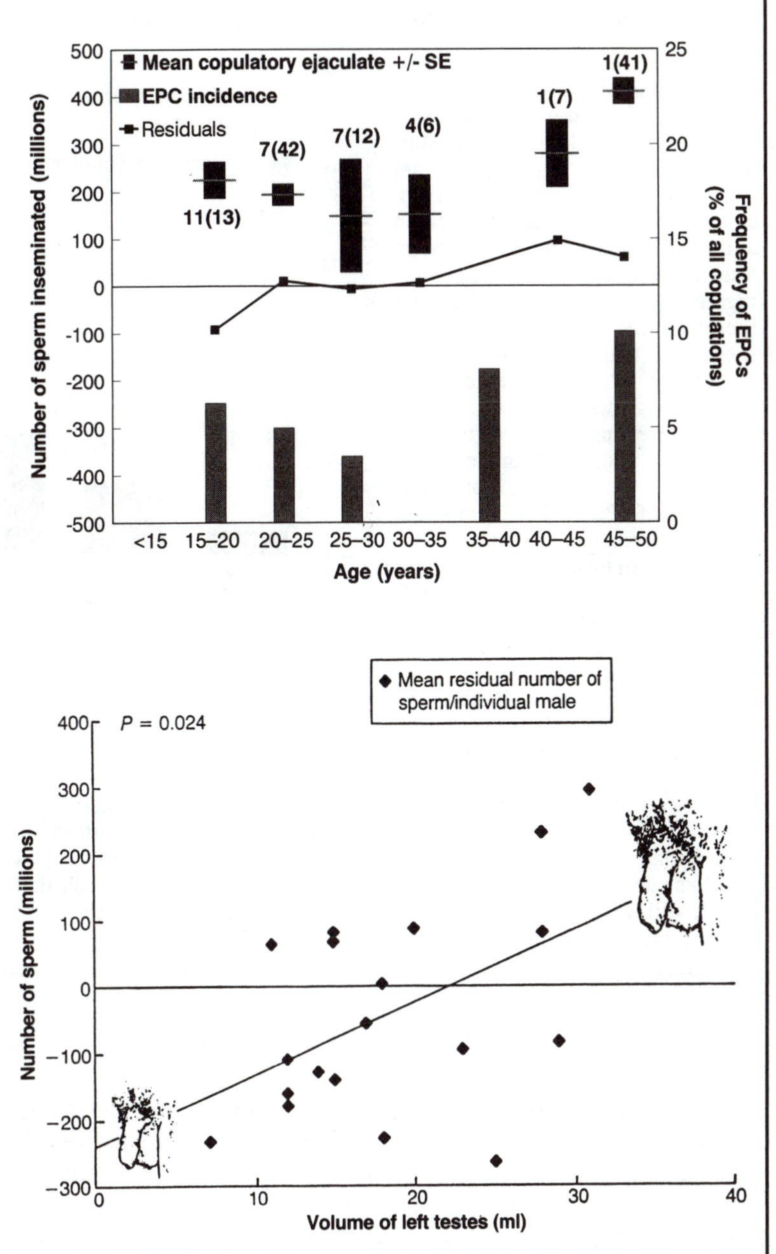

Testis size and sperm numbers

Males with larger testes inseminate more sperm during IPC and should thus be more successful in sperm competition (see also Chapter 5 and Box 9.4). Numbers expressed relative to average expected after allowing for percent time together, time since last copulation and time since last ejaculation.

as varying unpredictably due to circumstance. Sperm competition theory predicts that males should partition sperm numbers strategically between successive IPCs during the course of their partner's reproductive cycle.

There have been four models of the way in which this strategic partitioning of sperm numbers should proceed.

1. *Fixed inseminate model*: This assumes that males inseminate a relatively fixed number of sperm/IPC, no matter what the inter-IPC interval, and thus are relatively unconstrained by rate of sperm maturation. On this model, the number of sperm inseminated at each IPC is relatively constant and the total number inseminated

Box 9.3 Topping up the female: number of sperm inseminated during IPC varies with inter-IPC interval

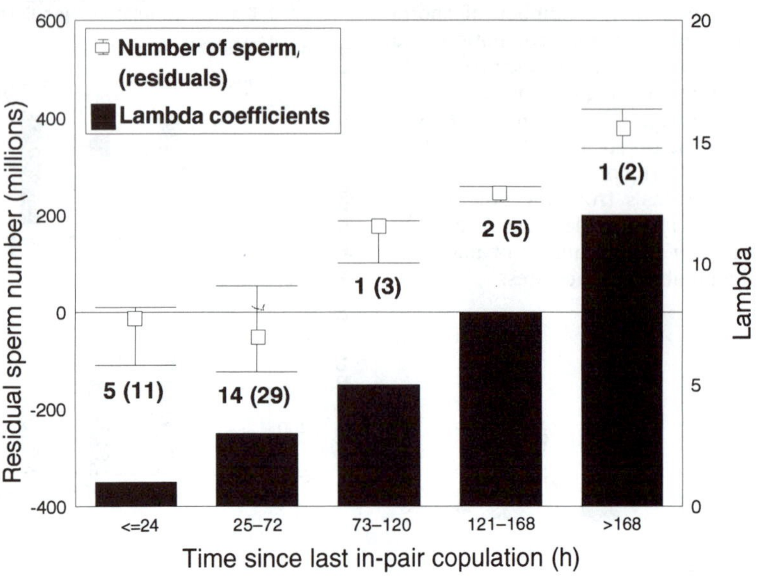

Variation in number of sperm ejaculated during in-pair copulation with time since last in-pair copulation. Occasions on which the male masturbated between copulations are excluded. The number of sperm ejaculated is expressed as the median and IQR residual number of sperm after correcting for percentage time together (Box 9.2). A value of zero is therefore the average number of sperm for that particular value of percentage of time together. The histogram shows the 'best fit' after also correcting for differences between males. Sample size expressed as: number of pairs (number of IPCs). The number of sperm inseminated during IPC increases with time since last IPC for up to at least 8 days (192 h). The male appears to be topping-up his female partner to a level determined by the risk of sperm competition (Box 9.2). Redrawn from Baker and Bellis (1993a).

Given the rate of sperm production in humans (12.5 million/h), our data on top-up rates at different levels of association between male and female suggest that a male should be able to inseminate: (a) one female with whom he spends 70% of his time while, during the other 30%, seeking occasional extra-pair copulations with other females with whom he spends relatively little time; (b) six females in a harem group such that he could associate with them all simultaneously and continuously (because of the 'topping-up phenomenon' this would be independent of intercopulation interval with each female); (c) two spatially separate females between whom he divided his time equally.

during a fixed interval is a function of the number of IPCs during that interval. With some qualification, this is essentially the model assumed by Ginsberg and Rubenstein (1990) for zebras (*Equus* spp.).

2. *Physiological constraint model*: This model assumes that, at each IPC, males inseminate all of the stored sperm mature enough to be ejaculated. On this model, number of sperm inseminated at each IPC will be a function of time since last ejaculation and the rates at which sperm mature (minus those which are shed or destroyed). Total number of sperm inseminated during a fixed interval, such as one menstrual cycle, will be the number of sperm matured during that interval.
3. *Parker's partitioning model*: Parker (1984) assumed that a given number of sperm are available for insemination during one fertile phase of the female partner. He then identified circumstances in which optimum strategy for the male would be to use these sperm during IPC in a series of smaller inseminations spread through the female's fertile phase rather than in a single large insemination at the beginning of that phase. His model predicts that in species, such as humans, with cryptic ovulation and a sperm life that is short relative to the interovulation interval, males should show multiple insemination. According to Parker's model, the total number of sperm inseminated during one fertile phase is

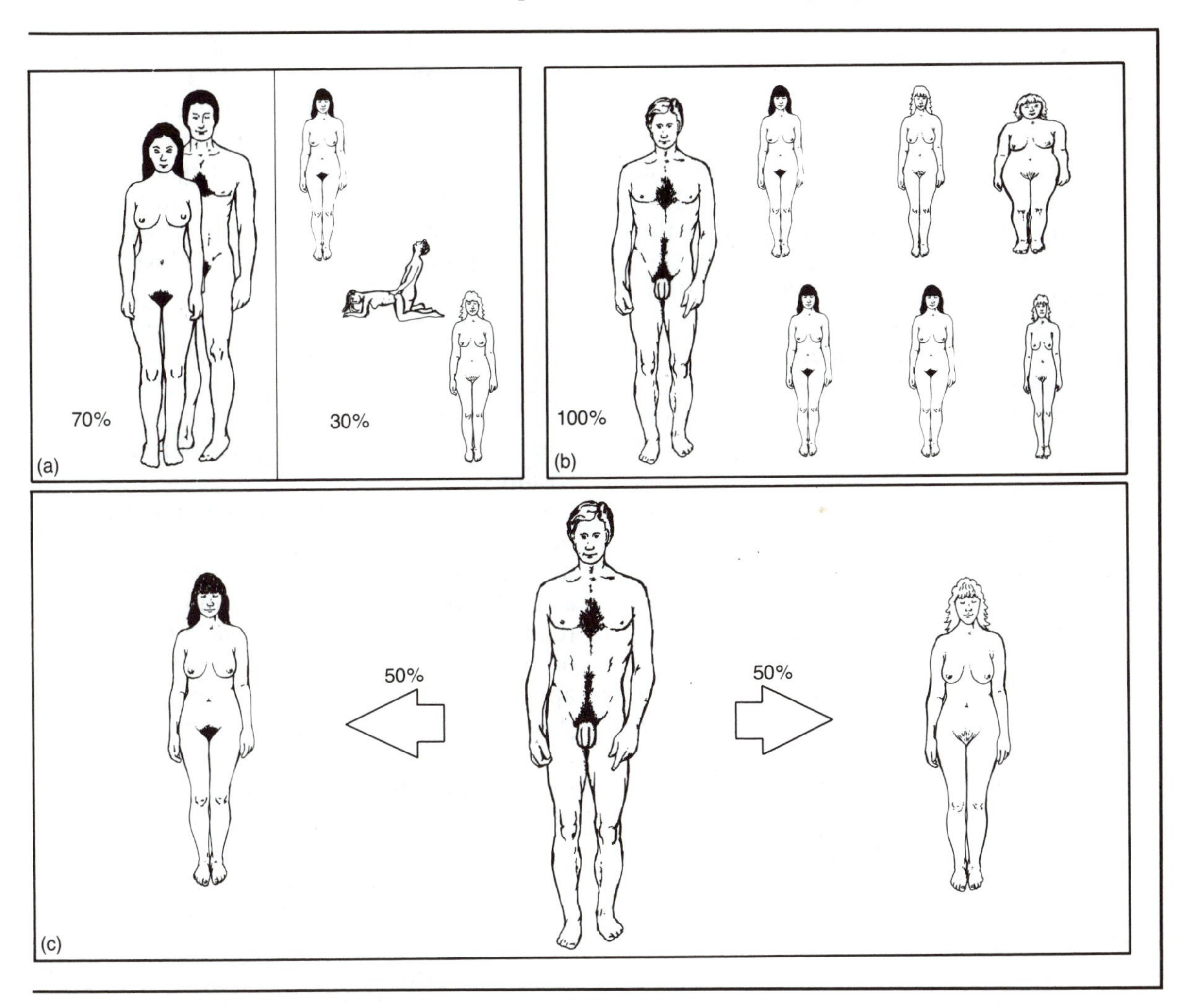

relatively constant and independent of the IPC rate. Number of sperm per insemination is given by the total number of sperm available divided by the number of IPCs during the phase. On average, therefore, there should be a positive association between inter-IPC interval and number of sperm inseminated.

4. *The 'topping-up' model*: According to our own 'topping-up' model (Baker and Bellis, 1993a), the total number of sperm inseminated during a given time interval is not fixed. Instead, males attempt to maintain an optimum-sized population of sperm in their partner's tract as a defence against sperm competition. Optimum size of sperm population will be a function of the risk of sperm competition. Successive IPCs thus become 'toppings-up'. The number of sperm inseminated is that necessary either simply to replace sperm lost through death and phagocytosis since last insemination and/or, if risk of sperm competition has changed, to adjust the total size of the sperm population. According to this model, number of sperm inseminated during IPC is a function of time since last IPC and risk of sperm competition. As in Parker's model, the total number of sperm inseminated during a fixed interval will be relatively independent of IPC rate but, rather than be fixed, will be a function of the risk of sperm competition.

Our study of the number of sperm inseminated during IPC has shown that inter-IPC interval and the proportion of time a couple spend together during that interval have significant but independent influences on the number of sperm inseminated at the next IPC (Baker and Bellis, 1993a). Males increase the number of sperm inseminated at the next IPC the longer the time since last IPC (Box 9.3). Presumably, the number of sperm inseminated would not increase indefinitely and at some time after last IPC the number inseminated would level off. However, levelling off does not seem to occur in the first eight days (Box 9.3).

Our data for humans are clearly inconsistent with the 'fixed ejaculate' model. Nevertheless, the form of the relationship (Box 9.1) means that the total number of sperm inseminated through IPC during a given time interval is slightly greater if there are more IPCs during that interval. Thus, if we assume male and female spend 100% of their time together, we can calculate that a male would inseminate 1372 million sperm during a complete (28 day) menstrual cycle of his female partner if the pair have four IPCs compared with 1440 million sperm if they have eight IPCs. Doubling the IPC rate thus increases the total number of sperm inseminated by 5%.

These data for humans provide some support for Ginsberg and Rubenstein's (1990) assumption for zebra that the number of sperm inseminated is a function of IPC rate. Ginsberg and Rubenstein may even be correct for zebras that adjustment of the IPC rate may be more important than any other adjustment the male might make. However, this is certainly not the case for humans. Thus, for humans (from Box 9.1), relative to four IPCs/menstrual cycle with 100% time together, halving the percentage of time together raises the total number of sperm inseminated per menstrual cycle by 50% compared with the 5% increase due to doubling the IPC rate.

The human data do not seem to fit the physiological constraint model. The relationship illustrated in Box 9.3 suggests an increase in number of sperm inseminated of only about 57 million sperm per day since last IPC. Yet adult human males probably manufacture nearly 300 million sperm per day (Johnson *et al.*, 1980; Neaves *et al.*, 1984; Box 5.8). Moreover, the number of sperm inseminated during IPC for each hour since the last IPC is not fixed but is significantly negatively correlated with the percentage of time together. Even the greatest rate of insemination (about 9.3 million sperm/h since last IPC when couples spend <25% of their time together) is lower than the estimated rate of sperm production by humans of 12.5 million/h (Johnson *et al.*, 1980). It thus seems likely that observed insemination rates are in some way strategic and not simply due to physiological constraint.

Of all the models, the human data seem to be most consistent with the 'topping-up' model. Thus, total number of sperm inseminated during a fixed time interval increases relatively little with an increase in the number of IPCs during that interval. Instead, males appear strategically to adjust the number of sperm with which they inseminate their partner according to time since last insemination. The number of sperm ejaculated during any given insemination increases with time since last IPC for at least the 8 days (192 h) that some active sperm are known to remain in the female tract (Austin, 1975). Any IPC less than 192 h since the previous IPC is effectively, therefore, a 'topping-up' to some particular level, rather than a complete insemination in its own right.

To what level the male tops-up the female depends on the risk of sperm competition. From the equation for IPCs in Box 9.1, if we assume an 8 day (192 h) inter-IPC interval (to allow all sperm from previous inseminations to die), males inseminate 389 million sperm/ejaculate when percentage of time with the female is 100% compared with 712 million/ejaculate when percentage of time with the female is only 5%.

Our data on top-up rates at different levels of association between male and female allow us to make one further calculation. Given a rate of sperm production in humans of 12.5 million/h (Johnson *et al.*, 1980), we can estimate the variety of situations in which a male can optimally inseminate different numbers of females. Our conclusions are illustrated in Box 9.3. In addition to the situations shown, a roving male with no female partner and little association with the females he inseminates should be able to produce an optimal ejaculate (about 700 million sperm) about every 2–3 days.

It is not yet known whether the 'topping-up' phenomenon is found in other animals in which male and female partners copulate intermittently over extended periods. Most monogamous birds begin to copulate some time (days) in advance of the first egg being laid, then mate with variable frequency until some point during egg-laying. In most monogamous birds, copulation frequency is controlled by the female and peaks a few days before the start of egg laying. It then declines markedly or ceases, despite the fact that females remain fertile until the day their penultimate egg is laid (Birkhead and Møller, 1992). However, in species with intense sperm competition, copulations continue until the end of the female's fertile period (Birkhead and Møller, 1993b). Males thus ensure a population of their sperm in the female tract throughout the female's fertile period.

In some species, pairs copulate very frequently in the production of a single clutch or litter. For example, American kestrels, *Falco sparverius*, copulate on average 690 times per clutch (Balgooyen, 1976) and Smith's longspurs, *Calcarius pictus*, copulate on average 350 times per clutch (Briskie, 1992). In this latter species, 99% of the copulations are preceded by female solicitation and it seems to be a general rule that when pairs have a high copulation rate copulation is solicited by the female more often than by the male (Hunter *et al.*, 1993).

Hypotheses to explain such female-driven, within pair, copulation behaviour have been formulated (e.g. Petrie, 1992) and predictions made (Hunter *et al.*, 1993). However, no critical tests have yet been made. A central factor could be the way in which the male partitions his sperm between copulations and whether the 'topping up' phenomenon described here for humans is also shown by the males of these other animals. If it is, then hypotheses involving female depletion of male sperm reserves may need re-appraisal.

9.2.3 Male response to female value

There has been minimal attempt to test the prediction that males should invest more in sperm competition when the female's reproductive value is higher. One initial problem is that such a response may sometimes be difficult to separate from other facets of sperm competition. For example, females of greater reproductive value may be inseminated by more males and thus offer greater risk of sperm competition.

Even so, there are some circumstantial data consistent with the prediction, at least for insects. Thus, male dung flies, *Scatophaga stercoraria*, copulate for shorter periods with females containing fewer eggs (Parker, 1970b). Male mormon crickets, *Anabrus simplex*, are less likely to transfer a spermatophore to lighter females containing fewer eggs (Gwynne, 1981). In both cases, it seems likely that, on average, females containing fewer eggs are inseminated with fewer sperm.

Female mammals may differ in their reproductive value in many ways. In this section we consider five: (1) body size; (2) sperm retention; (3) stage of menstrual cycle; (4) use of oral contraception; and (5) time of year.

(a) FEMALE BODY SIZE

In many animals heavier females contain more eggs (Halliday, 1983). Insofar as dizygotic twinning rates are higher in heavier human females (MacGillivray and Campbell, 1978) the same may, on average, be true for humans. In addition, heavier human females have reduced risk of miscarriage during pregnancy, faster growth of the fetus and heavier birth weight (Lechtig and Klein, 1981; Bongaats and Potter, 1983; Garn and LaVelle, 1983; Gray, 1983). In some societies, heavier women have higher offspring survival and, at any one time, more living children (Hill and Kaplan, 1988).

A male preference for larger females is reported to exist, both in humans (Ford and Beach, 1952; Borgerhoff Mulder, 1990) and other animals (Halliday, 1983) and male humans are reported to compete more for, and to show greater defence of, more fertile and fecund women (Flinn, 1988). Sperm competition theory would predict, therefore, that males inseminate larger females with more sperm.

(b) ORGASM PATTERN, PHASE OF MENSTRUAL CYCLE, AND CONTRACEPTION

The function of the female orgasm in humans is discussed in detail in Chapter 10. There we show a significant association between sperm retention and the occurrence and timing of the female orgasm relative to the timing of male ejaculation. It follows that, at any given copulation, the timing of female orgasm is likely to have some association with the reproductive value of the female to the male. We might expect males to vary the number of sperm inseminated accordingly.

Both stage of menstrual cycle and use and efficacy of contraception influence the probability of conception and hence, at any given copulation, the reproductive value of the female to the male. As far as contraception is concerned, the probability of conception (expressed as a percent of pregnant women in 100 fertile women years; i.e. 100 women for 1 year or 10 women for 10 years) is 5–30% when couples use a condom and <1% when the female uses an oral contraceptive (Box 7.1). When couples use both a condom and an oral contraceptive, the probability of conception should be correspondingly lower than when either method of contraceptive is used on its own. We might expect males to adjust the number of sperm inseminated in response to these altered levels of probability of conception.

(c) SEASON

The probability that a female will conceive in any one calendar month varies with the season, perhaps largely due to seasonal variation in the probability of ovulation (Box 4.10). Males should invest more in sperm competition, by inseminating more sperm, in those months that the probability of conception is greater. This prediction is evaluated separately in section 9.5.2.

(d) RESULTS

The prediction that the number of sperm inseminated should be a function of female body size was strongly supported (Box 9.4), female weight showing a slightly greater association with sperm number than female height. In contrast, no association emerged between the number of sperm inseminated and other measures of female reproductive value (Box 9.4). One possible explanation for this apparent failure of sperm competition theory is that whereas males can reliably judge female body size they cannot obtain reliable information concerning female orgasm and probability of conception.

As far as female orgasm is concerned, there is a major difference in sperm retention between a copulation involving no female orgasm and one with a female orgasm after male ejaculation (Chapter 10). Yet unless, on ejaculation without a female orgasm, males could predict whether the female will or will not experience orgasm before flowback, there would seem to be no opportunity for the male to adjust sperm number appropriately. Similarly, the difference in sperm retention between an orgasm during foreplay and an orgasm within a minute of ejaculation is also major. However, in this context, the time available for the male to adjust sperm number is very short (a few minutes or even seconds). Add to these difficulties the possibility of the female 'faking' an orgasm (Box 10.12) and the opportunity for the male to make appropriate strategic adjustments in sperm number in response to the occurrence and timing of female orgasm seems minimal.

The fact that males do not seem to adjust sperm number according to the stage of their partner's menstrual cycle is probably simply further testimony to the efficiency of female sexual crypsis in humans (section 6.6). Finally, given the recent origin of modern contraceptive techniques, it is perhaps not surprising that males may not have the psychophysiological repertoire to vary the number of sperm inseminated according to the use of oral contraceptives by their female partner. Even if male humans had such a repertoire, adjustment of sperm number may not be advantageous. Even if the female has

Box 9.4 Association between number of sperm inseminated during in-pair copulation and male and female stature

Table. Analysis of number of sperm inseminated during in-pair copulation by 25 pairs in relation to male and female stature

Parameter	*n*		Meddis' test	
	Pairs	Ejaculates	*z*	*P*
Female weight	24	114	4.437	<0.001
Male weight	25	116	3.462	<0.001
Female height	24	114	3.496	<0.001
Male height	25	116	1.560	0.059
Testis volume	19	110	2.465	0.007

Number of sperm is expressed relative to the average that would be expected after allowing for: percent time together since last in-pair copulation (IPC); hours since last IPC; whether the male ejaculated between IPCs and, if so, how long before IPC. Each data point is the mean number of sperm inseminated during IPC for a particular couple (based on 1–48 IPCs). The rank-order correlation with female body weight is very significant ($P = 0.004$). The results of a Meddis' specific test (which weights for sample size) for body weight and other measures of male and female stature are shown in the table. Female weight is the stature measurement that gives by far the strongest association with the number of sperm transferred from male to female during IPC.

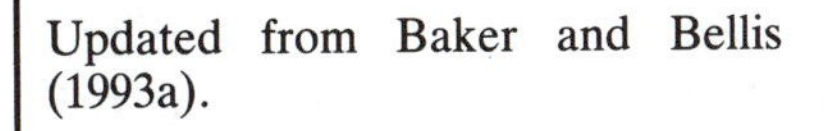

Updated from Baker and Bellis (1993a).

taken oral contraceptives up to the moment of copulation, the male cannot be certain that she will continue to do so for the next 5–8 days while the sperm he inseminates are competitive. Thus, males can never assume that the sperm being inseminated have no chance of entering competition and/or no chance of fertilization.

Perhaps, therefore, our failure to find any association between number of sperm inseminated and either orgasm pattern, stage of menstrual cycle, or use of oral contraceptives should not be considered a failure of sperm competition theory. Rather it could reflect the extent to which males are unable to obtain the information necessary for optimum adjustment of sperm number.

9.3 Number of sperm inseminated during extra-pair copulation (EPC)

Using a modelling technique, Parker (1990b) has investigated the competition between in-pair (IP) and extra-pair (EP) males to deter-

Box 9.5 Predicted and observed number of sperm inseminated during in-pair copulation, extra-pair copulation and double-mating

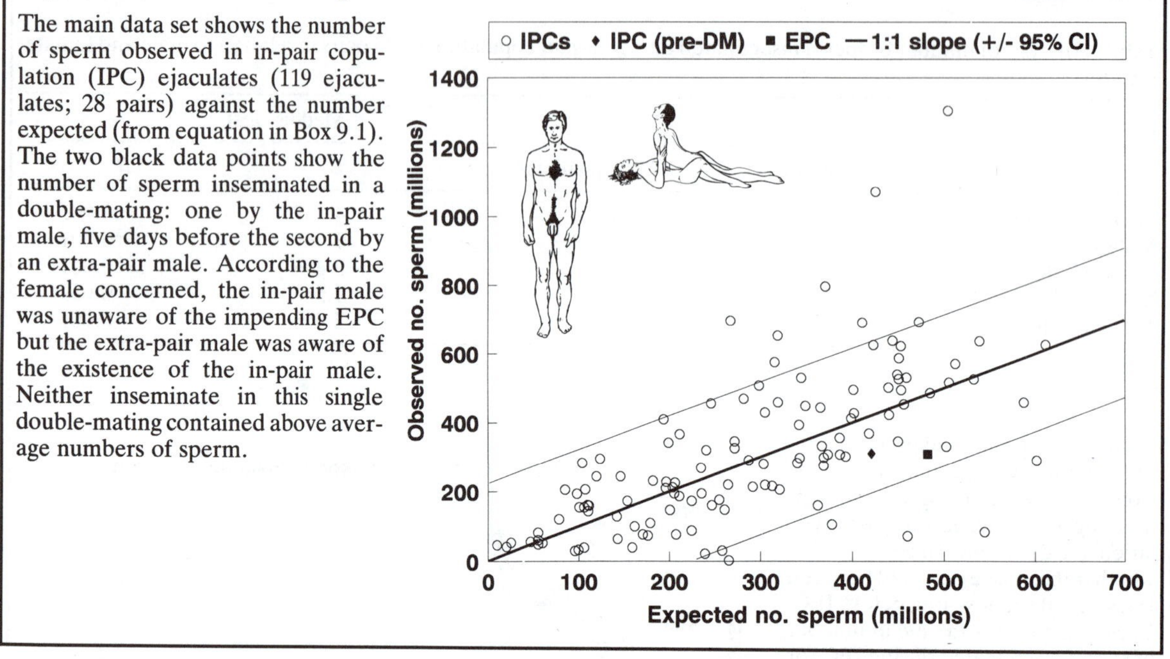

The main data set shows the number of sperm observed in in-pair copulation (IPC) ejaculates (119 ejaculates; 28 pairs) against the number expected (from equation in Box 9.1). The two black data points show the number of sperm inseminated in a double-mating: one by the in-pair male, five days before the second by an extra-pair male. According to the female concerned, the in-pair male was unaware of the impending EPC but the extra-pair male was aware of the existence of the in-pair male. Neither inseminate in this single double-mating contained above average numbers of sperm.

mine optimum strategy in terms of the number of sperm they should each inseminate. He concluded that a male should inseminate more sperm when performing an EPC than when inseminating his own female, unless he has detected an EPC by his female. In which case, he should again increase sperm numbers.

Our analysis suggests a more flexible scenario. It seems (Box 9.2) that male humans assess the probability of sperm competition in the female they are about to inseminate via the percentage time that they have spent with her. One possible implication of our data is that the concepts of EPC and IPC may be irrelevant to male strategies of sperm adjustment. For example, when a male has been away from his partner for any length of time (i.e. weeks or months) it is a moot point whether, on his return, he is an in-pair or extra-pair male.

Perhaps, when a male inseminates a female, his strategic concern should not be whether she is or is not his partner. Rather, the number of sperm inseminated may instead be determined simply by how long it is since he last copulated with that particular female (if ever) and/or the extent to which he has associated with her within the past 8 days (the active life of sperm she may already contain).

If this suggestion is correct, we should expect EPC inseminates to fit on the same line as IPC inseminates. Unfortunately, in our Manchester, UK, study (Box 3.8) we have whole ejaculates from only one (as far as we know) double-mating (or what would have been a double-mating had the two males not worn condoms). The female was not taking an oral contraceptive. Both IPC and EPC ejaculates were just within the range for IPC ejaculates but were towards the lower edge (Box 9.5). The meagre data give no indication that EP males (or the double-mated IP male) change their normal reaction to percentage time with the female by increasing the number of sperm inseminated.

9.4 Masturbation: features and function

The main characteristics of male self-masturbation and its interaction with other forms of male ejaculation were discussed at length in Chapter 5. The overwhelming impression was that, together with nocturnal emissions, the function of male self-masturbation related in some way to rate of sperm production and how long sperm are stored before being inseminated into a female. As frequency of ejaculation via copulation changes with age, so too does frequency of non-copulatory ejaculation (Box 5.1). The result is that, together, the two outlets maintain a total frequency of ejaculation that, as a male ages, relates closely to his rate of sperm production (Box 5.8).

This pattern, combined with the sequence and timing of successive copulations and masturbations (Box 5.5), means that females are rarely inseminated with sperm which have been 'in store' awaiting ejaculation for more than 2–4 days. It is tempting, therefore, to conclude that the function of masturbation, with nocturnal emission as some form of back-up, is in some way to influence the age and number of sperm with which the male next inseminates a female.

The apparent wastefulness of masturbation and spontaneous emissions would be resolved if, as a result of shedding sperm, a male produced a future ejaculate with a fitness enhanced beyond the level necessary to offset the numeric and energetic costs of the lost sperm. Previous authors have suggested that sperm have a limited 'shelf-life' and, if not used within a certain storage time, become suboptimal (e.g. Smith, 1984). Masturbation is thus suggested to be a mechanism for shedding suboptimal sperm and reducing the mean age of sperm inseminated at the next copulation.

The advantage to the male would be that younger sperm may be more acceptable to the female and/or may be better able to reach a secure position in the female tract. Moreover, once retained in the female tract, younger sperm could be more fertile in the absence of sperm competition and/or more competitive in the presence of sperm competition. Finally, if younger sperm live longer in the female tract, any enhanced fertility and competitiveness would also last longer.

9.4.1 Numbers and circumstance

Unlike IPC ejaculates, masturbatory ejaculates showed no association between the percentage of time a male had spent with his partner since their last IPC and the number of sperm ejaculated during masturbation (Baker and Bellis, 1989b, 1993a). Nor was there any difference in sperm numbers between: (1) self-masturbation in the absence of a female; and (2) masturbation (by self, partner or both) in the presence of a female. The non-significant trend was for more to be ejaculated when the female was present.

9.4.2 Time since last ejaculation

The number of sperm ejaculated during masturbation shows a highly significant correlation with hours since the male's last ejaculation, the number of sperm ejaculated continuing to increase linearly with time since last ejaculation, at least up to 9 days (216 h) after last ejaculation (Box 9.6). It made no difference whether masturbations were preceded by a masturbation or by an IPC.

It seems, therefore, that the primary factor associated with the number of sperm ejaculated by a given male during masturbation is the length of time since the male's last ejaculation. The circumstances of the last and current ejaculation have no significant influence.

9.4.3 Differences between males

There is a just significant heterogeneity between males in the number of sperm ejaculated during masturbation when time since last ejaculation is controlled. However, these differences were not a significant function of either male height, weight or volume of testes. As expected from Box 5.8, but in contrast to the pattern for the number of sperm inseminated during copulation (Box 9.2) there was a significant tendency for older males to ejaculate fewer sperm during masturbation (Box 9.6). This difference has already been discussed (section 9.2.1).

9.4.4 The function of masturbation

Masturbation between copulations reduces the number of sperm inseminated at the next IPC. The greatest reduction, however, occurs when masturbation immediately precedes copulation. With time since masturbation, the influence decreases until it has disappeared by 72 h since last masturbation (Box 9.7). Numerically, masturbation between IPCs reduces the number of sperm the male inseminates into his female partner at the next IPC by 228 million sperm minus 2.41 million sperm for every hour since last masturbation (up to 72 h).

We used the IPC equation (Box 9.1) to estimate the number of sperm that should have been inseminated into the female partner during all of the IPCs for which we obtained flowback (Box 3.8). Subtraction of the observed number of sperm ejected in the flowback from the predicted number inseminated provides a measure of the number of sperm retained on each occasion.

The number of sperm ejected and retained after IPC by females according to time since their partner's last ejaculation and whether that last ejaculation was the preceding IPC (IPC–IPC inseminates) or a masturbation (MAS–IPC inseminates) is shown in Box 9.7. Whether the male's last ejaculation was an IPC or a masturbation, the number of sperm in the flowback increases significantly with increase in time since last ejaculation. The number of sperm retained by the female also increases with time since the male's last ejaculation if that ejaculation was during IPC but not if the male's last ejaculation was masturbatory. The number of sperm retained does not change significantly with time since last masturbation though the tendency is for more sperm to be retained the more recent the masturbation.

When we compare the number of sperm retained by the female following IPC–IPC and MAS–IPC inseminations, no differences are significant. On average, therefore, males suffer no disadvantage from masturbation in terms of the number of sperm retained by the female at their next IPC, despite inseminating fewer (on average, by 228 million minus 2.4 million for every hour since last masturbation; Box 9.1). In fact, over all IPCs, the non-significant trend is for more sperm to be retained when the male masturbates between copulations.

Our study provides good support for the sperm age hypothesis. Assuming that the more recently the male has ejaculated, the younger the average age of sperm in the next ejaculate, there is a decline in residual sperm retention with age of sperm in both IPC–IPC and MAS–IPC inseminates (Box 9.7). Over all inseminates, the greater relative retention of younger sperm is highly significant.

Why younger sperm show better retention than older sperm is not yet known. It may be associated, however, with a greater ability of younger sperm to penetrate cervical mucus. Acrosin contained in/on the sperm acrosome has been implicated in the ability of sperm to penetrate cervical mucus (Schill *et al.*, 1979) and the level of acrosin activity is significantly higher in younger than older sperm (abstinence time since last ejaculation 1 day vs 7 days) (Schill *et al.*, 1988).

Our analysis of the dynamics and consequences of masturbation clearly supports the view that the behaviour is a functional strategy. The function, however, appears to be more to increase the 'fitness' of sperm retained by the female at the next IPC than to increase the number retained.

Spontaneous emissions by rats have different dynamics from non-copulatory emissions by humans but the function may well be the same. Rats shed smaller numbers of sperm/ejaculate during non-copulatory ejaculations than humans (10% of the number during copulation compared with about 63% for humans (slope of line; Box 9.9)) but do so more frequently (mean, 2/day; maximum c. 4/day) Ågmo, 1976). Moreover, the number shed/ejaculate is relatively constant no matter how many times the rat spontaneously ejaculates each day (Ågmo, 1976). However, after inseminating many sperm during copulation the rat takes longer to resume spontaneous ejaculation than after inseminating fewer and when nearby females are in oestrus the resumption of spontaneous ejaculation after a single copulation is hastened (Beach, 1976). Spontaneous emission by rats could thus be just as much an adjustment of the age of

Box 9.6 Variation in number of sperm ejaculated during masturbation in relation to sociosexual situation.

By far the most influential factor in determining the number of sperm shed by humans during masturbation is time since last ejaculation (data for 26 males, 88 ejaculates updated from Baker and Bellis, 1989b, 1993a).

Calculating residuals to control for time since last ejaculation, the second most influential factor is male age ($P = 0.003$), probably due to changes in daily sperm production (DSP) which peaks at around 25 years (DSP data based on autopsy; Neaves *et al.*, 1984).

Calculating residuals to control for both time since last ejaculation and male age (see equation; Box 9.1), no other factors (testis volume, right; percentage of time together; circumstances or sequence, see text) are significant. When weighted by sample size, testis volume shows a positive but not significant association with sperm number.

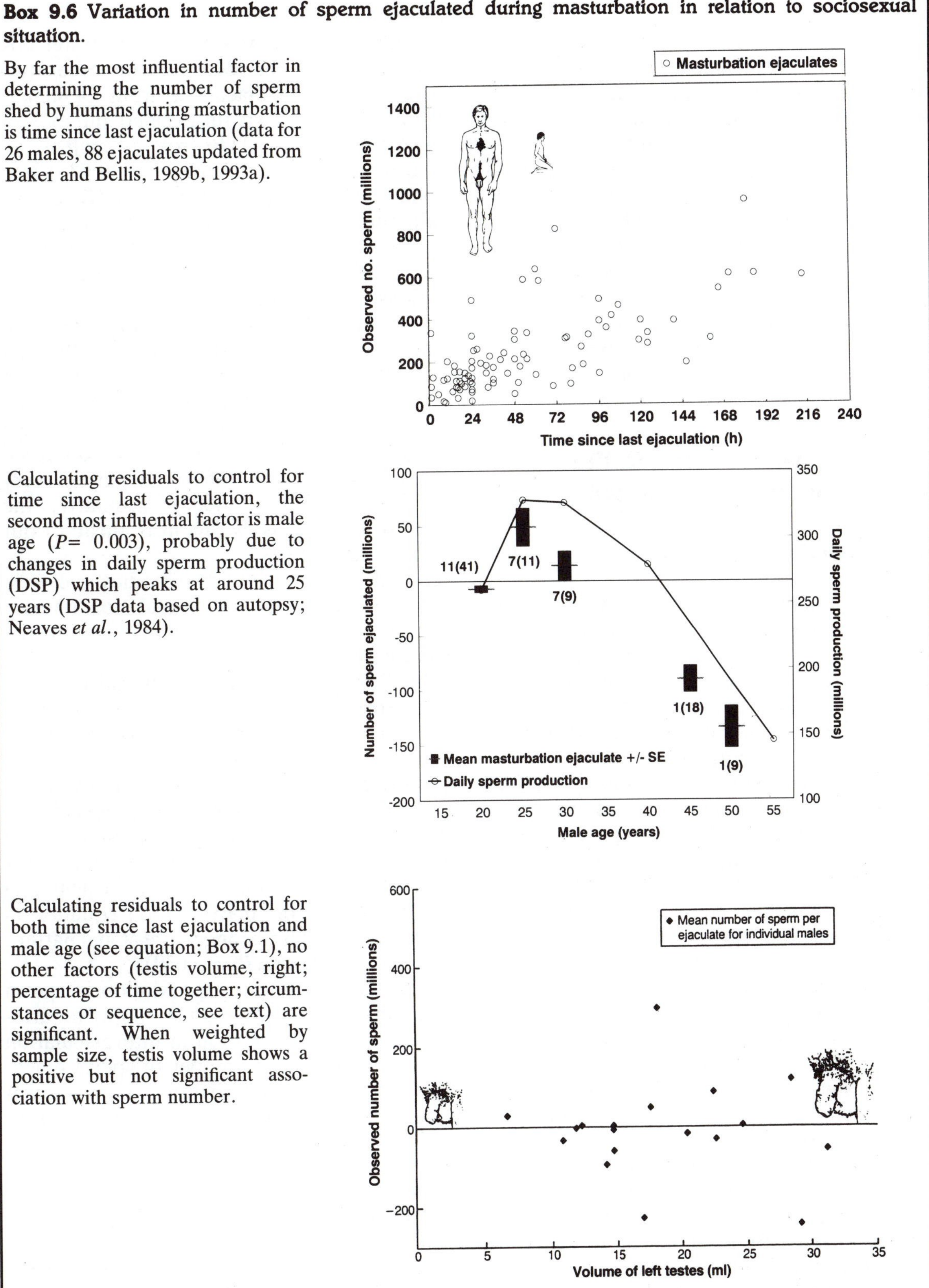

sperm to be inseminated into a female as is masturbation by humans.

Apart from our data on the higher retention rate of younger sperm, the only other empirical evidence concerning the advantage of younger sperm comes from sperm competition experiments. Aged cockerel sperm, when inseminated homospermically (i.e. in the absence of sperm competition), were capable of fertilization but when inseminated with younger sperm from another male achieved relatively few fertilizations. Similarly, rabbit sperm aged outside of the male rapidly became poorer at competing for eggs against sperm from freshly collected ejaculates (Roche *et al.*, 1968).

9.5 Seasonal variation in the number of sperm ejaculated

9.5.1 Masturbatory ejaculates

Even continuous breeders show seasonal peaks of reproductive activity which usually lead to seasonal variation in incidence of birth (section 4.4.3 and Box 4.10). Most often, the seasonal variation in conception and birth seems to be attributable to seasonal variation in probability of ovulation by the female. Some authors, however, have considered the possibility that there may be seasonal variation in the quality of the semen produced by males. Such variability has been claimed for humans (New York State, USA, Spira, 1984; Italy, Campaniello *et al.*, 1991) as well as for other species (e.g. Egyptian Buffalo; Shalash, 1981). In this section, we are concerned only with seasonal variation in total sperm count per ejaculate.

Both the USA and Italian studies used masturbatory ejaculates but did not control for time since last ejaculation. Both successfully demonstrated seasonal variation in sperm numbers (Box 9.8) when they analysed the total data set, the lowest sperm numbers being found in August in Italy and in September in New York. Spira (1984) also analysed his data for seasonal variation in the sperm count of individual males. Although the same pattern was obtained as for the total sample, the variation was no longer statistically significant.

Although our data set for Manchester UK was much smaller than for the other two studies (UK, 88 ejaculates from 24 males; Italy, 2405 from an unspecified number of males; New York, 1110 from 52 males), we also obtained a significant ($P = 0.035$) seasonal variation in sperm number (Box 9.8), minimum numbers being found in October. Moreover, our seasonal effect was significant ($P = 0.033$) when analysed within males as well as for the total data set.

The monthly pattern of variation was not quite significantly predicted ($P = 0.081$) by that observed by Spira (1984). However, if the New York months are moved on by one (i.e. New York August predicts UK September, and so on), Spira's observed pattern predicts our observed pattern very significantly ($P = 0.006$). Similarly, the pattern for Italy is two months ahead of our pattern for the UK.

Unlike the previous two studies, we were able to control for interejaculation interval which is such a strong influence on the number of sperm in masturbatory ejaculates (Box 9.6). Our data show seasonal variation in interejaculation interval that is significantly predicted both by our variation in sperm numbers ($P = 0.002$) and by the New York pattern of sperm numbers (shifted forward by one month) ($P = 0.002$). This association remains significant when analysed within males ($P = 0.012$). When we control for interejaculation interval by calculating residuals from the masturbation equation in Box 9.1), there is no longer a seasonal pattern of sperm numbers ($P = 0.630$).

The observed seasonal variation in sperm numbers can thus be seen as simply an artefact of seasonal variation in interejaculation interval. Nor is the number of sperm ejaculated during masturbation a significant function of seasonal variation in the probability of conception (Box 9.8).

9.5.2 Copulatory ejaculates

There is no automatic reason for a correlation between the number of sperm a male ejacu-

Box 9.7 Influence of male self-masturabation on the number of sperm inseminated (by the male) and retained (by the female) at the next copulation

(a) When a male masturbates between successive in-pair copulations (IPCs), the number of sperm he inseminates is reduced unless there is a gap of 72 h or more between masturbation and next IPC. In the figure, number of sperm inseminated are residuals against the number (standardized to zero) that would have been expected to have been inseminated if masturbation had not occurred (for calculation details, see Baker and Bellis, 1993a). Sample size in format number of pairs (number of copulations).

(b) On average, the more sperm a male inseminates at IPC, the more sperm are calculated to be retained by the female after flowback. However, on some occasions more than average are retained (positive residuals in (c)); on other occasions below average (negative residuals).

(c) The absolute number of sperm retained by the female depends strongly on time since last copulation (see Chapter 10) but not on time since the male's previous masturbation. Relative to the number inseminated (i.e. residuals as described in (b)), the number retained is greater when the time since the male's last previous ejaculation is shorter, no matter whether that ejaculation was an IPC or masturbation. For further details and statistics see Tables XIII and XIV in Baker and Bellis, 1993a).

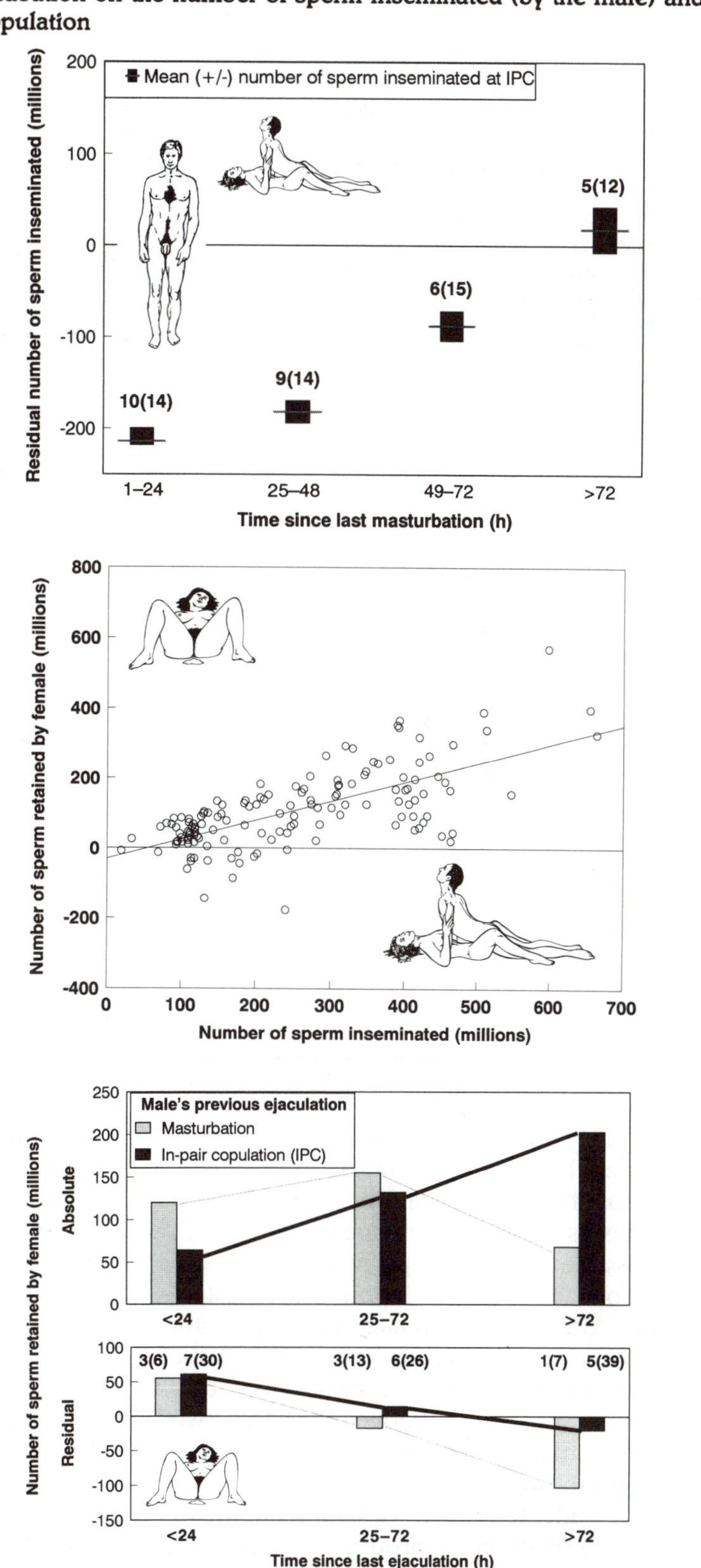

Box 9.8 Seasonal variation in number of sperm ejaculated during masturbation

(a) Three studies of masturbatory ejaculates have now shown seasonal variation in the number of sperm shed during masturbation (UK, Baker and Bellis, unpublished; Italy, Campaniello *et al.*, 1991; New York, USA, Spira, 1984). Unlike the previous two studies, our own UK study was able to show significant seasonal variation even when within-male variation was analysed ($P = 0.033$, $n = 26$ males, 88 ejaculates). All studies show a late summer/early autumn trough.

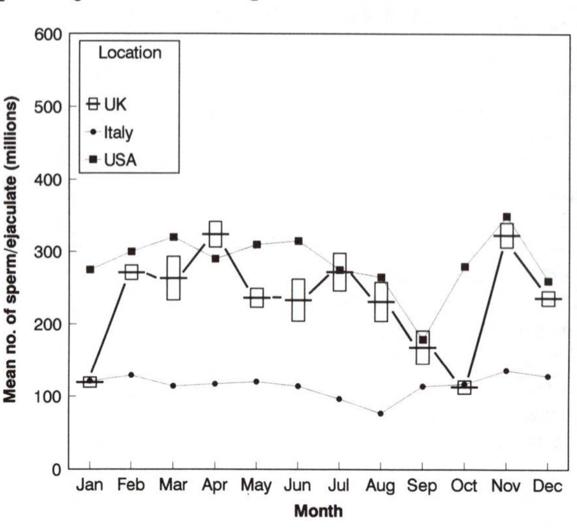

lates during masturbation and the number ejaculated during copulation. All associations seem possible. For example, if males with larger testes produce more sperm (Box 4.17) it is possible that more sperm will be ejaculated both during copulation and masturbation. Alternatively, males who ejaculate more sperm during masturbation may have fewer sperm to ejaculate during copulation and vice versa. In between these two extremes, there may be no association between the two.

In our study, testis size is a significant factor influencing daily sperm production rate (Box 4.17) and the number of sperm in copulatory ejaculates (Box 9.4) but not the number of sperm in masturbatory ejaculates (Box 9.6). Moreover, we find no significant correlation between the mean number of sperm produced by males in copulatory and masturbatory ejaculates (Box 9.9), especially when we control for other relevant factors. Finally, the number of sperm ejaculated during copulation and masturbation both show significant, but opposite, associations with male age (Boxes 9.2 and 9.6).

There is no reason, therefore, to expect seasonal variation in sperm numbers ejaculated during masturbation and copulation to be the same, and they are not. Most impor-

(b) The Italian and USA studies did not control for the possibility of seasonal variation in interejaculation interval. In our own study, there was considerable seasonal variation in this interval with a pattern that closely matched the variation in number of sperm shed during masturbation.

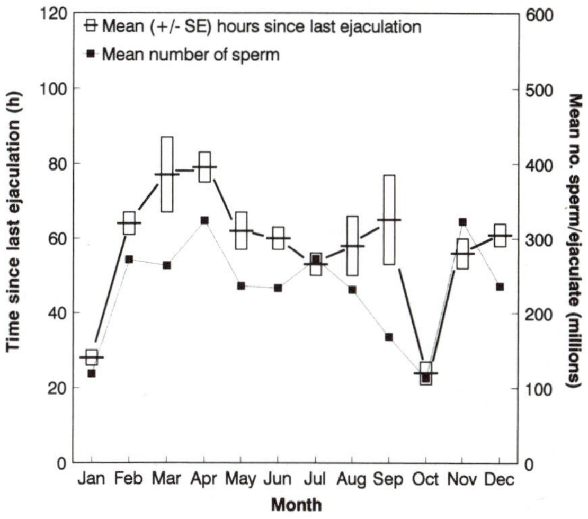

(c) When we controlled for inter-ejaculation interval (by calculating residuals from the masturbation equation in Box 9.1), the seasonal pattern disappeared ($P = 0.510$). In our study, therefore, the observed seasonal variation was the result of behavioural, not physiological (e.g. sperm production rate) variation.

There is no indication that either the number of sperm ejaculated or the residual number are in any way associated with seasonal variation in the probability of conception (numbers, $P = 0.614$; residuals, $P = 0.386$; analysis of within-pair variation).

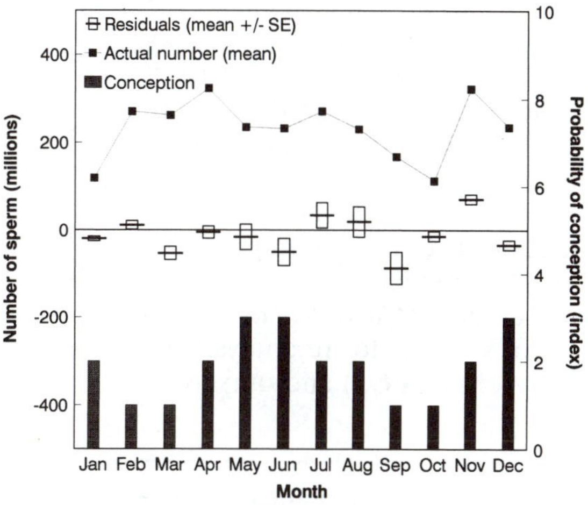

tantly, unlike masturbatory ejaculates, variation in the number of sperm inseminated during copulation is a significant positive function of the probability of conception (Box 9.10). This seems to be an example, therefore, of males varying their investment in sperm competition according to female value (section 9.2.3). As for masturbation, the seasonal pattern disappears when behavioural factors are controlled. Variation in sperm numbers, therefore, seems likely to be achieved behaviourally (e.g. by varying interejaculation interval) rather than physiologically (e.g. by varying sperm production rate). However, there is one final twist to this story (section 10.10.1).

9.6 Long-term changes in the number of sperm ejaculated (1938–1990)

In an analysis of sperm numbers produced during masturbation by men in various countries worldwide between 1938 and 1991, Carlsen *et al.* (1992) claimed that there has been a decline in semen quality during the past 50 years (Box 9.11). Although they restricted their analysis to males without a history of infertility, they suggest that their results may reflect an overall reduction in male fertility over the same period. They point out that there has been a concomitant increase in the incidence of genitourinary abnormalities such as testicular cancer, suggesting a growing impact of environmental factors with serious effects on male gonadal function (Chapter 7).

There are three main problems with this study. First, the data are for masturbatory, not copulatory, ejaculates. As we have seen several times in this chapter, patterns seen in one type of ejaculate cannot be extrapolated to the other. Second, analysis was rather superficial and no allowance was made for the possibility that interejaculation interval may also have changed over the same time period. Finally, exclusion criteria changed over the period analysed, subfertility being defined in terms of ejaculates containing <60 million sperm before 1965 but <20 million more recently. The result is to include more males with fewer sperm in the later data.

The data presented by Carlsen *et al.* (1992) have now been subject to re-analysis by Suominen and Vierula (1993) and ourselves. We both conclude that the decline suggested by Carlsen *et al.* was in fact an artefact of a relatively sudden drop in sperm numbers during masturbation in the 1960s. Suominen and Vierula (1993) show that, at least in Finland, there was a decline in interejaculation interval from pre-1960 (cf. Leikkola, 1955) to post-1970.

As with seasonal variation in sperm numbers, therefore, it seems likely that any long-term changes in the number of sperm in masturbatory samples are an artefact of long-term changes in interejaculation interval. Whether the decrease in interejaculation interval during the 1960s simply reflects a change in social attitudes to masturbation over that period or whether there is some more biological explanation is unknown.

9.7 Why do males show restraint in size of inseminate?

We have shown (section 9.2.2) that when a male inseminates a female he does not ejaculate all of the sperm available but instead inseminates a number that is a function of the risk of sperm competition inside that female.

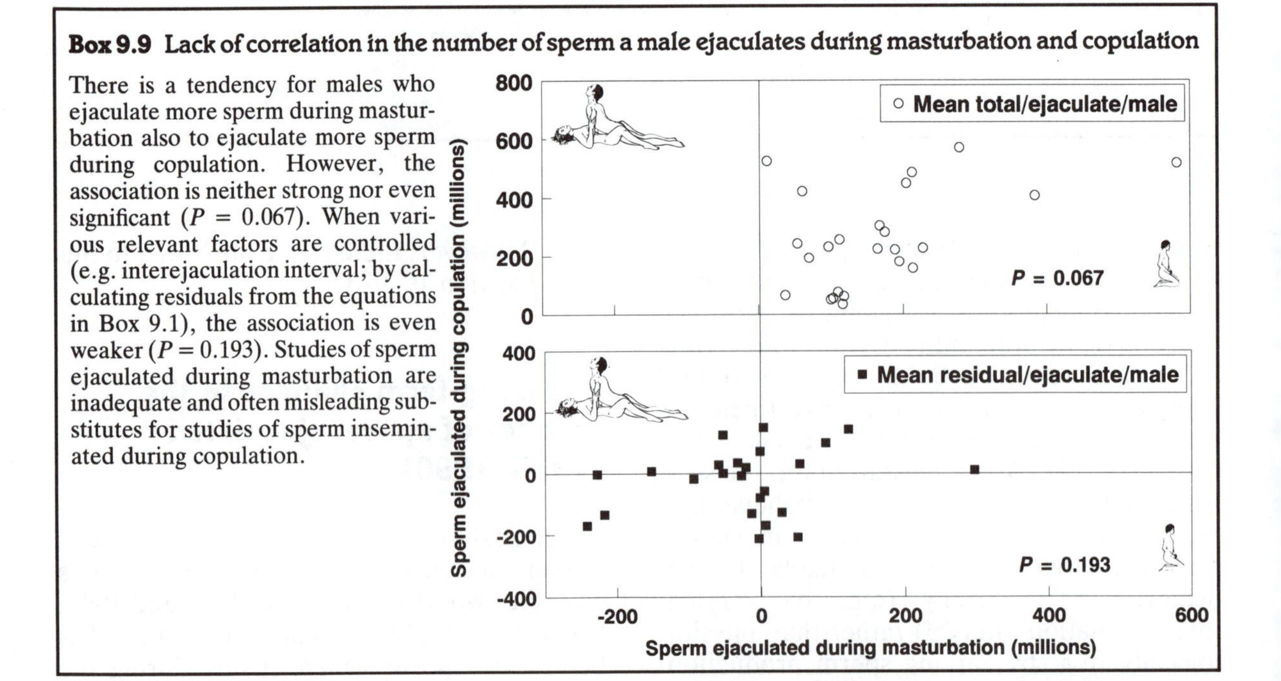

Box 9.9 Lack of correlation in the number of sperm a male ejaculates during masturbation and copulation

There is a tendency for males who ejaculate more sperm during masturbation also to ejaculate more sperm during copulation. However, the association is neither strong nor even significant ($P = 0.067$). When various relevant factors are controlled (e.g. interejaculation interval; by calculating residuals from the equations in Box 9.1), the association is even weaker ($P = 0.193$). Studies of sperm ejaculated during masturbation are inadequate and often misleading substitutes for studies of sperm inseminated during copulation.

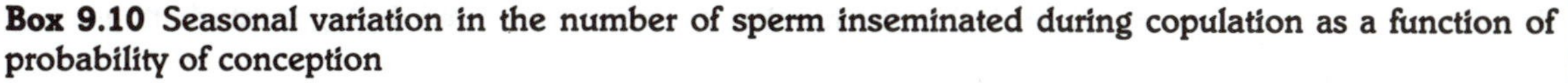

Box 9.10 Seasonal variation in the number of sperm inseminated during copulation as a function of probability of conception

Our study of sperm numbers in IPC ejaculates (28 pairs; 121 inseminates) found that monthly variation was a significant function of seasonal variation in the probability of conception (Box 4.10) even when controlled for differences between individuals (P = 0.012). When controlled for all known relevant factors (using residuals for equation, Box 9.1), the seasonal varation was no longer a function of probability of conception.

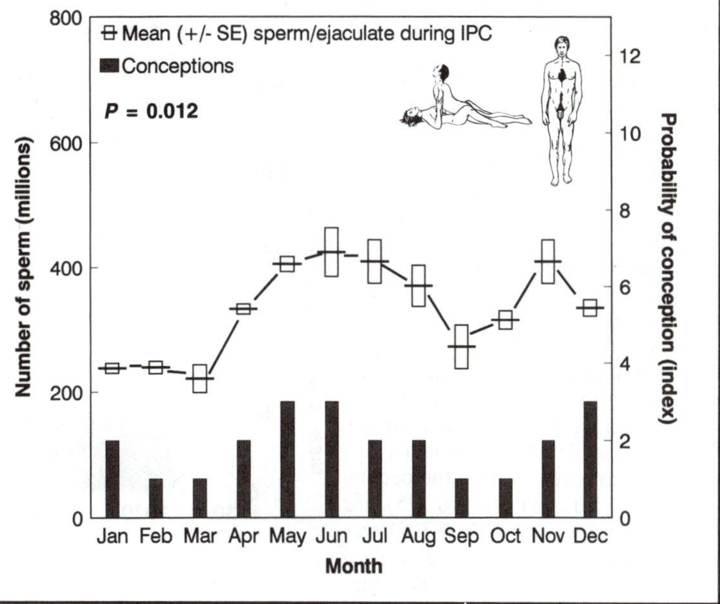

Thus, some sperm which could have been used are conserved and, we presume, move up on the 'conveyor belt' in the cauda epididymis and vas deferens to be next in line for ejaculation. The important question is why these sperm should be conserved instead of being inseminated at the current copulation. Essentially, interpretation rests on whether restraint generates: (a) a future advantage through use of the conserved sperm; and/or (b) an advantage or disadvantage at the current copulation.

9.7.1 Is there an advantage in saving sperm for a future copulation?

The sperm conserved by a male through restraint at the current copulation have four possible destinies. They may either be: (1) inseminated into the same female in a future copulation; (2) inseminated into a different female in a future copulation; (3) shed in a future masturbation or nocturnal emission; or (4) phagocytosed or voided in the male's urine.

Clearly, if the conserved sperm are eventually shed or voided, their conservation was of no future advantage. Indeed, any cost attached to postcopulation storage and then shedding renders their conservation a disadvantage. Phagocytosis may recoup some of this cost. However, only if the sperm are eventually inseminated into a female is there any real chance that their conservation could have been an advantage. In part, therefore, the value of conservation depends on the probability that the conserved sperm will be inseminated into a female rather than be shed or voided. In part, also, it depends on the value of the sperm for fertilization in a future copulation relative to what would have been their value in the present copulation.

We consider first the possible value of conserving sperm for a future copulation with the same female.

The rate at which sperm become suboptimal through waiting to be ejaculated at the storage end of the conveyor belt is unknown. It should be less than 12.5 million/h otherwise there could be no build-up of usable sperm waiting for ejaculation. According to the masturbation equation (Box 9.1), an extra 4.63 (say 5) million sperm are shed during masturbation for every hour since the male's last ejaculation. It is not unreasonable to

Box 9.11 Long-term changes (1938–90) in the number of sperm ejaculated during masturbation

Using historical data on sperm concentration in ejaculates produced by masturbation, Carlsen *et al.* (1992) concluded that there has been a genuine decline in semen quality over the past 50 years. Their data are here presented graphically.

We have re-analysed their data using the Meddis rank-sum test (Meddis, 1984) employed throughout this book (Box 3.8). For this analysis, we blocked the data for the three geographical areas as indicated and weighted samples logarithmically for sample size. Our re-analysis confirmed the original claim that the data are a significant fit to a (in this case rank-order) linear model ($P < 0.001$). However, neither the decline from 1938 to 1964 ($P = 0.275$) nor from 1969 to 1990 ($P = 0.300$) was significant. The data are described best by the post-hoc model that sperm density post-1964 was lower than that before 1964 with, presumably, a rapid decline through the 1960s as illustrated by the histogram.

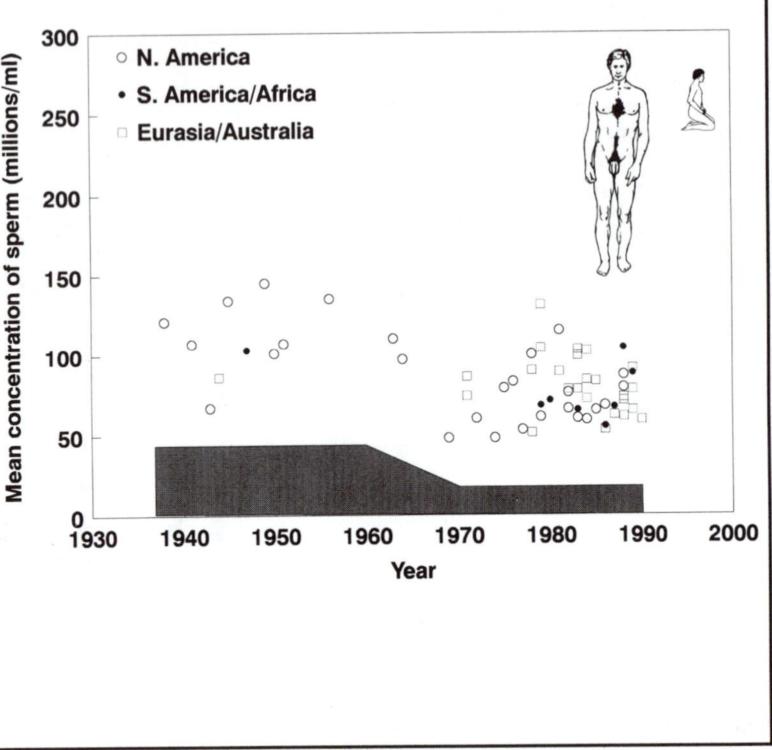

suppose that this approximates to the rate at which unused sperm become suboptimal while waiting to be ejaculated (though, to this number, should be added the unknown number that are phagocytosed and voided in the urine).

Thus, for each hour's delay before the next copulation, at least 5 million of the conserved sperm will become suboptimal and, even if inseminated into the female at the pair's next copulation, will probably be ejected in the flowback. At least these sperm, therefore, were of no future value. Our data show that males top-up their female by between 2 million and 9 million sperm (depending on the proportion of time the pair are together) for each hour between copulations. Yet new sperm are becoming available at the rate of 12.5 million/h, more than enough to top-up the female at the next copulation, no matter what proportion of time the pair have been together. It seems unlikely, therefore, that males conserve sperm at one copulation for use at some future copulation with the same female.

This conclusion could be complicated by the possibility that, if the female subsequently engages in extra-pair copulation, the male (as for birds; Birkhead and Møller, 1992) may need to show rapid 'unscheduled' IPC and inseminate large numbers of sperm to counteract the certainty of sperm competition. However, unlike birds which show a clear second male advantage (Birkhead and Møller, 1992) mammals may show first male (house mice), last male (prairie voles) and often no order effects (rats and swine) (section 8.4.2). In which case, it is by no means certain that the male does better to conserve sperm to inseminate after another male rather than before. This is especially so if any sperm conserved have, through greater age, a higher probability of being ejected in the flowback than if they had been inseminated earlier.

The only likely use for conserved sperm, therefore, is to increase the chances of securing fertilization of a different female. To be beneficial, the conserved sperm must increase the chances and/or value of fertilization of the second female by more than they would have increased the chances and/or value of fertilization of the first female. A major factor will be the relative probability of sperm competition in the two females. However, even when the conserved sperm would be more beneficial in the second female than the first, this greater benefit has to outweigh three other factors: (a) the cost of storage; (b) the conserved sperm would be younger when inseminated into the first female; and (c) the conserved sperm must be inseminated into the second female before they become suboptimal, are shed, or are inseminated into the first female. As an initial approximation, they have at most about 2–3 days (median inter-ejaculation interval at different ages; Box 5.1).

Without knowing the relative costs and benefits of these different parameters, no decision can be reached. Until it can, we have at least to entertain the possibility that conserved sperm are of no future advantage to the male and that the advantage of restraint may derive, not from economy and conservation, but from some direct benefit at the current insemination.

9.7.2 Is restraint an advantage or disadvantage at the current copulation?

There are several possible reasons why males could use restraint to gain maximum advantage from the current copulation, irrespective of what happens in the future to any sperm conserved.

Our analysis of ejaculate adjustment has identified two main elements to male strategy. First, successive IPCs are 'toppings-up' to some maximum level that is lower when risk of sperm competition is also lower. Second, masturbation seems to be a strategy to increase the fitness (perhaps longevity, competitiveness and/or fertility) of the sperm retained by the female at the next IPC without increasing the number retained. The overall impression is that there is some disadvantage to a male in placing too large a population of sperm in his partner's tract.

No data exist for any mammal on the relationship between large numbers of sperm in the female tract and the probability of fertilization. However, a known correlate of fertility impairment in humans is polyzoospermy (ejaculation of too many sperm; Wolf *et al.*, 1984). At present, clinical diagnosis places the lower level for polyzoospermy at 250×10^6 sperm/ml of ejaculate (when inter-ejaculation interval is about 72 h and ejaculates are collected during masturbation). With an average ejaculate volume of about 3 ml, this density is equivalent to an ejaculate containing about 750 million sperm. Why polyzoospermy should lead to impaired fertility is unknown. Apart from their concentration of sperm, polyzoospermic ejaculates have clinically normal parameters and, as long as concentration around the egg is controlled, their sperm fertilize eggs *in vitro*.

Our own data show that the more sperm a male inseminates, the more are retained (Box 10.1). It is also known that, in rabbits, the more sperm retained, the more are found at all positions throughout the female tract (Morton and Glover, 1974a,b). The implication is that, on average, the more sperm that are inseminated the more arrive at all positions in the female tract including, presumably, around the egg.

Studies *in vitro* show clearly that there is an optimum sperm:egg ratio in the vicinity of the egg, above which the probability of obtaining a viable zygote declines. Thus, in laboratory mice the probability of fertilization peaks (78%) at sperm:egg ratios of about 16 000:1 (Tsunoda and Chang, 1975). Increasing the sperm:egg ratio about tenfold decreases the probability of fertilization by more than 50%. The optimum sperm:egg ratio for humans *in vitro* is around 50 000:1 (Lee, 1988).

There are at least two potential disadvantages of a high sperm:egg ratio. First, above a certain ratio, enzymes released by sperm have an increasing chance of killing the eggs (Adams, 1969) (section 4.4.8). Second, the presence of too many sperm around the egg could lead to pathological polyspermic fertilization and a non-viable zygote (section 4.4.10).

Of course, there may also be a disadvantage in inseminating too few sperm, even in the total absence of sperm competition. However, just how low the optimum number may be for healthy ejaculates is not clear. It is now known that fertilization can occur even when no obviously healthy sperm can be seen in a male's ejaculate (Thomson *et al.*, 1993; section 6.4.2). Males who habitually inseminate low numbers of sperm (<60 million) range from being fully fertile to being infertile (Matson *et al.*, 1989). It may not be the low sperm numbers *per se* that is the problem for such men but other aspects of sperm subfertility that are often associated with the low numbers.

There will be an optimum number of sperm for fertilization in the absence of sperm competition. That optimum, however, may be lower than previously recognized.

9.7.3 Restraint: conclusion

We suggest that in the presence of sperm competition, a male's chances of fertilizing a female are increased by increasing sperm number but, in the absence of sperm competition, his chances are increased by decreasing sperm number to some optimum level. We

Box 9.12 Ejaculation and potential stages for adjusting the number of sperm in the ejaculate

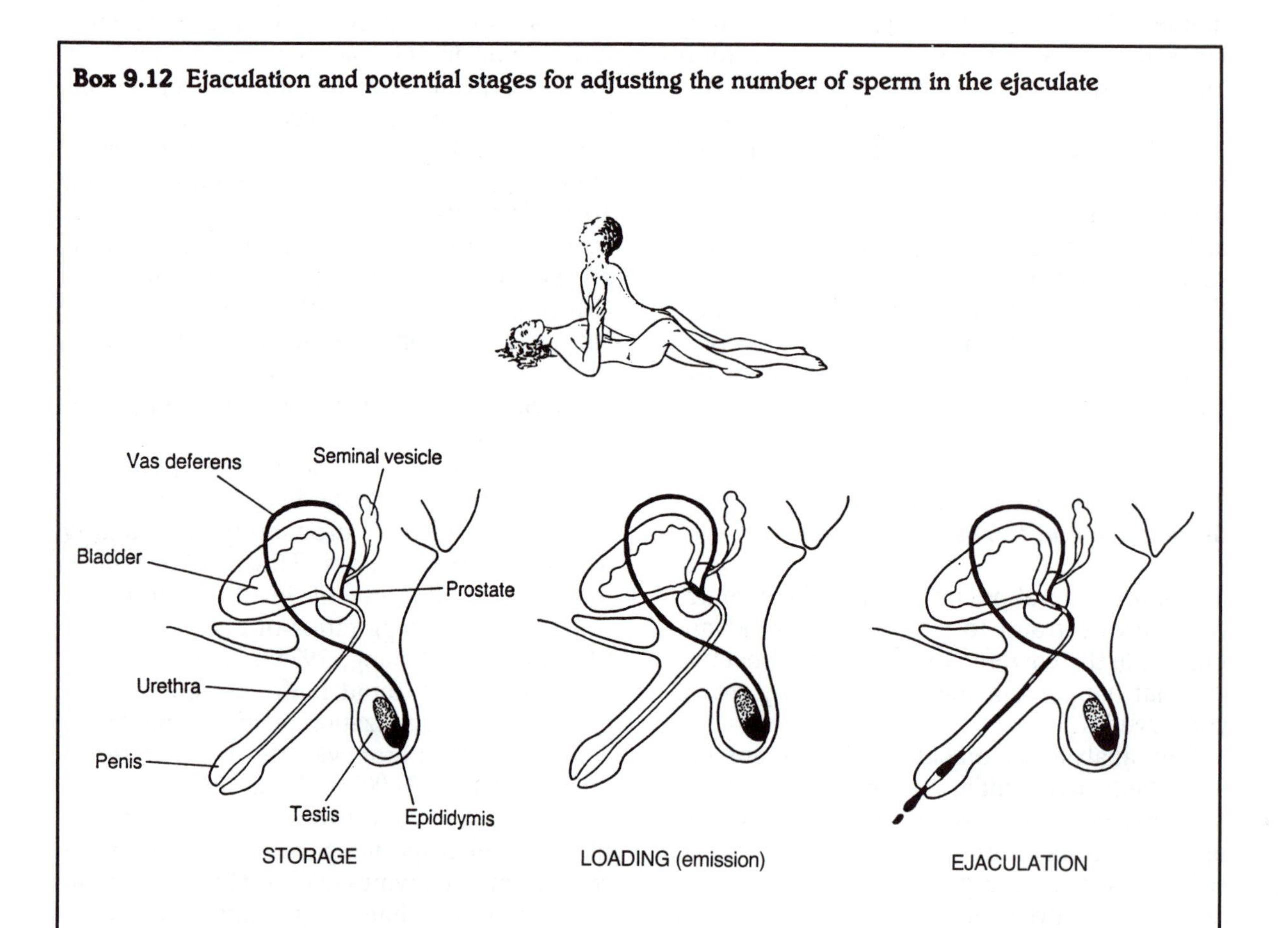

Storage When copulation begins, between 700 and 1000 million mature sperm (solid black colum) are in storage in the cauda epididymis and vas deferens.

Loading Towards the end of copulation, a proportion of the stored sperm are loaded into the urethra (= emission). A constriction in the prostatic urethra prevents sperm from entering the bladder.

Ejaculation First the prostate gland and then the seminal vesicles pour out fluids into the urethra to form the semen which is ejaculated as a series of spurts.

Adjustment Adjustment of the number of sperm in the ejaculate could occur at two stages:
1. during loading, by varying the proportion of the stored sperm that are loaded into the urethra; and/or
2. during ejaculation, by varying the proportion of the loaded sperm that are ejaculated. Any sperm retained in the urethra after ejaculation are voided at the next urination.

suggest further that this latter factor (the optimum sperm number in the female tract for fertilization in the absence of sperm competition) could be a major constraint to insemination of too many sperm at any given copulation. On this model, the primary trade-off in determining ejaculate size is between the probabilities that the male will fertilize the female if the inseminate does or does not encounter sperm competition. Optimum number of sperm is thus determined for each inseminate simply by the risk of sperm competition.

At present, we cannot determine the relative importance of this factor and any constraint on sperm numbers due to the cost of producing ejaculates (Dewsbury, 1982). Inevitably, ejaculate cost must have been a factor in the evolution of species-specific rates of sperm production. It is possible, however, that at least for mammals ejaculate cost could be less important than the factors discussed here in influencing restraint over the number of sperm inseminated on any given occasion.

9.8 When and how do males adjust sperm number?

Males may not look very sophisticated in the moments leading up to and during ejaculation but as this chapter has shown, some very sophisticated adjustments are taking place. First, masturbatory ejaculations are not simply substitutes for copulatory ejaculates. The number of sperm the male ejaculates is different in the two circumstances and adjustment for the risk of sperm competition only occurs in copulatory ejaculates. Second, during copulation, the male adjusts the number inseminated according to how long it is since he last inseminated that individual female and the risk of sperm competition.

Somehow, therefore, males are subconsciously making an adjustment to the number of sperm inseminated according to their sociosexual situation. Masturbations performed on the male by the female partner show the characteristics of self-masturbation, not the characteristics of copulation (Baker and Bellis, 1993a). Simple presence of a female is not, therefore, enough for the male to switch to a copulatory ejaculate. This implies that it is some psychophysiological stimulation during copulation that triggers the final adjustments.

Our studies of rats (Bellis *et al.*, 1990) show that a male can switch from inseminating a small to a large ejaculate within the 10 minutes or so that the female it has been guarding throughout a whole oestrous cycle is swapped for a novel female. Naturally, we have not been able to carry out the same experiments with humans, but the implication is that inseminate size is determined during copulation itself. The two main stages that adjustment could occur are shown in Box 9.12.

10 Ejaculate manipulation by females and the function of the female orgasm

10.1 Introduction

Until recently, models of sperm competition have tended to view the female tract as a passive receptacle in which males play out their sperm competition games. Females have the potential, however, to influence the outcome of the contest in several different ways (Thornhill, 1983; Bellis and Baker, 1990; Eberhard, 1990, 1991; Baker and Bellis, 1993b; Birkhead and Møller, 1993). The sequence and frequency with which the female mates with different males and the time interval between in-pair copulations and extra-pair copulations often have a major influence on the outcome of sperm competition (section 8.4.2). More directly, females eject sperm in the flowback (section 3.5.2).

Female humans eject up to 3 ml of fluid from their vagina after copulation. This fluid is the 'flowback', a mixture of sperm, seminal fluid, female tissue and female secretions. The flowback either (a) oozes or dribbles from the vagina while the female is lying down or walking; or (b) perhaps most often and most forcefully, is ejected as a discrete series of white globules when she urinates. Until our study (Boxes 3.8 and 3.9) no systematic attempt seems to have been made to determine the proportion of sperm ejected and retained by human females following normal copulation. Nor had any attempt been made to investigate variation in this proportion in relation to sociosexual situation or in relation to female physiological events, such as the female orgasm (sections 3.5.3 and 5.3).

Currently, there are two favoured hypotheses concerning the function of copulatory orgasms in females; (1) the 'poleaxe' hypothesis; and (2) the 'upsuck' hypothesis (section 3.5.3). The poleaxe hypothesis proposes that, as humans are bipedal, it is important for the female to lie down after copulation in order to reduce sperm loss. The orgasm thus functions to induce fatigue and sleep. The upsuck hypothesis proposes that the orgasm functions to suck up sperm during copulation (Fox *et al.*, 1970).

Our own hypothesis (Baker and Bellis, 1993b) is much more firmly based in behavioural ecology. It is that the timing of orgasm, both copulatory (section 3.5.3) and non-copulatory (Chapter 5) is the key feature of the mammalian female's armoury in male: female conflict and cooperation within the female tract. We proposed that nocturnal, masturbatory and copulatory orgasms are the primary mechanisms by which the female influences the ability of sperm in the next and/or current ejaculate to remain in, and travel through, her reproductive tract.

More specifically, our suggestion was that

Box 10.1 Separating male and female contributions to sperm retention

Measuring male and female contributions

The number of sperm retained by a female from any given in-pair copulation is essentially the product of two influences. In so far as the male inseminates a finite number of sperm, he places an upper limit on the number that may possibly be retained. The number actually retained then depends on what happens in the female reproductive tract. The behaviour of the sperm and/or the female's response to insemination may both influence the number of sperm actually retained.

In our study of flowbacks, we use the IPC equation shown in Box 9.1 to calculate the number of sperm inseminated into the female at each in-pair copulation. We then estimate the number of sperm retained by the female by subtracting the number of sperm observed in the flowback from the number calculated to have been inseminated.

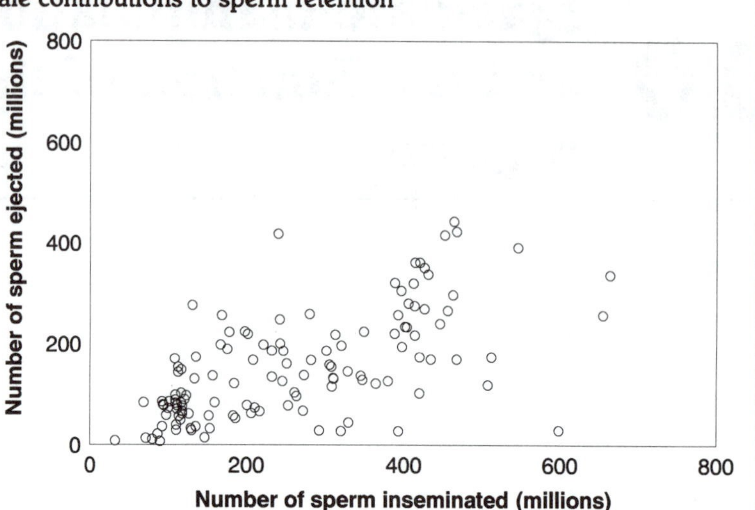

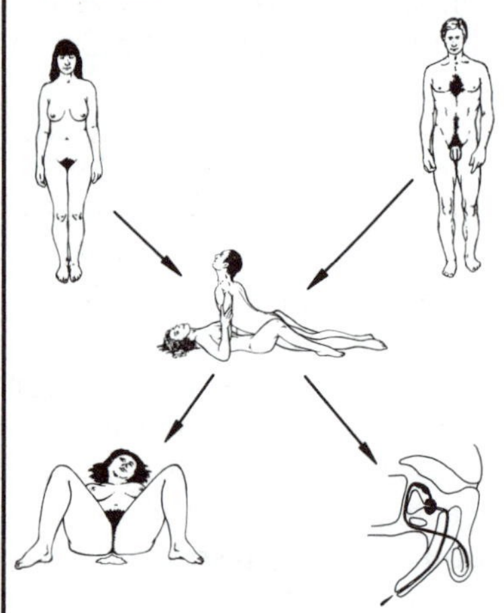

There is a significant positive relationship between the number of sperm inseminated and on the one hand the number of sperm in the flowback (see scattergram) and on the other the number of sperm retained (Box 9.7). We exploit this positive relationship to remove the influence of the male on the number of sperm retained. First, we calculate the least squares regression line for numbers of sperm ejected and retained against numbers inseminated (see Table).

We then apply these regression lines to each of the copulations in our total data set and calculate residuals. Variation in these residuals is thus independent of the variation in the number of sperm inseminated by the males and the residuals are, in effect, a measure of variation in the female's contribution to sperm retention.

Correcting for the age of sperm

Younger sperm are retained in the female tract in relatively, but not absolutely, greater numbers than older sperm but there is no further influence of sperm age once time since last ejaculation exceeds 72 h (Box 9.7).

It is a moot point whether the age of sperm is a male or female influence on sperm retention. Younger sperm may be more acceptable to the female (female influence) or they may be better able to attain a position in the female tract that makes them more resistant to ejection (male influence). Either way, before analysing contributions to sperm retention that are clearly female, such as timing of orgasm, it is necessary to correct the residuals from the above procedure to make them independent of sperm age. For full details, see Baker and Bellis (1993b).

Table. Relationships between the number of sperm inseminated into a female during in-pair copulation and: number of sperm ejected; number of sperm retained; and volume of flowback (analysis of only the first sample from each couple, $n = 11$ for sperm number, $n = 9$ for flowback volume)

	Sperm Number		Flowback volume ejected
	Ejected	Retained	
Correlation			
(r_s)	0.655	0.791	−0.113
($P_{1\text{-tailed}}$)	0.026	0.004	0.560
Regression			
Intercept	0.90	−0.90	
Slope	0.35	0.65	
% explained	24	58	

by altering the occurrence, sequence and timing of the different types of orgasm, the female can influence both the probability of conception in monandrous situations and the outcome of sperm competition in polyandrous situations. We predicted, also, that as part of female sexual crypsis (section 6.6), much of this influence will be cryptic to the male partner(s).

This chapter presents the main evidence for this general hypothesis and explores the dynamics of male:female conflict and cooperation over sperm.

10.2 Male and female influences on sperm retention

The number of sperm retained by a female from any given copulation is essentially the product of two influences: male and female. The way that male and female contributions to the number of sperm retained were separated during our study of flowbacks (Boxes 3.7–3.9) is outlined in Box 10.1.

One of the products of our work was a predictive equation (Box 10.2), based on both male and female contributions, for the calculation of number of sperm retained and ejected in different sociosexual situations.

10.3 Human flowbacks: general features

The main features of human flowbacks are described in section 3.5.2 and Boxes 3.10 and 10.3. Briefly, all of the flowbacks collected contained sperm and 94% of IPCs monitored subjectively for flowback volume were followed by noticeable flowbacks. On average, the more sperm inseminated, the more were ejected in the flowback (Box 10.1) and the greater volume of the flowback (Baker and Bellis, 1993b).

About 12% of all the IPCs for which we have flowback samples were followed by virtually 100% ejection of sperm (i.e. $<1\%$ retained) (Box 10.3). Females thus appear capable of total or near-total ejection of IPC ejaculates.

10.4 Oral contraceptives, pregnancy and flowbacks

The main results of this section are from Baker and Bellis (1993b).

Females taking oral contraceptives show no significant variation in retention or ejection during the 'menstrual' cycle ($P = 0.276$). Females not taking oral contraceptives, however, showed a significant decrease in retention during their most fertile phase (days 6–15) ($P = 0.016$). Finally, there was no significant difference in the number of sperm retained by females when they were taking oral contraceptives compared with when they were not ($P = 0.484$).

A longitudinal study of one female who collected flowbacks pre- and postpartum and during the first 6 weeks of pregnancy showed a significant reduction in sperm retention during pregnancy ($P = 0.016$). By 5–6 weeks postconception, virtually no sperm were being retained.

10.5 Are male strategies generally successful?

Our previous studies of whole ejaculates (Chapter 9) identified three male strategies involving adjustment of the number of sperm inseminated during IPC. First, the number of sperm inseminated during IPC is a positive function of time since last IPC. Adjustment of sperm numbers fits best a model based on the male 'topping-up' his partner to a particular level, inseminating only enough to make up the number of sperm likely to have died since last insemination. Second, males top-up larger females with more sperm. Third, males top-up females with more sperm when the male spends a lower proportion of his time with his partner and thus, on average, the risk of sperm competition is higher.

On any given occasion, it appears quite

Box 10.2 **Equation to calculate the number of sperm retained during in-pair copulation in different sociosexual situations**

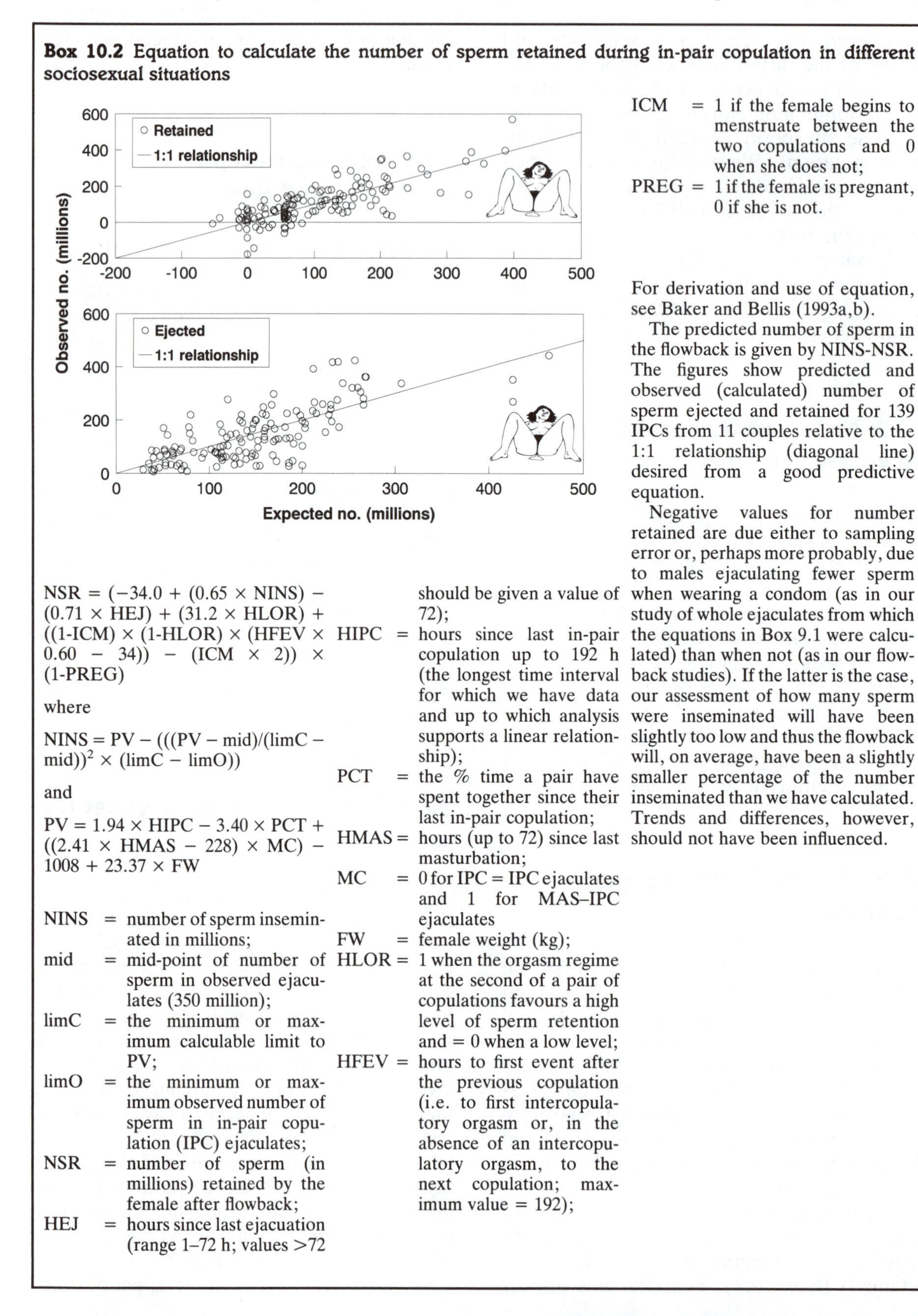

$$NSR = (-34.0 + (0.65 \times NINS) - (0.71 \times HEJ) + (31.2 \times HLOR) + ((1\text{-}ICM) \times (1\text{-}HLOR) \times (HFEV \times 0.60 - 34)) - (ICM \times 2)) \times (1\text{-}PREG)$$

where

$$NINS = PV - (((PV - mid)/(limC - mid))^2 \times (limC - limO))$$

and

$$PV = 1.94 \times HIPC - 3.40 \times PCT + ((2.41 \times HMAS - 228) \times MC) - 1008 + 23.37 \times FW$$

NINS = number of sperm inseminated in millions;
mid = mid-point of number of sperm in observed ejaculates (350 million);
limC = the minimum or maximum calculable limit to PV;
limO = the minimum or maximum observed number of sperm in in-pair copulation (IPC) ejaculates;
NSR = number of sperm (in millions) retained by the female after flowback;
HEJ = hours since last ejaculation (range 1–72 h; values >72 should be given a value of 72);
HIPC = hours since last in-pair copulation up to 192 h (the longest time interval for which we have data and up to which analysis supports a linear relationship);
PCT = the % time a pair have spent together since their last in-pair copulation;
HMAS = hours (up to 72) since last masturbation;
MC = 0 for IPC = IPC ejaculates and 1 for MAS–IPC ejaculates
FW = female weight (kg);
HLOR = 1 when the orgasm regime at the second of a pair of copulations favours a high level of sperm retention and = 0 when a low level;
HFEV = hours to first event after the previous copulation (i.e. to first intercopulatory orgasm or, in the absence of an intercopulatory orgasm, to the next copulation; maximum value = 192);
ICM = 1 if the female begins to menstruate between the two copulations and 0 when she does not;
PREG = 1 if the female is pregnant, 0 if she is not.

For derivation and use of equation, see Baker and Bellis (1993a,b).

The predicted number of sperm in the flowback is given by NINS-NSR. The figures show predicted and observed (calculated) number of sperm ejected and retained for 139 IPCs from 11 couples relative to the 1:1 relationship (diagonal line) desired from a good predictive equation.

Negative values for number retained are due either to sampling error or, perhaps more probably, due to males ejaculating fewer sperm when wearing a condom (as in our study of whole ejaculates from which the equations in Box 9.1 were calculated) than when not (as in our flowback studies). If the latter is the case, our assessment of how many sperm were inseminated will have been slightly too low and thus the flowback will, on average, have been a slightly smaller percentage of the number inseminated than we have calculated. Trends and differences, however, should not have been influenced.

Box 10.3 Some general features of flowbacks

The ejection of the flowback is a relatively discrete event. At some time after male ejaculation, the flowback fairly suddenly arrives in the vaginal vestibule. From there, it is either expelled with force during urination or else dribbles out slowly, either while the female is still down or when she begins to walk around.

(a) In our study, the time at which flowbacks were collectable after male ejaculation varied from 5 to 120 min. The commonest time interval, however, was between 20 and 40 min. A tendency for more sperm to be ejected when time to flowback was longer was not significant when controlled for differences between couples ($P = 0.108$).

(b) There was a tendency for time to flowback to be shorter when the number of sperm inseminated was either small or large but the variation was not significant when controlled for differences between couples.

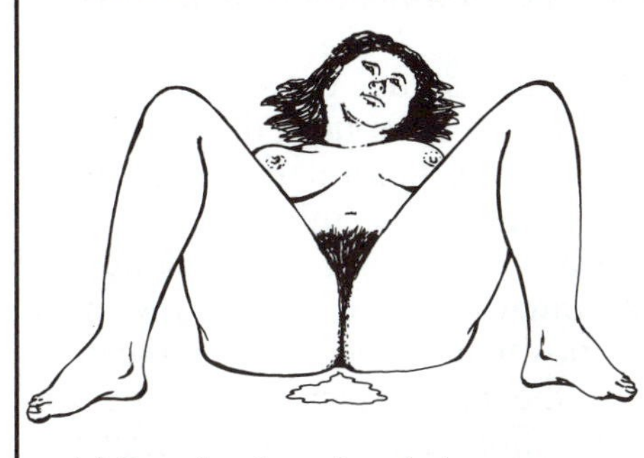

(c) Females show the whole range of sperm retention levels, from near zero to virtually total. Three modes are evident: <1% retained; 40–50% retained; and 60–70% retained.

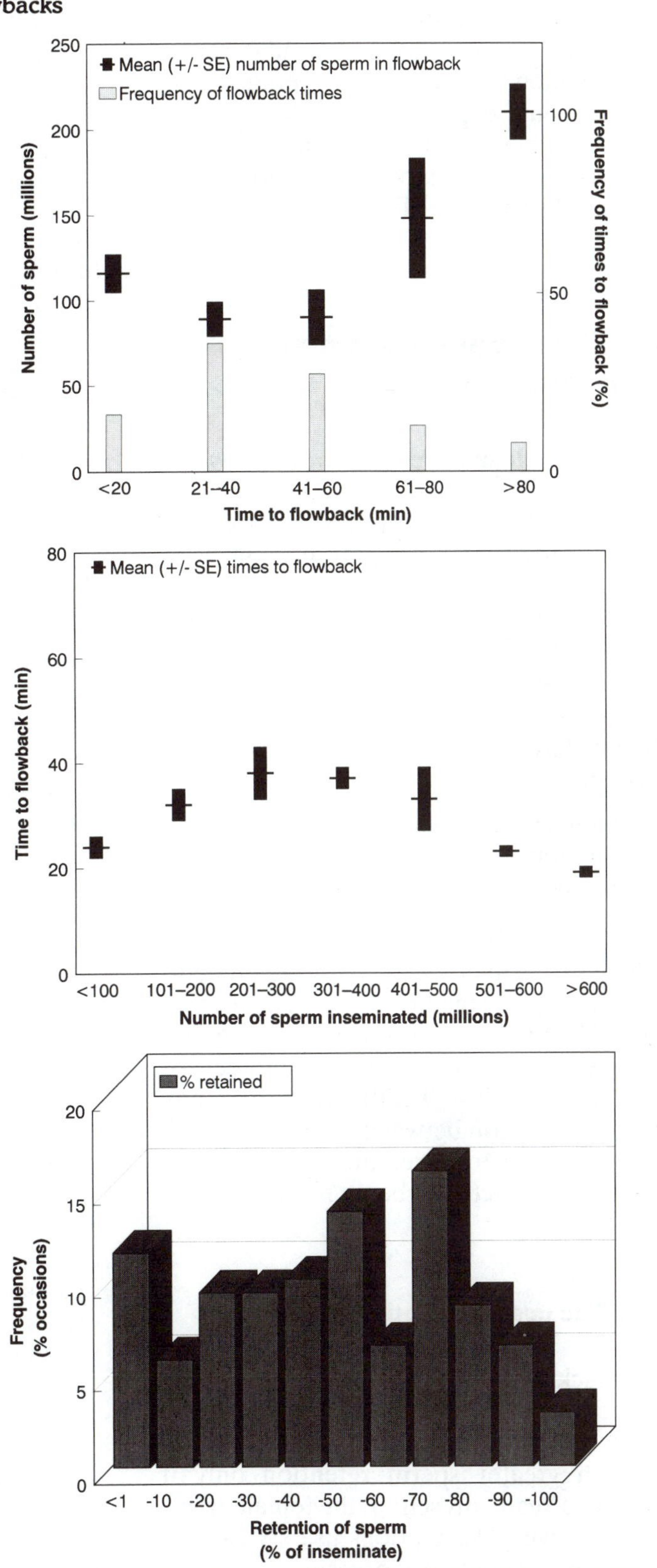

possible that females could override any attempt by the male to place more sperm in her tract in response to a particular socio-sexual situation (Box 10.3). On average, however, we should expect male strategies to be successful otherwise they would not have been promoted by natural selection. This appears to be the case (Box 10.4).

10.6 The female orgasm and sperm retention

10.6.1 The poleaxe hypothesis: falls down

Three predictions of the poleaxe theory have now been tested (Baker and Bellis, 1993b).

1. Time from male ejaculation to flowback should be longer if the female has a copulatory orgasm than if she does not.
2. More sperm should be retained if the female has a copulatory orgasm than if she does not.
3. The longer the time-interval from male ejaculation to flowback, the more sperm should be retained.

On the whole, there is no empirical support for the poleaxe hypothesis. Although time to flowback may be significantly longer after an orgasm ($P = 0.049$), there is no corresponding significant increase in the number of sperm retained ($P = 0.189$) nor is there any positive association between time to flowback and number of sperm retained ($P = 0.975$), the latter trend actually being negative.

10.6.2 The upsuck hypothesis: taken up

The upsuck hypothesis (Fox *et al.*, 1970) for the function of the female copulatory orgasm would predict that orgasm should be associated with greater sperm retention only if sperm are already present in the female tract.

The following descriptions use six categories of female copulatory orgasm:

1. CX = no orgasm between the beginning of foreplay and ejection of the flowback;
2. BC = orgasm before copulation (i.e. during foreplay; penis not in vagina);
3. BE = orgasm during copulation but before ejaculation (penis in vagina);
4. DE = during ejaculation (i.e. simultaneous climax; penis in vagina);
5. AE = after ejaculation (penis in vagina);
6. AC = after copulation but before flowback (penis not in vagina).

As already noted in relation to the poleaxe theory, whether the female has an orgasm at about the time of copulation (BC,BE, DE,AE or AC) or not (CX) has no significant influence on the number of sperm retained. However, if an orgasm does occur, the timing of the female climax relative to copulation and male ejaculation (i.e. comparison of BC,BE,DE,AE,AC) has a highly significant influence on the number of sperm retained. This influence depends on the relative timings of female climax and male ejaculation (Box 10.5).

In our flowback study, the earliest that a female experienced an orgasm during an IPC episode was 45 min before the male ejaculated and the latest was 45 min after (i.e. −45 min to +45 min, using negative numbers to indicate a female climax before the male ejaculates; positive numbers to indicate a climax after the male ejaculates). The frequency distribution of the timing of orgasm in relation to male ejaculation was shown in Box 3.12.

Climaxes earlier than one minute before the male ejaculates (i.e. −1 min) are associated with low sperm retention (Box 10.5). Maximum sperm retention is associated with climaxes from >0 to 1 min after the male ejaculates onwards. If the female climaxes earlier than −1 min, female influence on sperm retention is no better than if she fails to climax altogether. Only climaxes that occur between −1 min and emergence of the flowback (up to +45 min) are associated with a significant increase in the residual number of sperm retained compared with having no orgasm at all. During the 15 min or so period that the seminal pool is coagulated, however,

Box 10.4 The average success of male strategies

Males have three major strategies in terms of adjusting ejaculates: they vary the number inseminated according to (a) female weight, (b) time since last copulation, and (c) percentage time together. Essentially, they 'top-up' their partner to a level that is greater for larger females and when the risk of sperm competition is higher (see Chapter 9).

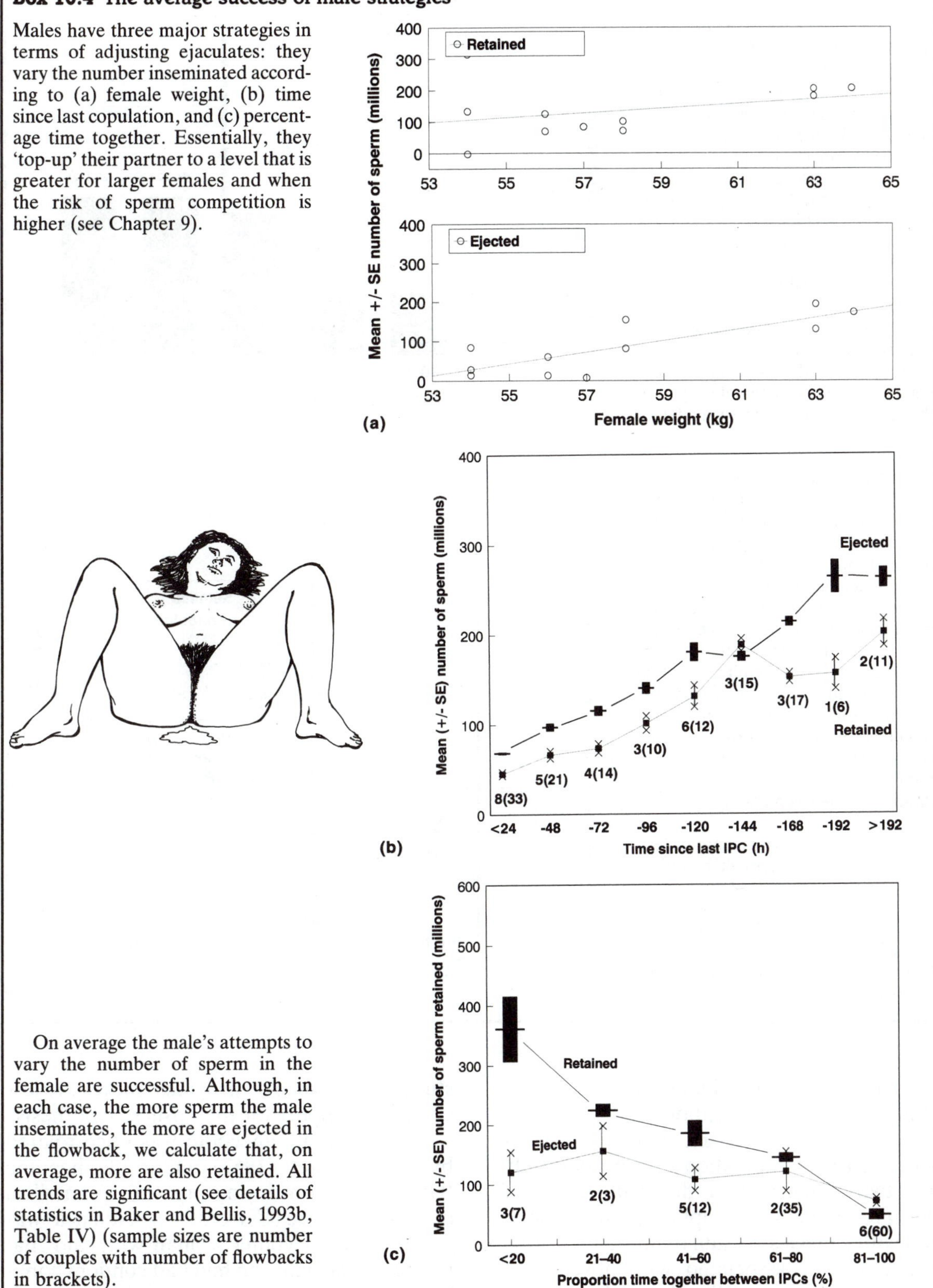

On average the male's attempts to vary the number of sperm in the female are successful. Although, in each case, the more sperm the male inseminates, the more are ejected in the flowback, we calculate that, on average, more are also retained. All trends are significant (see details of statistics in Baker and Bellis, 1993b, Table IV) (sample sizes are number of couples with number of flowbacks in brackets).

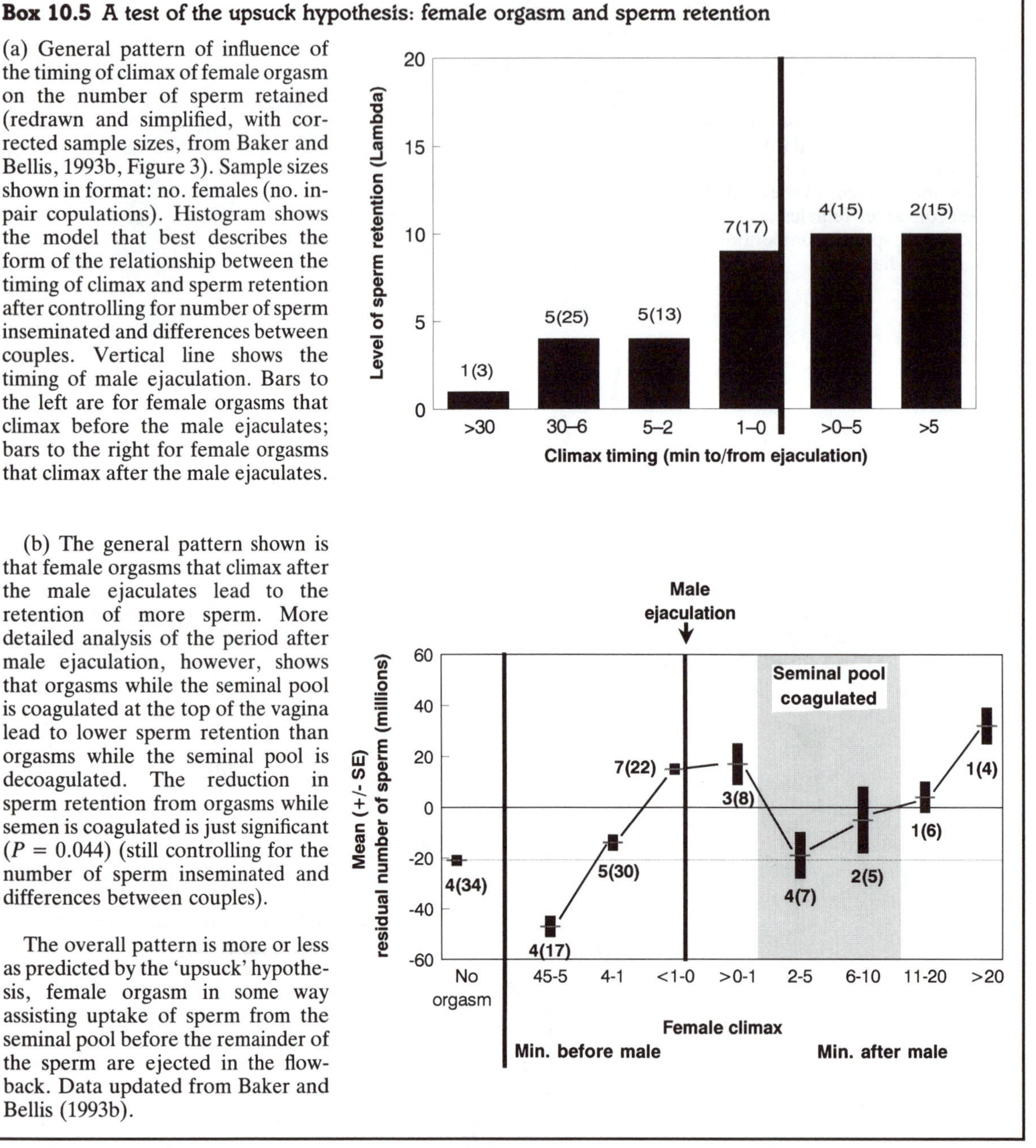

Box 10.5 A test of the upsuck hypothesis: female orgasm and sperm retention

(a) General pattern of influence of the timing of climax of female orgasm on the number of sperm retained (redrawn and simplified, with corrected sample sizes, from Baker and Bellis, 1993b, Figure 3). Sample sizes shown in format: no. females (no. in-pair copulations). Histogram shows the model that best describes the form of the relationship between the timing of climax and sperm retention after controlling for number of sperm inseminated and differences between couples. Vertical line shows the timing of male ejaculation. Bars to the left are for female orgasms that climax before the male ejaculates; bars to the right for female orgasms that climax after the male ejaculates.

(b) The general pattern shown is that female orgasms that climax after the male ejaculates lead to the retention of more sperm. More detailed analysis of the period after male ejaculation, however, shows that orgasms while the seminal pool is coagulated at the top of the vagina lead to lower sperm retention than orgasms while the seminal pool is decoagulated. The reduction in sperm retention from orgasms while semen is coagulated is just significant ($P = 0.044$) (still controlling for the number of sperm inseminated and differences between couples).

The overall pattern is more or less as predicted by the 'upsuck' hypothesis, female orgasm in some way assisting uptake of sperm from the seminal pool before the remainder of the sperm are ejected in the flowback. Data updated from Baker and Bellis (1993b).

orgasms have a significantly reduced influence on sperm retention.

This last detail apart, female orgasms associated with copulation divide essentially into two regimes with different levels of sperm retention: a high-uptake regime associated with climaxes -1 min or later; and a low-uptake regime associated with absence of orgasm or orgasms earlier than -1 min. The study of flowbacks provides more or less complete support for the upsuck hypothesis.

An irresistible anecdote was once related by De Carteret (1967) under the heading 'Uterine suction during orgasm'. A medical colleague of De Carteret's in Jamaica was consulted by a young woman who said that the previous night she had had condom copulation with a soldier and that on completion the condom was missing. The doctor did a speculum examination and found the blind end of the condom firmly held in the cervical canal.

We repeat this anecdote primarily to illustrate how not to interpret our work. There are facets to our and other workers' data that warn that the simplistic view of 'upsuck' inherent in De Carteret's story is probably far from the truth.

Orgasms associated with copulation influence not only the retention of sperm, but also the volume of the flowback (Baker and Bellis, 1993b). However, *contra* De Carteret, the primary effect is that, when an orgasm occurs, the volume of the flowback increases significantly. Volume is greatest when orgasm occurs during foreplay, presumably due to the increased flow of cervical mucus (Box 3.5). However, there is no other apparent influence of orgasm on flowback volume and, in general, orgasm regimes that favour sperm retention do not decrease flowback volume (Baker and Bellis, 1993b). However sperm are retained, therefore, retention is achieved without any gross retention of seminal fluid.

This conclusion is consistent with other evidence. Masters and Johnson (1966), for example, found no upsuck of radiopaque fluid across the cervix following orgasm and other workers could not detect any seminal plasma antigen in the cervix after copulation even though sperm were present in the cervical mucus channels. As a whole, these findings probably illustrate the unique rheological properties of the cervical mucus, which permit the forced permeation of small microparticles, such as sperm, but prevent bulk passage of fluids such as radiopaque media or seminal plasma (Katz and Overstreet, 1982; Overstreet, 1983).

It would also be a mistake to conclude that the female tract provides no assistance in sperm transport unless there is an orgasm (Overstreet, 1983). Even without female orgasm, inert carbon particles have been found in human oviducts 28–34 min after vaginal deposition, and sperm have been found in the oviducts within 5 min of artificial insemination (Settlage *et al.*, 1973).

It has long been known that female orgasm is not necessary for conception (Moghissi, 1977). Yet, females who copulate with their (infertile) partners within 4 h of artificial insemination of sperm from a (fertile) donor have a higher conception rate than females who abstain from copulation for 3 days (Kesserü, 1984). Even so, it makes no difference to their chances of conception whether females orgasm during this post-AID copulation or not.

In the absence of orgasm, some motility on the part of the sperm seems to be necessary for retention. In three cases in which the male partner had necrospermia, no sperm were in the cervical mucus 90–180 s after copulation. Yet of 44 women whose partners had motile sperm, 39 had active sperm in the cervical mucus within the same time after copulation.

Perhaps the best interpretation of our data, especially in view of the 'searching and dipping' behaviour shown by the cervix during orgasm (Box 3.5), is as follows. Orgasm facilitates the passage of sperm, probably via their own motility, from the seminal pool to the cervical mucus. It could do this in one or all of several ways: (1) dipping the cervix further into the seminal pool; (2) promoting greater mixing of cervical mucus and seminal fluid; (3) lengthening and/or increasing the number of seminal projections into the cervical mucus; and/or (4) lengthening the time that the cervix is dipped in the seminal pool. Whatever the mechanism, at the end of sperm transfer all traces of seminal fluid, perhaps still associated with cervical mucus, are extruded from the cervix into the vagina to form part of the flowback.

In summary, therefore, in the absence of orgasm, motile sperm enter and pass through the cervical mucus, then to be assisted by uterine contractions (stimulated simply by copulation) in passive, often rapid, transport through the female tract. In the presence of orgasm (of appropriate timing), more sperm avoid the flowback because their movement into the cervical mucus is facilitated by the behaviour of the cervix. The properties of the cervical mucus are such, however, that the seminal fluid is excluded whether the female has an orgasm or not. Finally, the increased retention of sperm has no influence on the probability of conception. It should, however, have considerable influence on the out-

Box 10.6 Intercopulatory orgasms and the retention of sperm at the next copulation

When only one intercopulatory orgasm occurs between two copulations, the type of orgasm (nocturnal, self-masturbatory, etc.) does not seem to make any consistent difference to the number or proportion of sperm retained at the next copulation. We assume, therefore, that the different types of intercopulatory orgasm are different ways of achieving the same result. However, the number of intercopulatory orgasms between any pair of copulations has a significant influence on the number of sperm retained at the second copulation of the pair ($P = 0.006$). This influence is independent of the time interval between the two copulations but is largely due to the time interval from the first of the pair of copulations and the first intercopulatory orgasm (Box 10.7). This time-interval is shorter when there are more intercopulatory orgasms.

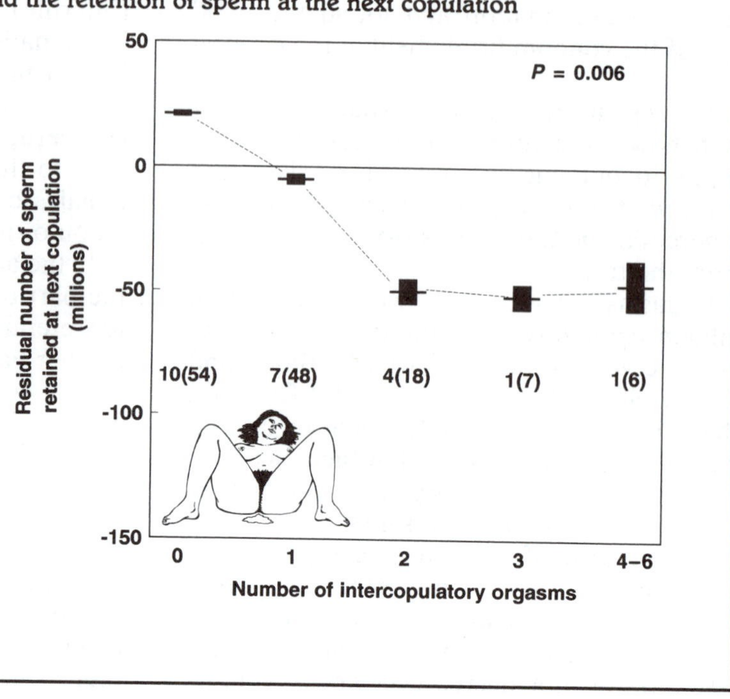

come of any sperm competition that may ensue.

10.6.3 Masturbatory, nocturnal and other intercopulatory orgasms

We define an intercopulatory orgasm (or ICO) as any female orgasm that occurs between the first withdrawal of the penis after male ejaculation at one copulation and the last insertion of the penis before male ejaculation at the next copulation. This section is concerned with whether intercopulatory orgasms have any influence on sperm retention at the second copulation and, if they do, what form that influence takes.

Initially, we recognize five categories of intercopulatory orgasm (see Chapter 5).

1. AC = during postplay at the first of a pair of copulations;
2. BC = during foreplay at the second of a pair of copulations;
3. NO = nocturnal/spontaneous;
4. SS = self-stimulation in absence of male;
5. PS = any stimulation, but not leading to copulation, in the presence of a male or female partner.

First, there is no significant heterogeneity between the five types of intercopulatory orgasm in terms of their influence on sperm retention at the second copulation ($P = 0.119$) (Baker and Bellis, 1993b). If intercopulatory orgasms have any influence on the residual number of sperm retained at the second copulation, it is independent of the type of intercopulatory orgasm. Henceforth, therefore, the five types of intercopulatory orgasm are considered as a single category. The number of intercopulatory orgasms has a very significant effect on the number of sperm retained at the next copulation (Box 10.6).

In the complete absence of either menstruation or an intercopulatory orgasm between two copulations, the number of sperm retained by the female at the second copulation depends on the time-interval between the two copulations. Females retain more sperm when the intercopulatory interval is longer (Box 10.7), with no clear indication of any plateauing of the relationship at least up to 192 h (8 days). It is as if one copulation sets up a block to sperm retention

Box 10.7 The influence of sperm from one copulation on the retention of sperm from the next copulation

(a) One copulation influences sperm retention at the next copulation. The diagonal line shows the change in sperm retention at the second copulation as a function of time since the previous copulation. Sperm retention at the second copulation improves with time since the previous copulation for a period of at least 8 days. The first intercopulatory orgasm, however, fixes the level of sperm retention at the next copulation at the level it had reached according to time since the last copulation. Level of retention at the next copulation is thus fixed by the time interval from the previous copulation to the first subsequent sexual event; either an intercopulatory orgasm or the next copulation.

(b) The critical factor in this influence of one copulation on another seems to be the sperm from the previous copulation. When no sperm are inseminated (through use of a condom) none of the above effects are seen. Moreover, on occasions when flowbacks were collected from two consecutive copulations, it was possible to calculate (crudely) how many sperm were likely still to be present in the female tract from the first copulation. Over these occasions, sperm retention at the second copulation seemed to be a function of the number of sperm remaining from the first.

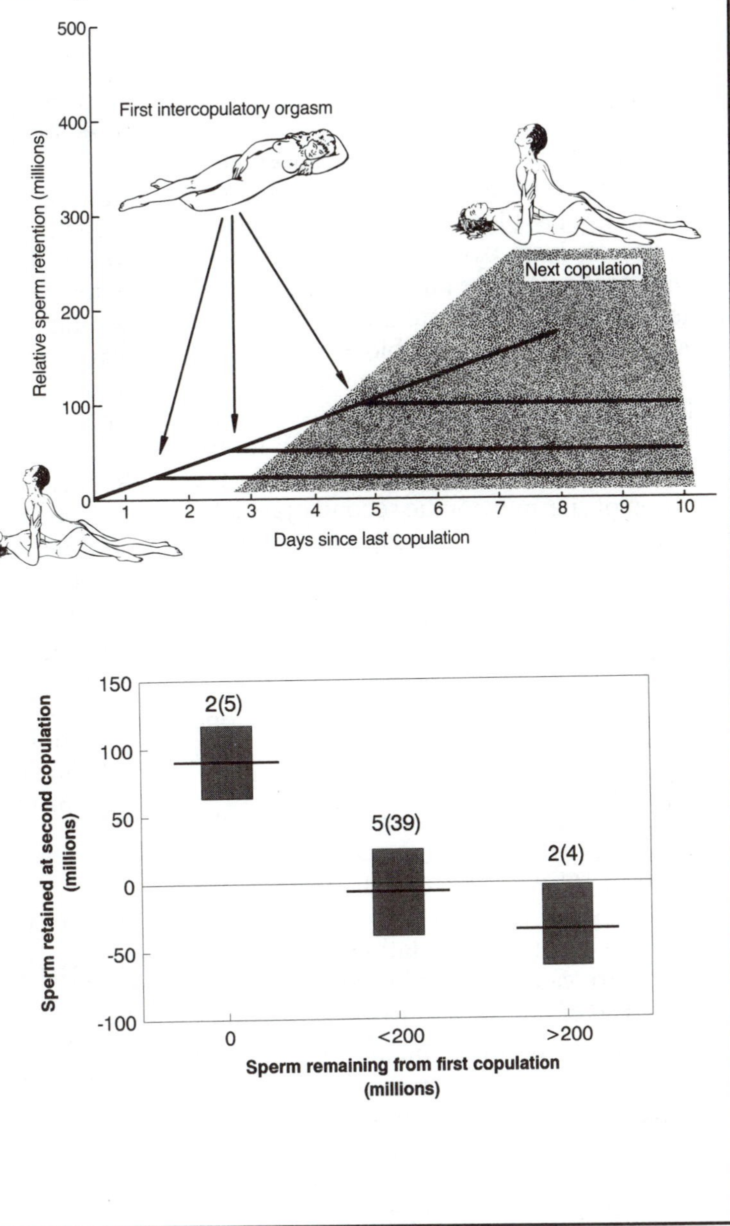

Figures redrawn from, or based on, data and conclusions in Baker and Bellis (1993b).

at the next, the block persisting with declining effectiveness for up to at least eight days.

Even a single intercopulatory orgasm, however, changes this pattern. The simplest interpretation of the data is that when an intercopulatory orgasm occurs it effectively fixes the level of sperm retention at the next copulation. The level at which sperm retention is fixed is more or less the same as the level determined by the efficiency of the block established by the previous copulation (Box 10.7).

Combining these two effects, it seems that the level of sperm retention at any given copulation is primarily a function of the time interval from the previous copulation to the next sexual event. The level of retention is the same whether this first sexual event is the next copulation itself or the first intercopulatory orgasm.

10.6.4 Blocking sperm?

So, one copulation apparently produces some form of block to sperm retention at the next copulation. This block gradually declines in efficiency, at least over the next eight days, the time that sperm are known to remain alive in the cervix. An intercopulatory orgasm fixes the level of efficiency of the block at about the level that it has reached during normal decline. The obvious implication is that this putative block is in some way due to the sperm from the first ejaculate hindering the retention of sperm from the second ejaculate. Three further observations support this interpretation.

1. When no sperm are inseminated at the first copulation (because the male either wore a condom or withdrew the penis before ejaculation) the normal influence of one copulation on sperm retention at the next is not found (Baker and Bellis, 1993b).
2. When the number of sperm retained by the female at the first copulation can be assessed (from a collected flowback), the more sperm that are calculated to be retained at this copulation, the fewer sperm are calculated to be retained at the next (Baker and Bellis, 1993b).
3. The more sperm estimated still to be present in the female tract from the previous copulation (assuming a linear decline in numbers from that copulation to zero, 8 days later) the fewer sperm are retained at the next copulation (Box 10.7).

It seems, therefore, that sperm transferred at one copulation interfere with sperm retention at the next copulation. As sperm numbers decline with time since copulation, their level of interference with sperm retention at the next copulation also declines. Intercopulatory orgasms in some way use those sperm which remain to fix the level of interference at roughly the point it had reached in its normal decline.

The putative involvement of sperm in blocking retention at a later copulation is discussed further in section 11.3.1

10.6.5 Undoing the block: the basis of female strategy

In the context of sperm competition, the benefit to a female of being able to use the sperm from one male to block or hinder the later retention of sperm from another male would be considerable. The benefit would be even greater if she could also facultatively remove that block so as to allow a high level of sperm retention at the next copulation whenever such high retention is advantageous. It seems that females do have such an ability.

The block established by one copulation and fixed by an intercopulatory orgasm only operates if the female has a low uptake response at her next copulation. The block can be totally negated by a high uptake response (Box 10.8).

In effect, female orgasm patterns can be partitioned to create four orgasm regimes which generate three levels of sperm retention (Box 10.8).

10.7 Menstruation and sperm retention

If the number of sperm retained by the female at any given copulation is in part influenced by sperm remaining in her cervix from a previous copulation, we should expect menstruation also to have some influence. Menstruation not only clears the cervix of any sperm present in the cervical mucus (but not necessarily of all those in the cervical crypts), it also, at least temporarily, replaces them with various female tissues and cellular debris. The result should be not only a change in the normal influence of one copulation on another but also, perhaps, a tendency for sperm retention to be reduced.

Both of these effects are observed (Box 10.8) but again, a high retention orgasm response during copulation can undo the block generated by menstruation. If a female

Box 10.8 Orgasm regimes and sperm retention levels

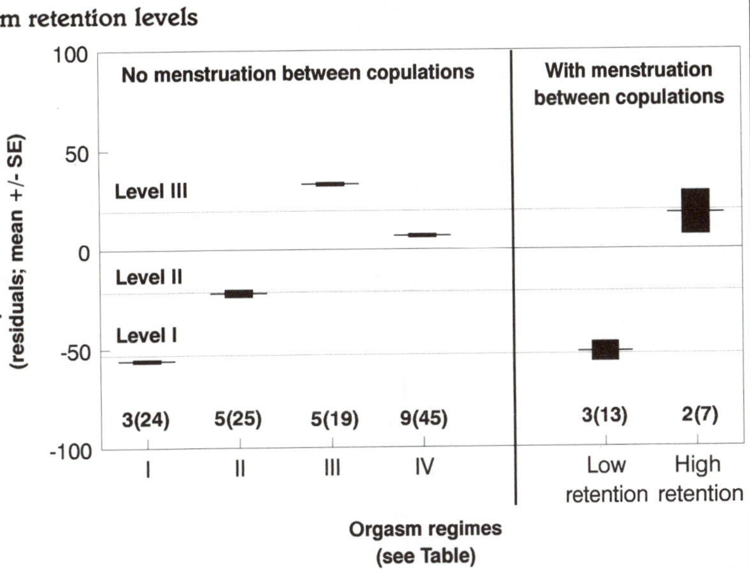

The first sexual event for a female after copulation is defined as her first intercopulatory orgasm or her next copulation, whichever comes first (Box 10.7).

When there is no menstruation between copulations and only a low retention orgasm regime associated with the next copulation (Box 10.5), sperm retention at that copulation is very significantly affected by time interval from the previous copulation to the first sexual event: the longer the time from first copulation to first sexual event, the more sperm are retained at the next copulation ($P = 0.002$). This trend is still evident, but no longer significant ($P = 0.198$), even when there is a high retention orgasm regime at the second copulation. Menstruation between the copulations destroys or even tends to reverse, these influences of one copulation (+ intercopulatory orgasms) on sperm retention at the next ($P = 0.896$). A high retention orgasm at the second copulation, however, still raises sperm retention significantly, even after menstruation ($P = 0.023$).

On current data, the most parsimonious model is as shown here and in the table. There are four orgasm regimes leading to three levels of sperm retention, Regimes III and IV producing levels of sperm retention (= Level III) that are not significantly different ($P = 0.320$).

Definition of orgasm regimes and levels of sperm retention

Orgasm regime	Time to FEV (h)	Retention level at second copulation	Number retained (millions above or below average)	Level of retention
I	≤24	L	−56	I
II	>24–≤72	L	−22	II
III	>72	L	33	III
IV	any	H	7	III
Menstruation	any	L	−52	I
Menstruation	any	H	18	III

FEV = first sexual event after previous copulation (i.e. either the first intercopulatory orgasm or, in the absence of an intercopulatory orgasm, the next copulation);
L = low retention level at the second of a pair of copulations (no copulatory orgasm or a climax earlier than one minute before the male ejaculates);
H = high retention level (climax some time between one minute before the male ejaculates and flowback).

menstruates between two copulations, the first copulation no longer influences sperm retention at the second) Baker and Bellis, 1993b). If the female shows a low retention response at her next copulation, then menstruation induces a low level of retention equivalent to level I (Box 10.8). A high retention response, however, removes all influence of menstruation and induces a high level of retention equivalent to level III.

10.8 Female orgasm, sperm retention and sperm competition: the female strategy

The study of flowbacks suggests that the combination of copulatory and non-copulatory orgasms, combined with variation in intercopulation interval, endows females with a flexible and powerful ability to influence sperm retention. This ability could, in principle, give them the power to influence the outcome of sperm competition by selectively favouring the retention of sperm from one of the competing males. In this section, based on the analyses in Baker and Bellis (1993b), we consider the evidence that they make use of this power.

In our UK, Nationwide survey, the majority of subjects provided enough information relating to their last copulation and intercopulatory orgasm to allow us to estimate the level of sperm retention (in terms of levels I–III; Box 10.8). In evaluating female behaviour, however, we also consider the extent to which the level of retention shown is a function of behaviour overt to her male partner (i.e. overt copulatory orgasms) or cryptic (intercopulatory orgasms that are either nocturnal, self-masturbatory, or encouraged by a female partner). We assume that, as concluded in section 10.6, an increase in cryptic intercopulatory orgasms leads to a decrease in sperm retention at the next copulation unless it is counteracted by a high retention copulatory orgasm.

All discussion concerns only hormonally 'normal' subjects (i.e. we exclude all females who are under an unusual hormonal regime, either oral contraception, depo-provera injection, or who have undergone hysterectomy). Our main conclusions are summarized in Box 10.9.

10.8.1 In-pair copulations and monandry

A total of 1207 hormonally 'normal' females provided information on their last IPC at times that they claimed they had no other sexual partners. We refer to these as monandrous females.

A slight, but not significant, tendency for monandrous females to show an increase in high retention copulatory orgasms during the main fertile phase of their menstrual cycle (days 6–15) compared with the remainder of the cycle is counteracted by a very significant increase in cryptic intercopulatory orgasms during this phase (cf. section 5.5). The result is no change in retention level during the menstrual cycle.

Twenty-one of our monandrous females were pregnant. These showed no difference in orgasm regime, either in overt copulatory orgasms or cryptic intercopulatory orgasms relative to non-pregnant monandrous females.

10.8.2 In-pair copulations and polyandry

Seventy-five hormonally 'normal' females provided information on their last IPC at times that they also had one or more other male sexual partners. We refer to these as polyandrous females.

Compared with monandrous females, polyandrous females show a significantly lower level of sperm retention when copulating with their main partner. They achieve this change, however, not by any change in overt copulatory orgasms but by a very significant increase in the frequency of cryptic intercopulatory orgasms. Median rate rises from about one per week during monandry to about four per week during polyandry.

We know of no other study of variation in the frequency of female self-masturbation under monandry and polyandry. Indirect evidence of changes in the incidence of nocturnal orgasms by females in relation to polyandry was provided by Winokur *et al.* (1959). They found a higher incidence of nocturnal orgasm in women who were neurotic, psychotic, divorced and separated than in control women (albeit attending surgical and medical clinics) who were neither neurotic nor psychotic and were either married or widowed. It is possible that the females with a higher incidence of nocturnal orgasms were in fact in a more polyandrous situation than those with a lower incidence.

10.8.3 Extra-pair copulations

A further 75 hormonally 'normal', polyandrous females provided information on their last copulation when the latter was an extra-pair copulation. Behaviour during extra-pair copulation for these females can be compared with behaviour during IPC for the 75 other

Box 10.9 Female sperm retention strategy in monandry and polyandry: data from a UK Nationwide survey

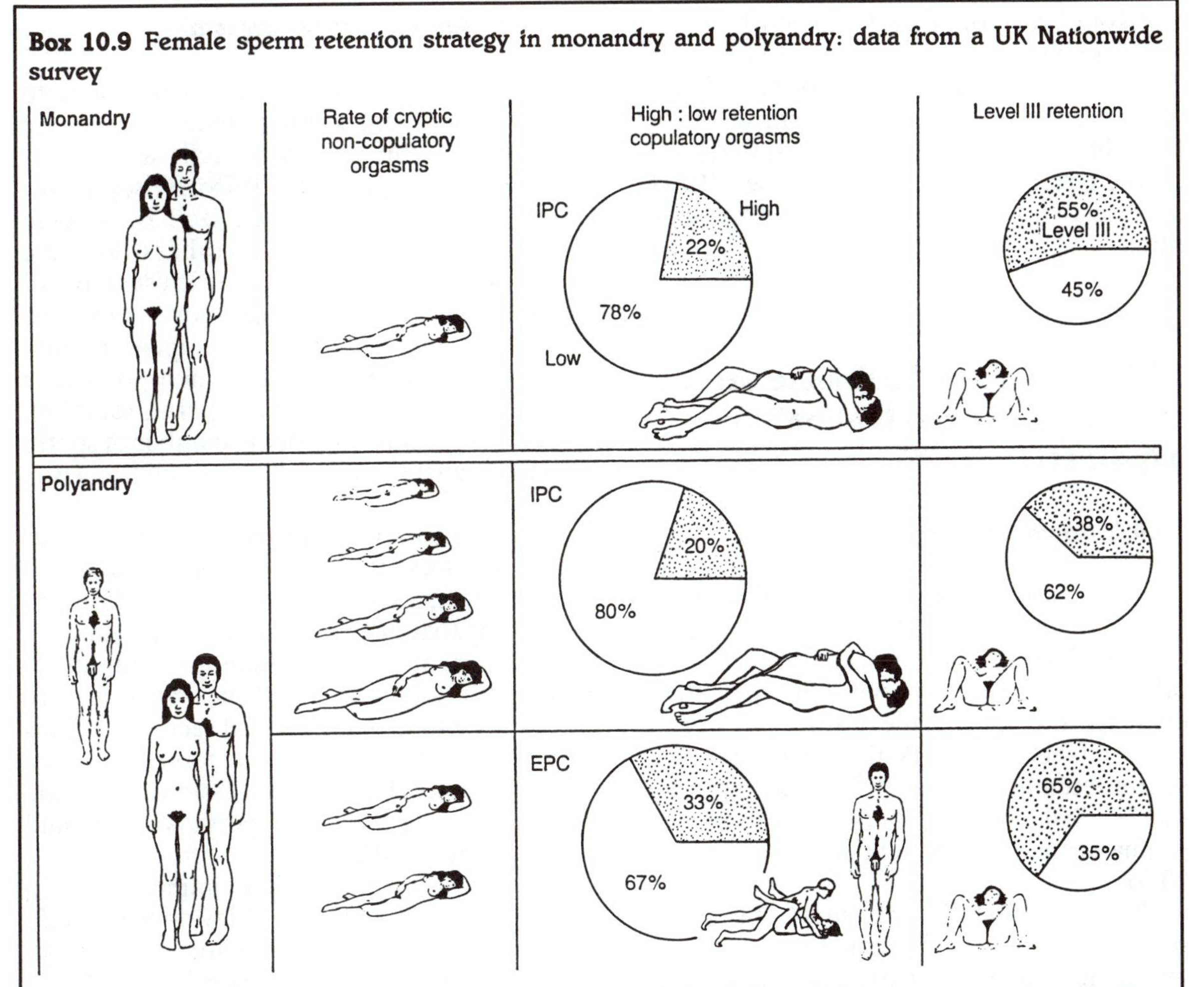

On average, monandrous females (defined here as women with just one male sexual partner) showed level III retention of sperm (Box 10.8) on 55% of their in-pair copulations (IPCs). In part, this was achieved by showing a high retention copulatory orgasm regime (Box 10.5) during 22% of their copulations. Rate of non-copulatory orgasms cryptic to their male partner (i.e. nocturnal, self-masturbatory or lesbian orgasms; Chapter 5) averaged about 1/week (1/170 h).

Polyandrous females (defined here as women with a main male sexual partner but who also had one or more other sexual partners) show a reduction in level III retention to 38% during IPCs with their main partner. They achieve this not by reducing the incidence of high retention copulatory orgasms but by increasing the frequency of cryptic non-copulatory orgasms. In contrast, during extra-pair copulations (EPCs) they show a higher incidence of level III retention of about 65%. They achieve this difference in part via a higher proportion of EPCs (33%) than IPCs (20%) being accompanied by high retention copulatory orgasms and in part by preceding EPCs with a lower rate of cryptic non-copulatory orgasms than IPCs (EPCs, 1/84 h; IPCs, 1/48 h).

(Summary of data and analyses in Baker and Bellis, 1993b.)

polyandrous females from the previous section.

Level of sperm retention was significantly higher during extra-pair copulation than during IPC (Baker and Bellis, 1993b). This increase was achieved primarily by a difference in overt copulatory orgasms when females were copulating with their extra-pair male(s) compared with their partner. However, there was also a (non-significant) tendency to lower the frequency of cryptic intercopulatory orgasms before copulating with the extra-pair male.

Essentially, therefore, females use cryptic

intercopulatory orgasms to set up blocks to sperm retention during periods of polyandry. They are then more likely to release that block with a high uptake response during EPC than during IPC. The result is that levels of sperm retention tend to favour the EP male over the IP male in any ensuing sperm competition.

10.9 How does the female orgasm influence sperm retention? A hypothesis

10.9.1 Copulatory orgasms

Our flowback data provide no support for the 'poleaxe' hypothesis of the female copulatory orgasm (Morris, 1967; Levin, 1981). On the other hand, they do provide the first direct evidence in support of the 'upsuck' hypothesis mooted by Fox *et al.* (1970). Highest levels of sperm retention (about 70%) occurred when the seminal pool was present in the upper vagina at the time the female climaxed (Box 10.5).

The fact that improvement in sperm retention is observed when the female climaxes up to a minute before the male ejaculates suggests that the upsuck mechanism continues to function for at least a minute after the female subjectively first experiences the climax. In rabbits, uterine contractions continue for 2–5 min after mating (Fuchs, 1972) and passive sperm transport to the oviduct may continue during this interval (Overstreet and Cooper, 1978).

We observed an increase in volume of the flowback if the female experienced an orgasm at any time during a copulation episode, the volume being greatest from orgasms during foreplay. The pattern suggests that the female adds some material, probably cervical mucus, to the seminal fluid in forming the flowback and that more is added if the female climaxes than if she does not. Climaxes during foreplay thus either add more material or add the same amount but without 'sucking' it back up to the same extent as when it is mixed with seminal fluid.

10.9.2 Non-copulatory orgasms

The penetrability of the cervical mucus to sperm from the next inseminate is a negative function of the density of cells and debris (Parsons and Sommers, 1978; Belsey *et al.*, 1987). Pregnancy, characterized by a large population of leucocytes and other cells in the cervical mucus (Davey, 1986) irrespective of copulation and intercopulatory orgasms, should be a time of maximum impenetrability (section 10.4). Our results for non-pregnancy flowbacks also show that sperm retention seems to be a function of the number of sperm remaining from the previous copulation (Box 10.7).

If the upsuck mechanism applies to copulatory orgasms, it is likely also to apply to intercopulatory orgasms. The difference would be that instead of sucking-up a mixture of cervical mucus and seminal fluid, intercopulatory orgasms would suck-up a mixture of cervical mucus and vaginal secretions. This would lower the pH of the cervix and thus have far-reaching repercussions on the mobility and survival of sperm which would be consistent with our data.

Sperm are immediately immobilized in acidic environments at pH levels below 6.5. Even at a pH of 7.0 the ability of sperm to penetrate cervical mucus is minimal. Penetration is 'normal' at pH 7.5 and above normal at pH 8.25 (El-Banna and Hafez, 1972). The pH of seminal fluid is normally in the range 7.0–7.8 (Raboch and Skachova, 1965) and buffers the sperm from the vaginal environment which is acidic, and thus hostile, having a pH of between 5.8 and 3.5 (Duerden *et al.*, 1987). Once sperm are no longer buffered by seminal fluid, they cannot live for more than a maximum of 10–12 hours at the hostile pH levels found in the vagina (Vander Vliet and Hafez, 1974). The pH of the cervical mucus can vary considerably from a favourable 7.4 to a hostile 4.0 (Kroeks and Kremer, 1977), at least in part perhaps depending on the extent to which it has been mixed with material from the vagina.

Mixing cervical mucus and vaginal secretions via intercopulatory orgasms should lower the pH of the cervix towards the acidic end of its observed range. This will slow down, or even immobilize, any sperm which subsequently enter the cervical mucus from reservoirs such as the cervical crypts. More may also die before reaching the uterus. Any hindrance to passage through the cervical mucus into the uterus may lead to the gradual accumulation of millions of sperm in the cervical mucus (Sagiroglu and Sagiroglu, 1970). Here they attract an even larger population of scavenging, female-produced leucocytes. The result could be that instead of declining with time since the most recent copulation the population of sperm, leucocytes and debris in the cervical mucus could remain constant or even increase, at least for as long as sperm continue to leave the crypts.

We suggest, therefore, that an intercopulatory orgasm lowers the pH of the cervical mucus. In the context of the manipulation of sperm, the function is to slow down or prevent further passage of sperm already present in the cervix, promote a build up of cells and debris in the cervical channel, and thus to reduce sperm retention at the next copulation. As with all such blocks, however, this can perhaps be circumvented by a high retention orgasm at the next copulation which may take sperm directly to their storage sites, bypassing mucus penetrability.

Most medically significant infections of the female tract also prefer a more-alkaline environment (Duerden *et al.*, 1987) and any increase in acidity due to intercopulatory orgasms could have an antibiotic effect. In the absence of copulation (i.e. while still virgin or during temporary periods without insemination) and during pregnancy, the function of nocturnal, masturbatory and other non-copulatory orgasms could be as an 'antibiotic' mechanism aimed at combating cervical infection.

10.10 The cryptic mating game: male–female conflict and cooperation over sperm

Although our data are consistent with the upsuck hypothesis as a mechanism for the female orgasm, both copulatory and non-copulatory, they also suggest a behavioural significance far more complex than simple assistance in sperm retention. The relatively high incidence of non-copulatory orgasms, copulations without orgasm, and copulatory orgasms which climax before the male ejaculates (Chapter 5) cannot simply be ignored as they have in most previous discussions of the female orgasm. Far from assisting sperm retention, all of these intercopulatory events reduce sperm uptake at the next copulation, perhaps using the mechanisms just discussed. Yet their influence can be overridden by a high uptake copulatory orgasm at the next copulation. The whole pattern hints strongly at a female strategy to influence sperm retention differently from different copulations.

We are not suggesting, of course, that this strategy is necessarily conscious or that, like any other strategy, it is infallible and omnipotent. As in any other arms race, we should expect male and female strategies to interact and it is by no means inevitable that one sex will always prevail. Nevertheless, the occurrence, pattern and timing of female orgasms emerge from our analysis as part of a female strategy to influence sperm retention from any given copulation. As such, the strategy should influence both the probability of conception and the outcome of sperm competition contests between males.

10.10.1 Ejaculate manipulation by females in the absence of sperm competition

Singer (1973) proposed that orgasms could be typed by whether they involved contractions of the uterus as well as the vagina. Uterine orgasms were suggested to facilitate conception whereas non-uterine orgasms did not. Singer further proposed that by a systematic shift from one type of orgasm to another a female could, consciously or subconsciously, influence the probability of conception.

In Chapter 9, we suggested that, in the absence of sperm competition, both male and female partner may benefit from improved chances of viable conception if fewer sperm are taken into the female tract. The advantage of fewer sperm may either be through reduced risk of debilitating the egg or through reduced risk of pathological polyspermy, eggs fertilized by more than one sperm failing to develop into viable embryos (section 4.4.10).

In this context, females who produce the optimum rate of arrival of sperm at the oviduct by manipulating both the number of sperm which enter the cervical crypts and the rate and length of time that sperm migrate through the cervical mucus will benefit both themselves and their male partner. We assume, however, that the more sperm the male inseminates beyond a certain optimum and/or the more are retained after flowback, the more difficult it is for the female to produce the optimum traffic of sperm (section 4.4.8).

Our flowback studies have produced several lines of evidence to support the suggestion that the retention of fewer sperm is a female strategy to aid conception.

(a) LOW SPERM RETENTION AT CONCEPTION

Lack of orgasm during copulation or a climax >1 min before the male ejaculates is associated with low sperm retention. It has long been known that copulatory orgasm is not essential for conception (Moghissi, 1977). In our data we had a situation in which date of conception could be estimated to within ±24 h and was attributable to one or other of two IPCs, about 156 and 120 h before conception (Box 8.1). The number of sperm inseminated on both occasions could be estimated (Box 9.1) and both flowbacks were collected. The amount of information available for this natural conception is probably unique. Neither of the relevant IPCs was associated with a copulatory orgasm and, summed for the two copulations, −6 ± 48 million sperm (i.e. near zero) were estimated to have been retained from a total insemination of 545 million sperm. Two non-copulatory orgasms occurred approximately 28 and 22 h before conception. This unique data point illustrates that neither a copulatory orgasm, nor any more than minimal sperm retention, is necessary for conception.

(b) LOW SPERM RETENTION DURING THE FERTILE PHASE OF THE MENSTRUAL CYCLE

Sperm retention was significantly reduced during the fertile phase of three flowback donors who were not taking oral contraceptives, two of whom conceived. This reduction was primarily the result of the timing of intercopulatory orgasms, menstrual variation in retention no longer being significant once timing of intercopulatory orgasm was statistically controlled. A significant increase in cryptic intercopulatory orgasms during the fertile phase was also found not only in our Nationwide survey (Box 5.7) but also by Harvey (1987). Nor, overall, was there any indication in our Nationwide survey of behaviour associated with an increase in sperm retention (measured in terms of levels I–III; Box 10.8) by monandrous females during their fertile phase.

Finally, there is a tendency for a reduction in sperm retention in hormonally normal females, both in our Nationwide survey and in our flowback analysis, compared with those taking oral contraceptives.

(c) LOW SPERM RETENTION AT SEASONS OF PEAK PROBABILITY OF CONCEPTION

Even though males inseminate more sperm at seasons when the probability of conception is higher (Box 9.10), females retain fewer sperm at these seasons (Box 10.10). This seasonal variation is no longer significant once orgasm pattern has been controlled though, as yet, we have been unable to identify which feature of the orgasm pattern is responsible for the seasonal variation.

There is no conflict between the strategies employed by males and females if our interpretations in this book are correct. Males inseminate more sperm at seasons of peak conception because: (a) female value is higher; and hence (b) the risk of sperm competition may be greater. Females retain fewer sperm in order to aid conception. If the proportion of sperm retained by the female from inseminates from different males is the same, males still benefit (in terms of sperm

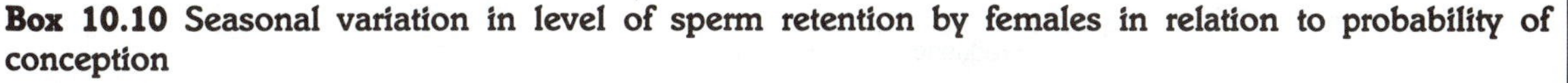

Box 10.10 Seasonal variation in level of sperm retention by females in relation to probability of conception

In our flowback study (Box 3.8), the number of sperm retained by females varied seasonally as a negative function of the probability of conception in the Manchester region (blocked by female to produce an analysis of within female variation). When controlled for orgasm pattern (residuals from equation; Box 10.2), variation was no longer significant. The implication is that females vary sperm retention seasonally through variation in orgasm pattern. However, none of our measures of orgasm pattern (number of intercopulatory orgasms; hours to first sexual event; proportion of high retention orgasms associated with orgasm; or even orgasm regime) showed a significant correlation with probability of conception.

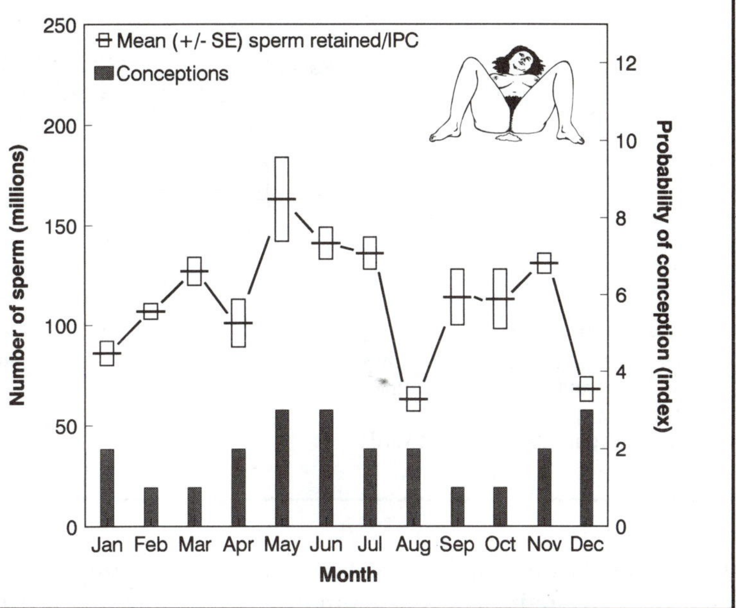

competitiveness) from inseminating more sperm, even if fewer are retained.

There is thus no indication in our data that monandrous females favour higher sperm retention when conception is more likely. On the contrary, the impression is that the favoured strategy associated with conception in a monandrous situation is to reduce the number of sperm retained, at least down to some optimum level.

We are not suggesting, of course, that high retention copulatory orgasms never lead to conception. Any negative relationship between the number of sperm retained and the probability of conception (Baker and Bellis, 1993a,b) will be quantitative, not absolute (i.e. there is always some chance of conception, no matter how many sperm are retained).

10.10.2 Ejaculate manipulation in the presence of sperm competition

Although a male's probability of fertilization from any given copulation may be increased by decreasing sperm number (to an optimum) in the absence of sperm competition (Chapter 9), it is generally accepted that it is increased by increasing sperm number in the presence of sperm competition (section 2.5).

The current explanation for concealed ovulation and continuous receptivity in some female mammals, such as humans, is that it allows the female to confuse the male over which copulations are likely to lead to paternity (section 6.5). In this way, females may manipulate the timing of inseminations from different males to suit their own priorities. Continuing copulation well into pregnancy maintains this deception by concealing the date of conception.

If females use copulation as a form of mate confusion, it is important that no aspect of their behaviour changes predictably with levels of infidelity and chances of conception. We have demonstrated that polyandrous females tend to favour the extra-pair male (or males) not only by the timing of extra-pair copulations (Box 6.10) but also by their use of contraceptives (Box 8.3) and now by their relative level of sperm retention (Box 10.9). When females switch from a monandrous to a polyandrous situation (but still retaining a main partner), they reduce the level of sperm retention from their partner's inseminates but

Box 10.11 Females faking orgasms

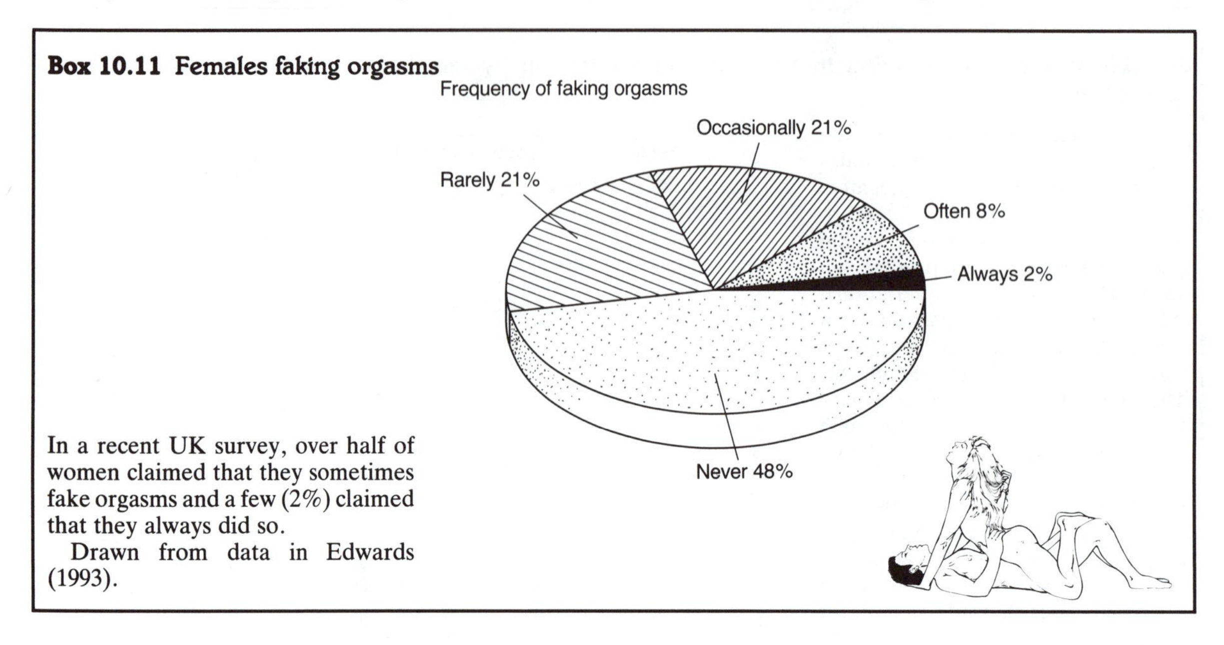

In a recent UK survey, over half of women claimed that they sometimes fake orgasms and a few (2%) claimed that they always did so.

Drawn from data in Edwards (1993).

show a significantly higher level of retention from the extra-pair male's inseminates.

Our data show that females achieve the change in retention with their partner by varying the frequency of intercopulatory orgasms cryptic to their partner while maintaining the same level of overt copulatory orgasms (Box 10.9). They achieve the difference in level of retention between males by varying the pattern of overt copulatory orgasms. The female response is exactly the one expected if the key factor in her behaviour were the deception of her primary partner. Add to these changes in pattern of real orgasms the possibility of females faking or hiding orgasms (Box 10.11) and the opportunity for males to detect any underlying female strategy is greatly reduced.

10.10.3 Sperm retention and mobilization in the absence of fertilization

There are three situations in which manipulation of ejaculates by both males and females at first sight seems redundant: (1) during the infertile phase of the female menstrual cycle; (2) when either male or female is using contraceptives; and (3) during pregnancy.

A male's sperm remain fertile for at least five days after copulation (Chapter 8). In addition, our data suggest that his sperm, simply by being present in the cervix, can influence retention of sperm at the female's next copulation for up to eight days (cf. Austin, 1975). This in turn could mean that the sperm could indirectly influence retention at the copulation after that, and so on. Thus, sperm from one insemination could in principle have some influence on fertilization for up to at least 13 (8 + 5) days and perhaps even longer. Rarely can males, or even females, predict whether or not the female will become fertile at some time during the next 13 or so days.

In a sense, with this time-scale in mind, discussion over why ejaculate manipulation should occur at times that conception is unlikely is largely unnecessary. The general answer is that it is very rare that any copulation has absolutely no chance of influencing fertilization. No form of contraception is perfect (Box 7.1), no phase of the menstrual cycle is absolutely infertile (Box 6.10), and even during pregnancy, spontaneous abortion followed by ovulation could occur within the period of influence of sperm.

As far as the male is concerned, a female has to be many months into pregnancy before her symptoms are unequivocal. Pregnancy, like the menstrual cycle, is a phase when the female may benefit from male confusion (section 6.6.7). During pregnancy, the vast majority of the sperm are apparently ejected in the flowback, no matter what the orgasm regime. Overt copulatory orgasms, therefore, seem to be functionless for sperm retention and we presume the main reason for overt regimes not to change during pregnancy is to maintain the efficiency of confusion. However, the frequency of cryptic orgasms also remains unchanged during pregnancy. Perhaps, as suggested for virgins, the function of these cryptic orgasms during pregnancy is largely to alter the cervical environment.

10.10.4 Individual variation in female strategies

The orgasm pattern shown by women differs considerably between individuals (Chapter 5). This variation is probably adaptive and retained through selection for a balanced polymorphism (section 5.6). Different orgasm patterns impart different levels of ability to manipulate ejaculates but perhaps also different levels of crypsis.

The greatest flexibility of ejaculate manipulation is shown by females who are capable of a varied range of orgasm patterns, both copulatory and intercopulatory, including at times having no orgasm at all.

According to our Nationwide survey, 84% of women, by the time they have had 500 copulations (henceforth, 'experienced' women), are using the whole range of ejaculate manipulation. Their strategy is 'mixed', involving copulatory and intercopulatory orgasms, as well as lack of orgasm. Only 2% of experienced women claimed never to have experienced any orgasm and thus to be using a 'no orgasm' strategy.

Of the experienced women in our survey 16% claimed not to experience copulatory orgasms though the majority of these (88%) were experiencing intercopulatory orgasms. Women with such an 'intercopulatory orgasm dominated' strategy are less flexible in ejaculate manipulation than women with a fully 'mixed' strategy only in so far as they cannot convert a level I retention into a level III (Box 10.8) during copulation itself. They can do so, however, by an orgasm after copulation (but before flowback), stimulated either by the male partner or by the female herself (either overtly or cryptically).

Our analysis of flowbacks suggest that no matter whether a woman uses a 'no orgasm', 'intercopulatory orgasm dominated', or a 'mixed' strategy, all are capable of manipulating sperm retention across the whole range from level I to level III. Indeed, what 'no orgasm' and 'intercopulatory orgasm dominated' strategies may lose in flexibility, they may recoup, at least partially, through greater crypsis.

11 Sperm polymorphism and the Kamikaze Sperm Hypothesis

11.1 Introduction

When the risk of sperm competition is low, both males and females contrive to reduce the number of sperm that will be retained in the female tract (Chapters 9 and 10). All indications are that, in the absence of sperm competition, the fewer sperm there are in the female tract (down to some low optimum), the greater the probability of conception. When the risk of sperm competition is higher, however, males contrive to place more sperm in the female tract. Females cooperate with some males but oppose others and, on average tend to favour the extra-pair male, both in terms of the timing of copulation during the menstrual cycle and by changes in orgasm pattern that are cryptic to her partner.

All of the male–male and male–female conflict and cooperation over sperm that we have considered so far has been numerical and has concerned the whole ejaculate. The evidence supports Parker's (1990a) assertion that, no matter what the mechanism by which sperm competition proceeds (section 2.5), the male who inseminates most sperm will be at an advantage. Nevertheless, there were indications that sperm competition is more complex than a simple game of numbers. The sperm themselves influence the level of sperm retention at the next insemination and female strategy is to use or circumvent this influence.

So far, we have discussed ejaculates and sperm competition as if all sperm were the same and equal. However, all sperm in an ejaculate are not the same (section 4.6). Sperm are polymorphic. In the past, 'non-normal' sperm have been considered to be unwanted passengers in the ejaculate; unavoidable deformities that are a hindrance to conception (Harcourt, 1991). Our Kamikaze Sperm Hypothesis (KSH; Chapter 2) argues otherwise. Each sperm morph has a part to play in the whole process of sperm competition and fertilization.

According to the KSH, the whole ejaculate should be considered as a living entity with differentiated cells and a division of labour. The mutual goal of these cells is to overcome the female's defences, outcompete similar entities from other males, and to secure the prize of fertilization for one of the cells (sperm) in that entity. The success of the entity, however, depends on the efficacy with which each cell type carries out its function.

The various sperm morphs differ in their size, shape and behaviour. In this chapter, we consider first the general evolution of sperm size and shape, then describe the characteristics and behaviour of some of the major sperm morphologies in the human ejaculate. Finally, we consider how the different morphs might be manipulated by the male so as to integrate their activities and produce the competitive, living entity that is the human ejaculate.

11.2 The evolution of sperm size and shape

11.2.1 The functional anatomy of human sperm

The structure and development of human sperm was introduced in section 2.4.3. A few more details are now needed for this chapter.

Briefly, mammalian sperm consist essentially of two structural parts: the head and the tail (= flagellum).

The head is capped by the acrosome, a modified lysosome, the enzymic contents of which have a clear anticellular/antitissue function (Schill *et al.*, 1988). Among the enzymes eventually released by the acrosome are: **hyaluronidase** (Johnson and Everitt, 1988); and **acrosin** (Polakoski and Siegel, 1986). The role, or lack of role, of these enzymes in fertilization are discussed in detail in Chapter 12 (Box 12.1). Acrosin has also been suggested to facilitate cervical mucus penetration and intrauterine sperm migration by releasing kinins from kininogen (Schill *et al.*, 1979; Kaneko *et al.*, 1986).

Apart from supporting the acrosome, the sperm head is primarily the carrier of the nuclear package. Most sperm contain a single (haploid) set of the paternal chromosomes. These chromosomes are highly condensed, being folded in such a way that the physical space they occupy is minimized (Johnson and Everitt, 1988). The head, which also contains some cytoplasm, is joined to the tail by a neck.

The powerhouse of the sperm is the front part of the tail, the mid-piece. A winding sheath of mitochondria surrounds and provides energy for the tail's axoneme which in turn provides the motive force for the sperm's movement. This axoneme consists of two central and nine outer microtubules and extends from its energy source in the mid-piece throughout the length of the flagellar tail (Johnson and Everitt, 1988). In the principle piece of the tail, the mitochrondrial sheath is replaced by a fibrous sheath. Dense fibres running between the outer microtubules are present throughout the principle piece of the tail but not in the end piece.

The whole sperm cell is covered by a cell membrane which provides the surface through which the sperm interacts with its environment. The molecular topography of this membrane forms a series of biochemical domains (Eddy, 1988). Each of these surface biochemical domains plays a most important part in particular interactions between the sperm and its environment, being involved in phenomena as diverse as agglutination (i.e. sperm aggregation) and avoidance of phagocytosis while in the female (Chapter 4).

11.2.2 Is sperm size, shape and behaviour determined by the male or the sperm?

Mammalian sperm are produced by a diploid organism (the male) but carry a haploid set of chromosomes. Evolutionarily, it becomes an important question whether sperm behaviour is dictated by the diploid or haploid chromosomes (Parker, 1993; Parker and Begon, 1993).

Genetically, a male germ cell is first subjected to a diploid effect, partly via its own premeiotic diploid chromosomes and partly via the diploid genotype of the surrounding body cells (Beatty, 1970). Potentially, it will also be subject after meiosis to a haploid effect due to its own haploid gene content. As sperm vary in their haploid genotype (due to chromosomal segregation, crossing over and other forms of genetic variation), the haploid effect would vary between individual sperm which might then behave correspondingly differently from each other.

One thing that should be established immediately is that the diploid organism does not simply hand over control to the haploid genome immediately segregation occurs and may even not do so at all. There are two ways by which the diploid genome may keep control of the haploid: 'environmental' and 'instructional'.

Throughout the time that sperm are passing through the epididymis, they are exposed to 'environmental' diploid influence. As they travel, they continue to mature and change structurally (Bedford *et al.*, 1972). They continue to be exposed to the diploid environment up to and through ejaculation and at least until they lose final contact with the seminal fluid and its contained cells.

Even then, it could be a mistake to view sperm as independent entities. Genetic instructions released in the diploid, premeiotic germ cell can lay dormant in the postmeiotic, haploid cell, then to become active at some preprogrammed stage in the sperm's life (Beatty, 1974). The most graphical illustration of this continuing diploid influence throughout a sperm's life is provided by the behaviour of sperm without a nucleus (e.g. the pinheaded sperm of humans and other mammals). These continue to behave and survive in a way that seems no less structured than that of their nucleated companions in the ejaculate. Even headless bovine sperm show normal (or at least near normal) metabolic activity and are capable of penetrating through cervical mucus (Mann and Lutwak-Mann, 1981).

The genetics of the size and shape of the sperm of mammals and other animals was extensively reviewed by Beatty (1970, 1971, 1974). In a detailed consideration of strain differences and the results of breeding experiments, particularly of mice, rabbits and cows, he concluded that almost all variation in sperm size and shape was orchestrated by the diploid genome, not the haploid.

Studies of sperm polymorphism in wild and captive cats (clouded leopard (*Neofelis nebulosa*), cheetah (*Acinonyx jubatus*), and lion (*Panthera leo*)), provide indirect evidence that variation in the diversity of sperm morphs in different lineages is under diploid control. In general, cat populations with greater diploid homozygosity have sperm with a greater coefficient of diversity (i.e. a more even spread of different sperm morphs) (Wildt *et al.*, 1983, 1986, 1987a,b). The converse would have been expected if polymorphism were the result of postmeiotic haploid differences. Further, there is direct evidence from selective breeding studies that the proportion of individual sperm morphs (e.g. pyriform sperm in bulls; Salisbury and Baker, 1966) also responds to selection on the diploid organism.

On the basis of these lines of genetic evidence, it seems a reasonable conclusion that the level of polymorphism is the result of selection on the diploid organism. Consistent with this conclusion is the biochemical evidence that the major part of sperm size, shape and function is directly the result of diploid expression (Beatty, 1970). Nevertheless, few people would now deny that there are at least a few unequivocal examples of sperm phenotype being the result of haploid expression.

Perhaps surprisingly, the least conclusive example concerns the possibility that sperm may differ phenotypically depending on whether they contain an X or Y sex-chromosome. Claims that it is possible to separate male- and female-producing sperm on the basis of their size, longevity or behaviour have for decades been the subject of controversy (e.g. Bedford and Bibeau, 1967; Beatty 1970; Ericsson *et al.*, 1973). Still, however, the evidence that such genotypically different sperm also differ in their properties and behaviour is unconvincing (Gledhill, 1987). Occasional positive results are obtained (e.g. Jaffe *et al.*, 1991) and the debate continues (Anon, 1990).

In contrast, heritable sex ratio distortion in insects (e.g. the mosquito, *Aedes aegypti*), inheritance being through the male, provides a clear example of postmeiotic haploid influence (e.g. Wood, 1961). So-called 'meiotic drive' leads males with the 'distorter' gene to produce an excess of males in their offspring. The distorting gene occurs on the Y chromosome and produces its effect by inducing breakage of the X chromosome during meiosis, as the chromosomes are segregating. The result is an absolute reduction in the number of sperm (Newton *et al.*, 1976) and an excess of Y-bearing sperm. The extent to which the sex ratio is distorted depends on the sensitivity of the X-chromosome to breakage. At least eight categories exist, ranging from resistant to highly sensitive (Wood and Newton, 1976).

The best known example of haploid influence in mammals, which may proceed via mechanisms very similar to those used by the sex ratio distorter gene of insects, is that due to genetic factors at the 'tailless' (T) locus in mice (the tail is that of the mouse, not the sperm) (Braden, 1958). There are several T-locus, non-normal, alleles (t) and most show the same characteristic: whereas females transmit t alleles to their offspring in accordance with Mendelian expectation, males do not. Instead, males produce an excess of offspring with a t allele as if t-bearing sperm have some advantage in fertilization. There is a complex genetic interaction between sperm and egg, determined by their respective genetic content at the T-locus (e.g. Bateman, 1960; Braden, 1960). Nevertheless, the general point remains: after meiosis and segregation, t- and T-bearing sperm seem to have different properties and/or behaviour. It is even possible that there is a visible difference between t and T sperm (Bryson, 1944; Rajasekarasetty, 1954).

The overall conclusion is that haploid effects are seen in mammalian gametes but are relatively rare (Beatty, 1970) and detailed examples of the transcription of RNA from the haploid genome (Kierzenbaum and Tres, 1975, 1978) have accumulated only slowly (Denny and Ashworth, 1991). However, we now know of tens of haploid-transcribed messenger RNAs and many characteristic spermeotelic proteins (Hecht *et al.*, 1990). A few are found in acrosomal sites (e.g. Wright *et al.*, 1990).

One of the major developments in the study of haploid influence involves detecting the early manifestation of a number of oncogenes (genes associated with various forms of carcinoma) which are now known to express themselves at least briefly during spermatogenesis (e.g. Propst *et al.*, 1988). In the mouse, some proto-oncogenes express themselves predominantly before or during meiosis; others are expressed only in postmeiotic haploid spermatids.

Cohen (1967, 1969) has long argued that postsegregation gene action occurs not rarely (cv. Beatty, 1970) but on a gigantic scale. 'Inexact' chiasma formation during meiosis is suggested to be a common phenomenon leading, at innumerable loci, to small interpolations and deletions that affect the morphology and behaviour of gametes and render most of them infertile or subfertile. Moreover, Cohen proposes that the sperm honestly 'announce' their imperfection to the female immunologically and as a result are the targets for phagocytosis (see Chapter 4 for more detailed discussion).

In the above discussion, we have concentrated on the traditional question of which elements of the sperm phenotype, both morphological and behavioural, are the product of diploid and which are the product of haploid expression. In other words, is this or that aspect of the sperm's phenotype due to messenger RNA physically transcribed from a diploid or haploid genome. Evolutionarily, however, this is not really the important question. What is important is the extent to which even haploid expression is itself under diploid control. In other words, to what extent is the haploid expression orchestrated by 'instructions' seeded in the sperm by the diploid organism.

We suggest that, in the absence of postmeiotic mutation, all haploid expression is actually under diploid control. After all, not only has the haploid genome been handed on to the developing sperm by the diploid organism but it also functions within structures and alongside organelles that were also handed on by the diploid. Diploids which produce haploids which express themselves maladaptively will be selected against in favour of diploids which produce haploids which express themselves more adaptively. In general, therefore, we should expect all haploid expression to be the result of selection on the diploid to produce haploids which express themselves optimally.

The key question is what happens when there is a postmeiotic mutation in the haploid genome so that it no longer 'obeys' the diploid instructions in the way that it would have done otherwise. Most often, of course, because the diploid instructions should have evolved to be optimal, honed by natural selection, we should expect the renegade haploid to behave maladaptively and any diploid to which it gives rise to be at a selective disadvantage. On the other hand, if it shows an improvement in behaviour, any diploid organism to which it gives rise should itself be at a selective advantage through producing haploids that behave in this improved way.

There are two further points that it is important to bear in mind. First, the vast majority of haploid genomes have inherited a sperm body (from the diploid) that is physically incapable of getting anywhere near the site of fertilization, no matter what mutation occurs postmeiotically. Only postmeiotic mutations to the few egg-getters (Chapter 12) in the ejaculate have repercussions in subsequent generations. Simplistically, these mutations will either increase or decrease the success of future diploid generations and be influenced by natural selection accordingly.

Second, varying the DNA content and/or nature of haploid expression between gametes could well be one mechanism by which the diploid could programme sperm to behave differently, as required for example by the Kamikaze Sperm Hypothesis. There is some evidence for a number of mammals (horse, cow, pig, sheep, goat, dog), using males of normal fertility, that different sperm morphs may have different DNA content (Leidl *et al.*, 1972). Even a gene that distorts sex ratios could be advantageous to the diploid if circumstances happen to favour such a distorted sex ratio (e.g. depending on the status of the male parent; Box 4.14). The widespread occurrence of, for example, mouse t-alleles and insect SD (sex distorting) genes in natural populations (Beatty, 1970; Wood and Newton, 1976) suggest that they do have some advantages to balance their apparent disadvantages.

In conclusion, therefore, we shall assume in the remainder of this chapter that sperm behaviour and sperm polymorphism: (a) are ultimately under diploid control; and (b) have evolved via natural selection on the diploid to produce haploid sperm which behave in an optimal, but not necessarily all in an identical, way.

11.2.3 Sperm size

Parker (1982, 1984) modelled the trade-offs involved in both the optimum proportion of energy that a male should invest in the production of ejaculates and the way in which that energy should be partitioned in terms of sperm size and number. According to these early models, the optimum proportion of energy the male should put into the ejaculate rather than in mate-searching or guarding is a function of the risk of sperm competition. Then, given this optimum level of energy investment in the ejaculate, the energy can be partitioned between fewer, larger sperm or smaller, more numerous sperm.

In Parker's (1982, 1984) models, the opposing factors in the trade-off were: (a) the benefit of increasing sperm size in order to contribute to the energy reserves of the zygote; and (b) the disadvantage of increasing sperm size in terms of reduced success in sperm competition. Doubling the size of the sperm would have a negligible influence on the viability of the zygote but would halve the number of sperm and hence the chances of success in sperm competition. Most often, therefore, animals should produce sperm that are the minimum size necessary to carry and deliver their package of DNA.

There is a compelling logic about Parker's hypothesis that until recently has led to its general acceptance. In particular, it receives support (see Chapter 2) from the existence of those few animals, such as self-fertilizing hermaphrodites, in which there is no sperm competition and in which sperm are relatively large, even being produced in about the same numbers as the eggs. However, Parker's interpretation cannot be the whole story.

Although sperm are generally tiny, some are more tiny than others. Size variation between species is considerable (e.g., among primates, from the 50 μm sperm of the marmoset, *Callithrix jacchus*, to the 94 μm sperm of the lesser mouse-lemur, *Microcebus myoxinus*; Cummins and Woodall, 1985). Even within a single ejaculate, some sperm are considerably smaller than others. The smallest sperm in a human ejaculate, for example, can have a volume at least 14 times smaller than the largest sperm (Laufer *et al.*, 1978).

Sperm are not, therefore, simply the minimum size necessary to carry and deliver the genetic package. Moreover, far from sperm size being a negative function of intensity of sperm competition, it is often a positive function (e.g. butterflies, Gage, unpublished; rodents, Roldan *et al.*, 1992; primates, Roldan *et al.*, 1992). Some qualification of Parker's original hypothesis is therefore necessary and a number of authors have recently re-examined the main selective forces acting on sperm size (e.g. Møller, 1988a,b); Gomendio and Roldan, 1991, 1993; Roldan *et al.*, 1992; Parker, 1993; Parker and Begon, 1993).

Female adaptations in mammals (Chapter 4) have created an arena for sperm competition that has three critical characteristics: (1) the length and complexity of the female tract, including storage areas and narrow channels; (2) the length of the menstrual cycle and the predictability of ovulation; and, on any given occasion, (3) the presence or absence of sperm from another male or males.

In principle, sperm size could influence sperm longevity, motility and, in the event of direct sperm–sperm interaction, competitiveness. Larger sperm with greater energy reserves could live longer but not if the energy needed to move the larger body at a given speed is disproportionately greater. Larger sperm could be faster or slower, depending on the relationship between total size and the motile force generated by the tail. Larger sperm could be more competitive if the critical factor were, say, the displacement of other sperm but could be less competitive if the critical factor were, say, the ability to pass through narrow channels.

In mammals, these biophysical factors seem to have resulted in longer sperm being faster (Gomendio and Roldan, 1992). Longer female tracts are associated with slower swimming sperm, perhaps following the same optimization principle which dictates that long-distance runners among human athletes complete the course fastest if on average they run slower than short-distance runners (Roldan *et al.*, 1992).

Larger female body size, which is often associated with a female tract of larger volume, is associated with more numerous sperm (Short, 1981) and, in some groups, with smaller sperm (e.g. cetaceans and rodents, Cummins and Woodall, 1985). In other families, larger females are associated with larger sperm (e.g. chiroptera, Cummins and Woodall, 1985) and in others there is no association (e.g. primates and muroid rodents, Cummins and Woodall, 1985; Gomendio and Roldan, 1991).

Finally, longer menstrual cycles are associated with longer-lived sperm (Parker, 1984) and an increased intensity of sperm competition is associated with larger sperm, at least in primates and muroid rodents (Gomendio and Roldan, 1991).

The fact that increased sperm competition can favour both more numerous and larger sperm should not be too much of a surprise. Presumably the cost of sperm production is such a small yet important part of the male's total energy budget that it rarely if ever constrains males into producing only more or larger sperm. If success in sperm competition can be increased by increasing the proportion of energy expended on sperm production, rather than on mate-guarding or searching, then males who do so will be favoured (Parker, 1984). Then, if sperm size and number have independent optima, in principle the threat of sperm competition can drive for sperm that are both larger and more numerous.

11.2.4 Sperm shape

Mammalian sperm fall roughly into three categories with regard to shape: (1) the worm-like form characteristic of monotremes which mammals inherited from their reptilian ancestors; (2) round/oval paddle-shaped forms characteristic of the majority of species, including humans; and (3) the hook-headed shape characteristic of rodents and some prosimian primates (Box 4.19).

New World marsupials produce paired sperm which swim together until near the egg when they separate and become independent (Bedford, 1974). It is possible that, despite the two heads, the presence of two tails provides a mechanical advantage imparting greater speed and efficiency. However, if this is so, two tails might be expected to be more common. Many species of eutherian mammals have a two-tailed morph within their sperm menagerie but it is usually one of the least common sperm in the ejaculate (Box 4.19).

Very little is known of the biophysics of sperm shape. The widespread occurrence in mammals of the round/oval paddle-shaped form with a short mid-piece and long, uniflagellate tail rather suggests that it is near to some form of optimum in terms of swimming efficiency within the female mammalian tract. The hook-shaped head of the sperm of rodents and lorisoid primates seem to be independently derived from the more general oval-headed form (Roldan *et al.*, 1992).

So far, discussion of variation in the shape of sperm heads in the literature has been restricted to consideration of how head morphology might affect the sperm's ability to penetrate the outer vestments of the egg during fertilization. Bedford (1983, 1991) suggested that several traits of mammalian sperm may have evolved in response to the thick layer of vestments surrounding the ova, which is a unique mammalian feature. He argued that these traits would include a large acrosome (but see Chapter 12), the presence of hyperactivated motility and enhanced stability of the sperm head, all of which would improve the physical thrust needed to penetrate the ova vestments.

McGregor *et al.* (1989) failed to find a positive relationship between the complexity of the sperm head (i.e. a presence of either one or three hooks) and the thickness of the zona pellucida or the structure of the surrounding vestments. Their study was limited, however, to only three species of Australian conilurine rodents.

11.2.5 Sperm polymorphism

Previous authors have discussed the evolution of sperm size and shape in the context of sperm competition as if there were only a single optimum; the best trade-off between opposing and contrasting selective pressures (Gomendio and Roldan, 1991; Roldan *et al.*, 1992). Thus, larger sperm are best suited to 'sprinting' over short distances; smaller sperm are best suited to slower travel over longer distances. Optimum sperm size for any given species is thus somewhere in between large and small!

The irony is that there is more variation in the size of sperm within a single human ejaculate (Belsey *et al.*, 1987) than there is in the mean sizes of sperm over all of the primates (Cummins and Woodall, 1985). Thus, the mean head length of primate sperm varies from about 4.5 to 8.0 μm whereas the head length of sperm within a human ejaculate varies from <1 μm (pinheads) to >7 μm (some macros and tapering heads). Yet this intraejaculate variation has always been ignored.

In all other such situations analysed by behavioural ecologists, it is readily accepted that there is often more than one modal optimum: e.g. large gametes and small gametes (Box 2.4); males and females (Box 2.4); monandrous and polyandrous females (Box 2.8); and so on. So, why should the same not be true for sperm in an ejaculate? It seems obvious that the most effective ejaculates will be polymorphic. In the above instance, for example, the ejaculate could consist of a certain proportion of large, fast sperm and a proportion of smaller, slower sperm. Selection will then act to produce the optimum ratio of the two morphs, in just the same way as it acts to produce the optimum ratio of males and females (Box 2.5). It is this principle of multimodal optima that is the basis of our Kamikaze Sperm Hypothesis (KSH) (section 2.5.3).

In this perspective, our KSH argues that sperm competition has generated many different ways in which an ejaculate from one male can compete with an ejaculate from another male. Each facet of competition favours a different modal optimum of sperm size, shape and behaviour. Selection acts to produce the optimum number and proportion of each morph within the ejaculate. This optimum may change according to circumstance (e.g. according to whether sperm competition is likely and, if so, whether it is more likely with sperm already in the female, with sperm yet to be inseminated, or both). In essence, there should be a division of labour between the sperm (and other cells and seminal fluids) within the ejaculate, selection acting to favour males who produce the most effective ejaculates with the most effective division of labour between the different elements of the ejaculate.

Seen in this perspective, the first step towards an interpretation of the functioning of the ejaculate is to examine the different roles and activities which might benefit from delegation to a specialized type of sperm. If the multimodal, KSH, scenario is correct, each sperm morph within the ejaculate should have an individuality of role that could never have been predicted if all non-fertilizing sperm were simply errors of production (Harcourt, 1991); 'deformities' which are unwanted passengers on the carousel of healthy sperm.

In a recent review, Gomendio and Roldan (1993) suggested that 'the most serious drawback' of the KSH 'is the lack of specific predictions concerning which sperm morphs perform the various functions that have been proposed.' In the next section, we respond to this comment by considering in detail four of the major hypotheses we have developed from the original kamikaze sperm idea.

11.3 Kamikaze sperm: four hypotheses

In our original formulation of the kamikaze sperm hypothesis (KSH) we suggested that the sperm in an ejaculate were of two major types: (1) egg-getters; and (2) kamikaze (Baker and Bellis, 1988, 1989b). In turn, kamikaze sperm divided into two types: (a) blockers; and (b) seek-and destroy. Our discussions in this book have led us to postulate two further kamikaze elements: (c) a division of labour based on size; and (d) sperm involved in 'spiteful' inseminations as part of male family planning.

A consideration of egg-getter sperm and fertilization is delayed until Chapter 12. In this chapter, we concentrate on the evidence and factors relevant to kamikaze behaviour.

11.3.1 Sperm with coiled tails: an ageing block

Our studies of flowbacks have provided evidence that sperm from one copulation can hinder the retention of sperm at the next copulation (Box 10.7). This apparent block to retention persists with declining effectiveness for up to eight days, approximately the length of time after insemination that sperm have been found in human cervical mucus.

To what extent it is the sperm themselves which hinder the retention of later sperm and to what extent it is the leucocytes and debris that their presence promotes is not yet known. Nor is it clear whether the sperm in the cervical mucus derive only from the initial invasion or whether their stocks are continually replenished by sperm from the cervical crypts. For present purposes, it is sufficient to consider that some sperm may be programmed to seek out and lodge in the channels in the cervical mucus, there to hinder the passage of later sperm until removed by phagocytosis or by the flow of mucus into the vagina.

(a) BLOCKING: COILED CREDENTIALS

Ideally, a sperm programmed to block the channels in the cervical mucus should have the following characteristics:

1. sufficient initial mobility to leave the seminal pool and penetrate the cervical mucus, some to lodge in the mucus channels relatively soon after penetration, others perhaps to swim on to the cervical crypts;
2. low upward, or indeed any, mobility thereafter except perhaps for any sperm which reached the cervical crypts eventually to leave and re-invade the cervical mucus; and
3. a physical configuration that most efficiently blocks the mucus channels.

The sperm which meet all these criteria (Box 11.1) are those which coil their tails. Even while alive, coiled tail sperm show the slowest upward mobility (Bellis *et al.*, 1990) which our subsequent studies have shown is

Box 11.1 Swimming performances of four different tail morphs

Four measures have been made of the swimming performance of different sperm morphs:

1. *Speed*: the track of an individual sperm across a microscope slide (as observed at ×1200 using phase-contrast microscopy through the grid of a micrometer eye-piece) was drawn by hand on a paper grid. The length of the track was then measured (with cotton). Speed was calculated by dividing total track by the period of observation (at least 15 s).
2. *Straightness*: straight-line distance of the above tracks measured from the beginning to end of observation and divided by total track length. This gives an index of straightness of track with limits of 1.0 (straight) to 0.0 (ending where it began).
3. *Forward speed*: straight-line distance of the above track divided by the period of observation to give a measure of speed of forward motility.
4. *Upward speed*. Two preparations were assessed for sperm morphology: an aliquot of the raw ejaculate; and a sample of the swim-up sperm used for fertilization in IVF treatment (Box 11.3). Comparison of the percentage change of each morph in the two samples gives an index of upward mobility (Bellis *et al*., 1990).

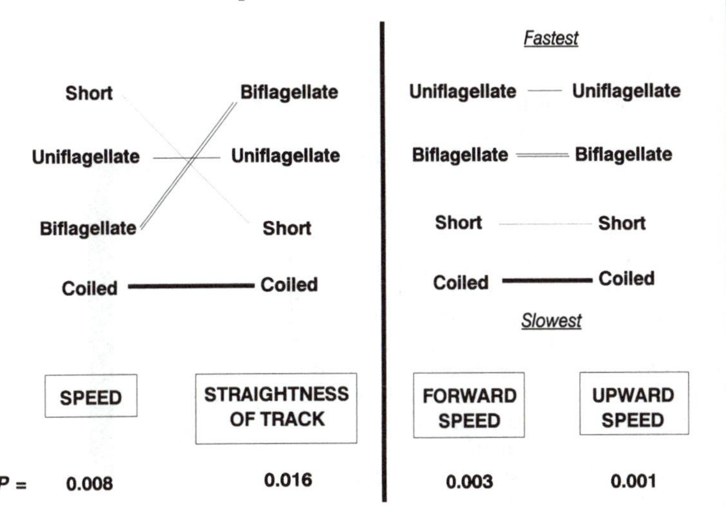

All measures are of sperm swimming in Human Tubal Fluid Medium (HTFM) under laboratory conditions. Measures (1)–(3) were made by a single observer (NM) from 28 ejaculates and 260 sperm (at least 10 of each morph). Measure (4) was also made by a single observer (JC) from 29 ejaculates and 11 600 sperm. Rank orders and significance levels calculated using Meddis' specific test, blocking by ejaculate to control for interejaculate differences.

A short tail seems to allow a sperm to swim fast but tortuously and hence is associated with sperm which make little forward or upward progress.

Two tails (= biflagellate sperm) seem to be an adaptation to allow a sperm to swim straighter but more slowly than sperm with a single but 'normal' tail. Nevertheless, sperm with two tails make good forward and upward progress.

Coiled tails are associated with sperm which make little forward or upward progress. Sperm with coiled tails in swim-up samples are almost certainly sperm which swam up, then coiled their tails.

due to a combination of slow swimming speed and a tortuous track. In addition, coiled-tail sperm have the ideal physical shape and cross-sectional size to lodge in and block mucus channels (Box 3.5).

Other authors have not commented specifically on the proportion of coiled tail sperm in the different spurts of split ejaculates but most have observed that motility, 'normality' and the proportion of sperm which appear to be alive are all reduced in later spurts (Eliasson and Lindholmer, 1972; Lindholmer, 1973; Marmar *et al.*, 1979). As described below, this almost certainly indicates a higher proportion of coiled tails in later spurts.

(b) COILED TAIL SPERM AS BLOCKERS: EMPIRICAL EVIDENCE

Males vary in the proportion of coiled tail sperm they inseminate (Bujon *et al.*, 1988) and females vary in their level of sperm retention (Chapter 10). This variation allows an empirical test of the hypothesis that coiled tail sperm are the primary morph responsible for blocking sperm retention. If the hypothesis is correct, we should expect females with the lowest levels of sperm retention to be those paired to males who inseminate the highest proportion of coiled-tail sperm. More specifically, the relationship should be most

Box 11.2 An empirical test of the hypothesis that the main 'blocker' sperm are sperm with coiled tails

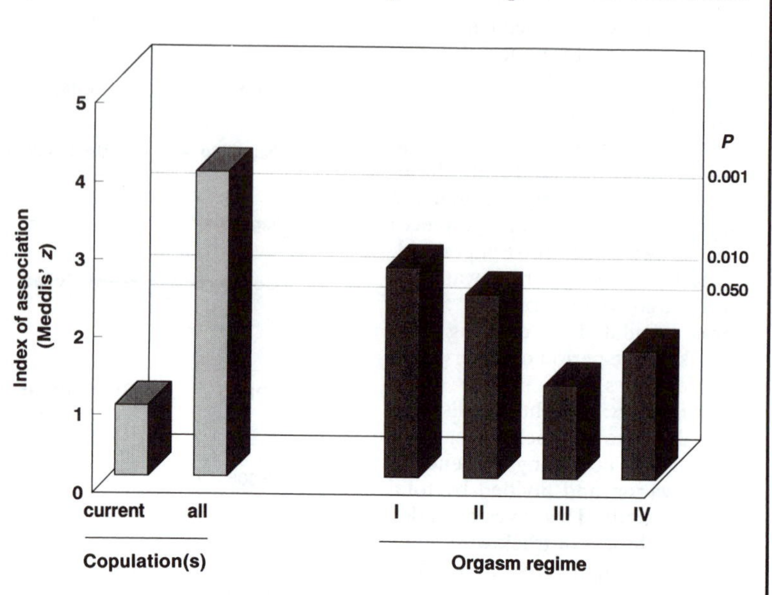

The level of sperm retention was calculated for individual females over all the flowbacks for which sperm morphology had been typed (9 females; 99 flowbacks; 15 600 sperm). Variation in retention between females was then compared using the mean proportion of sperm with coiled tails in copulatory ejaculates provided by their male partner as lambda coefficients in Meddis' specific test. The figure shows the z-values calculated from this test under various circumstances.

The hypothesis that the level of sperm retention by a female over a succession of copulations is a function of the proportion of sperm with coiled tails inseminated by her partner is supported: (a) over all copulations; and (b) in copulations with orgasm regimes I and II (Box 10.8). It is not supported: (c) for orgasm regimes III and IV. In all cases the tendency is for fewer sperm to be retained by females whose partners inseminate a higher proportion of coiled tail sperm.

As a control for the possibility that inseminates with a higher proportion of coiled tail sperm simply show lower levels of sperm retention, the level of retention at the current copulation was tested against the proportion of coiled tail sperm at that copulation (as determined from the flowback). The z-value is shown as the left-hand bar in the figure.

The analysis suggests that the proportion of coiled tail sperm in the inseminate does not influence sperm retention at that copulation but does influence sperm retention at subsequent copulations.

All analyses were controlled for variation in the number of sperm inseminated, age of sperm, orgasm regime, and pregnancy by calculating residuals from the equation in Box 10.2. Analysis was also controlled for observer bias by blocking by observer (seven observers: HF, HN, JB, JO, KT, MB, PC).

pronounced under orgasm regime I (in which retention is maximally influenced by the putative block) but should not be evident under orgasm regime IV (in which the female circumvents the block with a high retention orgasm during copulation) (Box 10.8). Our study of flowbacks produced results totally consistent with these predictions (Box 11.2).

An alternative interpretation of our data could be that males with a higher proportion of coiled tail sperm in their ejaculates produce sperm which are generally less able to attain a secure position in the female tract before being ejected in the flowback. Then, only if they receive female assistance are they capable of the same level of retention as sperm produced by other males. Our analysis controlled for factors that we know influence sperm retention (age of sperm, female weight, and total number of sperm inseminated) but this does not rule out the possibility of an influence of some other, as yet unidentified, factor associated with proportion of coiled tail sperm.

If this alternative explanation were correct, we should expect retention at the current (as opposed to future) inseminations to be lower when the proportion of coiled tail sperm in the ejaculate is higher. However, this is not the case. Sperm retention at the current insemination is not a significant function of the proportion of coiled tail sperm in the inseminate (calculated from flowback sperm, still controlling for sperm age, female weight, and total number of sperm inseminated) (Box 11.2).

In the absence of a more parsimonius explanation, therefore, we conclude that males who inseminate more coiled tail sperm set up a more efficient block to sperm

Box 11.3 The swim-up process and Human Tubal Fluid Medium (HTFM) used in experiments on live sperm

Swim-up procedure
Males masturbated and ejaculated into a wide-mouthed container which then stood at room temperature for 10–20 min until the ejaculate deco-agulated. The semen was then poured into a large Falcon tube. Two ml of Human Tubal Fluid Medium (HTFM) was added and mixed. The mixture was centrifuged for 5 min at 500 *g*, after which the supernatant was discarded. Three ml of HTFM was then added to the sperm pellet and mixed. Centrifugation, discarding of the supernatant, and re-mixing was repeated three times. After the third centrifugation, the sperm were re-suspended in 0.25 ml of HTFM.

The swim-up process was initiated by pouring further HTFM into the tube, ensuring a clean interface. The tube was then left at 37°C for 1 h for sperm to swim up and out of the mixture at the base of the tube into the clear HTFM. The top layer of supernatant, containing the most upwardly mobile sperm, was then removed with a Pasteur pipette.

For full details, see Matson *et al.*, 1989).

Human Tubal Fluid Medium (HTFM)
All of our observations on live sperm have been carried out while the sperm were in HTFM. Most recently, the medium used was Medi-Cult © Serum-free IVF Medium. This consists of:

EBSS
Synthetic serum replacement (SSR2)
Na-pyruvate 0.8 mM
$NaHCO_3$ 2.2 g/l
Penicillin 50U/ml
Streptomycin 50 μg/ml
Human serum albumin (HSA) 1% (10 mg/ml)
Osmolality: 280 mOsm/kg

retention at the next copulation. How this relates to male strategy is discussed below.

(c) BLOCKING: A SEDENTARY ROLE FOR GERIATRIC SPERM

Blocking is a sedentary activity that, once the sperm is in position, can even be continued after the sperm is dead. For this reason, blocking would be an ideal function for old and dying sperm. Previous authors have commented that tail 'defects' (Tyler and Crockett, 1982), specifically coiled tails (Mortimer *et al.*, 1982) are more common among dead and senescent sperm.

In our study of sperm from IVF swim-up samples (Box 11.3) (58 males; 5800 sperm), a mean of 22.0 ± 1.3% of apparently dead sperm had coiled tails compared with only 1.3 ± 0.3% of live sperm. Moreover, once sperm have been ejaculated, the proportion which have coiled their tail increases significantly with time from 5% after one day to 27% after 6 days (sperm maintained in Human Tubal Fluid Medium (HTFM) at room temperature and lighting; the longest we have seen a sperm remain active under such conditions is 14 days) (Box 11.4).

An influence of sperm age on the proportion of sperm with coiled tails is evident for time before as well as for time after ejaculation. The time that sperm have been in store in the cauda epididymis and vas deferens awaiting ejaculation has a very significant influence on the proportion of coiled tails already present at ejaculation. This is true whether the ejaculate is masturbatory or copulatory, is collected after IVF swim-up procedure, or is collected as a flowback after some time in the female (Box 11.4). The longer the time since last ejaculation, the greater the proportion of coiled tail sperm present.

The suggestion is, therefore, that as sperm wait to be ejaculated from the vas deferens and perhaps cauda epididymis, they lose energy reserves. With time, an increasing proportion of stored sperm become physically more suited to a sedentary role as blockers than to any more active role. The proportion suited to such a role should be greatest in those oldest sperm (at the urethral end of the vas deferens) and decrease towards and up through the younger and younger sperm in the vas deferens into the cauda epididymis. If this scenario is correct, for any given average age of sperm (i.e. as measured by time since last ejaculation), the larger the number of sperm ejaculated, the greater the recruitment of these younger sperm into the ejaculate. For any given time since last ejaculation, therefore, larger ejaculates should contain a lower proportion of sperm which coil their tails. This is just the pattern found (Box 11.4).

(d) RESPONDING TO THE ENVIRONMENT

In the same way that many sperm coil their tails when they die through running out of energy reserves, they also coil their tails when exposed to osmotic stress (Drevius, 1975). We also have investigated this response. In our own studies, the more dilute the medium, the faster and the more sperm coil their tails. Also the older the sperm (whether due to longer interejaculatory interval or longer time since being ejaculated) again the faster and the more sperm coil their tails.

Osmotic stress may not be the only chemical challenge to trigger at least some sperm to coil their tails. The spermicidal nature of seminal vesicle fluids could be enough to increase the proportion of coiled tail sperm towards the back of the ejaculate (cf. Lindholmer, 1973) even though the sperm concerned may be younger than those nearer the front of the ejaculate.

In view of the propensity of at least some sperm to coil their tails when exposed to a chemical challenge, we expected to see a higher proportion of coiled tail sperm in the flowback than in the ejaculate. We also expected to see an increase in the proportion of coiled tail sperm in the flowback the longer the time interval from insemination to flowback and hence the longer the sperm were in the vagina. However, neither of these expectations have materialized in our flowback studies.

Two explanations seem possible: (1) the flowback sperm have been sufficiently buffered by seminal fluid (section 3.3.2) not to be challenged by the hostile female vaginal fluids during the 5–120 min between insemination and flowback; or (2) many sperm do coil their tails while in the vagina but to such an extent that the tail membranes rupture and the tail straightens once more. We have not as yet investigated either of these possibilities further except to show that the latter does happen when sperm are exposed to prolonged or extreme osmotic stress.

The sensitivity of older sperm to chemical challenge in terms of coiling their tails should perhaps be viewed as part of blocking strategy rather than as pathological. The requirement of a blocking sperm is to retain motility until it is in a position to block a channel, then to coil its tail, lose motility, set up a block, and perhaps even die. The sudden change in chemical environment from male vas deferens to seminal fluid and cervical mucus that occurs at ejaculation and insemination could well be the chemical signal that triggers the oldest sperm to coil their tails. Younger sperm, eventually destined also to coil their tail, may well penetrate the cervical mucus further and perhaps even reach the cervical crypts. Only later, when older and/or returned from the crypts to the mucus, should they coil their tails. In this way, the block could be extended through both time and cervical space.

(e) RETENTION OF COILED TAIL SPERM

We can find no evidence in our flowback studies that coiled tail sperm are retained any less by the female than other sperm. In fact, we estimate that, on average, 50 ± 2% of coiled tail sperm are retained compared with 45 ± 2% of sperm with 'normal' tails (Box 11.5).

In possible contrast, is the observed drop in the incidence of unspecified 'tail defects' from a mean level of about 6% in the (masturbatory) ejaculates of males from nine couples undergoing infertility investigations to about 2% in the cervical mucus about 1–6.5 h after copulation (Mortimer et al. 1982). This could imply that coiled tail sperm had a lower retention rate than sperm with 'normal' tails. However, interpretation is not that straightforward. Even apart from the problem of comparing masturbatory and copulatory ejaculates, it is unclear whether coiled tail sperm were included in the category of tail defects.

In our flowback studies, coiled tail sperm are assumed to be hindered by the putative block to sperm retention due to sperm from a previous copulation (Box 10.7) in just the same way as are other sperm. All else being equal, significantly fewer ($P = 0.010$) coiled tail sperm are retained by the female under a level I orgasm regime than a level III (Box 10.8). Moreover, a high retention orgasm at copulation significantly increases the retention of coiled tail sperm ($P = 0.032$),

Box 11.4 The proportion of sperm with coiled tails increases as sperm age

(a) When sperm are ejaculated (by masturbation) and stored in HTFM (Box 11.3) under laboratory conditions, the proportion of coiled tail sperm increases significantly with time since ejaculation. The proportion that coil their tails seems to level off at about 27% after 5 days. (Two observers: (DH, SP), 65 males, 65 ejaculates, 20 600 sperm, analysis blocked by ejaculate and observer to produce a within ejaculate analysis over time, controlling for interejaculate and interobserver differences.)

(b) The proportion of sperm with coiled tails in the ejaculate also varies with the time that the sperm were in storage in the vas deferens/cauda epididymis (as measured by time since last ejaculation). As time since last ejaculation increases, more sperm have coiled tails at ejaculation (copulatory and masturbatory ejaculates). The same trend is evident even after selection (e.g. by IVF swim-up, Box 11.3; by the female tract, in the flowback). (*Swim-up*: as for (a). *Masturbation*: eight observers (AN, HN, JB, JO, KT, LH, RO, MB), nine males, 68 ejaculates, 10 900 sperm, blocked by observer and male to produce a within male ejaculate analysis over time, controlling for intermale and interobserver differences. *Copulation*: seven observers (HN, JB, JO, KT, LH, PC, RO), 11 males, 116 ejaculates, 13 500 sperm, analysis as for masturbation. *Flowback*: observers, etc. as in Box 11.2, analysis as for masturbation.)

(c) For any given time since last ejaculation, the greater the number of sperm in the ejaculate, the lower the proportion of sperm with coiled tails. This suggests that there is a gradient of coiled tails in sperm being stored awaiting ejaculation in the vas deferens and epididymis. This gradient should be such that older sperm near the urethra have a higher proportion of sperm with coiled tails than the younger sperm in the epididymis. (Data controlled for time since last ejaculation by calculating residuals. Analysis of masturbatory and copulatory samples as above, blocked by observer, male, and type of sample (masturbatory or copulatory).)

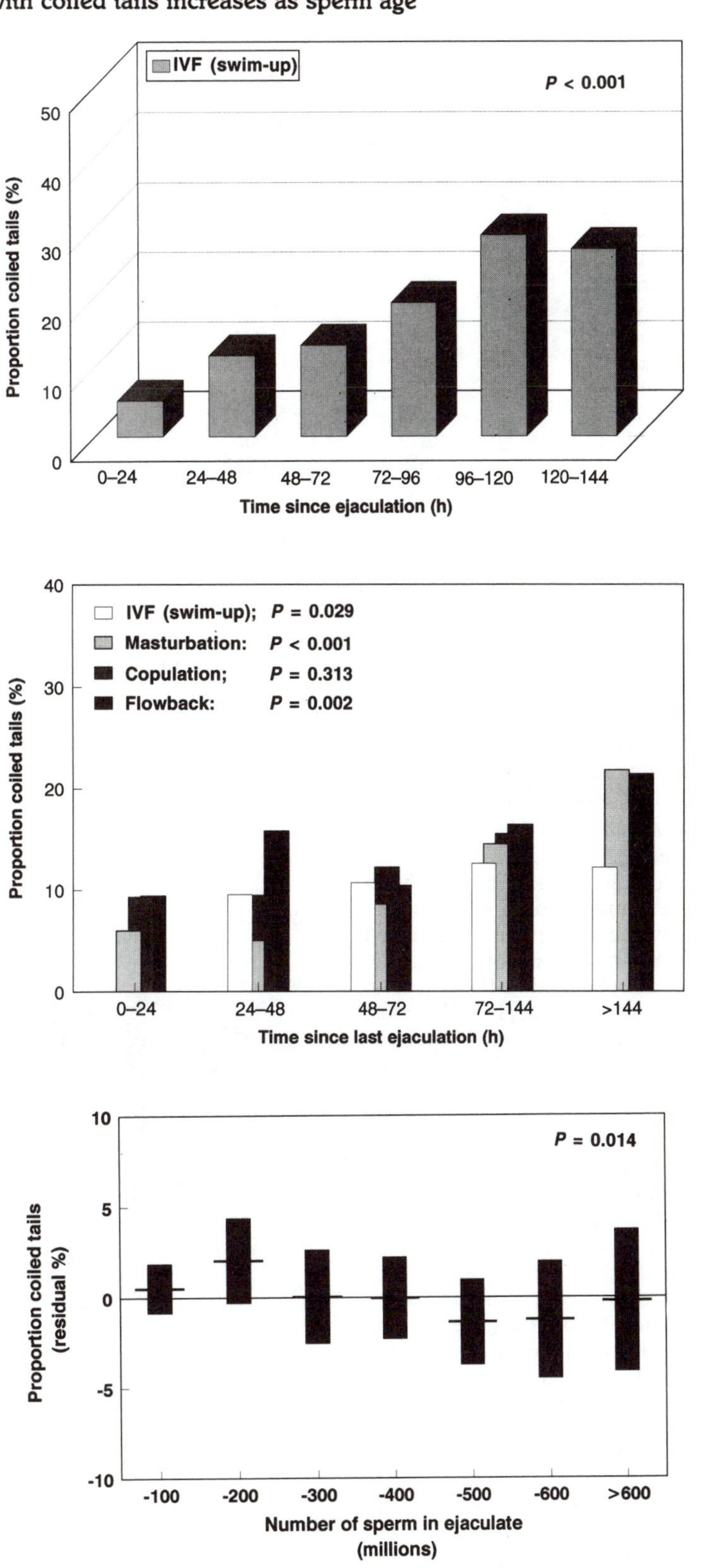

Box 11.5 Variation in retention by the female of sperm of different morphologies

Formulae for the calculation of the number of sperm of each morphology inseminated by males during copulation

Morph	Intercept	HEJ	Total sperm	% time together	Male age	% explained
Oval heads						
Macros	11	−0.027	0.051***	−0.124*	−0.041	38
Modals	−47	0.123	0.794****	0.340	1.517**	95
Micros	3	−0.020	0.038****	0.012	−0.226*	38
Other heads						
Amorphous	7	−0.028	0.029**	−0.033	−0.319*	16
Tapering	14	0.022	0.026*	−0.100*	−0.498*	20
Pyriform	10	−0.024	0.024**	−0.081*	−0.266*	28
Round	1	−0.039**	0.037****	−0.003	−0.157	46
Tail morphs						
'Normal'	42	−0.209**	0.887****	−0.208	−2.620****	97
2-tailed	2	0.005	−0.001	−0.016*	−0.018	3
Short	0	0.023*	0.006	−0.005	0.058	18
Coiled	−39	0.184	0.094***	0.182	2.740****	65
Mid-piece morphs						
Cytoplasmic droplets	−4	−0.001	0.084****	0.110	−0.944***	46
Bent	8	−0.017	0.044****	−0.061	−0.509**	33

Multiple regression equations calculated from all copulatory samples (Box 11.4). HEJ, time (h) since last ejaculation; Total sperm, total sperm in ejaculate; % time together, proportion of time male and female spent together since previous copulation and is a negative index of risk of sperm competition; Male age, absolute difference in years between male age and age 25 (Box 11.6). Asterisks indicate significant multiple regression coefficients: *, $P<0.05$; **, <0.02; ***, <0.01; ****, <0.001. % explained indicates the proportion of the variance explained by the equation. The regression coefficient for the Total sperm parameter indicates the morph's average proportion in the ejaculate. The presence of the Total sperm term in the equation means that all other coefficients relate to changes in proportion of the morph with respect to that parameter.

Retention of different morphs by the female

Females retain different morphs in different proportions. Microcephalous oval-headed sperm are retained in the highest proportions among the head morphs and 2-tailed sperm among the tail morphs. The most common sperm characteristics in the ejaculate (modal oval-headed and 'normal' tails) are actually retained in the lowest proportions. (The method used for calculating the proportion retained is the same as for total sperm (Chapter 10)). Equations shown in the table are used to calculate the number of sperm of each morph inseminated at each copulation. Then the number ejected in the flowback is subtracted to calculate the number retained. When forced through the intercept, the regression of number retained against number inseminated over all copulations then gives the mean proportion of that morph retained by the female. All standard errors are less than 5% of the mean. Details of observers, number of samples etc. as in Box 11.2.)

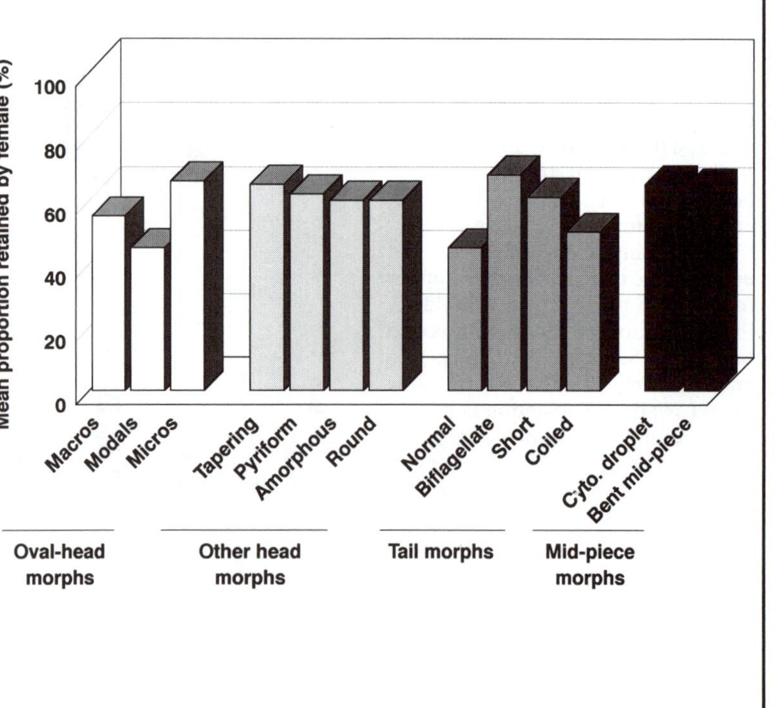

thus circumventing the block. Females thus have as much ability to influence the retention of these relatively sedentary sperm as they do the more mobile sperm in the ejaculate.

(f) SPERM COMPETITION AND BLOCKING STRATEGY

Strategically, males are in a difficult position in terms of how to deploy blocking sperm. There is no obvious value in inseminating more blockers just because the risk of the female already containing sperm from another male is greater. In contrast, there should be some value in inseminating more blocking sperm if the risk of the female double-mating in the near future is greater. However, the male would need to be able to make some assessment of that future risk.

As we have shown (Box 10.7), sperm hinder retention of the male's own sperm at the next copulation as well, presumably, as hindering retention of another male's sperm. The critical trade-off in the male's assessment, therefore, should be the probability that the female's next copulation will be with him or with a different male. In the absence of suitable information, optimum strategy may well be to inseminate enough blocking sperm successfully to block the channels in the female's cervical mucus to a level appropriate to the average risk of the female's next copulation being with another male. This should lead the male: (1) to inseminate more blocking sperm the longer it is since last copulation with the female (to 'top-up' the block; section 9.2.2); but (2) for any given time since last copulation, to inseminate a relatively fixed number of blockers. In at least humans and rats, this does seem to be the male's strategy.

Males show the same 'topping up' phenomenon with coiled tail sperm as they do with total sperm, apparently replenishing the block by a number of coiled tail sperm that depends on how long it is since last copulation ($P < 0.001$). The proportion of sperm with coiled tails is unaffected by time since last copulation ($P = 0.498$).

In rats, the blocking sperm are those which help to form the copulatory plug (Baker and Bellis, 1988). These sperm have significantly smaller heads and are more likely to shed their heads than sperm which progress to the uteri (Bellis *et al.*, 1990). The sperm in the copulatory plugs of bats are also more likely to shed their heads (Fenton, 1984). We have shown (Bellis *et al.*, 1990) that male rats inseminate blocking sperm in numbers that are independent of the risk of the female already containing sperm from a different male. If the coiled tail sperm of humans are also blockers, then male humans seem to show a similar strategy.

Even though male humans, like male rats (Bellis *et al.*, 1990) inseminate more sperm in total when the risk that the female already contains sperm from another male is high (Box 9.2), the number of coiled tail sperm inseminated does not vary in the same way (Box 11.6). This pattern is achieved by the male significantly increasing the proportion of coiled tail sperm inseminated when the risk of the female already containing sperm from a previous male is low (i.e. a high percentage time with the female since last in-pair copulation; Box 2.11). Nor does the male vary the number of coiled tail sperm according to the risk of the female collecting sperm from another male in the days after copulation (i.e. according to the percentage time together between this copulation and the next; $P = 0.553$).

It would seem, therefore, that from one copulation to the next, males of both humans and rats invest in blocking sperm independently of variation in the immediate past and future risk of sperm competition. The assumption is that they aim for a fixed level of blocking efficiency in part determined by the number and volume of channels to be blocked in the female tract and in part by the average level of risk of sperm competition.

To achieve the same level of efficiency of block in different females, males may need to inseminate more coiled tail sperm into larger females than smaller. Our data show that human males do inseminate significantly more coiled tail sperm into larger females ($P < 0.001$). However, this is part of the general insemination of more sperm into larger females (Box 9.4) and does not involve a change in the proportion of sperm with coiled tails.

Box 11.6 Variation in the number of coiled tail sperm inseminated during copulation in relation to risk of sperm competition

(a) Males respond to increased risk that the female already contains sperm from another male (low % time together since their last copulation) by increasing the number of sperm inseminated (Chapter 9). The number of coiled tail sperm inseminated, however, does not increase. This is achieved by there being a higher proportion of coiled tail sperm in smaller ejaculates (shown by figures on the histogram bars; see also Box 11.4). (Data from copulatory samples; see Box 11.4. Analysis is of within-pair variation, controlled for observer).

(b) The proportion of coiled tails varies with male age in both copulatory and masturbatory ejaculates in a way that is a positive function of the probability of extra-pair copulation (EPC) and double-mating by female partners of comparable age. (Ejaculate data from copulatory and masturbatory samples, 28 males, 203 ejaculates, 22 500 sperm. Analysis blocked by observer to give inter-male differences controlled for inter-observer differences.)

(c) The number of coiled-tail sperm in copulatory inseminates also varies seasonally as a function of the probability of conception in the Manchester region (Chapters 9 and 10). (Data from copulatory samples; eight males, 108 ejaculates, otherwise as Box 11.4. Analysis is of within-pair variation between months, controlled for observer.)

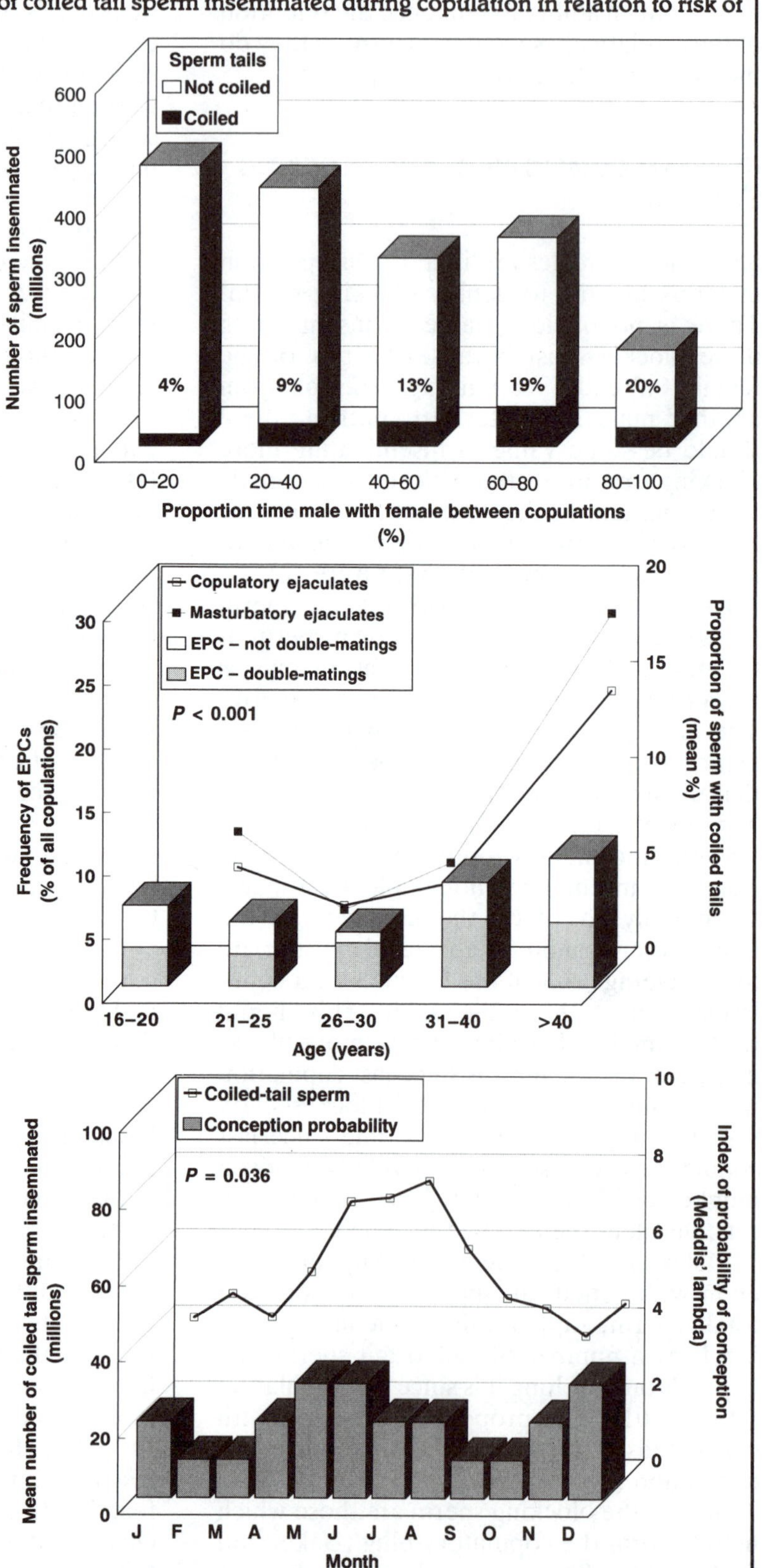

Level of efficiency of the block should evolve to be a general optimum based on the average probability that the next insemination will be by the male himself rather than by some other male. Just because a male does not, or cannot, assess variation in this probability from one copulation to the next, does not mean that a more general variation does not occur.

Bujon *et al.* (1988) have shown for humans that the proportion of sperm with coiled tails in masturbatory ejaculates varies with male age, coiled tails being least common in the ejaculates of 25–30-year-old males and most common in the ejaculates of older males. Our own data (Box 11.6) show a pattern in both masturbatory and copulatory ejaculates that is almost identical to that in Bujon *et al.*'s studies. This pattern bears a striking similarity to the variation with age in the probability of double-mating by females. As a result, males on average inseminate more coiled tail (putative blocker) sperm into females who are on average more likely to double-mate.

The fact that males show the same pattern of age-related variation in coiled tails in both masturbatory and copulatory samples suggests that the variation is largely due to a change in testicular function rather than to behavioural changes such as, for example, variation in masturbation rate. Support for this view derives from the continuing significance of the age-related changes ($P = 0.001$) even when variation in ejaculation interval, ejaculate size, and percentage time with the female partner are controlled.

The implication is that although from copulation to copulation males cannot assess the risk of sperm competition via their partner's infidelity in the immediate future, the more general age-related risk of sperm competition is counteracted by age-related variation in the production of coiled tail sperm. Moreover, the fact that the variation is physiologically linked to male age suggests that the age-related variation in female polyandry (Box 8.2) may owe more to the age of the male partner than to the age of the female herself. Perhaps, irrespective of her own age, a female may be most faithful to a male aged around 25 and least faithful to older males. As yet, we have no data with which to test this prediction.

Age-related variation in risk of sperm competition is not the only general and hence predictable variation to which the male could be adapted. Seasonal variations fit into the same category. We have shown earlier that males alter the number of sperm inseminated with time of year such that more sperm are inseminated when the probability of conception, and hence female value, is higher. We have no data on whether the risk of sperm competition is also higher at these times, but we predict that it will be. If risk of sperm competition is higher when the probability of conception is greater then, like age-related variation, season-related variation also permits the male some opportunity to vary his investment in blocking sperm.

Seasonal variation in sperm morphology has been reported for various mammals (e.g. Indian buffalo; Pant and Mukherjee, 1972; dairy bulls, Sekoni and Gustafsson, 1987) as well as humans (Spira, 1984) but the published data do not allow assessment of whether the number of coiled tails varies relative to the risk of sperm competition. Analysis of our own data in a longitudinal study of eight males (100 ejaculates) shows that seasonal variation in the number of coiled tails inseminated is positively associated ($P = 0.036$) with the probability of conception and hence, perhaps, to the risk of sperm competition. However, when we control for behavioural factors (intercopulation interval; percentage time together; masturbation rate; and female weight) the association disappears ($P = 0.201$). The implication is that, unlike age-related response which seems to involve a change in the testicular production of coiled tail sperm, season-related changes are achieved by behavioural mechanisms, primarily inter-ejaculation interval (Box 9.8).

Although males inseminate more sperm at seasons when the probability of conception is high (Box 9.10), females retain absolutely fewer sperm (Box 10.10). This discovery was consistent with our general hypothesis (Chapters 9 and 10) that conception is more likely in the presence of fewer sperm (down to some low optimum). This implies, however, that females should only discriminate against sperm destined to reach the site of fertilization (egg-getters and seek-and-destroy sperm; Chapter 12). There is no advantage and may even be a disadvantage in reducing the number of blocking sperm retained during seasons when conception is favoured.

We should not expect, therefore, to see a reduction in the number of coiled tail sperm retained by women during seasons of peak conception, and in our longitudinal study of five females (91 flowbacks) none was found ($P = 0.347$). Retention of a constant number of blockers while retaining fewer total sperm, meant that females retained a higher proportion of coiled tail sperm during seasons of peak conception ($P = 0.036$; controlled for sperm age as well as number of sperm inseminated).

(g) COILED TAILS: CONCLUSIONS

In summary, therefore, sperm from one copulation hinder sperm retention at the next copulation and the most effective blocks appear to be found in females paired to males with a high proportion of coiled tail sperm in their ejaculates. Such sperm have physical attributes and show behaviour that would suit them to a blocking role. Moreover, (a) their variation in different splits of the ejaculate, (b) their lack of variation in numbers from copulation to copulation in relation to past risk of sperm competition, (c) their variation in numbers with age of the male and the season, and (d) their seasonal pattern of retention by females, are all consistent with a role as blocking sperm.

The fact that coiled tail sperm tend to be older sperm suggests that at least some sperm in the ejaculate are programmed to switch from more active roles to a more sedentary blocking role as they age. *In vivo*, at least some sperm do not coil their tail until they reach the uterus and/or oviduct (Mortimer et al. 1982). Such sperm may well be programmed to block the narrow channel through the uterotubal junction. As such, of course, they should not influence levels of retention but could influence the number of sperm which gain access to the oviducts.

During the passive phase of rapid sperm transport (section 10.6.2), the uterotubal junction appears to offer little resistance to sperm passage. Thereafter, however, during the subsequent phases of active sperm migration, the flagellar activity of the sperm cells seems to be required to negotiate this barrier (Overstreet, 1983). This difference could be due to the presence of other sperm during the latter phase but not at the first.

Motile sperm have a definite advantage over immotile cells in crossing the uterotubal barrier (Gaddum-Rosse, 1981). Even dye does not cross the junction, suggesting that passive transport by uterine contractions is minimal (Gaddum-Rosse, 1981). The potential for the passage of sperm through such a restrictive channel to be hindered still further by blocking sperm seems considerable.

In mice, sperm head morphology is different across the uterotubal junction (Krzanowska, 1974; De Boer *et al.*, 1976). Whether this indicates that: (a) some sperm head morphs are prevented from crossing the uterotubal junction; (b) some sperm head morphs are programmed not to attempt to cross the junction; or (c) head morphology changes after crossing the junction (section 11.4.2), remains to be investigated.

Blocking behaviour, whether in the cervical mucus channels or at the uterotubal junction, may thus characterize a phase of a sperm's life rather than be a function of a particular morph. However, the observation that even the oldest ejaculates and populations of dead sperm do not contain, on average, more than 25% of coiled tail sperm could indicate that there is a particular group of sperm programmed to coil their tail and behave as blockers as they age.

Although we consider that the above evidence strongly indicates that sperm with coiled tails play a major role in hindering the retention of later sperm, we do not suggest that they are the only sperm type to be involved in this activity. For example, bicephalous (2-headed) sperm also have low upward mobility (Bellis *et al.*, 1990), a physical shape that should suit them to a blocking role (Baker and Bellis, 1988), and do not vary in number as a function of risk that the female already contains sperm from another male (P = 0.188). However, as yet we have too few data for this relatively (though with about 300 thousand in an average human ejaculate, not absolutely) rare morph to carry out a detailed analysis.

11.3.2 Macros and micros: hares and tortoises in the human ejaculate?

When compared across mammalian species, larger sperm (i.e. sperm with longer tails) are faster and smaller sperm are slower (section 11.2.3). In section 11.2.5 we hypothesized that even within a species, there could be some benefit in producing some sperm which are fast and others which are slow. In this section we consider the evidence that the observed variation in sperm head size within the human ejaculate represents just such a polymorphic adaptation.

(a) SIZE POLYMORPHISM AMONG OVAL-HEADED SPERM

The vast majority of sperm in the human ejaculate have a head with a roughly oval outline when viewed from 'above' but are narrow and pointed when viewed from 'the side' (i.e. they are paddle-shaped). Within this general category of head shape, however, there is a considerable range in head size (Box 4.19).

For convenience, we shall refer to the different ends of the spectrum of sizes of oval-headed sperm using the traditional terminology of 'macros' and 'micros'. To avoid giving the impression that there is something special about sperm with the most common, medium sized, head, however, we shall avoid use of the traditional term of 'normal' and refer to them instead as 'modal' (Box 4.19). In reality, however, there is probably a more or less continuous size distribution from macros at one end to micros at the other (Laufer *et al.*, 1978).

(b) SWIMMING SPEED

For the first 24 h after ejaculation, there is a significant positive association between sperm head size and both upward and forward velocity (Box 11.7), at least when observed in Human Tubal Fluid Medium (HTFM) (Box 11.3). It is not, however, that the macros are moving faster than the smaller sperm but rather that they are moving in a straighter line. From 48 h after ejaculation onwards, the roles are reversed. Macros are now moving forward the slowest. The change in rank-order owes more, however, to an increase in the speed and straightness of track of the smaller sperm than to any change in the swimming behaviour of the macros.

(c) SURVIVAL

At first sight and in contrast to upward and forward mobility, sperm survival appears to be a negative, not a positive function of sperm size (Box 11.7). Sperm collected after swim-up (Box 11.3), all of which are therefore alive at the start of the experiment, appear to show a significantly lower survival of macros 24 h later. The survival advantage of smaller size remains significant for a further four days. Beyond that time, the influence of size is evident but no longer significant and eventually, of course, all sperm are dead and size has no influence.

Other authors have also obtained data which are consistent with modal and micro morphs showing better survival after ejaculation than macros (Fredricsson, 1978; Tyler and Crockett, 1982). The situation, however, is not that simple and there is an important final twist to the story (section 11.4.2).

(d) RETENTION BY THE FEMALE

Our measurements of swimming performance, survival and development suggest that, on being inseminated, the larger oval-headed sperm make the most rapid progress through the female tract. In contrast, the

Box 11.7 Oval-headed sperm: influence of head size on swimming performance, longevity and retention by the female.

(a) *Swimming performance*
Oval-headed sperm differ in their swimming performance as a function of head size. The modal size appears to be optimum for swimming speed but larger heads seem better able to maintain a straighter track. The combined result is a rank-order that is a positive function of head size and is identical for both forward and upward progression, despite the different methods of estimation. (Data collected as in Box 11.1. Rank-orders only for sperm <24 h after swim-up.)

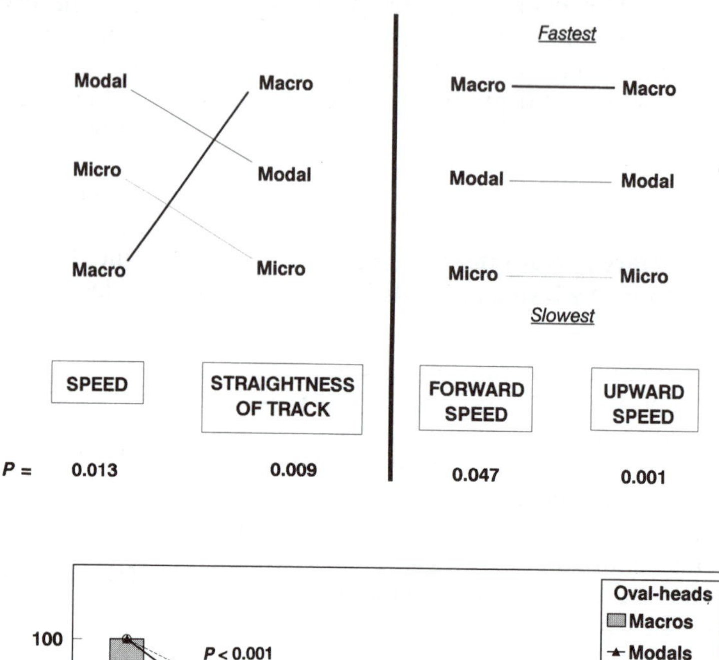

(b) *Apparent survival*
We used swim-up to select out only live, upwardly mobile sperm from masturbatory ejaculates, and then examined oval-headed morphs for longevity as a function of head size. Within 24 h of swim-up, survival is a negative function of head size, sperm with larger heads having a higher death rate. Thereafter, until all sperm are dead, macros have a consistently higher proportion dead than the other morphs. The negative relationship between head size and survival continues until 5 days (120 h) after swim-up, but the main contributory factor is the difference between macros and the others rather than differences between the smaller morphs. Sperm were categorized as dead or alive on the basis of movement, not by staining.)

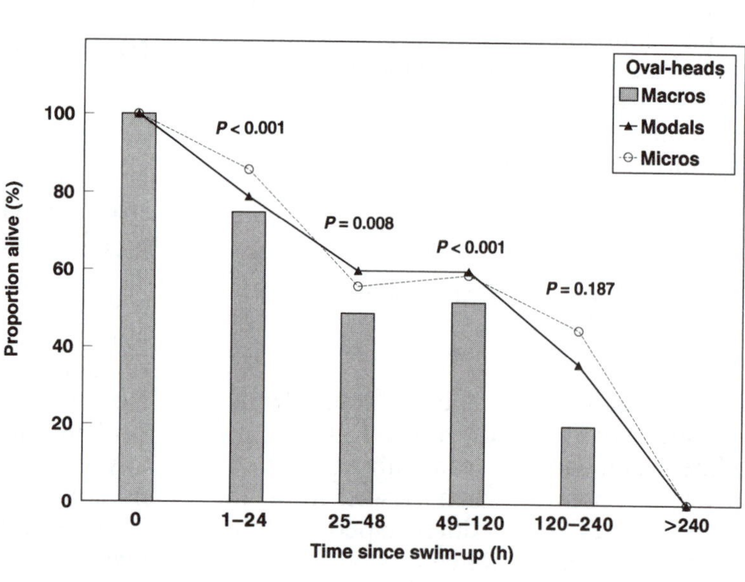

smaller oval-headed sperm at insemination make delayed and slower progress through the tract. Our studies of flowbacks, however, have identified a further consequence of size polymorphism: smaller sperm are much less hindered by the cervical block created by (coiled tail?) sperm from a previous copulation (Box 11.7).

When the sperm block is most powerful (orgasm regime I; Box 10.8), there is a very significant influence of sperm size on level of retention by the female, smaller sperm being retained in much higher proportions than larger sperm (Box 11.7). With decreasing efficiency of the block (orgasm regimes II–III), all sizes of oval-headed sperm are retained by the female in significantly greater numbers, but the greatest increase is shown by the larger sperm. When the sperm receive female assistance in retention (orgasm regime IV), the advantage of smaller sperm size on the proportion retained by the female, although still significant, is much reduced.

There is a strong indication, therefore, that

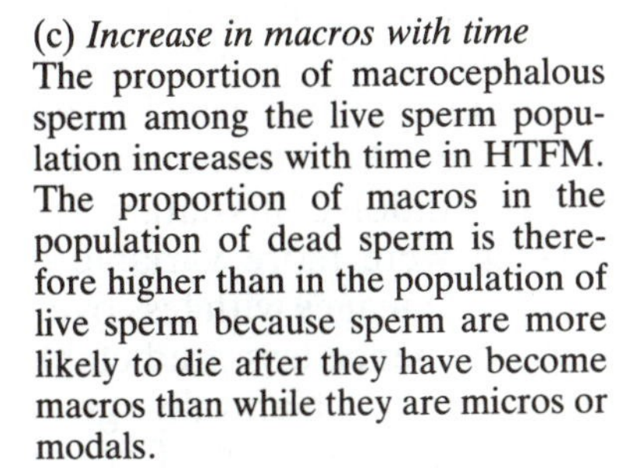

(c) *Increase in macros with time*
The proportion of macrocephalous sperm among the live sperm population increases with time in HTFM. The proportion of macros in the population of dead sperm is therefore higher than in the population of live sperm because sperm are more likely to die after they have become macros than while they are micros or modals.

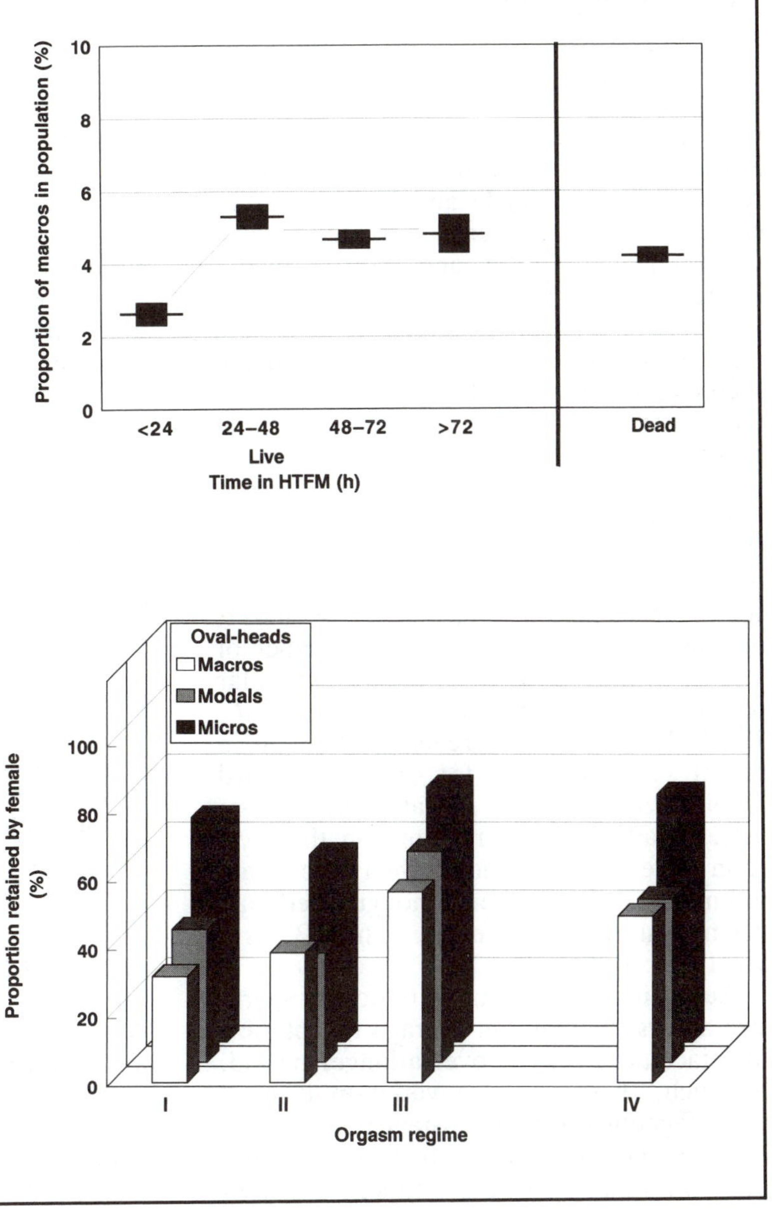

(d) *Retention by the female*
All three sizes of oval-headed morphs benefit in terms of higher retention from more favourable orgasm regimes (Box 10.8) on the part of the female. However, micros, which show the highest mean retention of all head morphs (Box 11.5), benefit least (macros, $P = 0.001$; modals, $P < 0.001$; micros, $P = 0.018$). Under all orgasm regimes, however, retention is a significant negative function of head size ($P < 0.001$).

(Data details as for flowbacks, Box 11.2. Analysis blocked by female and by observer to provide a within-female analysis controlled for inter-female and interobserver differences.)

smaller sperm are, at least in part, a male adaptation for sperm retention on occasions that retention is not assisted by the female. Quite possibly the small head size of such sperm allows them to gain access to the channels in the cervical mucus even when these channels are blocked by other sperm, leucocytes and debris.

It thus seems that on some occasions (with female assistance) those sperm which have larger heads at insemination are taken up and should be the first to begin their activity whereas on other occasions (without female assistance) the smaller sperm are the most likely to be retained. Given this variation over which sperm size gains most from first access to the cervical mucus, there is no clear reason for there to be a gradient of sperm sizes from early to late spurts of the ejaculate. Nor did our study of split ejaculates find any such gradient, a tendency for sperm in the earlier spurts to have larger heads than those in the later spurts failing to reach significance ($P = 0.120$).

(e) A STRATEGY FOR SPREADING ACTIVITY THROUGH TIME AND SPACE

Size variation in sperm seems to allow the male to counter a greater variety of situations. Smaller headed sperm are more likely to be retained by the female and live longer. In contrast, the larger sperm, if retained, carry out their tasks faster and more directly. Larger sperm thus seem to be adapted to occasions when sperm retention is female assisted; smaller sperm are adapted to make their own way through the female tract over a longer period, perhaps in the face of blocking sperm from another male.

When the risk of the female already containing sperm from another male is higher (low percentage time together; Box 9.2), male humans increase the number of larger sperm inseminated (macros, $P = 0.021$; modals, $P = 0.006$) but not the number of smaller sperm (micros, $P = 0.813$). A similar response is shown by male rats (Bellis *et al.*, 1990). The result is a greater investment in sperm programmed for rapid upward mobility and activity. In rats, many more larger-headed sperm are found in the uteri soon after insemination when the male inseminates a female he has not guarded than when he inseminates a female he has (Bellis *et al.*, 1990).

According to our data, variation in sperm size allows the male on average to optimize his ejaculate's activity over a longer period. As such, therefore, size variation is most advantageous when circumstances are unpredictable and particularly when the male has little information over the timing of either ovulation or any previous inseminations by another male. This conclusion suggests a further prediction that the coefficient of sperm size variation should be greatest in species in which the timing of ovulation and female polyandry is most unpredictable (e.g. species with female crypsis) and least in species in which they are more predictable (e.g. species with conspicuous sexual swellings) (Chapter 6).

Harcourt (1991) has analysed the coefficient of size variation for the sperm of a variety of primate genera and interpreted his results in terms of sperm competition. He concluded that if there was any trend in his data it was for lineages less exposed to sperm competition to produce sperm which were more variable in size. An alternative but equally viable interpretation of his data would be that sperm which were more variable in size were produced by lineages with less overt signs of ovulation and hence less predictable timing of female polyandry. Naturally, given the almost inevitable higher level of sperm competition in species with sexual swellings (Chapter 6), it is difficult to separate these two factors.

11.3.3 Oval-headed sperm: seek-and-destroyers?

We have argued that kamikaze sperm fall into two main categories: (1) blockers, programmed to hinder the passage of sperm from any later male to inseminate the female (section 11.3.1); and (2) seek-and-destroyers, programmed to wander through the female track, locate sperm already present from another male, and destroy or at least incapacitate them.

(a) SEEK-AND-DESTROY CREDENTIALS

Numerically, the seek-and-destroy role must be the most demanding. Sufficient numbers have to be able to pass through any cervical block established by a previous male (and the female) and survive phagocytosis to be able to find and destroy any sperm present from a previous male. Perhaps even then there needs to be enough sperm to spare still to patrol the female tract to be a hindrance to any sperm which might arrive from a later male.

We have always anticipated, therefore, that the seek-and-destroy morph(s) should be the most common sperm in the male's menagerie (Baker and Bellis, 1989b). In humans, other primates, and many other mammals (Chapter 4), this expectation immediately targets the macro-micro range of paddle-headed (in humans oval-headed) sperm as being the most likely candidates for a seek-and-destroy role.

(b) EVIDENCE FOR SEEK-AND-DESTROY BEHAVIOUR

Our evidence that sperm possess seek-and-destroy behaviour comes from mixing ejaculates from different males (Box 11.8). The *in vitro* behaviour of sperm in such heterospermic mixes is then compared with the behaviour of sperm in the two control homospermic mixes in which the sperm from one male are divided into two populations which are then mixed back together.

Within 3–6 h of being mixed with sperm from another male, the sperm in heterospermic mixes are more likely to have coiled their tails (P = 0.021) and to be immobile (P = 0.002), apparently dead, than sperm in homospermic mixes. Even those which are still motile are moving significantly more slowly (P = 0.007) and are making significantly slower forward progress (P = 0.011). All of these effects are still significant after 24–48 h.

A decrease in motility has also been observed in heterospermic mixes of sperm from bulls (Campbell and Jaffe, 1958; Dott and Walton, 1958; but see Hess *et al.*, 1954).

(c) SEEK-AND-DESTROY MORPHS?

In humans, the main morphs to suffer this loss of vitality in the presence of sperm from another male are the larger sperm, particularly the larger oval-headed sperm (macros, P = 0.018; modals, P = 0.017). The smaller sperm seem to be relatively unaffected (micros, P = 0.904; pinheads, P = 0.500).

Coiled tail sperm are not the only category significantly to increase in number in the

Box 11.8 A study of sperm behaviour and survival when in the presence of sperm from another male

Pairs of males (A and B) produced masturbatory ejaculates which were subjected to standard IVF swim-up procedure in Human Tubal Fluid Medium (HTFM) (Box 11.3). After swim-up, the sample from each male was divided into three equal portions and placed in 0.5 ml Ependorff tubes. Two of the portions were mixed with each other to produce two homospermic mixes (AA and BB). The remaining two portions, one from each male, were mixed to produce a heterospermic sample (AB). The three mixes (AA, AB, BB) were cryptically labelled by a third party before being examined by an observer for sperm morphology, behaviour, motility and survival. In our early experiments, sperm remained in a tube and portions were taken out at intervals to examine change in behaviour with time. This risked breaking up aggregations and sampling live and dead sperm differently. In our recent experiments, subsamples (20 μl) were placed under a cover slip on a microscope slide immediately after mixing and different slides were examined at different time intervals thereafter. This avoided the problems just noted but such slides could usually only be examined once before being discarded. (The data in subsequent boxes are based on 66 mixing experiments using sperm from 132 males. Six observers (KE, CD, MS, JB, DJ, DH) have now examined the fate and behaviour of 61 200 sperm. Data from both tube and slide experiments are included in analyses, no data having been discarded. All analyses are blocked by individual experiment and hence observer to produce a within-experiment comparison of homospermic (AA and BB) vs heterospermic (AB) mixes, controlled for interexperiment and interobserver differences. Only one time interval is used for any given analysis to avoid pseudoreplication of data.)

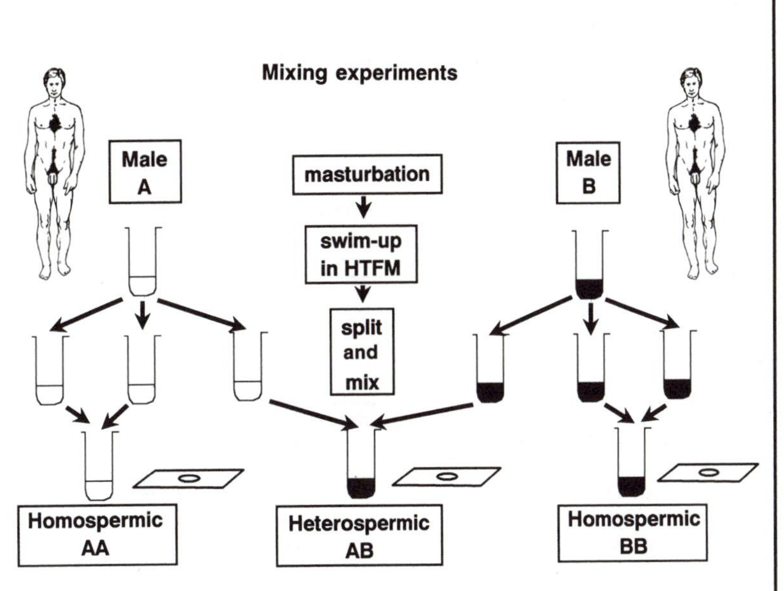

presence of sperm from another male. Round-headed sperm also increase significantly (Box 11.9). At 24 h after mixing they show a 31% increase (from 2.89 ± 0.49% of the ejaculate in homospermic mixes to 3.79 ± 0.48% in heterospermic); 48 h after mixing they show a 45% increase (3.55 ± 0.55% in homospermic; 5.22 ± 0.89% in heterospermic). This can only happen if at least some of these sperm are the result of the transformation of another type of sperm and in our mixing experiments the increase in round heads is matched by a significant decrease in oval-heads, particularly macros. We are forced to conclude that some oval-headed sperm become round-headed sperm and that this is more likely to occur in the presence of sperm from another male.

Two types of round-headed sperm have been reported previously (Singh, 1992). Both are in some way acrosome deficient. Type I lack an acrosome (Schirren *et al.*, 1971; Afzelius, 1981), its intrinsic enzymes (Schill *et al.*, 1988) and a postacrosomal sheath (Singh, 1992). Occasional males produce ejaculates consisting entirely of such sperm and seem to be absolutely infertile. Type II round-headed sperm may possess remnants of the acrosome and may be capable of fertilization (Singh, 1992).

Detached acrosomes are often seen floating in the medium along with sperm and we have assumed that the increase in round heads is the result of oval heads shedding their acrosome or undergoing the acrosome reaction (Box 11.10). Even after losing the acrosome, however, some additional change in head shape would also be necessary before an oval-headed sperm could transform into a round-headed sperm (H. Moore, personal communication; P. Matson, personal communication).

It has been postulated that type-I round-headed sperm may be produced during spermatogenesis by the failure of the acrosome to develop and hence, by mechanical factors, to taper the nucleus to a conical extremity as in oval-headed sperm (Holstein, 1975). In which case, loss of the acrosome could conceivably lead to oval-headed sperm coming to resemble a type II round-headed sperm.

We cannot yet resolve the question of whether all, some or any of the round-headed sperm we see in heterospermic mixes are

Box 11.9 Summary of results from mixing experiments: the response of sperm to encountering sperm from another male

Comparison of sperm in heterospermic and homospermic mixes (Box 11.8) should indicate the response of sperm to encountering sperm from another male. Swim-up procedure removes most, if not all, seminal fluid constituents except for those chemically bound to the sperm surface. Changes in behaviour should primarily, therefore, be the result of encountering sperm from another male, presumably identified via the molecular domains on the sperm plasma membrane (section 11.2.1).

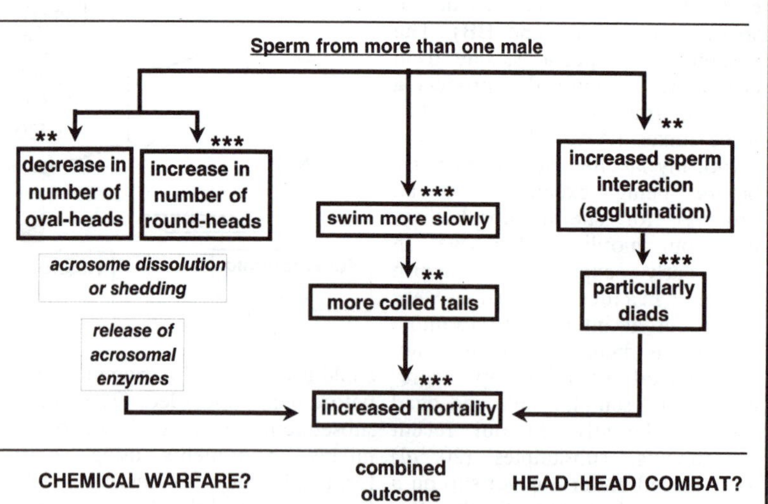

Responses surrounded by solid lines are observed as significant differences between heterospermic and homospermic mixes. Level of significance is indicated by asterisk: *, $P < 0.05$; **, < 0.02; ***, < 0.01; ****, < 0.001. Comments surrounded by dotted lines are suggested mechanisms by which observed responses might occur. It is suggested that sperm have two main methods of attempting to incapacitate sperm from another male. These are akin to chemical warfare and head-to-head combat as indicated.

Box 11.10 The acrosome reaction

The acrosome reaction involves multiple fusions between the outer acrosomal membrane and the overlying plasma membrane, which enables the contents of the acrosome to escape through the fenestrated membranes. Once sperm are capacitated and receive an appropriate stimulus, they undergo the acrosome reaction rather quickly, within perhaps 2–15 min or so.

In vivo, a few acrosome reacted sperm may be found at any point throughout the female tract but the vast majority of sperm swimming freely within the ampulla at the site of fertilization have not yet undergone the acrosome reaction. Those in the cumulus have either unreacted, reacting or reacted acrosomes, and nearly all those on the zona surface have reacting or reacted acrosomes. For detailed review and references, see Yanagimachi (1988). Figure redrawn and simplified from Nagae *et al*. (1986).

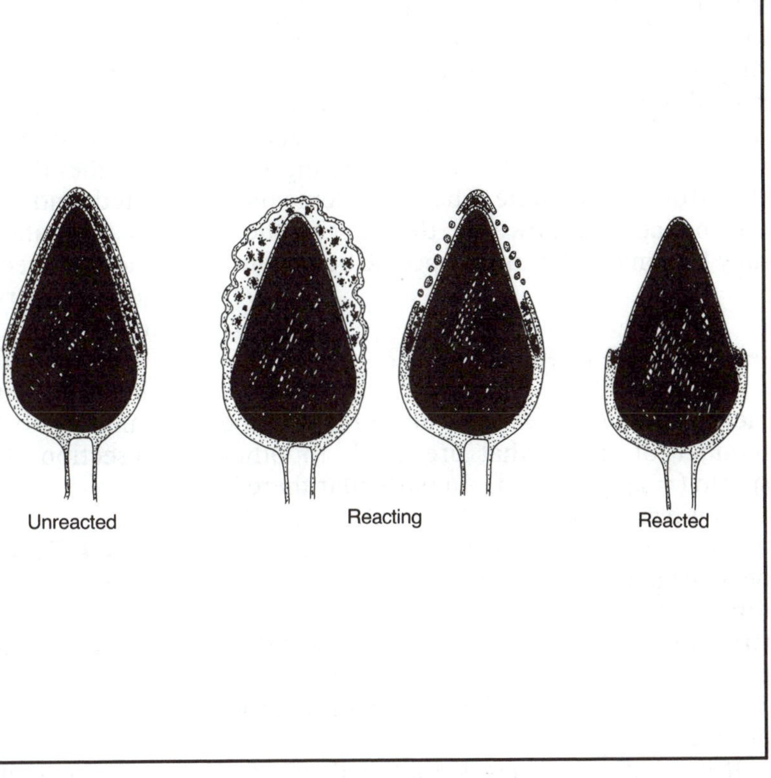

oval-headed sperm which have shed their acrosome and changed their head shape. However, our data are relatively unequivocal that when sperm from one male encounter sperm from another male round-headed sperm increase in number and oval-headed sperm decrease.

(d) AN ACROSOME FOR CHEMICAL WARFARE

John Cohen once commented to us that by carrying around an acrosome, sperm were in effect carrying a bomb on their heads. The acrosome contains enzymes which are released by a physiological event known as the acrosome reaction (Box 11.10). The enzymes contained in the acrosomal package are powerful enough, if enough sperm surround the egg, actually to dissolve the egg itself (section 4.4.8). There can be little doubt that, once released into the surrounding medium, these enzymes could promote morbidity and death in the sperm themselves. We have observed sperm appear, in effect, simply to drop dead after having swum past another sperm. Perhaps chemical warfare is a better analogy than a bomb.

Although other authors do not seem to have investigated acrosome reactions in heterospermic mixes, there is some supporting evidence for our hypothesis from other authors from studies of homospermic mixes. Although the incidence of spontaneous acrosome reaction (i.e. release of the acrosome contents and acrosome dissolution in the absence of follicular fluid) is low in such conditions, it is nevertheless shown significantly more often by modal oval-headed sperm than by other head morphs (9% as opposed to 4% after 24 h incubation; Fukuda *et al.*, 1989).

The effects we have just described, as observed in our heterospermic mixes, suggest that up to a point seek-and-destroy behaviour is relatively non-specific. The results imply that, whenever oval-headed, seek-and-destroy sperm encounter sperm from another male, they release acrosomal enzymes into the surrounding medium, perhaps even losing or shedding the whole acrosome. The result is that they slow down and perhaps kill all sperm in the vicinity irrespective of by which male they were produced. However, if a sperm does this in response to a sperm from another male, it is likely that the enzymes it releases will affect more sperm from a competing male than from its own male. There are elements in our mixing data, however, that suggest there may be some rather more specific behaviour.

(e) HEAD-TO-HEAD COMBAT?

Heterospermic mixes show an increase in the number of sperm that are stuck to other sperm (= agglutinated). In particular there is a significant increase ($P = 0.009$) in the number of diads, pairs of sperm joined at the head which, at least when dead, can sometimes be mistaken for biflagellate sperm (frontispiece). In our experiments, the difference is of the order of 42% after 6 h (8.02 ± 3.1 in homospermic; 11.38 ± 3.8 in heterospermic).

It is tempting to suggest that diads involve sperm joining at the head as part of a mechanism by which one sperm attempts to incapacitate another. Invariably the point of attachment is the equatorial region of one of the sperm but can be any part of the acrosome of the other sperm.

We have not yet observed the process of diad formation so cannot confirm that both sperm are alive as they join. We have, however, followed diads (in heterospermic mixes). One of the two sperm often appears to be dead by the time the diad is first seen, so either that sperm was already dead at diad formation or, perhaps most likely, died soon after. We have seen diads separate after as short a time as 10 min, the live sperm swimming off apparently normally (frontispiece). However, we have also followed diads for over an hour until both sperm appear to be dead.

We have so far been unable to determine whether diads are more likely to form between sperm from different males. Diads are seen in homospermic mixes, so they do occur when the sperm are from the same male, but they are more common in heterospermic mixes. One hypothesis we are evaluating is that sperm form diads in homospermic mixes but separate sooner with both sperm being more likely to be alive at separation than in heterospermic mixes.

An important evolutionary consideration is whether the surviving sperm, having incapacitated and separated from one sperm, can incapacitate further sperm. If so, this is yet another mechanism by which one seek-and-destroy sperm can remove from competition more than one sperm from another male. Such an ability would be of considerable theoretical importance to the evolution of kamikaze behaviour and is discussed further in section 11.6.

11.3.4 Tapering and pyriform sperm, male stress and spiteful insemination

(a) STRESS AND FAMILY PLANNING

Female mammals, with almost universal maternal care, have a wide range of mechanisms for family planning ranging from ovulation avoidance to infanticide (section 4.4). As female circumstance is translated into family planning, the physiological syndrome most often manifest is one or other of the forms of stress reaction.

We have argued (sections 6.4 and 7.2.1) that the similar evolution of paternal care in a few mammalian lineages has led to the evolution of male family planning. These adaptations seem to take three forms: (1) avoiding copulation; (2) copulation without insemination; and (3) spiteful insemination. As with the female, at least some of these should be mediated via a stress response. This section is concerned with the possibility that sperm polymorphism is an important element in spiteful insemination (i.e. insemination intended to reduce the chances of fertilization).

The expectation would be that, if any sperm morphs are involved in spiteful insemination, they should show an increase in proportion in the ejaculate in response to male stress. Moreover, an increase in proportion of these same morphs should result in a temporary reduction in the fertility of the male's ejaculates. Why a male should continue to inseminate his female partner while attempting to avoid fertilization has already been discussed (section 6.4).

(b) STRESS AND SPERM POLYMORPHISM

A number of studies of humans have now shown that two sperm morphs increase in proportion in the male's ejaculate after periods of stress (MacLeod, 1951, 1970; Poland *et al.*, 1986; Giblin *et al.*, 1988; Gerhard *et al.*, 1992). The main morph to show an increase is the sperm type with a tapering head but MacLeod also reported an increase in sperm with 'amorphous' heads. This 'amorphous' category used by MacLeod will have included sperm that are now more normally classified as sperm with pyriform and round heads, as well as sperm still categorized as amorphous (Belsey *et al.*, 1987).

In a longitudinal study of six males (section 11.4.1), we have also observed a significant increase in the proportion of tapering sperm associated with stress ($P = 0.003$). In the general 'amorphous' category identified as responsive by MacLeod, we observed a significant increase in the proportion of pyriform ($P = 0.016$) and amorphous sperm ($P = 0.019$) but not in the proportion of round ($P = 0.925$) or of any other head morph. Among the tail morphs, sperm with normal ($P < 0.001$) and short ($P = 0.004$) tails also increased with increased stress but neither coiled nor biflagellate tails showed an increase.

We discuss the time interval between stress and the observed change in the ejaculate in section 11.4.1. Our concern here is whether any of these morphs could be responding to stress as part of a male strategy for family planning via spiteful insemination. The test is whether any of these morphs are associated with a decrease in fertility of the ejaculate.

(c) MORPHS AND FERTILITY: THE MTP RATIO

Our studies of the fertility of ejaculates used in both artificial insemination by donor (AID) and *in vitro* fertilization (IVF) treatments have demonstrated a significant association between a higher proportion of both tapering ($P = 0.005$) and pyriform ($P = 0.035$) sperm in the ejaculate and a lower probability that the female will become pregnant (Box 11.11). There was no such association for sperm with amorphous heads ($P = 0.118$) or sperm with normal ($P = 0.702$) or short ($P = 0.917$) tails. When combined together, the proportion of tapering plus pyriform sperm in the ejaculate showed a very strong negative association with the probability of the female becoming pregnant ($P = 0.001$). These two morphs, therefore, have all the characteristics necessary for an involvement in spiteful insemination. However, there is one further twist.

If an increase in the proportion of one or two sperm morphs in the ejaculate has a significant negative association with ejaculate fertility, the proportion of some other morph is likely to show a significant positive association. In our study, this morph was the macrocephalous oval-headed morph ($P = 0.001$) (Box 11.11). No other morph, including the modal oval-headed morph ($P = 0.474$), showed an association with ejaculate fertility. The result is that the numerical ratio of macro:(tapering+pyriform) sperm in the ejaculate provides a very significant indication of whether a given ejaculate will lead to pregnancy in AID and IVF treatment. This conclusion clearly has important implications for assisted conception methods of infertility treatment (Chapter 12). For present purposes, however, the main points of interest are the ramifications of the conclusion for morph function and male contraceptive strategy.

In our study of the fertility of ejaculates in AID and IVF treatments, there was a just significant association between the proportion of women who became pregnant following artificial insemination by donor and the macro: (tapering+pyriform) ratio (MTP ratio) of the ejaculate used ($P = 0.042$). Such women, of course, have no physical association with the donor either before or after insemination. Thus, there can be no artefact in the correlation due to stress levels common to both male and female which could influence both the male's MTP ratio and the

Box 11.11 The MTP ratio, male fertility, stress and male family planning

(a) *The macro: (tapering + pyriform) (= MTP) ratio and fertility in in vitro fertilization (IVF) and artificial insemination by donor (AID) treatments*
Sperm were produced and processed as in Box 11.8. At one of two stages (before or after swim-up), ejaculates were subdivided, part being used for IVF or AID and part for examination of sperm morphology. Morph typing was 'blind' as it was carried out before the outcome of IVF or AID was known. Once the results of IVF or AID were available, the outcome was analysed with respect to the MTP ratio of the ejaculate (Box 12.2). Levels of association between MTP ratio and fertility are indicated by asterisks: *, $P < 0.05$; **, < 0.02; ***, <0.01; ****, <0.001.

(b) *The MTP ratio and lifetime fertility: re-analysis of data collected by Bostofte et al. (1985)*
Over 1000 males who had attended a sperm analysis clinic in Denmark between 1950 and 1952 were surveyed 20 years later by questionnaire with respect to the number of living children they had produced. The results are presented in sufficient detail by Bostofte *et al.* to allow re-analysis of fertility with respect to MTP ratio. Such an analysis would be justified as long as the long-term (perhaps genetic) differences in MTP ratio between males were great enough to show through the 'noise' of within-male variation (e.g. due to individual stress levels and other factors at the time of sperm analysis in 1950–52). Analysis reveals a significant association between MTP ratio and fertility over almost a lifetime.

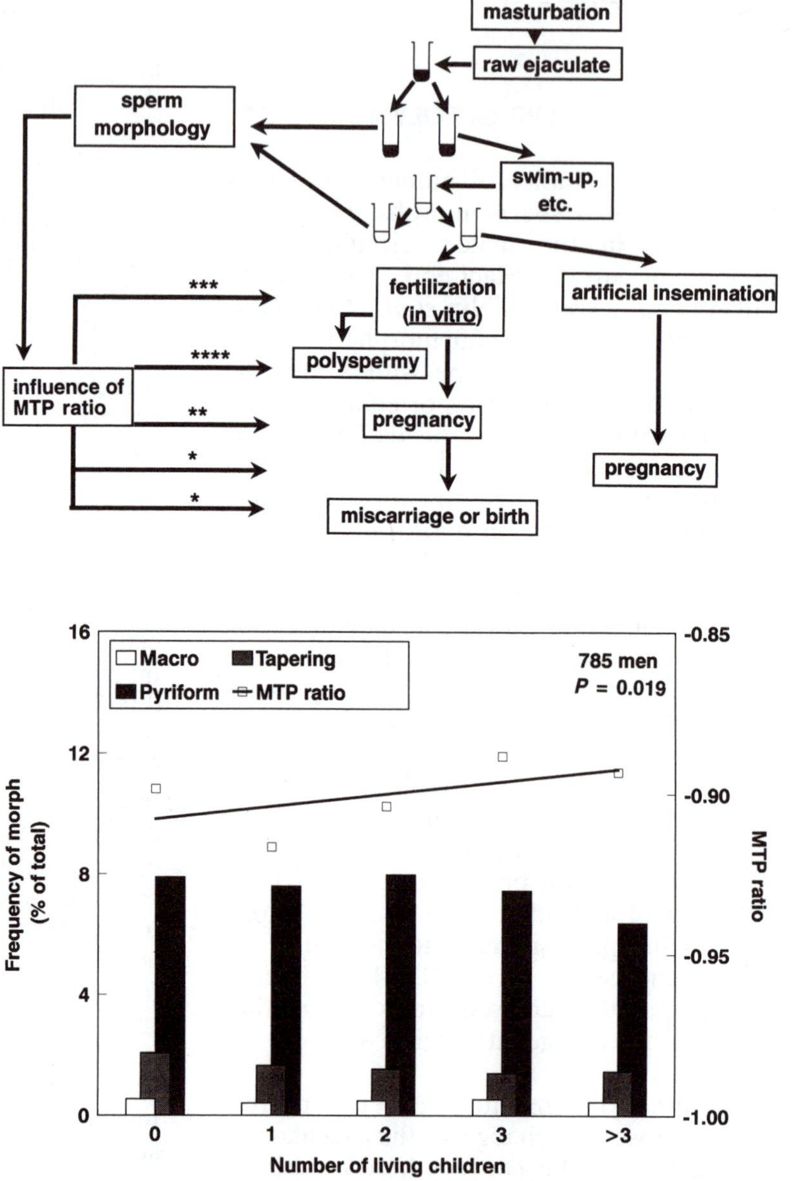

female's chances of conceiving and implanting.

In IVF treatment, the primary manifestation of the MTP ratio was with respect to the percentage of eggs fertilized *in vitro* ($P = 0.007$). All three of the morphs involved in the MTP ratio showed a separate significant influence (tapering and pyriform a negative influence each at $P < 0.001$; macros a positive influence at $P = 0.013$). The MTP ratio also influenced the probability of fertilization being polyspermic ($P = 0.001$), higher MTP ratios being associated with greater probability of polyspermy. As with our artificial insemination by donor study, there should again be no artefact from stress levels common to both the male and the female as the effect was manifest *in vitro*.

At the next stage of IVF treatment, however, there was the possibility of an artefact.

(c) *Stress, MTP ratio and male family planning*
Summary of the proposed mechanism (via stress and the MTP ratio) for the translation of environmental conditions unsuitable for the male to reproduce to a reduced probability that his partner will conceive. Events in boxes surrounded by a solid line are significant in our data. The comment surrounded by a dotted line is a proposed mechanism not yet investigated directly.

The sequence refers to a situation in which there is no sperm competition. If sperm competition occurs, the reduced MTP ratio gives the male more chance of removing his opponent's macros from competition. Responding to unfavourable (to him) conditions by spiteful insemination, rather than by either not copulating or by not inseminating sperm, gives the male some defence against his partner's infidelity and, if she is going to conceive anyway (by another male if not by him), some chance of the offspring being his. Full discussion of spiteful insemination can be found in Chapters 4 and 7.

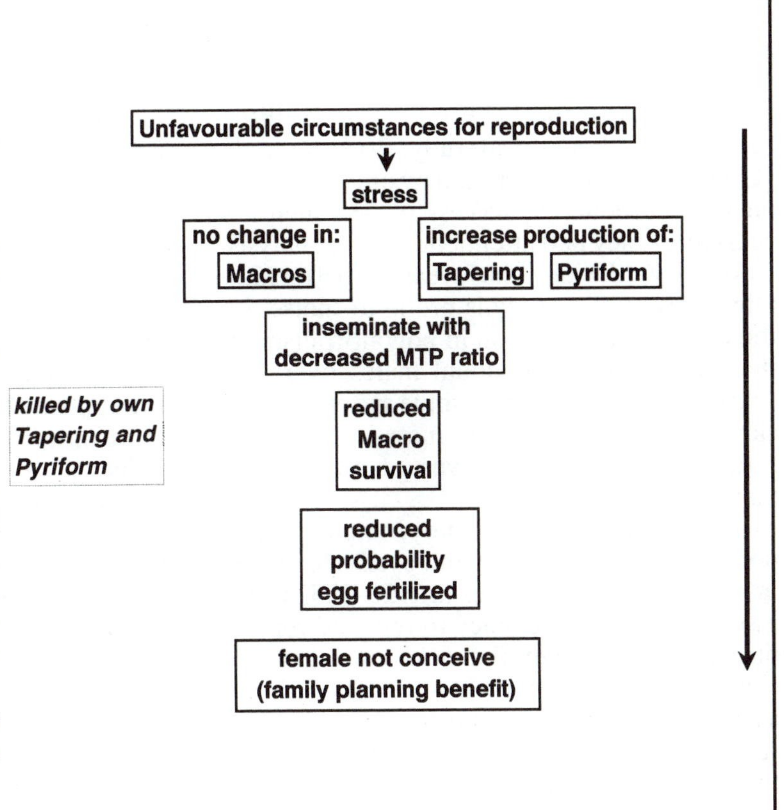

Analyses for (a) based on 283 ejaculates (36 500 sperm examined) for which the outcome of AID (96 cases: 12 pregnancies, 84 failures) or IVF (187 cases: 57 pregnancies, 130 failures) was known. Sperm morphed by an observer for AID (JJ) and three observers for IVF (TD, ES, DJ). All analyses blocked by observer, producing an analysis of within-observer variation in association between MTP ratio and fertility of ejaculates, thus controlling for interobserver differences.)

We again found a significant association between the MTP ratio of the ejaculate and the proportion of females who successfully implanted once the fertilized egg(s) were returned to their tract. Pregnancy was more likely if the male's ejaculate had a higher MTP ratio ($P = 0.018$). In this case, however, an artefact due to stress levels common to the male and female is perhaps a more parsimonious explanation than the more interesting alternative: that ejaculates with a high MTP ratio produce zygotes more likely to implant and develop in the female.

Finally, there was a just significant association between the MTP ratio of the ejaculate and the proportion of those women who became pregnant who went on to give birth, rather than miscarrying ($P = 0.040$). Again, however, this could be an artefact due to common stress levels for male and female partner. However, there may be parallels here with the association between ejaculate fertility and offspring viability reported for adders (section 2.5.6).

Apart from our own analysis of the association between the MTP ratio and fertility, there is one other critical study. Bostofte *et al.* (1985) collected information by questionnaire on the reproductive success of 785 men 20 years after they had attended a sperm analysis laboratory during the years 1950–1952. Obviously such a study is prone to 'noise' in the data due to variation in polymorphism in the ejaculates of individual males over the years. Nevertheless, the sample size was sufficiently large for real differences in the MTP ratio between males to average out and to manifest an effect.

In Bostofte *et al.*'s study, higher proportions of both tapering and pyriform heads in masturbatory ejaculates in 1950–1952 were associated with a tendency to reduced reproductive success over the next 20 years. In contrast, proportion of macros showed no association with lifetime reproductive success. Re-analysis of these data show that there is a significant ($P = 0.019$) relationship between the MTP ratio and subsequent reproductive success (Box 11.11). The mean MTP ratio for males with three or more living children was 34% higher than that for males with only one living child.

(d) SPITEFUL INSEMINATION: A MECHANISM?

The study on which our conclusions concerning MTP ratio are based was correlational and as such gives no direct evidence of cause and effect. For example, it would be possible for younger ejaculates to happen to have a higher MTP ratio than older ejaculates, perhaps due to selective phagocytosis of macros in the vas deferens (section 11.4.3). In which case, high MTP ratios could be associated with higher ejaculate fertility, not because of any function of the macro, tapering and pyriform morphs themselves but simply because ejaculates with higher MTP ratios contain younger sperm.

In fact, we can rule out this particular possibility because, at least in our whole ejaculate study (Box 3.8), younger ejaculates (shorter time since last ejaculation) do not have a higher MTP ratio ($P = 0.868$). However, we cannot rule out every possible third factor that could generate an artefactual association between MTP ratio and ejaculate fertility. Even so, while acknowledging that the MTP ratio may be an artefact, we consider there is some value in briefly discussing the possibility that the MTP ratio has a direct influence on ejaculate fertility and that manipulation of the MTP ratio provides the male with a mechanism for family planning.

Suppose that it is no coincidence that the three morphs which seem to have the greatest influence on ejaculate fertility are those with considerable upward mobility (macros the fastest; tapering the third fastest; pyriform the fourth; Bellis *et al.*, 1990) and the three largest. All three morphs almost certainly also reach all parts of the female tract, including the site of fertilization in the ampulla of the oviduct (Mortimer *et al.*, 1982). Given that macros enhance and tapering and pyriform reduce the chances of fertilization, it is not unreasonable to postulate that there is some direct interaction between the three morphs in normal homospermic (i.e. no sperm competition) insemination.

An excess of macros is associated with increased rates of fertilization. Not only does a high MTP ratio increase the proportion of eggs fertilized ($P < 0.001$), it also increases the proportion which are fertilized polyspermically ($P = 0.001$). Suppose, for simplicity, that at least a proportion of macros are egg-getting sperm (Chapter 12) but that they are hindered (or even killed) by tapering and pyriform sperm from their own ejaculate (hence their higher initial mortality; section 11.3.2). A high proportion of tapering plus pyriform sperm in an ejaculate not only reduces the proportion of eggs fertilized ($P < 0.001$), they also reduce the proportion fertilized polyspermically ($P < 0.001$). Moreover, there is a highly significant association between the proportion of tapering plus pyriform sperm in the ejaculate and the survival of macros during the first six hours after ejaculation. The higher the proportion of tapering plus pyriform morphs, the lower the survival of macros after ejaculation ($P < 0.001$).

Under normal (i.e. favourable environment; no sperm competition) circumstances, the MTP ratio is sufficiently high for enough macros to avoid tapering and pyriform sperm and reach the oviduct for fertilization to occur. In poor environments, however, stressed males inseminate more tapering and pyriform sperm and reduce the chances of any of their macros fertilizing the egg, thus meeting the dictates of family planning.

Why males should inseminate a female when conception is disadvantageous has already been discussed (section 6.4). By so doing, the male ensures that, if the female is inseminated by another male, he at least has some 'defence' inside the female and some chance of being the one to fertilize her egg(s) if she might conceive anyway. Moreover, the excess of tapering and pyriform sperm are present to hinder not only the male's own macros but also macros from the other male, thus also reducing the other male's chances of fertilization.

We might suppose that, ideally, in the presence of sperm from another male tapering and pyriform sperm should target the 'foreign' macros more than their 'own' macros. One of the biggest effects we observe when we mix sperm from two different males is a decrease in the MTP ratio among live sperm ($P = 0.004$), though a corresponding increase in ratio among dead sperm is nowhere near significance ($P = 0.335$).

It might appear as though macros are a primary target for the seek-and-destroy activity of foreign sperm. There is, however, an alternative interpretation: after the initial death of macros in the ejaculate, fewer sperm transform into macros in the presence of sperm from another male (section 11.4.2).

(e) THE MTP RATIO ACROSS SPECIES: A PREDICTION

One prediction that derives from our model is that tapering and pyriform sperm should be most common in species which show male parental care. Only in such species does the male benefit from having a mechanism to reduce the fertility of his own ejaculate.

Among humans and great apes, for example, we should predict that tapering and/or pyriform sperm should be more common in the ejaculates of humans and gorillas with relatively high levels of paternal care than in the ejaculates of the orangutan and both species of chimpanzee, which show little paternal care. Such data as exist indeed suggest that this is the case (see data in Seuanez *et al.*, 1977 and Box 4.19).

A systematic cross-species investigation is now needed to test our prediction.

(f) SPITEFUL INSEMINATION AND SPERM WITH CYTOPLASMIC DROPLETS: A SURPRISE

One of the surprises in our analysis of spiteful insemination was the apparent absence of a role for sperm with cytoplasmic droplets. As the cytoplasmic droplets themselves are rich in lysosomal enzymes (Mann and Lutwak-Mann, 1981), a role in seek-and-destroy behaviour for sperm carrying such droplets was anticipated. However, as an increase in the proportion of sperm carrying cytoplasmic droplets attached to their mid-piece is associated with a significant decline in ejaculate fertility in cows in artificial insemination (e.g. Wood *et al.*, 1986), a specific seek-and-destroy role in spiteful insemination also seemed a possibility. Yet no indication of such a role emerged in our studies.

First, sperm with cytoplasmic droplets declined rather than increased in abundance in response to stress ($P < 0.001$). Second, although in our AID and IVF studies an increase in this morph tends to be associated, as in cows, with a decline in the probability of the female becoming pregnant, the tendency is not significant ($P = 0.139$).

11.3.5 Kamikaze sperm: a summary

Our final synthesis, therefore, is as follows.

1. Macros are the morph primarily responsible for ejaculate fertility. Either they, or a proportion of them, are egg-getters or

they in some way enhance the performance of sperm that are egg-getters (Chapter 12).

2. Tapering and pyriform sperm are rather specific seek-and-destroy morphs the primary targets of which are macros, both from their own ejaculate and from the ejaculates of other males. The reasons for having two morphs in this niche are not yet clear.
3. Modal oval-headed sperm are generalized seek-and-destroy morphs which react primarily on encountering sperm from another male. Size polymorphism from larger to smaller heads is an adaptation to spread seek-and-destroy activities through time and space. Round-headed sperm, at least in part, are oval-headed sperm which have shed their acrosome during seek-and-destroy activity.
4. As they age, any of these sperm, or perhaps only a programmed proportion (25%?), may lodge in the channels of the cervical mucus and/or perhaps uterotubal junction, coil their tails, and become blockers.

This summary does not cover all of the morphs in the human ejaculate. For those not so far covered, either (a) we have no *a priori* hypothesis (e.g. for amorphous, pinheaded and two-tailed), (b) our data are too few (e.g. for two-headed) or totally lacking (e.g. for diploid sperm), or (c) no clear pattern has yet emerged from our analyses (e.g. for sperm with cytoplasmic droplets or bent mid-pieces). Nor have we yet considered different interactions between head, tail and mid-piece morphs (e.g. the roles of macros with and without cytoplasmic droplets; tapering sperm with short and normal tails).

Although sperm are the most obvious cells in the ejaculate, a number of non-sperm cells are also present and often abundant. These include undifferentiated germ cells and a range of phagocytes. We suspect that these may also have a direct role in sperm competition. Undifferentiated germ cells may well act as energetically cheap decoys, distracting other males' sperm and/or the female's phagocytes. The male's own phagocytes, of course, could play a much more direct role in attacking sperm from other males. We have begun a study of these cells to complement our investigation of sperm polymorphism but as yet have insufficient data to discuss in this book.

11.4 How do males adjust level of polymorphism?

One of the most intriguing aspects of our analysis of polymorphism is that even within successive ejaculates from individual males, the proportion of different sperm morphs varies. Moreover, the proportion varies in a way that relates to circumstances such as the risk of sperm competition and the dictates of family planning. The implication is that males customize their ejaculate to have the optimum proportions of the different types of sperm for the circumstances surrounding the copulation. The mechanisms by which this customization might be achieved deserve further discussion.

In fact, it is not necessary to postulate very complex mechanisms, five simple elements alone being enough to generate all the variation we have observed. These are: (a) morph differentiation during spermatogenesis and maturation; (b) ontogenetic transformation; (c) selective reabsorption or phagocytosis; (d) varying the size of the ejaculate; and (e) environmentally induced transformation.

11.4.1 Morph differentiation during spermatogenesis and maturation

Through examining the timing of the influence of antiandrogen drugs on a male sex-offender, Fredriccson (1978) concluded that the different head morphologies of sperm were differentiated during spermatogenesis, probably at the spermatid stage.

A less drastic study of the influence of illness-induced stress on sperm morphology by MacLeod (1951) showed that an increase in tapering and amorphous sperm first appeared in ejaculates 20–40 days after the illness (chickenpox or pneumonia) and disappeared about 20–40 days later. This implies that stress (in this case perhaps simply high body temperature) acted to promote the production of tapering and amorphous sperm during the later stages of spermatogenesis but before the sperm arrived in the epididymis (section 2.4.3).

In our longitudinal study of six healthy males, we were able to calculate the level of association over different time intervals between: (a) self-assessed stress level (recorded daily on a scale of 1–5 as a diary); and (b) the proportion of each morph in masturbatory and copulatory ejaculates. We analysed morph proportions relative to stress: (a) on the day of ejaculation; (b) 1–5 days before ejaculation; (c) 6–15 days before ejaculation; and (d) 16–72 days before ejaculation.

These timings more or less correspond to the times that sperm were, respectively: (a) about to be ejaculated; (b) in store in the vas deferens and perhaps cauda epididymis; (c) undergoing maturation during passage through the epididymis; and (d) undergoing spermatogenesis in the seminiferous tubules. To allow for missing diary data, we used the means of daily stress levels for categories (b), (c) and (d). Of course, such an approach can only identify the time of differentiation of those morphs which are influenced by stress levels.

The results are summarized in Box 11.12 and, for the morphs concerned, are more or less in total agreement with the more traumatic studies by MacLeod (1951) and Fredricsson (1978). As a result, we suggest that differentiation into tapering, pyriform, amorphous, pinhead and oval-headed morphs is initiated during early spermatogenesis. The differentiation of all tail morphs, bent mid-piece and bicephalous morphs is initiated towards the end of spermatogenesis. The differentiation of oval-headed sperm into macros, modals and micros, however, does not seem to take place until maturation in the epididymis.

Variation in the proportion of the different morphs with male age (Bujon *et al.*, 1988; Box 11.6) is probably achieved by long-term changes in differentiation during spermatogenesis. However, just because the differentiation of many of the morphs is initiated in the testis does not mean that the proportion of sperm cannot change with respect to circumstance. The very fact that stress has such far reaching influence on the proportion of the different morphs (section 11.3.4) shows this clearly.

11.4.2 Ontogenetic transformation

Andrologists have long interpreted sperm with cytoplasmic droplets to be an ontogenetic phenomenon (e.g. Mann and Lutwak-Mann, 1981). To this example, we can now add two more: (1) sperm with coiled tails; and (2) the size polymorphism of oval-headed sperm.

(a) SPERM WITH CYTOPLASMIC DROPLETS

When a sperm enters the epididymis, the neck region is still surrounded by the collar of cytoplasm it has had since the end of spermatogenesis in the testis (Mann and Lutwak-Mann, 1981). This cytoplasm contains characteristic flattened vesicles derived from the Golgi region. On entering the epididymis, some of the cytoplasm is lost leaving a cytoplasmic droplet attached to the front end of the mid-piece. As the sperm travels along the epididymis and maturation takes place, the cytoplasmic droplet slowly migrates along the mid-piece from head end to tail end. Then, before the sperm enters storage, the droplet is most often shed. Some sperm, however, retain their cytoplasmic droplet through storage, ejaculation, and passage in the female (at least into the cervical mucus).

Our stress study suggested that the shedding or persistence of cytoplasmic droplets through ejaculation was determined during maturation in the epididymis (Box 11.12). However, neither Fredricsson (1978) nor ourselves found any indication that frequent ejaculation increased the proportion of cytoplasmic droplets in the human ejaculate. It is not simply, therefore, that the sperm do not have time to shed their droplets before ejaculation.

Box 11.12 Male control of sperm polymorphism: the time and the place

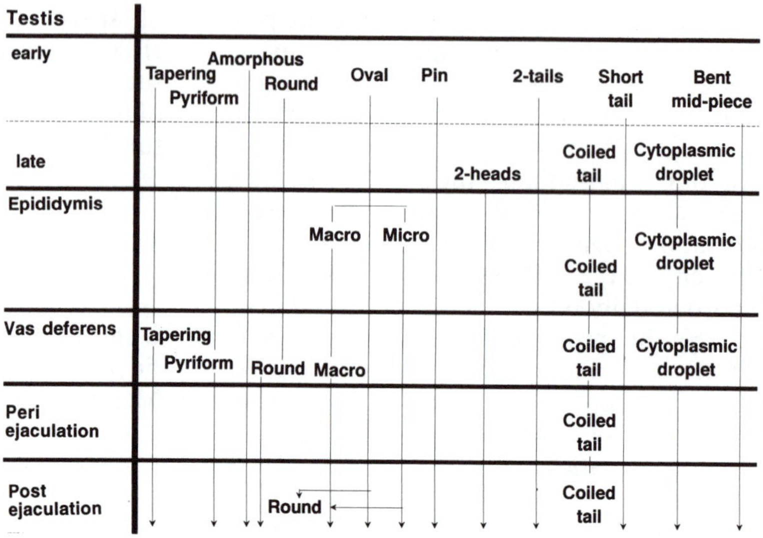

Summary of the locations in the male tract where the proportions of the different sperm morphs may be influenced. For full description of evidence for the different events see main text.

The relative abundance of most of the different morphs seems first to be determined during early spermatogenesis in the testis. Either specific germ cells are programmed to produce particular morphs or the variable (e.g. hormone profile) aspects of the testicular environment influences the proportion of germ cells or spermatogonia that will develop along particular morphological lines. Some lines of development (two-heads, coiled tails, persistent cytoplasmic droplets) are more influenced by their testis environment later in spermatogenesis, perhaps during spermatogenesis.

Testis
early
Amorphous
Tapering
Pyriform
Round
Oval
Pin
2-tails
Short tail
Bent mid-piece
late
2-heads
Coiled tail
Cytoplasmic droplet
Epididymis
Macro
Micro
Coiled tail
Cytoplasmic droplet
Vas deferens
Tapering
Pyriform
Round
Macro
Coiled tail
Cytoplasmic droplet
Peri ejaculation
Coiled tail
Post ejaculation
Round
Coiled tail

The size distribution of oval-headed sperm seems to be determined during maturation in the epididymis. The probability of sperm eventually coiling their tail and shedding their droplet is also influenced during maturation.

The proportion of sperm in the ejaculate which will have a cytoplasmic droplet or will coil their tails (or will have coiled tails), is again influenced during the sperm's period of residence in the vas deferens, awaiting ejaculation. So, too, is the proportion of macros, pyriform and round-headed sperm. The most likely mechanism is differential phagocytosis.

Only the proportion of coiled tail sperm seems to be influenced by circumstances around the time of ejaculation. Finally, after ejaculation, sperm may coil their tails and oval-headed sperm may transform into round-headed sperm, depending on the environment in which they find themselves.

Nor does the ontogenetic history of cytoplasmic droplets end with ejaculation. Our studies of 132 ejaculates in HTFM (Box 11.3) show a significant decline in the proportion of sperm with cytoplasmic droplets with time after ejaculation ($P < 0.001$). The proportion remains constant for the first 24 h then declines sharply over the next 48 h. However, there is no increase in the rate of loss of cytoplasmic droplets in heterospermic as opposed to homospermic mixes (Box 11.8) and so far, our studies give no indication of the nature of any postejaculation role for sperm with cytoplasmic droplets.

(b) SPERM WITH COILED TAILS

Previous authors have already concluded that coiled tail sperm are senescent sperm and that tail coiling, if it occurs, does not take place until the sperm are at least as near the vas deferens as the epididymis (Fredricsson, 1978). Our stress study also suggests that tail coiling, if it occurs before ejaculation, takes place in the epididymis and vas deferens (Box 11.12). This is supported by the observed influence of interejaculation interval on the proportion of coiled tails (Box 11.4), the ejaculate containing a higher proportion if abstinence time since last ejaculation is longer.

In this book, we have presented evidence that tail coiling is in fact part of an ontogenetic transition from a mobile sperm with an active role to a more sedentary sperm with a blocking role (section 11.3.1). Two major questions remain: (1) do all sperm have the potential to become coiled tail sperm as they age; and (2) if not, what is their active role before coiling their tail?

Our own data suggest that at least 25% of

sperm are programmed to transform into coiled tail forms as they age and that the number increases with the male's age beyond 25 years. Our data also suggest that the older sperm near the urethral end of the vas deferens are more likely to have coiled tails than younger sperm towards the epididymis. However, our stress study suggests that the number of sperm which will coil their tails is first influenced by conditions while the sperm are undergoing spermatogenesis. On the whole, therefore, our data suggest that a proportion of sperm are preprogrammed to coil their tails under appropriate circumstances. Whether this proportion is really as low as 25% or is much higher we cannot yet determine.

(c) SIZE POLYMORPHISM OF OVAL-HEADED SPERM

The final ontogenetic effect to emerge in our data is perhaps the most interesting and important. It was first noticed but misinterpreted by Fredricsson (1978) who suggested that a higher proportion of macros were found among dead sperm, not because they had lower survival than other oval-headed sperm (Box 11.7) but because the heads of such sperm swell after death.

If Fredricsson's hypothesis were correct, not only would the proportion of macros be greater among dead sperm, the proportion of macros in the ejaculate would gradually increase with time after ejaculation. We have checked this possibility in ejaculates from 132 males (20 300 sperm; maintained in HTFM after washing and swim-up) and found a very strong tendency for such an increase ($P < 0.001$). Contrary to Fredricsson's hypothesis, however, the increase with time occurred only in the population of live sperm ($P = 0.004$) not in that of apparently dead sperm ($P = 0.453$). Moreover, the increase was complete by 24 h after ejaculation, there being no further increase in proportion nor an increase from older live sperm to dead sperm (Box 11.7).

Such an increase is only possible if some other head morph in the ejaculate converts into the macrocephalous morph with time. The only morph in our data to show a corresponding decrease in proportion is the microcephalous morph ($P < 0.001$). However, at present we think it is more likely, but cannot yet prove, that micros convert into modals and some modals convert into macros rather than that micros convert directly into macros. In other words, we suggest that there is a gradual increase in the head size of oval-headed sperm with time after ejaculation. This now requires investigation using actual micrometric measurements to supplement our current morphometric approach.

In our present data set, the most sensitive measure of this ontogenetic increase in head size of oval-headed sperm is the ratio of macros to micros (the macro:micro ratio). We have used this ratio to investigate further this apparent morphological maturation of human sperm.

There is a significant increase in macro: micro ratio ($P = 0.036$) in response to stress levels experienced by the male during the period that sperm are in the epididymis (section 11.4.1) suggesting that the proportion of any given cohort of sperm that matures into macros before ejaculation is first determined during epididymal maturation. However, there is no influence of stress levels while the sperm are in the vas deferens ($P = 0.810$). Nor is there any influence of the length of time the sperm are in the vas deferens (measured by abstinence time since last ejaculation). This is so whether the ejaculates examined are masturbatory ($P = 0.130$) or copulatory ($P = 0.504$). In fact, in both types of ejaculate the non-significant trend is for older ejaculates to have a lower macro: micro ratio. Finally, there is no indication that older sperm (i.e. longer in the vas deferens) convert into macros faster, once ejaculated, than younger sperm. Once again, any tendency is for the opposite to occur and, after 4 days in HTFM, the trend is significant ($P < 0.001$) Four days after ejaculation, sperm populations which spent less time in the vas deferens actually contain a higher proportion of macrocephalous sperm.

There is thus no indication in our data that the conversion of smaller, oval-headed sperm into macros is in any way part of senescence (cv coiled tail sperm; section 11.3.1). In which case, the conversion could be some form of postejaculatory maturation. However, such maturation, apparently begun in the epididymis but suspended in the cauda epididymis and vas deferens, does not seem to resume while in the seminal plasma. Flowbacks show no increase in macro:micro ratio with time in the female vagina between insemination and flowback ($P = 0.773$), the non-significant trend actually being in the opposite direction.

We do not observe a change in macro: micro ratio, therefore, unless or until sperm are placed in HTFM. This means that the change in head size that we observe is either an artefact of the chemical environment provided by HTFM or that, *in vivo*, it does not begin until sperm encounter an environment in the female that is mimicked by HTFM (i.e. the uterus and/or oviduct). Evidence in support of the latter is provided by our studies of rats (Bellis *et al.*, 1990) in which the mean head length of sperm in the uterus was significantly greater than that of sperm in the vagina and cervices.

11.4.3 Selective reabsorption and/or phagocytosis

Any change in polymorphism during spermatogenesis, even when supplemented by post-ejaculation ontogenetic change, can only be adaptive to long-term factors, such as stress, family planning and age- and season-related factors. This is because there is a 72 day or so gap between the beginning of spermatogenesis and ejaculation of the sperm produced. The later in spermatogenesis that polymorphism is influenced, the faster the response to environmental circumstance can be, but even the fastest response can only be as short as 2–3 weeks. Any response to short-term factors (e.g. risk of the female containing sperm from another male) require mechanisms to change the polymorphism first established during spermatogenesis at much shorter notice such as while the sperm are in the cauda epididymis and vas deferens awaiting ejaculation.

Many more sperm are produced by the testes than are normally ejaculated (Barratt and Cohen, 1986) and there is now strong evidence from studies using immunocytochemical staining techniques that phagocytosis of sperm occurs and is selective (Tomlinson *et al.*, 1992).

Leucocytes are present throughout the male reproductive tract and are found in almost every human ejaculate (Tomlinson *et al.*, 1990). Phagocytic cells predominate. Three types of seminal phagocytic cells have been found to contain sperm (Tomlinson *et al.*, 1992): small polymorphonuclear leucocytes (c. 10–12 μm), monocytes of similar size and much larger macrophages (30–40 μm) capable of engulfing multiple sperm heads.

In the study by Tomlinson *et al.* (1992) there was a strong association between the abundance of each of these cell types and the relative abundance of different sperm head morphs but not with the abundance of the different sperm tail and mid-piece morphs. These data, along with repeated observation of sperm or sperm fragments within phagocyte cells, support the hypothesis that leucocytes have a role in the differential removal of sperm head morphs while the sperm are waiting in the vas deferens to be ejaculated.

Apart from an increase, presumably 'ontogenetic', in the proportion of coiled tails with increased time in the vas deferens (Boxes 11.5 and 11.6), the only morph to show a significant change in proportion with abstinence time is the round-headed morph which decreases in proportion the longer the sperm are in the vas deferens (Box 11.5). The implication is that some round-headed morphs are present as each cohort of sperm move from the epididymis into the vas deferens but that, as the cohort waits for ejaculation, the round heads are selectively phagocytosed. This again implies that such sperm have no further role (sections 11.3.3. and 11.4.5).

Although neither the macro:micro ratio nor the MTP ratio changes as a function of time (= abstinence time) that the sperm are in the vas deferens, the ratios do vary as a function of the proportion of time that the male spends with his female partner. Both ratios decrease, the macro:micro ratio significantly ($P = 0.007$), the MTP ratio not quite significantly ($P = 0.062$). Perhaps the most likely explanation of these changes is that when males spend more time with their partner there is selective phagocytosis of the macrocephalous among oval-headed sperm. This selection is nearly matched by selective phagocytosis of tapering and pyriform sperm (Box 11.5), leading to a much less-marked decrease in MTP ratio than macro:micro ratio. The result is an ejaculate adaptive to the risk that the female partner already contains sperm from another male.

Our various sources of data suggest that there is selective phagocytosis of round, macro, tapering and pyriform head morphs under at least some sociosexual circumstances while the sperm are waiting in the cauda epididymis and vas deferens. Overall, therefore, there should be a positive correlation between the number of phagocytes deployed in the vas deferens in the production of any given ejaculate and the proportion of modal oval-headed sperm but no correlation with mid-piece and tail morphs. This is precisely the pattern found by Tomlinson *et al.* (1992).

11.4.4 Varying the size of the ejaculate

Differential production of sperm morphs during spermatogenesis allows the male to customize his ejaculates according to long-term and enduring changes in circumstance, such as those related to stress, age, etc. Differential phagocytosis allows the male to customize his ejaculate to circumstances that develop between copulations. Neither system, however, would seem to allow the male to customize his ejaculate to circumstances that change suddenly at about the time of copulation.

When a male spends little time with his partner between copulations, there is probably little difference between the ejaculates that would be optimum for his partner and for an extra-pair female (except perhaps for more coiled tail blocking sperm in the latter because he is less likely to be the next male to inseminate her). The biggest problem in ejaculate optimization for the male, however, arises when, having spent a high proportion of his time with one female, he suddenly has the opportunity to inseminate a female with whom he has had little association. In such circumstances, the morph distribution that is optimum for the latter (i.e. high macro:micro and MTP ratios and, because he is unlikely to be the next male to inseminate her, a high proportion of coiled tails) is the opposite of the morph distribution that would be optimum for his partner.

In fact, we have no direct evidence for humans that the male is able to make such an adjustment. Our data for rats, however, suggest that the male can make a rapid adjustment in sperm polymorphism depending on whether he is offered a female which he has guarded or one that he has not (Bellis *et al.*, 1990). We assume that male humans have a comparable ability.

The secret of such rapid adjustment seems to lie with the number of sperm that the male inseminates. When the risk that the female already contains sperm from another male is higher, the male inseminates more sperm (Box 9.2). For simplicity of description, let us assume that small ejaculates are derived entirely from sperm in the vas deferens whereas larger and larger ejaculates in addition recruit progressively more and more sperm from the cauda epididymis.

Suppose that the sperm in the tail of the epididymis exist in the morph ratio appropriate for high risk of sperm competition. Suppose further that those in the vas deferens, after selective phagocytosis, exist in the ratio appropriate for insemination of the female with whom the male has spent most of his time since his last copulation. All potential coiled tail sperm are in the vas deferens so that, no matter how many sperm are inseminated in total, the number of coiled tail sperm stays constant (Box 11.6). By altering the number of sperm loaded and inseminated (Box 9.12), the male then automatically inseminates an ejaculate with a morph distribution appropriate to the risk of sperm competition.

So far, we have considered masturbation only as a mechanism for reducing the age of sperm in the vas deferens (section 9.4.4). In particular, masturbation will influence the proportion of sperm with coiled tails. However, the above model for adjustment of sperm polymorphism suggests an additional function for masturbation. Phagocytosis is suggested to produce a morph distribution in the vas deferens appropriate for insemination into the female with whom the male spends most time and into whom, therefore, he would inseminate fewest sperm if it were her he next inseminated. Change of circumstance (e.g. sudden separation from that female or beginning to associate with another female) renders the morph distribution in the vas deferens inappropriate. Masturbation allows the male to shed that cohort and begin selective phagocytosis on a new cohort which can be customized to suit current circumstances.

Such a system, in combination with differentiation during spermatogenesis, epididymal maturation and differential phagocytosis in the vas deferens/cauda epididymis as already described accommodates all of the ejaculate adjustments encountered in our investigations so far (e.g. 'reaching further back' for bigger ejaculates when the risk of sperm competition is high).

11.4.5 Environmentally induced transformation

Our analyses have provided three apparent examples of environmentally induced transformation: (1) sperm coiling their tails in response to chemical challenge; (2) the staggered change in size polymorphism of oval-headed sperm on encountering uterine or oviductal conditions; (3) round-headed sperm being produced in the presence of sperm from another male.

(a) COILED TAIL SPERM

It has been known for some time that at least some sperm coil their tails when exposed to a variety of chemical challenges, particularly osmotic stress (Drevius, 1975). Older sperm are more likely to respond and respond faster. This is probably part of the mechanism by which sperm at the back of the ejaculate are induced to coil their tail (a response to secretions from the seminal vesicle) and at appropriate locations in the female tract (a response to local osmolarity, pH and chemistry, etc.).

(b) OVAL-HEADED SPERM

The postejaculatory increase in size of the head of oval-headed sperm which we have inferred from our morphometric studies seems to be a primarily ontogenetic development that is initiated in the epididymis. Nevertheless, it is possible that after development has been suspended in the vas deferens, seminal plasma and vagina, further development is not triggered until the sperm encounter uterine and/or oviductal fluids. In which case the phenomenon is partly a reaction to the sperm's environment.

(c) ROUND-HEADED SPERM

Round-headed sperm increase in number in the presence of sperm from another male. If, as we suggest, at least some of these sperm are produced by oval-headed sperm shedding their acrosome, they are an environmentally induced transformation. It is possible, however, that having shed their acrosome the sperm's role is finished and that the round-headed sperm that they have become has no further part to play in sperm competition.

11.5 The kamikaze sperm hypothesis: an evaluation

So far, two authors have critically sought to evaluate the kamikaze sperm hypothesis (KSH), using their own interpretation of its predictions (Harcourt, 1991; Møller, 1988c).

Møller (1988c) considered four aspects of an ejaculate: total number of sperm; ejaculate volume; the proportion of 'normal' sperm; and the proportion of motile sperm. He argued that all four characteristics were positively associated with fertility, both in the presence and absence of sperm competition. His analysis of mammalian ejaculates suggested that all four characteristics change together during evolution and Møller argues that the implication is that these different measures of ejaculate quality in mammals have been improved simultaneously, apparently by a common selective force, presumably sperm competition.

As far as the KSH is concerned, Møller concludes that if the 'normal' sperm are fertilizing sperm, his analysis provides evidence that is inconsistent with the hypothesis. On the other hand, as we have argued previously (Baker and Bellis, 1989a) and do so again in Chapter 12, if these so-called 'normal' sperm do not have a fertilizing function, his analysis provides evidence consistent with the KSH.

Harcourt (1991) compared various aspects of primate ejaculates (proportion of 'normal' sperm; variance in sperm size; motility; and absolute size) with levels of sperm competition. He concluded that mammalian sperm are adapted to fertilize and solely to fertilize and do so through scramble competition, not contest competition.

Unfortunately, Harcourt's tests of KSH are based on his own misinterpretations of our hypothesis. The most unfortunate misinterpretation means that, unlike Møller (1988c), he does not allow the possibility that the vast majority of 'normal' sperm are actually kamikaze sperm (Baker and Bellis, 1989a; see also section 11.3.3 and Chapter 12). Had he done so then, like Møller, he would have been forced to acknowledge that his analyses may well support, rather than contradict, the KSH. Indeed, one of his predictions (that, if the KSH were correct, species with a higher risk of sperm competition should have larger sperm) for which he had minimal data for primates, has since been shown to be the case for both primates and rodents (Gomendio and Roldan, 1991).

If (as we argue in section 11.3.3) the so-called 'normal' sperm in mammalian ejaculates are the most common kamikaze morph, probably with a generalized seek-and-destroy function, then the papers by Harcourt (1991) and Møller (1988c), in combination with the findings of Gomendio and Roldan (1991) actually support the KSH.

In this book, two further, but more direct, sets of evidence for the KSH have been presented. First, there is a strong indication that sperm from one insemination 'block' and thus hinder sperm retention at the next insemination. Second, when ejaculates from different males are mixed, there is a significant increase in the morbidity and levels of agglutination of sperm, as would be expected if seek-and-destroy behaviour had been triggered.

11.6 The ejaculate as a male organ

Cartoon caricatures of sperm see each as a hopeful fertilizer, frantically racing through the female tract, each attempting to fertilize the egg before another. Even in the absence of intermale sperm competition, most people view an inseminate as a hot-bed of intraejaculate sperm competition. It has even been said that the reason for the competition is so that only the fittest sperm succeed, thus ensuring the fitness of the population.

Even if intramale sperm competition did exist, selection for fast sperm is only that, selection for fast sperm. There is no reason to suppose that fast sperm have any more or less beneficial genes for all other aspects of anatomy, physiology and behaviour than slower sperm. Besides which, if selection had acted on intramale competition, it would be even more difficult to explain the 40% of slow, lame and virtually immobile sperm that still make up the human ejaculate after millions of years of intensive natural selection.

The alternative view is that the ejaculate is not a collection of competing cells but instead is one of the male's organs, fully comparable to the liver, the testes or, perhaps an even better analogy, to the circulating part of the immune system. Admittedly, the ejaculate is an unusual organ. It operates outside of and detached from, the male's body. However, like the circulating part of the immune system, it consists of cells and fluids that are continuously changing their spatial relationships with each other. Also like the immune system, it becomes interspersed with cells and fluids from another individual (the female into whom it is inseminated or the ejaculate of another male with whom it competes). Nevertheless, it remains a male organ with a single purpose, to prevail over both female defences and male opposition in such a way as to have the female's egg fertilized by one of the male's cells.

To achieve this purpose with maximum efficiency then, just like the testis or circulating immune system, it has evolved to consist of a number of cell types, each specialized to a different role. The development and differentiation of the cell types is under the genetic control of the male (section 11.2.2) and so too is the proportion of the different cell types in any one inseminate (section 11.4).

Interpreting the human ejaculate as an organ leaves little room for the concept of intraejaculate sperm competition. Kamikaze and egg-getter sperm do not compete with each other to be fertilizers any more than Sertoli and seminiferous cells compete to be seminiferous cells in the testis, somatic and germ cells compete to be germ cells in the embryo, or monocytes and macrophages compete to be macrophages in the immune system. In each case, the fate of each cell is determined by the diploid instructions which the parent organism itself has inherited from its ancestors.

There is, of course, one sense in which the ejaculate differs from all other organs in the body. Whereas eyes, the liver, kidneys etc. of a diploid organism consist of cells that are all diploid and genetically identical (apart from local mutations), most of the cells in the ejaculate (the sperm) are haploid and genetically different. In this sense, the ejaculate is more comparable to a colony of ants.

Even this analogy is not perfect because the individual workers and soldiers of an ant colony, although non-reproductive (like kamikaze sperm) and although genetically different (like kamikaze sperm) are diploid, not haploid. Nevertheless, the different individuals within a colony do not compete with each other (e.g. for food or survival) nor do they (usually) become reproductive. Their development and role is mapped out for them by the genetic programme they inherit from their parents and the environment in which they find themselves just as, we assume, are the development and roles of sperm.

This does not mean that there has been no selection on the prowess of egg-getters at fulfilling their role, but the selection has acted on the male, not on the sperm themselves. Males who produce egg-getters which are inadequate in some way, either in the absence or presence of sperm competition, will be selected against. Males whose programme produces a set of egg-getters only a minority of which are competent will be less successful than males who produce egg-getters a greater proportion of which are competent. Whether the egg-getter performance of a male's lineage can be selectively enhanced by the fact that it is only the most competent of his egg-getters which fertilizes the egg is an interesting genetic and evolutionary question that must await future analysis for an answer.

Any suggestion, though, that a male's ejaculate would be more efficient if he produced all egg-getters is no more reasonable than suggesting that the immune system should produce only macrophages, the testis should consist only of seminiferous cells, or indeed a male's entire body should consist only of testes. In each case, selection acts on the optimum proportion of different cell types, not the maximum number of one particular type, no matter how important that type may seem to be.

One simple reason why, beyond a certain point, it may be more efficient to produce more kamikaze sperm than more egg-getter sperm is indicated in both sections 11.3.1 and 11.3.3. This is that one kamikaze sperm may be able to take out from competition many more than one sperm from another male. For example, a few blockers in the cervical mucus channels can potentially keep at bay innumerable egg-getters and kamikaze sperm subsequently inseminated by another male (section 11.3.1). Similarly, seek-and-destroy sperm may be able to incapacitate several, or at least more than one, egg-getter (or kamikaze sperm) from another male (section 11.3.3). In both cases, beyond the production of a certain minimum number of egg-getter sperm, a male may always do better to produce more kamikaze sperm than more egg-getter sperm.

12 Human fertility and infertility: the kamikaze perspective

12.1 Introduction

We have concluded (Chapter 11) that the ejaculates of humans and other animals should be conceived as a male organ made up of fluids and various 'tissues' of motile cells (e.g. different sperm morphs, germ cells, phagocytes). The overall function of the organ is to enhance the probability that one of its own cells succeeds in fertilizing the egg(s) of the female into which it has been transplanted. We suggest that, like any organ, the ejaculate's various tissues have evolved to interact in an optimum way and that, again like any organ, optimum interaction will depend on the circumstances in which the ejaculate finds itself.

In the presence of competition from a similar organ from another male, the full diversity of function of the various tissues are brought into play, with attempts at the blocking and destruction of the opponent's tissues. In the absence of competition from another male, much of the organ is redundant. Even then, however, the egg-getter sperm do not have a clear run. We postulate that they are attacked by spiteful 'family planning' sperm (tapering and pyriform) from their own organ (Chapter 11). They are also hindered by the architecture and immunological defence system of the female tract (Chapter 4).

The egg-getter sperm constitute the one tissue within the ejaculate that we have not yet considered. Despite their obvious importance, however, they are surprisingly the sperm morph about which we know the least. In part, this may be because, at any one time, they are one of the least common morphs in the ejaculate.

12.2 How many egg-getters?

12.2.1 What is an egg-getter?

An egg-getter sperm is not simply a sperm which, when placed within an egg, activates development. Such activation does not need a sperm. Some animals show spontaneous egg activation and, in parthenogenetic species, such activation produces a normal individual. In others, a simple pin-prick is sufficient for activation and even human eggs may be activated in this way (Laws-King *et al.*, 1987). Calcium ions injected into the egg of most mammals will activate development (Yanagimachi, 1988) even occasionally to the beating heart stage (Johnson and Everitt, 1988). As an intriguing possibility, maybe even a haploid nucleus from one calcium activated, 'meiosis complete', egg, when injected into another egg, could even achieve activation and at least partial zygote development.

We should not be surprised, therefore, to find that the act of microinjecting any sperm into the egg can often activate development. Such techniques are becoming increasingly used in the context of assisted conception. Originally, the technique was to inject single sperm into the perivitelline space, between the zona pellucida and the vitelline membrane (Laws-King *et al.*, 1987; Lassalle *et al.*, 1987; Yamada *et al.*, 1988). The sperm then only had to penetrate the vitelline membrane to achieve fertilization. However, even this step was more than many sperm seemed able to achieve. More recently, single sperm have been injected directly into the egg cytoplasm. The result has been an increased success at both fertilization (51.0% intracytoplasmic vs 14.3% subzonal; Van Steirteghem *et al.*, 1993) and implantation rates (Van Steirteghem *et al.*, 1993).

Nevertheless, in the current context, just because almost any sperm may fertilize if fertilization is forced, or made sufficiently easy, does not mean that all sperm are egg-getters. To be an egg-getter, a mammalian sperm must have the following abilities. It must be able to:

1. avoid ejection in the flowback by escaping the seminal pool in the vagina;
2. enter the narrow channels in the cervical mucus;
3. avoid or by-pass any block created by sperm (from a previous insemination by the same or another male) and phagocytes (of either male or female origin);
4. enter the uterus (either directly or after a sojourn in the cervical crypts);
5. traverse the uterus (either by its own locomotion or with assistance from uterine contractions);
6. pass through the uterotubal junction (again by-passing any block from a previous insemination);
7. travel the length of the oviduct to the ampulla, perhaps after a second sojourn in the oviductal isthmus;
8. find the egg and pass through the surrounding cumulus;
9. penetrate the egg's last defence, the zona pellucida, and finally achieve the climax of fertilization.

Moreover:

10. throughout this whole journey, the successful egg-getter has to avoid the seek-and-destroy attention of both sperm from other males and even 'family planning' sperm from within its own ejaculate.

Any one of these stages might separate egg-getters from non-egg-getters. Most recent attention, however, has been directed at differences between sperm in their ability to: (1) orient towards; (2) swim to; and (3) penetrate the three outer final layers of the mammalian egg (i.e. the cumulus, zona pellucida, and finally the vitelline membrane, Box 4.6).

When human sperm are placed in a vertical multi-well micro-chemotaxis chamber, more accumulate in wells containing follicular fluid than in control wells containing buffer (Ralt *et al.*, 1991). Moreover, more sperm accumulate in wells containing fluid from the more fertilizable eggs. However, only a very small proportion of sperm from the original ejaculate show such an orientation response though the proportion is increased by selection through swim-up. The implications are that only egg-getter sperm respond to chemical signals from eggs and that these sperm are only a tiny fraction of the original ejaculate.

Working with hamsters and mice, respectively, Bavister (1979) and Thadani (1982) found that eggs stripped of their surrounding cumulus could be fertilized by sperm at very low sperm:egg ratios, sometimes approaching unity (about 4:1). Experiments on the same species with the cumulus intact, however, showed that many of the sperm which arrived at the cumulus failed to penetrate through to the zona pellucida (Tsunoda and Chang, 1975; Siddiquey and Cohen, 1982; Talbot *et al.*, 1985).

In both mouse and human, the number of sperm bound to the zona pellucida is related directly to the chances of fertilization *in vitro* (Mahadevan *et al.*, 1987; Morroll *et al.*, 1993). This could suggest that only egg-getters actually bind to the zona. However, differences in ejaculate fertility, perhaps due to differences in the number of egg-getters, are still evident even when the sperm are presented with zona-free eggs (Morroll *et al.*, 1992).

It appears, therefore, that without assistance (such as microinjection), only programmed egg-getters can achieve fertilization even when confronted with only the vitelline membrane.

12.2.2 So how many human sperm in an ejaculate are egg-getters?

Experiments on mice, using eggs with the cumulus intact, show that *in vitro*, 1000 sperm from the epididymis in 1 μl of medium will fertilize all of the 4–5 eggs with which they are presented (Siddiquey and Cohen, 1982). In such a small volume, collision rate may not be a limiting factor and all sperm should contact an egg. The same preparation also fertilizes 4–5 eggs out of 15 and 4–5 eggs out of 30 (Siddiquey, 1982; Siddiquey and Cohen, 1982). The implication, therefore, is that at any one time 4–5 sperm per 1000 (say 1 in 200) sperm from the epididymis is an egg-getter. This figure is similar to that obtained by Tsunoda and Chang (1975) who also used mouse eggs with cumulus intact and concluded that only one in every few hundred sperm is an egg-getter.

If we extrapolate from mice to humans this figure of one egg-getter for every 200 sperm in the epididymis, we obtain a figure of about 1.5 million egg-getters in the ejaculate (assuming that the proportion of egg-getters does not change between the end of maturation in the epididymis and the moment of ejaculation). This figure may actually be on the high side as Quinn and Marrs (1991) showed that thousands of human sperm were needed for fertilization. However, they were using moderately large volumes of medium and collision rate between sperm and egg may have been a limiting factor. As a nice round working figure, 1 million egg-getter sperm at ejaculation seems reasonably consistent with the available evidence.

12.2.3 How many egg-getters reach the oviduct?

The vast majority of inseminated sperm never reach the oviduct. From a human ejaculate of about 300 million sperm, only about 20 000–30 000 (c. 1 in 10 000) ever pass through an oviduct (Chapter 8), perhaps most of these ending up in the female's body cavity. At any one time only 2000–5000 are present in the whole oviduct and only about 300 (c. 1 in a million from the original ejaculate) are at the site of fertilization in the ampulla. Such proportions seem about typical for mammals (Johnson and Everitt, 1988).

We could calculate the number of egg-getters which reach the site of fertilization simply by assuming that the sperm which arrive are a random selection of the sperm which were inseminated. Both intuition and evidence, however, argue that this procedure would underestimate egg-getter success.

In an elegant study, Cohen and McNaughton (1974) allowed a male rabbit with a particular genetic coat marker to inseminate a female. About 10–1000 sperm from this male were then removed from the female's oviduct. They were then mixed with 10^5–10^7 washed sperm from another male (with a different genetic coat marker) and artificially reinseminated into the uterus of another female. Over a series of trials, sperm from the oviduct of the first female succeeded in obtaining some fertilizations in the second female. Sperm from the first female's oviduct achieved absolutely fewer, but proportionately more, fertilizations than randomly selected sperm which had not previously travelled through a female tract.

This study demonstrates a number of points. However, the two that concern us here are: (1) the population of sperm which reaches the oviduct contains a higher proportion of egg-getters than the initial population in the ejaculate; and (2) potential egg-getters which reach the oviduct are by no means 'exhausted', being quite capable of completing the journey, at least from the uterus, all over again in another female.

An increase in the proportion of egg-getters as sperm migrate from vagina to oviduct has also been demonstrated for mice.

In contrast to sperm from the epididymis (one egg-getter in every 200 sperm), sperm taken from the oviduct appear to consist of egg-getters in the proportion of about one sperm in seven (Siddiquey and Cohen, 1982). By inference, therefore, sperm in the oviduct consist of non-egg-getters in the proportion of about six sperm in seven.

The kamikaze sperm hypothesis would not expect all the sperm which reach the oviduct to be egg-getters. The oviduct and beyond, in the body cavity, should be an arena ripe for the attention of seek-and-destroy sperm. What better place for an egg-getter sperm to wait for the female to ovulate than as near to the site of egg-release as possible, perhaps even to follow the egg back into the oviduct to the ampulla until the egg is ready to be penetrated? Where more important, therefore, for a seek-and-destroy sperm to search for and attempt to neutralize any rival egg-getters than in the body cavity between the ovary and the fimbria, as well as in the ampulla and the remainder of the oviduct?

Extrapolating the mouse data to humans, we can suggest that of the 20 000–30 000 sperm which pass through the oviduct, about 4000 (1/7) are likely to be egg-getters. Moreover, of the 300 found in the ampulla at any one time, about 40 will be egg-getters. Again, these figures receive some independent support. Lee (1988), in discussing the influence of successive IVF swim-ups on the fertility of human sperm, concluded that there may be as few as 2000 egg-getter sperm available for fertilization.

For present purposes, therefore, let us assume that out of an ejaculate of 300 million sperm, about 1 million start as egg-getters and that about 4000 egg-getters (i.e. about 1 in every 250 sperm inseminated) will pass through the oviduct. These deceptively simple figures, however, are just the beginning of the discussion, not the end.

12.2.4 Which morph is the egg-getter: the macro credentials

In Chapter 11, we presented evidence that the fertility of an ejaculate was a function of the proportion of macrocephalous oval-headed sperm it contained. More specifically, we argued that fertility was a strong function of the ratio of macros to what we have termed the 'family planning' sperm, those with tapering or pyriform heads.

While acknowledging that the MTP ratio may well simply reflect the action of some other factor, we presented some evidence in Chapter 11 that there could actually be a direct interaction between macros on the one hand and tapering and pyriform sperm on the other. The most parsimonious interpretation of that interaction was that macros were egg-getter sperm and that they were the specific target for the seek-and-destroy activity of the 'family planning' sperm in their own ejaculate.

On average, about 0.3% of sperm in the human ejaculate are macrocephalous (Belsey *et al.*, 1980), giving a figure of about 1 million in an ejaculate of 300 million, the same as calculated for egg-getter sperm. Moreover, macros have the highest upward mobility (Bellis *et al.*, 1990b) of all sperm morphs. They are also, as would be expected for egg-getters, the primary targets of other sperm in heterospermic mixes.

We propose here that the macro morph, or at least a temporal subset of the macro morph population (see below), is the human egg-getter.

12.3 Maturation, ejaculation, capacitation and fertilization: three views of sperm ontogeny

12.3.1 The standard view: a steady trickle of ripe sperm

While developing in the testes, sperm are incapable of fertilization. As they pass through the epididymis, the ability to fertilize slowly develops, as do the first signs of motility (Yanagimachi, 1988). The first sperm to become fertile as the sperm mature do so either in the head (caput) of the epididymis (e.g. pigs), main body (corpus) of the epididymis (e.g. rats), or, in most mammals, in the tail (cauda) of the epididymis. Almost certainly, in all species, the process is one of gradually more and more sperm becoming capable of behaving as egg-getters. As we have seen for mice, however, even in the cauda epididymis, only about 1 sperm in 200 is immediately capable of behaving as an egg-getter.

In most mammals, there is no further increase in the proportion of sperm immediately capable of behaving as egg-getters once they have reached the cauda epididymis (Yanagimachi, 1988). At ejaculation, however, the proportion of sperm immediately capable of behaving as egg-getters drops dramatically. In rats, for example, in one test 59% of eggs were fertilized by epididymal sperm but only 21% by freshly ejaculated sperm (Shalgi *et al.*, 1981). In pigs and cats recorded decreases are from 75% and 100% respectively to zero (Hammer *et al.*, 1970; Nagai *et al.*, 1984; Niwa *et al.*, 1985).

This decline is generally thought to be due to the sperm being coated with large molecules (antigens) while in the seminal plasma (Yanagimachi, 1988). These large molecules serve the function of stabilizing the acrosome and reducing the likelihood of spontaneous acrosomal reaction (Chapter 11) thereby temporarily reducing the sperm's ability to penetrate an egg.

As sperm migrate through the uterus, pass through the uterotubal junction, and then migrate along the oviduct, first the uterine environment, then the oviductal, strips away these large molecules, once again preparing the sperm to show the acrosome reaction. This preparation is capacitation (Chang, 1951; Austin, 1951), the process of rendering sperm capable of showing the acrosome reaction (Austin, 1967; Bedford, 1970).

Towards the end of capacitation, before showing the acrosome reaction, sperm of at least some mammalian species (e.g. hamster, mouse, bat, rabbit, dolphin, dog, sheep, cattle, pig, rhesus monkey, human; see references in Yanagimachi, 1988) begin to show a particular swimming action, known as hyperactivation. The behaviour involves a unique movement characterized by a very vigorous whiplash-like beating of the tail, while the sperm head traces an erratic figure of 8. This non-progressive 'dancing' movement is interspersed with brief linear 'dashing' movements.

In vivo, sperm begin to show hyperactivation some time after leaving the oviductal isthmus but before entering the ampulla. Hyperactivated motility can be seen through the wall of the ampulla about the time of fertilization (Yanagimachi, 1970; Katz and Yanagimachi, 1980) and hyperactivated sperm have been collected from the ampullae of sheep, rabbits, and guinea pigs (Yanagimachi and Mahi, 1976; Cooper *et al.*, 1979; Cummins, 1982).

A number of studies have attempted to measure the minimum exposure to the female tract necessary for the capacitation of sperm of various mammals. Some figures produced are 5–6 h for rabbits, 2 h for hamsters and 1 h for mice (Yanagimachi, 1988). However, some epididymal and freshly ejaculated sperm, never exposed to the female tract, fertilize eggs in artificial media. Moreover, human sperm can show the acrosome reaction immediately on encountering an appropriate glycoprotein (ZP3) from the zona pellucida (Barratt, personal communication). Similarly, hamster sperm can capacitate immediately on contact with cumulus cells from around an egg without any contribution from the female tract (Gwatkin *et al.*, 1972).

Yet, even in hamsters, when sperm do contact the cumulus cells around the egg, the majority, at least *in vitro*, cannot penetrate and become trapped in the cumulus and corona. A small proportion, however, presumably mainly egg-getters, simply swim straight through to the zona (Talbot *et al.*, 1985).

When sperm reach the zona pellucida, having negotiated the cumulus, they can only penetrate further if they no longer have an acrosome. The fact that sperm must complete the acrosome reaction before entering the zonas which was first noticed by Austin and Bishop (1958), has been confirmed by all subsequent investigators (see review by Yanagimachi, 1988), and seems to be one of the few absolutes in sperm biology. Clearly, the egg is adapted to exclude sperm still carrying the main bulk of acrosomal material.

Just how the sperm penetrates the zona pellucida to achieve the climax of fertilization

Box 12.1 Two hypotheses for the nature of egg penetration in eutherian mammals

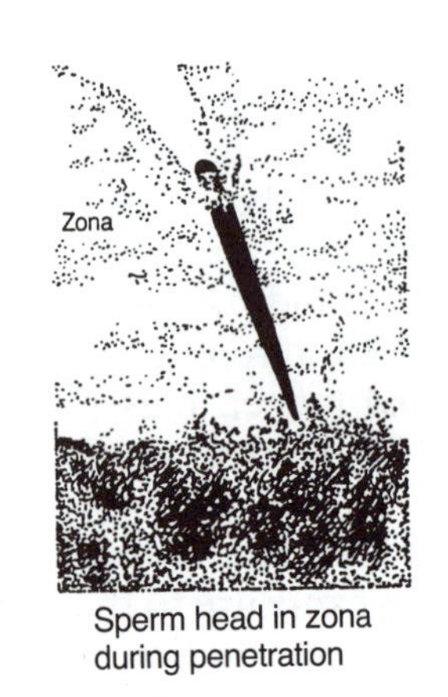

Sperm head in zona during penetration

When a sperm of a eutherian mammal arrives at an egg, it has in most species three more obstacles to negotiate before it can fertilize. It has to: (1) penetrate the cumulus; (2) penetrate the zona; and (3) fuse with the vitelline membrane. There are two major hypotheses as to how the sperm might achieve steps (1) and (2). These are: (a) the mechanical hypothesis; and (b) the enzymatic hypothesis.

The mechanical hypothesis
Sperm penetration through the egg vestments is purely mechanical. The sharply pointed perforatorium (the inner acrosomal membrane underlined by the subacrosomal material), which is exposed by the acrosome reaction, cuts open the zona mechanically as the sperm beats its tail vigorously and moves its head from side to side.

The enzymatic hypothesis
Every step of sperm passage through the egg vestments is enzyme dependent. Surface hyaluronidase and/or acrosomal hyaluronidase released during sperm passage through the cumulus lubricates the sperm as it swims. Some other enzymes on the sperm surface bind the sperm to the surface of the zona. Most of the acrosomal enzymes are released by the sperm onto the zona surface and some of the acrosomal hyaluronidase digests the hyaluronic acid in the outer region of the zona. Acrosin, although it does not dissolve the zona, softens the outer surface. Some of the acrosin and perhaps some other acrosomal enzymes remain bound on the inner acrosomal membrane. These enzymes cleave zona molecules as the inner acrosomal membrane is pushed forward by vigorous movement of the sperm.

Evidence for the mechanical hypothesis
1. The force exerted by the sperm can, in theory, be as great as 100 pN. This is enough to induce stress relaxation of zona glycoproteins which can behave as a viscoelastic fluid.
2. The penetration slit the sperm leaves in the zona has very sharply defined borders as if the zona has been 'cut' rather than 'dissolved'.
3. Rabbit zonae treated to be resistant to trypsin and acrosin digestion are still readily penetrated by sperm.
4. Proteinase inhibitors block binding of the sperm to the zona, but once binding is established they cannot prevent sperm from passing through.

Evidence against the enzymatic hypothesis
1. Sperm lacking hyaluronidase are nevertheless able to pass through the cumulus.
2. The enzymes released by the acrosome on the zona surface might possibly be enough to soften the zona through its entire depth when the zona is thin (e.g. 1–2 μm, as in marsupials and birds) but is unlikely to be enough when

is still far from clear (Box 12.1). At present, the weight of evidence seems to favour a purely mechanical mechanism, with perhaps no other enzymatic involvement than initially to bind the sperm head to the zona and perhaps soften the outer zona surface.

12.3.2 Cohen's view: a tale of two sperm

According to the standard view, differences in rate of transport through the female tract and differences between sperm in capacitation time are adaptive to achieve a steady trickle of precisely 'ripe' sperm from each copulation as the female tract awaits ovulation (Braden and Austin, 1954a; Hunter, 1991). Any morphologically normal sperm can progress through the changes involved in capacitation, hyperactivation, acrosome reaction, and fertilization if it arrives at the right place at its precisely ripe time. It is clear, however, that sperm life is not that straightforward. Sperm are not equally programmed to travel through the female tract and achieve fertilization. Differences exist.

A focal piece of evidence comes from the work by Talbot *et al.* (1985) on the ability of some hamster sperm, but not others, to penetrate the cumulus. Traditionally, it was thought that sperm penetrate the cumulus by using the acrosomal enzyme, hyaluronidase, released by the acrosome reaction, to separate the cell matrix. Hence the sperm which

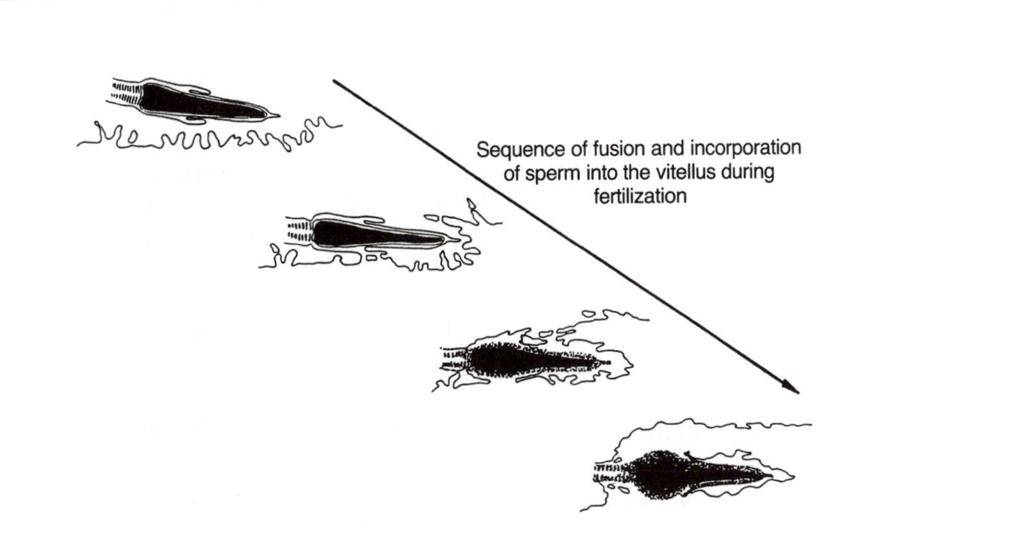

the zona is thick (e.g. 5–27 μm, as in eutherian mammals).

Sequence of fusion and incorporation of sperm into the vitellus during fertilization

At present, the balance of evidence is in favour of a mechanical penetration of cumulus and zona by the sperm of eutherian mammals, with some chemical involvement during the initial binding of the sperm to the zona before penetration.

Sperm–egg fusion
After passing through the zona, the sperm head crosses the perivitelline space and becomes attached to the vitellus, into which it gradually becomes incorporated. In marsupials, as in marine invertebrates, lampreys, amphibians, and birds, sperm–egg fusion begins between the inner acrosomal membrane and the egg plasma membrane. In eutherian mammals, it is plasma membrane over the equatorial region on the sperm head, not the inner acrosomal membrane, that first fuses (as shown) with the egg plasma membrane. This equatorial plasma membrane can only fuse with the egg plasma membrane if the sperm has undergone the acrosome reaction. With a few exceptions, entry of the sperm head into the egg is followed by the gradual incorporation of the entire sperm tail.

For full review and references, see Yanagimachi (1988). Figures redrawn and simplified from Bedford and Cooper (1978) and Huang and Yanagimachi (1984).

Talbot *et al.* observed to swim straight through the hamster cumulus were likely tobe those which arrived at precisely the right moment to be able to release their hyaluronidase. Yet Talbot *et al.* (1985) showed convincingly that hyaluronidase was not involved. Sperm from the toad *Xenopus*, which lack acrosomal hyaluronidase, penetrated the hamster cumulus with ease. So, too, did sea urchin sperm and even cells of the alga *Chlamydomonas*, both of which also lack hyaluronidase.

Talbot *et al.*'s important paper suggests strongly that the cumulus is keeping out sperm with some positive characteristic but letting through almost anything which is neutral for that characteristic. Cohen (1992) suggests that such neutrality is the primary characteristic of egg-getting sperm. Moreover, he suggests that neutrality is not only the secret of their success in penetrating the cumulus, it is also the secret of their success in arriving at the cumulus in the first place.

In at least one sense there are at any one time two types of sperm in the mammalian ejaculate (Cohen and Adeghe, 1986): those that the female coats with immunogammaglobulin (IgG) and those which she does not (Chapter 4 for details). In rabbits and mice, sperm from the oviduct contain a far higher proportion of sperm which are not coated with IgG than sperm from the uterus (Cohen and Werrett, 1975). Moreover, there is some evidence of a link between the IgG-coat-ability of sperm and their ability to function as egg-getters (Cohen and Tyler, 1980). There is also evidence that, during phagocytosis, females do not reduce the number of egg-getters (Taylor, 1982).

The sperm receptor for IgG is similar to, and may be derived from, prostatic secretory protein in the seminal fluid (Liang *et al.*, 1991). Sperm may thus differ initially in their affinity for prostatic protein and secondarily in IgG coating. Cohen's (1992) hypothesis is that sperm coated with IgG are selectively phagocytosed by the female, those neutral sperm without IgG having relatively safe passage through the leucocytic minefield to the oviduct. When a selection of sperm reaches the cumulus, the female again bars passage to those coated with IgG but allows through those not so coated.

Cohen (1984) suggests that the function of female selectivity is to keep out sperm which suffered errors of crossover during meiosis and which thus would be unsuitable for fertilization. He postulates that such sperm honestly advertise these errors via chemical signatures on the surface of their plasma membrane. As a result, they become coated with IgG once in the female and are then selectively phagocytosed and, if they survive, barred from the zona by being prevented from penetrating the cumulus.

One problem with Cohen's hypothesis derives from the existence of the SCID mouse. This strain is immunologically deficient, lacking T-cells, B-cells, IgGs and NK cells (Croy and Chapeau, 1990); yet it breeds normally. If female selection against mutant sperm were so important, the SCID mouse should not exist (Cohen, 1992).

According to the standard view, any selection of sperm that takes place in the female tract is necessary simply because a precisely ripe population of sperm may need to be selected from a vast juvenile or senile majority of otherwise equal sperm (Hancock, 1978; Eisenback and Ralt, 1992). According to Cohen (1971, 1984), female selection occurs because sperm are of two different types, and most sperm should not be allowed through. Our kamikaze view combines both of these scenarios.

12.3.3 The kamikaze view: a job for all ages

So far, our analyses (Chapter 11) appear to have identified two major families of sperm types within the human ejaculate: (1) the family planning sperm (tapering and pyriform); and (2) the oval-headed sperm. Whether amorphous, pinheaded and bicephalous sperm are part of one or other of these 'families' or have as yet unidentified affinities remains to be investigated (though bicephalous sperm may be a 'blocking' morph; section 11.3.1).

Crudely, the oval-headed sperm subdivide into three morphs (macros, modals and micros) but in reality they form a continuous distribution with a positive skew towards larger heads (Laufer *et al.*, 1978). More importantly, head size is not constant but seems to increase, first during maturation in the epididymis, next in the presence of tubal (and perhaps uterine) fluid (Chapter 11).

Our data suggest that head size is associated with different roles and abilities: (1) micros are better able to achieve retention in the female; (2) modals are the major generalized seek-and-destroy sperm during sperm competition; and (3) (some) macros are egg-getters. Our model for the ontogeny of oval-headed sperm is thus as follows.

(a) THE ONTOGENY OF OVAL-HEADED BEHAVIOUR: FROM DESTROYER TO FERTILIZER

During maturation in the epididymis, some oval-headed sperm mature faster than others so that, by the time they enter the vas deferens, a proportion (1 in 200) are macros, immediately ready to behave as egg-getters. No further development takes place while in the vas deferens, though differential phagocytosis may alter the size distribution as appropriate to the risk that the next female to be inseminated will already contain sperm from another male (Chapter 11).

On ejaculation, depending on conditions in the cervix and the assistance or resistance from the female's orgasm regime that the sperm encounter, macros may either be retained in high or low numbers. Those that are retained are the first to progress through the female tract towards the oviduct (Overstreet, 1983). They require little or no capacitation and could achieve a rapid fertilization in the event of early ovulation.

The 'immature' micros are retained in larger proportions (Box 11.7) but may require a longer period of capacitation before they can function as seek-and-destroy sperm. Some modal sperm are already capacitated and ready to show a spontaneous acrosome reaction if they encounter sperm from another male. The remainder capacitate more slowly, increasing in head size as they do so. Micros become modals, becoming capacitated and ready to act as seek-and-destroy morphs. Some modals become macros and change their behaviour to that appropriate for egg-getters, thereafter reserving the acrosome reaction for when and if they encounter cumulus cells around an egg. These egg-getters swim through the oviduct and, eventually, in the absence of an egg, through into the body cavity.

Our earlier caution over interpreting just what the calculated number of egg-getters in an ejaculate really means should now be clear. If our model is correct, egg-getters are a temporal and temporary morph. Experiments on the number of egg-getters in any given population of sperm can only estimate the number that are egg-getters at that time.

Our conclusion that the human ejaculate contains about one million egg-getters on ejaculation and that about 4000 pass through the oviduct does not necessarily mean therefore that these 4000 are a subset of the original million. Of course, the first wave of egg-getters should be sperm from the ejaculate which made rapid progress to the oviduct after insemination (Chapter 8). The remainder of the 4000 will be sperm which attained egg-getter status over a period of time after ejaculation. Equally, many of the initial million will become senescent before ever reaching the oviduct.

The fact that sperm from the oviduct of one rabbit can re-trace their journey in a second rabbit and still perform successfully as egg-getters (Cohen and McNaughton, 1974) does not invalidate our ontogenetic model. As yet, we do not know how long an individual sperm can function as an egg-getter. In any case, the subset of sperm (1 in 7) which could have fertilized eggs in the first female may well be different from the subset which fertilizes eggs in the second female.

One question which we cannot yet resolve is whether all oval-headed sperm have the potential to pass through all ontogenetic phases and thus briefly during their lives to behave as egg-getters. The alternative is that only a portion show the full ontogenetic sequence. The remainder, curtailing their development at the seek-and-destroy phase, may be destined to wander forever in search of foreign sperm until they either die or are killed by phagocytosis.

In so far as micros and modals, although underrepresented among dead sperm (Tyler and Crockett, 1982), nevertheless do die, it would seem that not all oval-headed sperm are programmed to develop into macros before they die. It is still possible, however, that all have the potential but that some environmental trigger (e.g. presence of sperm from another male, the breakdown products of dead macros) is necessary before many begin to change their shape and role.

(b) COMPARISON WITH OTHER VIEWS

Our model retains the paradigm from the standard view of a few precisely 'ripe' sperm amidst a larger population of senile and immature sperm. It also retains the image of a steady trickle of ripe sperm past the site of fertilization. It differs, however, in that it proposes that egg-getters have a morphological characteristic (larger head size) as well as a behavioural characteristic. It also differs in that it proposes a role (seek-and-destroy) for sperm during their phase of immaturity. Capacitation is seen as preparation for an acrosome reaction which is used in response to foreign sperm. Finally, our model also differs from the standard view in proposing a role (as blockers) for senescent or even dead, sperm (section 11.3.1).

Our model retains Cohen's notion of two populations: one gametic, one not. Our non-gametic population, however, is not simply a motley assemblage of unwanted and potentially dangerous sperm as in Cohen's model. It is instead a subpopulation of sperm with active and important roles to play in the success of the ejaculate.

Cohen's evidence for the relative fertility of sperm which are and are not coated with IgG in the female tract appears convincing and we see no reason why our category of egg-getter sperm should not be Cohen's non-IgG coated sperm. In which case, we should predict that, on average, sperm not coated with IgG should have a larger head than sperm coated with IgG. If the situation were really that simple, we should also expect that, as oval-headed sperm pass through their ontogenetic phases from micro to modal to macro, they should at some point lose their IgG coat as they become egg-getters. Such stripping of the coat could well be part of capacitation. Cohen and Tyler (1980) have shown that uncoated sperm stay uncoated but did not investigate a loss of coat by coated sperm.

One potential confounding factor in such examination is the possibility that not all oval-headed sperm may be programmed to develop into egg-getting macros. In which case, there may well be a large population of sperm, destined never to progress beyond seek-and-destroy activity, which might never lose their IgG coat. If so, the prediction will need to be tested by careful statistical analysis of changes in proportions rather than by gross impressions.

(c) ACROSOMES AND FERTILIZATION: SOLUTION OF A MYSTERY?

In one review, Yanagimachi (1988) agonized over the conundrum that despite extensive research there was still no convincing evidence that acrosomal enzymes were involved in fertilization in eutherian mammals. On the contrary, there is evidence that they are not (e.g. penetration of cumulus, Talbot *et al.*, 1985; penetration of zona, Bedford and Cross 1978; Box 12.1).

Yanagimachi (1988) thus asks

> If mammalian spermatozoa can pass through the egg vestments without any assistance from enzymes, why do they have acrosomes loaded with such powerful enzymes as hyaluronidase and acrosin, which are capable of hydrolysing the matrix of cumulus and the zona? If the acrosome and acrosomal enzymes are relics of evolution, we should encounter at least some mammalian species in which the spermatozoa totally lack acrosomes or acrosomal enzymes. Since this is not the case, acrosome and its enzymes must exist to perform very important functions.

Our kamikaze sperm hypothesis offers a solution: the acrosomes and acrosomal enzymes are adaptations to sperm competition.

We propose that acrosomal enzymes are used: (a) by 'family planning' sperm, specifically against egg-getters in their own and other ejaculates; and (b) by modal oval-headed sperm for general release in the presence of sperm from another male.

(d) THE SPERM AND THE ZONA

The fact that sperm cannot penetrate the zona unless they undergo the acrosome reaction (Austin and Bishop, 1958) is not evidence that the acrosomal enzymes are used to penetrate the zona. An equally viable alternative is that the zona only allows through sperm which have shed their acrosomal enzymes. These enzymes are dangerous to eggs (Chapter 4) and the zona could be simply a vestment designed to keep acrosomal enzymes away from the egg's (vitelline) surface.

During the course of eutherian evolution the zona has increased in thickness. How long sperm should delay abandoning their acrosomal load before attacking the zona will depend on how late they may be needed for antisperm activity. Mouse and hamster sperm, for example, undergo the acrosome reaction on the zona surface (Saling and Storey, 1979; Barros *et al.*, 1984) but guinea-pig and rabbit sperm often undergo the acrosome reaction before arriving at the zona (Huang *et al.*, 1981; Kusan *et al.*, 1984). The fact that the thickness of the zona seems to correlate with where the sperm abandon their acrosomal contents may be less an adaptation on the part of the sperm to the thickness of zona to be penetrated (Yanagimachi, 1988) and more a female adaptation against the quantity of enzymes to which they are likely to be exposed.

(e) ACROSOMES: A BRIEF PHYLOGENY

In contrast to the situation for eutherian mammals, the acrosome of birds and marsupials seems more likely to play a part in zona penetration (Yanagimachi, 1988). Acrosomal contents released by an acrosome-reacted sperm dissolve the thin zona locally to produce a hole through which the sperm then swims (Okamura and Nishiyama, 1978; Rodger and Bedford, 1982). The major zona lysins in these species are most probably trypsin-like enzymes (Langford and Howarth, 1974; Rodger and Bedford, 1982; Talbot and DiCarlantonio, 1984). Even so, we predict that in birds and mammals the acrosome has a dual role, partly for seek-and-destroy activity and partly for zona penetration.

A role for the acrosome in fertilization probably has a long evolutionary history, stretching back to the earliest days of external fertilization (Chapter 2). A role in seek-and-destroy behaviour probably expanded or even began with the evolution of internal fertilization (Chapter 3), seek-and-destroy activity utilizing, in an evolutionary sense, the powerful antitissue enzymes that had evolved first for a role in fertilization.

In eutherian mammals, females evolved a thicker and thicker zona pellucida around the egg. This was perhaps partly to give more control over polyspermy as a contraceptive device (Chapter 4) and partly to keep acrosomal enzymes away from the vitelline membrane. Eventually, sperm were only allowed through the zona if they had lost most or all of their antitissue enzymes by an earlier acrosome reaction which was then used primarily, or only, as an antisperm device.

(f) PREDICTIONS OF THE MODEL

Our model of the ontogeny of oval-headed sperm is a good model in at least one sense: it generates a variety of easily testable hypotheses and is hence falsifiable. These predictions are as follows.

1. The heads of sperm higher up the female tract should, *on average* be larger than the heads of sperm lower down the tract, in the ejaculate, and in the epididymis.
2. The average size of heads in the oviduct (and perhaps in the uterus) should increase with time since insemination.
3. The average size of heads around the zona should be greater than that of heads around the cumulus.
4. The average size of heads of sperm not coated with IgG should be greater than that of heads of sperm coated with IgG.
5. In at least some types of subfertility, the average size of heads of less fertile males should be smaller than that of more fertile males.

For humans and most other mammals, these predictions relate to the oval/round-headed sperm. The 'family planning' sperm (tapering and pyriform) and those with as yet unknown roles (amorphous, pinhead and bicephalous) should not be included in the measurements. For other mammals (e.g. rodents) the predictions should apply to the size distribution of the most common head shape.

None of these predictions has yet been tested deliberately. However, we have already shown (Bellis *et al.*, 1990a) for rats that the average size of sperm heads in the uterus is greater than that of heads in the cervix and vagina (prediction 1). Moreover, Laufer *et al.* (1978) have shown that the average volume of an oval-headed sperm (most of which is in its head) is smaller in oligozoospermic males (prediction 5).

12.4 Kamikaze sperm and differences in fertility

Males differ in the ease and efficiency with which their ejaculates fertilize females. In

Box 12.2 The MTP ratio of the ejaculate as a predictor of success in IVF and AID-assisted conception treatment

The ratio of macrocephalous to (tapering + pyriform) sperm (MTP ratio) shows a significant association with the success of artificial insemination by donor (AID) and of various steps in *in vitro* fertilization (IVF) treatments (Chapter 11). We calculate the MTP ratio as:

$$(M - (T+P))/(M+T+P)$$

This gives an index with limits from −1 to +1 in which values <0 indicate fewer macros than (tapering + pyriform) in the ejaculate and values >0 indicate more macros than (tapering + pyriform).

Experiments were carried out from 1989 to 1993. Different years involved different IVF protocols and different assistants. All analyses are blocked by year and assistant to give a within year and assistant analysis thus controlling for interyear and interobserver differences.

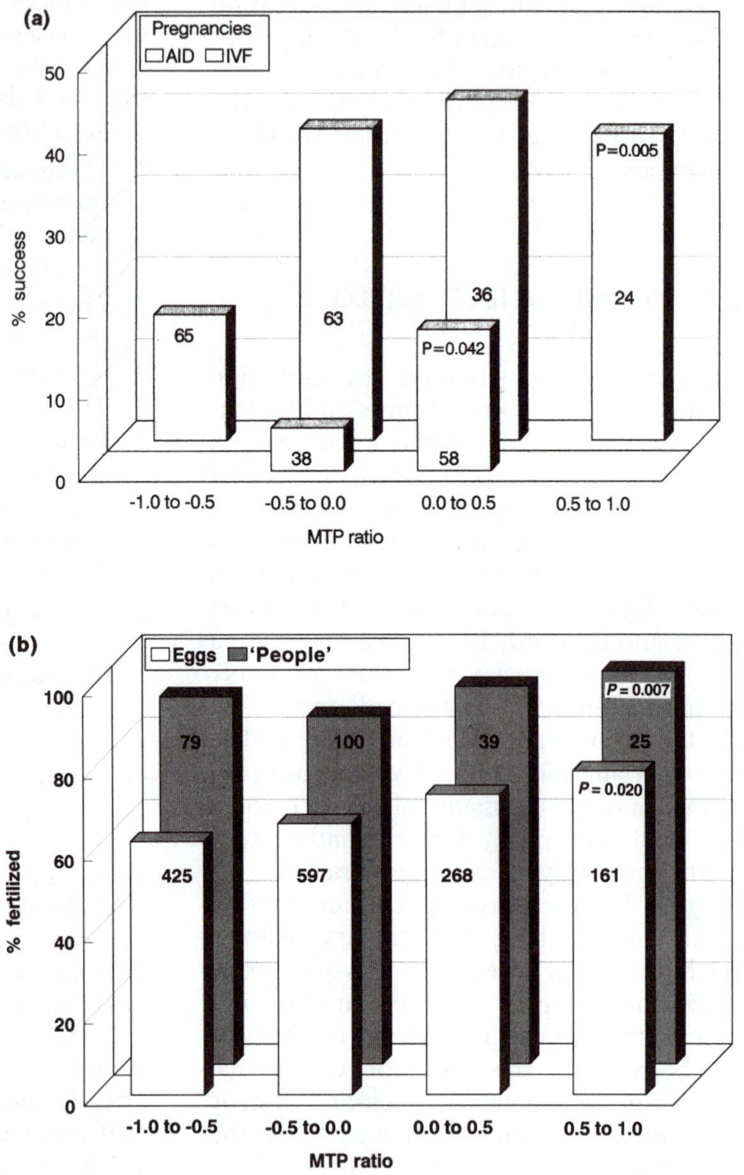

humans, even in the absence of sperm competition, some males never produce children, despite frequent sexual access to fertile partners. Others easily produce a succession of children. The existence of such differential fertilizing capacity (DFC) has been known for many years in animals as diverse as cockerels (Martin and Dzuik, 1977), mice (Finn, 1964), rabbits (Parrish and Foote, 1985), pigs (Boender, 1966), cattle (Beatty *et al.*, 1969) and humans (Bostofte *et al.*, 1985; Enginsu *et al.*, 1993). We suggest that many of the observed differences will only make sense when viewed in the perspective of the kamikaze sperm hypothesis.

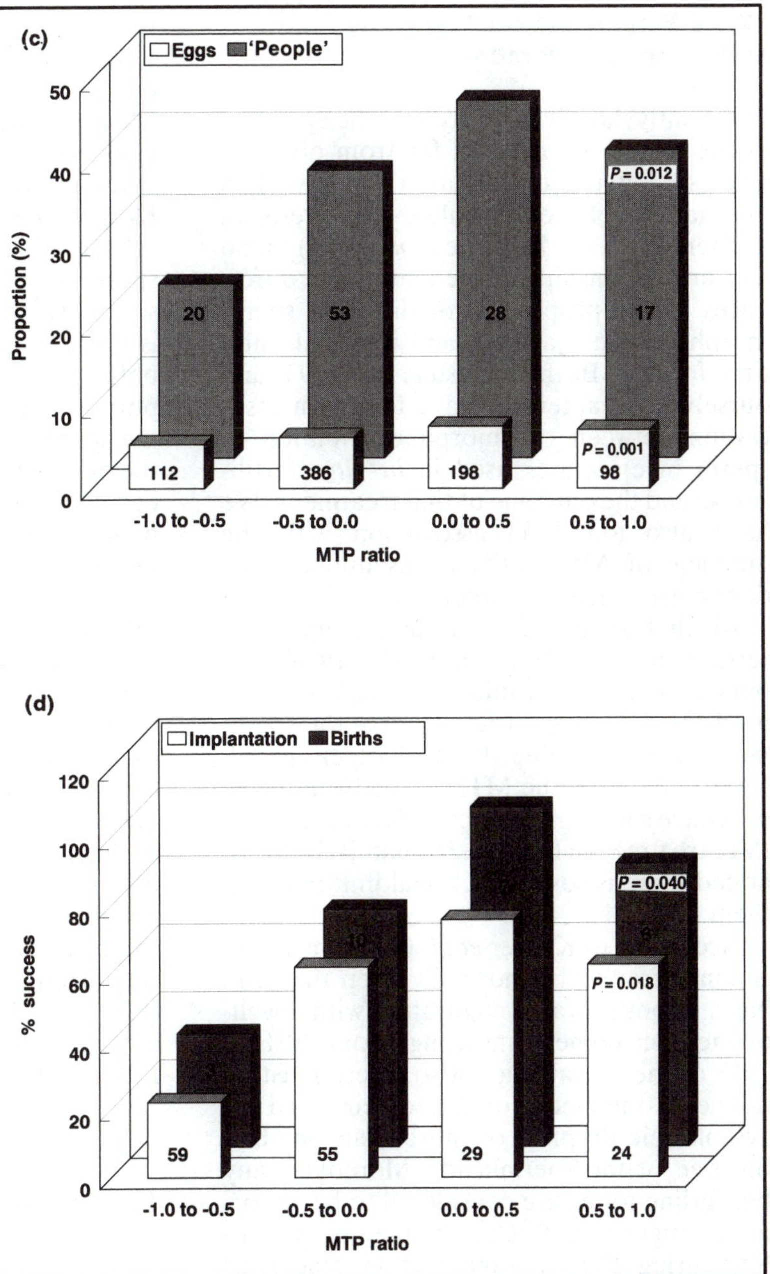

Higher MTP ratios in the male's ejaculate are associated with: (a) greater probability that the female will become pregnant in both IVF ($P = 0.005$) and AID ($P = 0.042$); (b) a higher proportion of eggs fertilized ($P = 0.020$) and a lower probability that no eggs will be fertilized ($P = 0.007$); as well as (c) a higher probability of polyspermy ($P = 0.001$); and (d) a higher probability of implantation of returned embryos ($P = 0.018$); and a higher probability that pregnancy will result in a successful birth rather than a miscarriage ($P = 0.040$).

12.4.1 Ejaculates and fertility: gross associations

Frequently, infertility or subfertility is clearly related in a gross and conspicuous way to the ejaculate that the male produces. For example, some males produce no sperm (azoospermia), others produce few sperm (oligospermia) and yet others produce too many sperm (polyzoospermia). Others produce sperm which are immotile, or relatively so, or which stick together (agglutinate) en masse. Yet others produce almost all pin heads or tapering heads.

In all such cases, a proximal link between the characteristics of the ejaculate and infertility or subfertility is clear, even if the medical condition (genetic; infection; accident; etc.) which led to such an ejaculate in each case may be less clear.

12.4.2 Fertility indices: 'normality' evaluation vs MTP ratio

Frequently, however, the role of the ejaculate in the male's infertility is far from obvious. Even so, detailed examination may reveal an association between ejaculate characteristics and fertility. Thus Bostofte *et al.* (1985) found a clear relationship between intermale differences in the proportion of different sperm morphs in the ejaculate and intermale lifetime fertility. Both Enginsu *et al.* (1993) and ourselves (Chapter 11) have found an association between the morph distribution of sperm in ejaculates used in *in vitro* fertilization and the outcome of that treatment. We have also found an association with the outcome of AID (artificial insemination by donor) treatment (Chapter 11).

At first sight, the criteria for ejaculate fertility used by Enginsu *et al.* (1993) and ourselves seem very different. Enginsu *et al.* used the so-called 'strict' criteria for sperm 'normality' introduced by Kruger *et al.* (1986). We used the MTP ratio. Both measures have such high predictability of success in IVF treatment (Box 12.2) that it is to be hoped there is some functional link between them.

According to Kruger *et al.* (1986), a sperm is considered to be 'normal' 'when the head has a smooth, oval configuration with a well-defined acrosome comprising about 40% to 70% of the spermhead. Also, there must be no neck, midpiece, or tail defects and no cytoplasmic droplets of more than one-half the size of the spermhead.' Moreover, any borderline forms are considered to be abnormal. Kruger *et al.* (1986) obtained peak fertility during IVF treatment from ejaculates with 31–45% 'normal' sperm, higher and lower proportions being associated with reduced fertility.

These criteria for normality are much stricter than the conventional WHO criteria. However, as the criteria involve a particular category of oval-headed sperm then, according to our model of the ontogeny of egg-getting in oval-headed sperm (section 12.3.3), the proportion of 'normal' sperm by Kruger *et al.*'s criteria could well be an index of the proportion of egg-getters currently in the sperm population. In this sense, at least, Kruger *et al.*'s 'normality' index is similar to our assessment of the proportion of macros, which should also indicate the proportion of egg-getters in the sperm population.

In principle, the criteria proposed by Kruger *et al.* and ourselves both exploit a subset of the population of oval-headed sperm to obtain an index of the proportion of sperm which are currently egg-getters. There is no reason, therefore, why they should not both give high predictability. For present purposes, it does not really matter whether the sperm being used as an index are actually the egg-getters or not (though we suggest that egg-getting represents a precise phase of being a macro). As far as utilization of an index in male screening for IVF is concerned, the only important parameter is the success of the index in predicting the chances of success of the treatment.

In one way, Kruger *et al.*'s index is better than ours in that it requires the observer to look for only one category of sperm and thus is perhaps less prone to observer bias. Our index, however, not only includes a factor perhaps positively related to the proportion of egg-getters in the ejaculate but also includes a factor related to the proportion of sperm (tapering and pyriform) which have a significant negative influence on fertility (Chapter 11). There is a need for the two indices to be compared in working trials by an IVF clinic.

12.4.3 Sperm morphology and polyspermy

An increase in the MTP ratio increases the chances of eggs being fertilized but it also increases the risk of polyspermy (Chapter 11). All else being equal, an increase in the proportion of 'family planning' sperm in the sperm population not only decreases the chance of fertilization but also decreases the chances of polyspermy.

It is difficult to see how these effects could occur unless macros as a category are (or contain) egg-getter sperm which in turn are hindered by the 'family planning' sperm in their own ejaculate.

12.4.4 Sperm for assisted conception: masturbation vs copulation

Universally, clinics for assisted conception (IVF and AID) use sperm produced by masturbation. We have stressed throughout this book that masturbation is not simply a substitute for copulation. Rather, it is preparation for copulation; a mechanism by which ageing sperm, perhaps with an unsuitable morph distribution for the male's current or developing sociosexual situation, are shed to make way for younger and/or more suitable sperm.

Viewed in this light, it seems obvious that techniques for assisted conception should use sperm produced during copulation rather than masturbation. Zavos and Goodpasture (1989) showed that, when provided by men with a variety of andrological symptoms, all conventional semen parameters were clinically better in copulatory than masturbatory ejaculates. Our own studies of apparently healthy males, some of proven fertility, have provided similar results (as presented at various points in this book).

It is, of course, far easier and cheaper for busy IVF and AID clinics to obtain masturbatory ejaculates from their patients than copulatory ejaculates. Whether any increase in success through using copulatory ejaculates would be sufficiently great to be cost effective would again seem to require practical investigation.

12.4.5 Kamikaze sperm in competition

If males differ in their fertility when their sperm are not in competition, it should be no surprise to find that they also differ when their sperm are in competition. Such intermale differences have been known for many years for animals such as cockerels (Martin and Dzuik, 1977), mice (Finn, 1964), rabbits (Parrish and Foote, 1985), pigs (Boender, 1966), and cattle (Beatty *et al.*, 1969). Naturally, no such information exists for humans.

Differences in fertility between males are much more conspicuous when their sperm are placed in competition (Beatty *et al.*, 1969) than when they are not and heterospermic insemination has been used for years to study the differential fertilizing capacity (DFC) of individuals of domestic animals.

Genetically inbred strains of a variety of domesticated animals often show strain-related differences in DFC, indicating a genetic basis for ejaculate fertility. Thus, sperm from Leghorn cockerels were more competitive than sperm from Columbian cockerels (Martin *et al.*, 1974). Similiarly, sperm from Long–Evans rats were more competitive than sperm from F344-strain rats (Dewsbury and Hartung, 1980) and sperm from Albino rats were more competitive than sperm from Norwegian hooded rats (Sharma and Hays, 1975). Within strains, some individual males have consistently more competitive sperm than others (e.g. rabbits, Dziuk, 1965; Parrish and Foote, 1985).

DFC between males exposed to sperm competition need not show the same rank-order as DFC for the same males in the absence of sperm competition. In theory, all combinations are possible.

1. Males who are most fertile in homospermic tests (i.e. more females conceive and/or females have larger litters) are also most fertile in heterospermic tests (sire more offspring than their competitors).
2. Males who are fully fertile in homospermic tests are rarely successful when their sperm are in competition.
3. Males who are relatively infertile in homospermic tests are very successful when their sperm are in competition.

We know of examples of (1) (e.g. cockerels and boars, Martin and Dziuk, 1977), (2) (e.g. bulls, Beatty *et al.*, 1969), and (3) (e.g. bulls, Beatty *et al.*, 1969). The last situation is particularly interesting. For example, polyzoospermic males who inseminate very large numbers of sperm are often infertile in the absence of sperm competition, perhaps because too many sperm reach the egg (Chapters 4 and 9). In the presence of sperm from another male, however, the seek-and-destroy behaviour of sperm from both males could reduce the total number which survive (from both males) to more fertile levels, yet the numerical superiority of the polyzoospermic male could promote success.

As in this theoretical example, there is absolutely no reason for kamikaze activity by sperm to lead to reduced fertility of the female under conditions of sperm competition. It is, of course, possible to envisage kamikaze sperm from both males totally neutralizing each other's ejaculates to the extent that no egg-getters from either male survive to reach the oviduct. However, so few egg-getters need to reach the site of fertilization for a normal litter to result (e.g. only one in humans) that such total devastation should be rare.

We rather envisage that kamikaze activity simply biases the number of egg-getters from each male that reach the site of fertilization. In any case, the total number that arrive at the site of fertilization may owe as much to female control of sperm transport as to the number of survivors from sperm competition (Chapter 4). Consequently, we do not see that the kamikaze sperm hypothesis can make any generalized prediction over whether the total number of egg-getters to arrive at the ampulla from the two males will be more than, equal to, or less than the number that would have arrived if only one of the males had inseminated the female.

Direct studies of the outcome of sperm competition (e.g. using fluorochrome-labelled sperm; Parrish and Foote, 1985) suggest, not surprisingly, that the DFC of the ejaculates of different males can be generated at just about every stage of sperm transport and fertilization. Thus, during sperm competition following artificial heterospermic insemination in rabbits, males differ in the proportion of their sperm that arrive at the site of fertilization and in the proportion of those that do arrive which successfully penetrate the egg vestments (Parrish and Foote, 1985). In part this may be due to differences in the speed and timing of capacitation (Dziuk, 1965; Parrish and Foote, 1985).

Although the outcome of competition between given pairs of males is relatively consistent, it is not invariable. On any one occasion the usually less competitive male's sperm may outcompete the other (Martin *et al.*, 1974). Sometimes, in rats, sperm competition can lead to different outcomes in the right and left uterine horns of the female reproductive tract (Špinka, 1988).

The whole arena for the study of kamikaze interactions in sperm competition is waiting to be studied, perhaps using the elegant techniques pioneered by Parrish and Foote (1985) for rabbits.

12.4.6 Kamikaze problems in fertility

One of the confusing facts about animal ejaculates has always been their apparent inefficiency at simple fertilization. Reproductive success at each generation is such a vital factor in determining which lineages dominate future generations that we could quite reasonably expect ejaculates to be near perfect at achieving fertilization when they get the chance. That they are not clearly shows that fertilization is not the only function for the ejaculate and that success at any one function may sometimes hinder success at another, such as fertilization.

We claim in this book to have demonstrated that male humans, in common with most primates and many other mammals, are trying to achieve a number of objectives with every single ejaculate. Together, the requirements of these different objectives interact to reduce the efficiency of each, including fertilization. Presumably, however, the ejaculate strikes the optimum compromise. These objectives are:

1. fertilization, but only in the context of male family planning (in species with paternal care);
2. spreading the ejaculate through space and time in the female tract to counter the unpredictability of ovulation;
3. taking advantage of any female assistance in sperm retention and transport while being able to counter a lack of assistance;
4. blocking passage to sperm if the female's next copulation is with a different male but allowing some passage if it is not; and
5. being able to compete if foreign sperm are already present or later arrive in the female while not hindering progress of

one's own egg-getters if they are not or do not.

In every case, an ejaculate adaptive to one circumstance may be less adaptive or even maladaptive to another. The ejaculate is thus continuously in a fine state of balance between helping and hindering fertilization and male success will often depend on the accuracy with which he predicts circumstance, often with very little reliable information.

Seen in this light, it is almost more surprising that ejaculates succeed in fertilization as often as they do than that they fail as often as they do. The mechanisms for compromise are so finely balanced (e.g. MTP ratio; macro: micro ratio; the proportion and placing of coiled tails) that it is not surprising that small perturbations can have big influences on male fertility.

The perturbations, such as intermale or interejaculate differences, should not necessarily be seen as maladaptive. If we are right that the males themselves exist in a balanced polymorphism for involvement in sperm competition (Chapter 5), we should expect this polymorphism to be reflected in the males' ejaculates.

For example, polygynous males programmed for an active, roving, role in sperm competition should not only produce many more sperm (Chapter 5), they should perhaps produce fewer family planning sperm, more coiled tail sperm, and more oval-headed sperm, particularly more macros. Monogynous males programmed for intensive mate-guarding, however, should produce more family-planning sperm, a wider spread of sizes of oval-headed sperm, and fewer coiled tails.

In clinical trials, the ejaculates of monogynous males may well appear less fertile than those of polygnous males because monogynous males are adapted to induce a slower rate of conception in their partner. Yet, as we have already argued (Chapter 5) over a lifetime in the right circumstances, both types of male should do equally well. In the wrong circumstances, however, both may seem maladaptive. Polygynous males, perhaps relying on their sperm being reduced in number in the female tract by foreign sperm, may verge on polyzoospermy in a monogynous situation. Monogynous males, relying on long-term access to a given partner for maximum efficiency of family planning, may appear subfertile.

13 Final thoughts

Approximately 1000 million years ago some of the first life forms on planet Earth evolved gametes of two modal sizes and two conspecific morphs, males and females. That single evolutionary step unleashed one of the most powerful selective forces to operate on subsequent generations in the millenia that followed: sperm competition. There cannot be any animal alive today the sexual anatomy, physiology and behaviour of which has not been shaped and moulded by this force at least at some time during its evolutionary past. In most species, including humans, sperm competition will have been an ever present factor.

In Chapters 2–6 we have argued that the major part of human sexual programming has evolved in response to the pressures of sperm competition as males struggled to outcompete males and females quietly and secretly manipulated and deceived them all, bending the cryptic mating game to suit their own priorities. Aspects of human sexuality as diverse as internal fertilization, forced copulation, penis shape, masturbation, homosexuality and female sexual crypsis were all initiated, crudely shaped and then finely honed by the force of sperm competition.

As adaptation followed adaptation, sperm competition rebounded from one to the other, shaping, reshaping and finely tuning each animal's sexuality according to the species' current status and ecology. In the human lineage, the end result was the subtle and sophisticated mating game that we enjoy today and in which we are all, consciously or subconsciously, active players. This game still involves sperm competition. Even in the modern sexual world of efficient contraceptives and lethal sexually transmitted diseases, sperm competition is still as ruthless a force as ever.

One of the facets of human sexuality to feel the full force of sperm competition was (not surprisingly) the male ejaculate. There is a naïve view among non-biologists (and many biologists) that evolution leads to greater efficiency and eventually perfection in the ability of animals to do what they have to do. This is not true. Nearly everything is a compromise. If the ejaculate had only one job to do, to fertilize eggs, and if both male and female interests were identical, then maybe evolution would have shaped the ejaculate into a near-perfect entity for fertilization. Unfortunately, male and female interests are rarely identical and as a result the ejaculate does have more than one job to do. It has to overcome female defences, outcompete sperm from other males, and then achieve fertilization. Adaptations that increase the efficiency of one job decrease the efficiency of another. It is probably impossible to shape an ejaculate that is perfect in everything and evolution seems to have failed to do so. It has, of course, made the best of an impossible job and come up with what we assume is the best compromise, but it has always to be remembered that it is a compromise.

Compromise raises no problems as long as all angles are clear and understood. There is a danger, though, that if the nature of the compromise is not appreciated errors of interpretation will occur. There is a natural tendency for people interested only in fertility to assume that the ejaculate is there solely for that function. Thus, all aspects are interpreted in terms of fertility. This is dangerous when hardly any obvious feature is related to fertility. To refer to the analogy in Chapter 1, the tree of fertility may be difficult to find in the forest of sperm competition.

It was Robert Smith, in his brave and provocative paper (Smith, 1984b), who first pointed out how important sperm competition might have been in the shaping of human sexuality. Frustrated by the lack of any real data, we have done little more than take up his mantle and run with it.

It is our wish that this book showing, as we hope it does, how easy and important it is to obtain real data on human sperm competition will stimulate a rapid expansion in research and an application to all of the relevant biomedical fields. If it does this, we shall be well pleased.

References

Abele, L. and Gilchrist, S. (1977) Homosexual rape and sexual selection in acanthocephalan worms. *Science*, **197**, 81–3.

ACSF (1992) AIDS and sexual behaviour in France. *Nature*, **360**, 407–9.

Adams, C. E. (1969) Intraperitoneal insemination in the rabbit. *J. Reprod. Fertil.*, **18**, 333–9.

Adler, M. W. (1991) Sexually transmitted diseases, in *Oxford Textbook of Public Health*, Vol. 3, (eds W. W. Holland, R. Detels and G. Knox), Oxford Medical Publications, Oxford, pp. 345–57.

Adler, N. T. (1969) Effects of the male's copulatory behaviour on successful pregnancy of the female rat. *J. Comp. Physiol. Psychol.*, **69**, 613–22.

Adler, N. T. and Zoloth, S. R. (1970) Copulation and the inhibition of pregnancy in rats. *Science*, **171**, 311–12.

Afton, A. D. (1985) Forced copulation as a reproductive strategy of male lesser scaup: a field test of some predictions. *Behaviour*, **92**, 146–67.

Afzelius, B. A. (1981) Abnormal human spermatozoa including comparative data from apes. *Am J. Primat.*, **1**, 175–82.

Ågmo, A. (1976) The number of spermatozoa in spontaneous ejaculates of rats. *J. Reprod. Fertil.*, **48**, 405–7.

Ahlgren, M. (1975) Sperm transport to and survival in the Human fallopian tube. *Gynecol. Invest.*, **6**, 206–14.

Ahlgren, M., Boxtrom, K. and Malmqvist, R. (1975) Sperm transport and survival in women with special reference to the fallopian tube, in *The Biology of Spermatozoa*, (eds E. S. E. Hafez and C. G. Thibault), Karger, Basle, pp. 63–73.

Alberts, B., Bray, D., Lewis, J. *et al.* (1983) *Molecular Biology of the Cell.* Garland Publishing, New York.

Alcock, J. (1987) Ardent adaptationism. *Nat. Hist.*, **96**, 4.

Alcock, J. (1989) *Animal Behaviour: An Evolutionary Approach*, 4th edn. Sinauer, Massachusetts.

Alexander, A. J. (1957) Courtship and mating in the scorpion, *Opistophthalamus latimanus. Proc. Zool. Soc. Lond.*, **128**, 529–44.

Alexander, R. D. and Noonan, K. B. (1979) Concealment of ovulation, paternal care and human social evolution, in *Evolutionary Biology and Human Social Behaviour*, (eds N. A. Chagnon and W. Irons), Duxbury Press, North Scituate, MA.

Allen, G. J., Bishop, M. W. H. and Thompson, T. E. (1974) Lysis of photographic emulsions by mammalian and chicken spermatozoa. *J. Reprod. Fertil.*, **36**, 249–52.

Allen, M. L. and Lemmon, W. B. (1981) Orgasm in female primates. *Am. J. Primat.*, **1**, 15–34.

Altmann, J., Hausfater, G. and Altmann, S. A. (1988) Sex ratio bias in baboons. in Reproductive Success, (ed. T. H. Clutton-Brock), University of Chicago Press, Chicago, pp. 403–18.

Amelar, R. D. and Hotchkiss, R. S. (1965) The split ejaculate: its use in the management of male infertility. *Fertil. Steril.*, **16**, 46–60.

Anderson, R. M., May, R. M., Boily, M. C. *et al.* (1991) The spread of HIV-1 in Africa; sexual contact patterns and the predicted demographic impacts of AIDS. *Nature*, **352**, 581–9.

Anon (1990) Are sex selection techniques fact or fiction? *Reprod. Tech. Update*, **2**, 117–21.

Archer, J. (1810) Facts illustrating a disease peculiar to the female children of negroe slaves. *Med. Reposit.*, **1**, 319–23.

Archunan, G. and Dominic, C. J. (1991) Oestrous cycle disruption in group-housed mice: evaluation of the involvement of tactile and pheromonal stimuli. *Acta Physiol. Hung.*, **78**, 275–82.

Aronson, L. R. (1949) Behaviour resembling spontaneous emissions in the domestic cat. *J. Comp. Physiol.*, **42**, 226–7.

Ashton, G. C. (1980) Mismatches in genetic markers in a large family study. *Am. J. Hum. Genet.*, **32**, 601–13.

Athanasiou, R. (1973) A review of public attitudes on sexual issues, in *Contemporary Sexual Behavior*, (eds J. Zubin and J. Money), Johns Hopkins University Press, Baltimore, MD, pp. 361–90.

Aroux, M. R., Jaques, L., Mathieu, D. and Auer, J. (1991) Is the sperm bacterial ratio a determining factor in impairment of sperm motility: an in-vitro study in man with *Escherichia coli*. *Int. J. Androl.*, **14**, 264–70.

Austin, C. R. (1951) Observations on the penetration of the sperm into the mammalian egg. *Aust. J. Sci. Res. (B)*, **4**, 581–96.

Austin, C. R. (1961) *The Mammalian Egg*, Charles C. Thomas, Springfield.

Austin, C. R. (1967) Capacitation of spermatozoa. *Int. J. Fertil.*, **12**, 25–31.

Austin, C. R. (1975) Sperm fertility, viability and persistence in the female tract. *J. Reprod. Fertil., Suppl.*, **22**, 75–89.

Austin, C. R. (1982) The egg, in *Reproduction in Mammals*, Vol. 1, (eds C. Austin and R. Short), Cambridge University Press, Cambridge, pp. 46–62.

Austin, C. R. and Biship M. W. H. (1958) Role of the rodent acrosome and perforatorium in fertilization. *Proc. R. Soc. Lond. [Biol.]*, **149**, 241–8.

Baccetti, B. and Afzelius, B. A. (1976) *The Biology of the Sperm Cell. Monographs in Developmental Biology* Vol. 10, S. Karger, London.

Bailey, J. M. and Pillard, R. C. (1991) A genetic study of male sexual orientation. *Arch. Gen. Psychiatry*, **48**, 1089–96.

Baker, R. R. (1978) *The Evolutionary Ecology of Animal Migration*. Hodder & Stoughton, London.

Baker, R. R. and Bellis, M. A. (1988) 'Kamikaze' sperm in mammals? *Anim. Behav.*, **36**, 937–80.

Baker, R. R. and Bellis, M. A. (1989a) Number of sperm in human ejaculates varies in accordance with sperm competition theory. *Anim. Behav.*, **37**, 867–9.

Baker, R. R. and Bellis, M. A. (1989b) Elaboration of the kamikaze sperm hypothesis: a reply to Harcourt. *Anim. Behav.*, **37**, 865–7.

Baker, R. R. and Bellis, M. A. (1993a) Human sperm competition: ejaculate adjustment by males and the function of masturbation. *Anim. Behav.*, **46**, 861–85.

Baker, R. R. and Bellis, M. A. (1993b) Human sperm competition: ejaculate manipulation by females and a function for the female orgasm. *Anim. Behav.*, **6**, 887–909.

Baker, R. R. and Parker, G. A. (1979) The evolution of bird coloration. *Phil. Trans. R. Soc. B*, **287**, 63–130.

Baker, R. R., Bellis, M. A., Hudson, G. *et al.* (1989) *Company*, September, 60–2.

Baldwin, J. D. (1970) Reproduction synchronization in squirrel monkeys (*Saimiri*). *Primates*, **11**, 317–26.

Balgooyen, T. G. (1976) Copulation rate in American kestrels. *Univ. Calif. Publ. Zool.*, **103**, 1–85.

Bancroft, J. (1987) Hormones, sexuality and fertility in women. *J. Zool., Lond.*, **213**, 445–54.

Barlow, D. (1979) *Sexually Transmitted Diseases: The Facts*. Oxford University Press: Oxford.

Barnard, C. J. (1983) *Animal Behaviour: Ecology and Evolution*, Croom Helm, London.

Barnett, S. A. (1971) *The Human Species: A Biology of Man*, 5th edn, revised. Harper & Row, London.

Baron, M. (1993) Genetic linkage and male homosexual orientation: reasons to be cautious. *Br. Med. J.*, **307**, 337–8.

Barratt, C. L. R. and Cohen, J. (1986) Fate of superfluous spermatozoal products after vasectomy, and in the normal male tract of the mouse. *J. Reprod. Fertil.* **78**, 394–411.

Barrett, J. C. and Marshall, J. (1969) The risk of conception on different days of the menstrual cycle. *Popul. Studies*, **23**, 455–61.

Barros, C., Jedliki, A., Bize, I. and Aguirre, E. (1984) Relationship between the length of sperm preincubation and zona penetration in the golden hamster: a scanning electron microscopy study. *Gam. Res.*, **9**, 31–43.

Bateman, A. J. (1948) Intra-sexual selection in *Drosophila*. *Heredity*, **2**, 349–68.

Bateman, N. (1960) High frequency of a lethal gene (t^e) in a laboratory stock of mice. *Genet. Res.*, **1**, 214–25.

Bavister, B. D. (1979) Fertilization of hamster eggs *in vitro* at sperm:egg ratios close to unity. *J. Exp. Zool.*, **210**, 259–64.

Beach, F. A. (1975) Variables affecting 'Spontaneous' seminal emissions in rats. *Physiol. Behav.*, **15**, 91–5.

Beach, F. A. (1976) Sexual activity, proceptivity, and receptivity in female mammals. *Horm. Behav.*, **7**, 105–38.

Beach. F. A. and Eaton, G. (1969) Androgenic control of spontaneous seminal emission in hamsters. *Physiol. Behav.*, **4**, 155–6.

Beams, H. W. and King, R.L. (1933) The sperm storage function of the seminal vesicles. *J. Urol.*, **29**, 95–7.

Bean, J. A., Leeper, J. D., Wallace, R.B. *et al.* (1979) Variations in the reporting of menstrual histories. *Am. J. Epidemiol.*, **109**, 181–5.

Beatty, R. A. (1970) The genetics of the mammalian gamete. *Biol. Rev.*, **45**, 73–91.

Beatty, R. A. (1971) *The Genetics of Size and Shape of Spermatozoan Organelles*. Proceedings of the International Symposium on The Genetics of the Spermatozoon. Edinburgh.

Beatty, R. A. (1974) The phenogenetics of spermatozoa. *J. Reprod. Fertil.*, **24**, 291–9.

Beatty, R. A., Bennett, G. H., Hall, J. G. *et al.* (1969) An experiment with heterospermic insemination in cattle. *J. Reprod. Fertil.*, **19**, 491–502.

Bedford, J. M. (1966) The influence of the uterine environment upon rabbit spermatozoa, in *Reproduction in the Female Mammal*, (eds G. E. Lamming and E. C. Amoroso), Butterworths, London, pp. 478–99.

Bedford, J. M. (1970) Sperm capacitation and fertilization in mammals. *Biol. Reprod. Suppl.*, **2**, 128–58.

Bedford, J. M. (1977) Evolution of the scrotum: the epididymis as the prime mover? in *Reproduction and Evolution*, (eds J. H. Calaby and C. H. Tyndale Boscoe), Australian Academy of Science, Canberra, pp. 171–82.

Bedford, J. M. (1983) Oocyte structure and the design and function of the sperm head in eutherian mammals, in *The Sperm Cell*, (ed. J. Andre), Academic Press, London, pp. 75–89.

Bedford, J. M. (1991) The co-evolution of mammalian gametes, in *A Comparative Overview of Mammalian Fertilization*, (eds B. S. Dunbar and M. G. O'Rand), Plenum Press, New York, pp. 3–35.

Bedford, J. M. and Bibeau, A. (1967) Failure of sperm sedimentation to influence the sex ratio of rabbits. *J. Reprod. Fertil.*, **14**, 167–70.

Bedford, J. M. and Cooper, G. W. (1978) Membrane fusion events in fertilization of vertebrate eggs, in *Membrane Surface Reviews* (*Membrane Fusion*), Vol. 5, (eds G. Poste and G. L. Nicolson), North-Holland, Amsterdam, pp. 65–125.

Bedford, J. M. and Cross, N. L. (1978) Normal penetration of rabbit spermatozoa through a trypsin and acrosin resistent zona pellucida. *J. Reprod. Fertil.*, **54**, 385–92.

Bedford, J. M., Cooper, G. W. and Calvin, H. I. (1972) Post meiotic changes in the nucleus and membranes of mammalian spermatozoa, in *The Genetics of the Spermatozoan*, (eds R. A. Beatty and S. Gluehksohn-Waelsch), Departments of Genetics of the University of Edinburgh and the Albert Einstein College of Medicine, Edinburgh and New York, pp. 69–89.

Beecher, M. D. and Beecher, I. M. (1979) Sociology of bank swallows: reproductive strategy of the male. *Science*, **205**, 1282–5.

Bell, G. (1978) The evolution of anisogamy. *J. Theor. Biol.*, **73**, 247–70.

Bell, J. A. (1991) The epidemiology of Down's syndrome. *Med. J. Aust.*, **155**, 115–17.

Bellis, M. A. and Baker, R. R. (1990) Do females promote sperm competition: data for humans. *Anim. Behav.*, **40**, 997–9.

Bellis, M. A., Baker, R. R., Hudson, G. *et al.* (1989) *Company*, April, 90–2.

Bellis, M. A., Baker, R. R. and Gage, M. J. G. (1990a) Variation in rat ejaculates is consistent with the kamikaze sperm hypothesis. *J. Mammal.*, **71**, 479–80.

Bellis, M. A., Baker, R. R., Matson, P. and Chew, J. (1990b) A guide to upwardly mobile sperm *Andrologia*, **22**, 397–9.

Belsey, M. A., Eliasson, R., Gallegos, A. J. *et al.* (1980) *WHO Laboratory Manual for Examination of Human Semen and Semen–Cervical Mucus Interaction*, 1st edn, Press Concern, Singapore.

Belsey, M. A., Eliasson, R., Gallegos, A. J. *et al.* (1987) *WHO Laboratory Manual for Examination of Human Semen and Semen–Cervical Mucus Interaction*, 2nd edn. Cambridge University Press, Cambridge.

Benjamin, B. (1965) Social and economic differentials in fertility, in *Genetic and Environmental Factors in Human Ability*, (eds J. E. Meade and A. S. Parkes) Oliver & Boyd, London, pp. 177–84.

Benshoof, L. and Thornhill, R. (1979) The evolution of monogamy and loss of estrus in humans. *J. Soc. Biol. Struct.*, **2**, 95–106.

Bercovitch, F. B. (1987) Reproductive success in male savanna baboons. *Behav. Ecol. Sociobiol.*, **21**, 163–72.

Bercovitch, F. B. (1992) Re-examining the relationship between rank and reproduction in male primates. *Anim. Behav.*, **44**, 1168–70.

Bernstein, D., Glezerman, M., Zejdel, L. and Insler, V. (1977) Quantitative study of the number and size of cervical crypts, in *The Uterine*

Cervix in Reproduction, (eds V. Insler and G. Bettendorf), Theime, Stuttgart, pp. 166–85.

Bertram, B. C. R. (1975) Social factors influencing reproduction in wild lions. *J. Zool.*, **177**, 463–82.

Betteridge, K. J. (1981) A historical look at embryo transfer. *J. Reprod. Fertil.*, **62**, 1–13.

Betzig, L. (1986) *Despotism and Differential Reproduction: A Darwinian View of History*, Hawthorne, New York.

Betzig, L. (1988) Mating and parenting in Darwinian perspective, in *Human Reproductive Behaviour: A Darwinian Perspective*, (eds L. Betzig, M. Borgerhoff Mulder and P. Turke), Cambridge University Press, Cambridge, pp. 3–22.

Betzig, L. (1992a) Roman polygyny. *Ethol. Sociobiol.*, **13**, 309–49.

Betzig, L. (1992b) Roman monogamy. *Ethol. Sociobiol.*, **13**. 351–83.

Betzig, L. and Turke, P. W. (1986) Parental investment by sex on Ifaluk. *Ethol. Sociobiol.*, **7**, 29–37.

Betzig, L., Borgerhoff Mulder, M. and Turke, P. (1988) *Human Reproductive Behaviour: A Darwinian Perspective*. Cambridge University Press, Cambridge.

Bickers, W. (1961) Sperm migration and the emotions. *Int. J. Fertil.*, **6**, 159–67.

Bielański, W., Dubek, E., Bittmar, A. and Kosiniak, K. (1982) Some characteristics of common abnormal forms of spermatozoa in highly fertile stallions. *J. Reprod. Fertil. Suppl.*, **32**, 21–6.

Bingham, H. C. (1928) Sex development in apes. *Comp. Psychol. Monogr.*, **5**, 1–165.

Birkhead, T. R. (1993) Avian mating systems: the aquatic warbler is unique. . . . *TREE*, **8**, 390–1.

Birkhead, T. R. and Møller, A. O. (1992) *Sperm Competition in Birds*. Academic Press, London.

Birkhead, T. R. and Møller, A. P. (1993a) Female control of paternity. *TREE*, **8**, 100–4.

Birkhead, T. R. and Møller, A. P. (1993b) Why do birds stop copulating while their partners are still fertile? *Anim. Behav.*, **45**, 105–18.

Birkhead, T. R., Hunter, F. M. and Pellatt, J. E. (1989) Sperm competition in the zebra finch, *Taeniopygia guttata*. *Anim. Behav.*, **38**, 935–50.

Björntorp, P. (1987) Fat cell distribution and metabolism, in *Human Obesity*, (eds R. J. Wurtman and J. J. Wurtman), New York Academy of Sciences, New York, pp. 66–72.

Björntorp, P. (1991) Adipose tissue distribution and function. *Int. J. Obesity*, **15**, 67–81.

Blackburn, D. G. (1991) Evolutionary origins of the mammary gland. *Mammal Rev.*, **21**, 21–96.

Blackwell, R.E. (1984) Detection of ovulation. *Fertil. Steril.*, **41**, 680–1.

Blower, S. M. (1993) Exploratory data analysis of three sexual behaviour surveys: implications for HIV-1 transmission in the U.K. *Phil. Trans. R. Soc. Lond. B*, **339**, 33–51.

Blurton Jones, N. G. (1989) The cost of children and the adaptive scheduling of births: towards a sociobiological perspective on demography, in *The Sociobiology of Sexual and Reproductive Strategies*, (eds A. E. Rasa, C. Vogel and E. Voland), Chapman & Hall, London, pp. 265–83.

Boender, J. (1966) The development of AI in pigs in the Netherland and the storage of boar semen. *World Rev. Anim. Prod.* Special Issue, 29–44.

Bohannan, P. (1960) *African Homicide and Suicide*, Princeton University Press, Princeton, NJ.

Bongaarts, J. and Potter, R. G. (1983) *Fertility, Biology and Behaviour: An Analysis of the Proximate Determinants*. Academic Press, New York.

Borgerhoff Mulder, M. (1989) Reproductive consequences of sex-biased inheritance, in *Comparative Socioecology of Mammals and Man*, (eds R. Foley and V.Standon), Basil Blackwell, London, pp. 405–27.

Borgerhoff Mulder, M. (1990) Kipsigis bridewealth payments, in *Human Reproductive Behaviour: A Darwinian Perspective*, (ed. L. Betzig, M. B. Mulder and P. Turke), Cambridge University Press, Cambridge, pp. 85–92.

Borgerhoff Mulder, M. (1991) Human behavioural ecology, in *Behavioural Ecology: an Evolutionary Approach*, 3rd edn, (eds J. R. Krebs and N. B. Davies), Blackwell, London, pp. 69–104.

Borgerhoff Mulder, M. and Milton, M. (1985) Factors affecting infant care in the Kipsigis. *J. Anthropol. Res.*, **41**, 231–62.

Bostofte, E., Serup, J. and Rebbe, H. (1985) The clinical value of morphological rating of human spermatozoa. *Int. J. Fertil.*, **30**, 31–7.

Boue, J., Boue, A. and Lazar, P. (1975) Retrospective and prospective epidemiological studies of 1500 karyotyped spontaneous human abortions. *Teratology*, **12**, 11–25.

Bounds, W., Guillebaud, J. and Newman, G. B. (1992) Female condom (Femidom). A clinical study of its use-effectiveness and patient acceptability. *Br. J. Fam. Plan.*, **18**, 36–41.

Bourgeois-Pichat, J. (1951) Evolution génèrale de la population française depuis le XVIIIème siècle. *Population*, **6**, 635–62.

Bowen-Jones, C. (1992) Twins who have different fathers. *Marie Claire*, no. 47, 47.

Braden, A. W. H. (1958) Influence of time of mating on the segregation ratio of alleles at the T locus in the house mouse. *Nature*, **181**, 786–7.

Braden, A. W H. (1960) Genetic influences on the morphology and function of the gametes. *J. Cell. Comp. Physiol.*, **56** (suppl. 1), 17–29.

Braden, A. W. H. and Austin, C. R. (1954a) The number of sperms about the eggs in mammals and its significance for normal fertilization. *Aust. J. Biol. Sci.*, **7**, 543–51.

Braden. A. W. H. and Austin, C. R. (1954b) Time regulations and their significance in the ovulation and penetration of eggs in rats and rabbits. *Aust. J. Biol. Sci.*, **7**, 543–51.

Breder, C. M. jr. and Rosen, D. E. (1966) *Modes of Reproduction in Fish*, The Natural History Press, New York.

Brillard, J. P. and Bakst, M. R. (1990) Quantification of spermatozoa in the sperm storage tubules of turkey hens and its relation to sperm numbers in the perivitelline layer of eggs. *Biol. Reprod.*, **43**, 271–5.

Briskie, J. V. (1992) Copulation rate of Smith's longspurs. *Auk*, **109**, 563–75.

Bronson, F. H. and Manning, J. M. (1991) The energetic regulation of ovulation: A realistic role for body fat. *Biol. Reprod.*, **44**, 945–50.

Brooke, M. and Birkhead, T. (1991) *The Cambridge Encyclopedia of Ornithology*. Cambridge University Press, Cambridge.

Broude, G. E. and Greene, S. J. (1976) Cross-cultural codes on twenty sexual attitudes and practices. *Ethnology*, **15**, 410–29.

Brownell, R. L. and Ralls, K. (1986) Potential for sperm competition in baleen whales. *Rep. Int. Whal. Commun.* Special Issue **8**, 97–112.

Brownmiller, S. (1975) *Against our Will: Men, Women and Rape*. Simon and Shuster, New York.

Bruce, H. M. (1960) A block to pregnancy in the mouse caused by proximity of strange males. *J. Rep. Fertil.*, **184**, 105.

Bryson, V. (1944) Spermatogenesis and fertility in *Mus musculus* as affected by factors at the T-locus. *J. Morphol.*, **74**, 131–79.

Buckley, T. (1982) Menstruation and the power of Yurok women: methods of cultural reconstruction. *Am. Ethnol.*, **9**, 47–60.

Buhrich, N. J., Bailey, J. M. and Martin, N. G. (1991) Sexual orientation, sexual identity, and sex-dimorphic behaviours in male twins. *Behav. Genet.*, **21**, 75–96.

Bujon, L., Mieusset, R., Mondinat, Ch. *et al.* (1988) Sperm morphology in fertile men and its age related variation. *Andrologia*, **20**, 121–8.

Bullough, V. and Bullough, B. (1978) *Prostitution: An Illustrated Social History*. Crown Publishing Inc., New York.

Burke, T. (1989) DNA fingerprinting and other methods for the study of mating success. *Trends Ecol. Evol.*, **4**, 139–44.

Buss, D. M. (1989) Sex differences in human mate preferences: evolutionary hypotheses tested in 37 cultures. *Behav. Brain Sci.*, **12**, 1–49.

Busse, C. D. and Estep, D. Q. (1984) Sexual arousal in male pigtailed monkeys (*Macaca nemestrina*): effects of serial matings by two males. *J. Comp. Psychol.*, **98**, 227–31.

Cameron, T. W. M. (1956) *Parasites and Parasitism*. Methuen, London.

Campaniello, E., Ucci, N., Ferrari, R. *et al.* (1991) Seasonal variations of human sperm count. *Fertil. Steril.*, **27**, 5–10.

Campbell, R. C. and Jaffe, W. P. (1958) The motility of mixed semen. *J. Agric. Sci.*, **50**, 64–5.

Cappieri, M. (1970) The racial homogeneity of the Andamanese: Part I. *Mankind Qt.*, **10**, 199–212.

Carayon, J. (1974) Insemination traumatique heterosexuelle et homosexuelle chez *Xylocoris maculipennis* (Hem. Anthocoridae). *C.R. Acad. Sci. Paris*, **D 278**, 2803–6.

Carlsen, E., Giwercman, A., Keiding, N. and Skakkebæk, N. E. (1992) Evidence for decreasing quality of semen during the past 50 years. *Br. Med. J.*, **305**, 609–13.

Carpenter, C. R. (1942) Sexual behaviour of free ranging Rhesus monkeys (*Macaca mulatta*) II. Periodicity of oestrous, homosexual, autoerotic and non-conformist behaviour. *J. Comp. Physiol.*, **33**, 143–62.

Carr, D. H. (1967) Chromosome anomalies as a cause of spontaneous abortion. *Am. J. Obstet. Gynecol.*, **97**, 283–8.

Carr, D. H. (1972) Chromosome anomalies in human foetuses. *Res. Reprod.*, **4**, 3–8.

Carre, D. and Sardet, C. (1984) Fertilization and early development in *Beröe ovata. Dev. Biol.*, **105**, 188–95.

Carrick, F. N. and Setchell, B. P. (1977) The evolution of the scrotum, in *Reproduction and Evolution*, (eds J. H. Calaby and C. H. Tyndale-Boscoe), Australian Academy of Sciences, Canberra, pp. 165–70.

Carr-Saunders, A. M. (1922) *The Population Problem*. Oxford University Press, Oxford.

Chadwick, A. (1977) Comparison of milk-like secretions found in non-mammals. *Symp. Zool. Soc. Lond.*, **41**, 341–58.

Chang, M. C. (1951) Fertilizing capacity of spermatozoa deposited in fallopian tubes. *Nature*, **168**, 997–8.

Chang, M. C. (1959) Fertilization of rabbit ova in vitro. *Nature*, **184**, 406.

Chang, M. C. (1965) Fertilization of rabbit ova and sperm longevity. *J. Exp. Zool.*, **158**, 87–92.

Chapman, R. F. (1982) *The Insects: Structure and Function*, 3rd ed, Hodder & Stoughton, London.

Charles, D. and Larsen, B. (1987) Infectious agents as a cause of spontaneous abortion, in *Spontaneous and Recurrent Abortion* (eds M. J. Bennet and D. K. Edmonds) Blackwell, London, pp. 149–67.

Cheney, D. L. and Seyfarth, R. M. (1977) Behaviour of adult and immature male baboons during inter-group encounters. *Nature*, **269**, 404–6.

Cheng, K. M., Burns, J. T. and McKinney, F. (1982) Forced copulation in captive mallards (*Anas platyrhynchos*): II. Temporal factors. *Anim. Behav.*, **30**, 695–9.

Chimbos, P. D. (1978) *Marital Violence: A Study of Interspouse Homicide*. R. and E. Research Associates, San Francisco.

Chretein, F. C. (1989) The saga of the human spermatozoa throughout the jungle of the female genital tract. *Dev. Ultrastruct. Reprod.*, **296**, 263–72.

Chu, S. Y., Berkelman, R. L. and Curran, J. W. (1992) Epidemiology of HIV in the United States, in *AIDS: Etiology, Diagnosis, Treatment and Prevention*, 3rd edn, (eds V. T. DeVita, S. Hellman and S. A. Rosenberg), Lippincott, Philadelphia, pp. 99–109.

Clark, A. B. (1978) Sex ratio and local resource competition in a prosimiam primate. *Science*, **201**, 163–5.

Clark, J. H. and Zarrow, M. X. (1971) Influence of copulation on time of ovulation in women. *Am. J. Obstet. Gynecol.*, **109**, 1083–5.

Clarke, A. L. (in press) Reproductive consequences of human dispersal in 19th century Sweden. *Behav. Ecol.*

Clegg, P. C. and Clegg, A. G. (1975) *Biology of the Mammal*, 4th edn. Heinemann, London.

Clutton-Brock. T. H. (1991) *The Evolution of Parental Care*. Princeton University Press, Princeton, NJ.

Clutton-Brock. T. H. and Harvey, P. H. (1976) Evolutionary rules and primate societies, in *Growing Points in Ethology*, (eds P. P. G. Bateson and R. A. Hinde), Cambridge University Press, Cambridge.

Clutton-Brock, T. H. and Vincent, A. C. J. (1991) Sexual selection and the potential reproductive rates of males and females. *Nature*, **351**, 58–61.

Clutton-Brock, T. H., Guinness, F. E. and Albon, S. D. (1982) *Red Deer: The Behaviour and Ecology of Two Sexes*. Chicago University Press, Chicago.

Clutton-Brock. T. H., Albon, S. D. and Guinness, F. E. (1984) Maternal dominance, breeding success and birth sex ratios in red deer. *Nature*, **308**, 358–60.

Cohen, J. (1967) Correlation between chiasma frequency and sperm redundancy. *Nature*, **215**, 862–3.

Cohen, J. (1969) Why so many sperms? An essay on the arithmetic of reproduction. *Sci. Progr. (Lond.)*, **57**, 23–41.

Cohen, J. (1971) The comparative physiology of gamete populations. *Adv. Comp. Biochem. Physiol.*, **4**, 267–380.

Cohen, J. (1973) Cross-overs, sperm redundancy and their close association. *Heredity*, **31**, 408–13.

Cohen, J. (1975) Gamete redundancy: wastage or selection, in *Gamete Competition in Animals and Plants*, (ed. D. Mulcahy), Pergamon Press, Oxford, pp. 99–112.

Cohen, J. (1977) *Reproduction*. Butterworths, London.

Cohen, J. (1984) Immunological aspects of sperm selection and transport, in *Immunological Aspects of Reproduction in Mammals*, (ed. D. B. Crighton) Butterworths, London, pp. 77–89.

Cohen, J. (1992) The case for and against sperm selection, in *Comparative Spermatology – Twenty Years After*, (ed. B. Bacceti), Raven Press, New York, pp. 65–87.

Cohen, J. and Adeghe, A. J. (1986) The other spermatozoa: fate and functions, in *New Horizons in Sperm Cell Research*, (ed. H. Mohri), Japan Science Society Press, Tokyo, pp. 125–34.

Cohen, J. and McNaughton, D. C. (1974) Spermatozoa: the probable selection of a small population by the genital tract of the female rabbit. *J. Reprod. Fertil.*, **39**, 297–310.

Cohen, J. and Tyler, K. R. (1980) Sperm populations in the female genital tract of the rabbit. *J. Reprod. Fertil.*, **60**, 213–18.

Cohen, J. and Werrett, D. G. (1975) Antibodies and sperm survival in the female tract of the mouse and rabbit. *J. Reprod. Fertil.*, **42**, 301–10.

Collins, W. P., Branch, C. M., Collins, P. O. and Sallam, H. N. (1981) Biochemical indices of the fertile period in women. *Int. J. Fertil.*, **26**, 196–202.

Comfort, A. (1971) Likelihood of human pheromones. *Nature*, **230**, 432–43.

Connor, J. M. and Ferguson-Smith, M. A. (1984) *Essential Medical Genetics*, 3rd edn. Blackwell Scientific Publications, Oxford.

Cook. L. M. (1991) *Genetic and Ecological Diversity: the Sport of Nature*. Chapman & Hall, London.

Cooper, G. W., Overstreet, J. M. and Katz, D. F. (1979) The motility of rabbit spermatozoa recovered from the female genital tract. *Gam. Res.*, **2**, 35–42.

Cords, M. (1987) Forest guenons and patas monkeys: male–male competition in one-male groups, in *Primate Societies*, (eds B. B. Smuts, D. L. Cheney, R. M. Seyfarth *et al.*), Chicago University Press, Chicago, pp. 98–111.

Corner, G. W. (1923) The problem of embryonic pathology in mammals with observations on intrauterine mortality in the pig. *Am. J. Anat.*, **31**, 53–8.

Coulson, J. C. (1966) The influence of the pair-bond and age on the breeding biology of the kittiwake gull, *Rissa tridactyla. J. Anim. Ecol.*, **35**, 269–79.

Cowgill, U. M. (1969) The season of birth and its biological implications. *J. Reprod. Fertil. Suppl.*, **6**, 89–103.

Craft, I. (1984) *In vitro* fertilization clinical methodology. *Br. J. Hosp. Med.*, **7**, 90–102.

Cramp, S. (ed.) (1983) *Handbook of the Birds of Europe, the Middle East and North Africa*, Vol. III. Oxford University Press, Oxford.

Crew, F. A. E. (1952) The factors which determine sex, in *Marshall's Physiology of Reproduction*, Vol. 2, 3rd edn, (ed. A. S. Parkes), Longmans, Green, London, pp. 741–84.

Crews, D. and Young, L. J. (1991) Pseudocopulation in nature in a unisexual whiptail lizard. *Anim. Behav.*, **42**, 512–14.

Crisp, A. H. (1980) *Anorexia Nervosa – Let Me Be*. Academic Press, London.

Croxatto, H. B., Faundes, A., Medel, M. *et al.* (1973) Studies on sperm migration in the human female genital tract. *INSERM.*, **26**, 165–82.

Croxatto, H. B. *et al.* (1975) Sperm migration, in *The Biology of Spermatozoa*, (eds E. S. E. Hafez and C. Thibault), Karger, Basel, pp. 56–62.

Croy, B. A. and Chapeau, G. (1990) Evaluation of the pregnancy immunotropism hypothesis by assessment of the reproductive performance of young adult mice of genotype scid/scid.bg/bg. *J. Reprod. Fertil.*, **88**, 231–9.

Cummins, J. M. (1982) Hyperactivated motility patterns of ram spermatozoa recovered from the oviducts of mated ewes. *Gam. Res.*, **6**, 53–63.

Cummins, J. M. and Woodall, P. F. (1985) On mammalian sperm dimensions. *J. Reprod. Fertil.*, **75**, 153–75.

Cummins, J. M. and Yanagimachi, R. (1982) Sperm–egg ratios and the site of the acrosome reaction during *in vivo* fertilization in hamsters. *Gam. Res.*, **5**, 239–56.

Curry, P. T., Ziemer, T., Van der Horst, G. *et al.* (1989) A comparison of sperm morphology and silver nitrate staining characteristics in the domestic ferret and the black-footed ferret. *Gam. Res.*, **22**, 27–36.

Curtsinger, J. W. (1991) Sperm competition and the evolution of multiple mating. *Am. Nat.*, **138**, 93–102.

Cutler, W. B. (1980) Lunar and menstrual phase locking. *Am. J. Obstet. Gynecol.*, **137**, 834–9.

Cutler, W. B., Preti, G., Erickson, B. *et al.* (1985) Sexual behaviour frequency and ovulatory biphasic menstrual cycle patterns. *Physiol. Behav.*, **34**, 805–10.

Cutler, W. B., Preti, G., Krieger, A. *et al.* (1986) human axillary secretions influence women's menstrual cycles: the role of donor extract from men. *Horm. Behav.*, **20**, 463–73.

Dagg, A. I. (1984) Homosexual behaviour and female–male mounting in mammals: a first survey. *Rev. Mammal.*, **14**, 155–85.

Dahl, J. F. (1986) Cyclic perineal swelling during the intermenstrual intervals of captive female pygmy chimpanzees (*Pan paniscus*). *J. Hum. Evol.*, **15**, 369–85.

Dalton, K. (1964) *The Premenstrual Syndrome*. Thomas, Springfield, Ill.

Dalton, K. (1966) The influence of mother's menstruation on her child. *Proc. R. Soc. Med.*, **59**, 1014.

Daly, M. (1979) Why don't male mammals lactate? *J. Theor. Biol.*, **78**, 325–45.

Daly, M. and Wilson, M. (1978) *Sex, Evolution and Behavior*, Duxbury Press, North Scituate, MA.

Daly, M. and Wilson, M. (1983) *Sex, Evolution and Behaviour*, 2nd edn. PWS Publishers, Boston.

Daly, M. and Wilson, M. (1984) A sociobiological analysis of human infanticide, in *Infanticide: Comparative and Evolutionary Perspectives*, (eds G. Hausfater and S. B. Hrdy), Aldine Press, Hawthorne, NY.

Daly, M. and Wilson, M. (1987) Evolutionary psychology and family violence, in *Sociobiology and Psychology: Issues, Ideas and Applications*, (eds C. Crawford, M. Smith and D. Krebs), Erlbaum, Hillsdale, NJ.

Daly, M., Wilson, M. and Weghorst, S. (1982) Male sexual jealousy. *Ethology and Sociobiology*, **3**, 11–27.

Darling, F. F. (1963) *A Herd of Deer*, Oxford University Press, London.

Darwin, C. R. (1871) *The Descent of Man and Selection in Relation to Sex*. Murray, London.

Daunter, B., Chantler, E. and Elstein, M. (1976) Scanning electron microscopy of cervical mucus. Normal menstrual cycle and pregnancy. *Br. J. Obstet. Gynaecol.*, **1**, 343–58.

Dauzier, L., Thibault, C. and Wintenberger, S. (1954) La fécondation in vitro de l'œuf de lapine. *C. R. Acad. Sci.*, **238**, 844–5.

Davenport, W. (1965) Sexual patterns and their

regulation in a society of the southwest pacific, in *Sex and Behaviour*, (ed. F. A. Beach), Wiley, New York, pp. 164–207.

Davey, D. A. (1986) Normal pregnancy: physiology and antenatal care, in *Dewhurst's Textbook of Obstetrics and Gynaecology for Postgraduates*, (ed. C. R. Whitfield), Blackwell Scientific Publications, London, pp. 126–58.

David, W. A. L. and Gardiner, B. O. C. (1962) Oviposition and hatching of the eggs of *Pieris brassicae* (L.) in a laboratory culture. *Bull. Entomol. Res.*, **53**, 91–109.

Davies, N. B. (1983) Polyandry, cloaca pecking and sperm competition in dunnocks. *Nature*, **302**, 334–6.

Davies, N. B. (1985) Cooperation and conflict among dunnocks, *Prunella modularis*, in a variable mating system. *Anim. Behav.*, **33**. 628–48.

Dawkins, R. (1976) *The Selfish Gene*. Oxford University Press, Oxford.

Dawkins, R. and Krebs, J. R. (1979) Arms races between and within species. *Proc. R. Soc. Lond. B*, **205**, 489–511.

De Boer, P., Van Der Hoeven, F. A. and Chardon, J. A. P. (1976) The production, morphology, karyotypes and transport of spermatozoa from tertiary trisomic mice and the consequences for egg fertilization. *J. Reprod. Fertil.*, **48**, 249–56.

de Carteret, R. J. (1967) Uterine suction during orgasm. *Br. Med. J.*, **1**, 761.

de Kretzer, D., Dennis, P., Hudson, B. *et al.* (1973) Transfer of a human zygote. *Lancet*, **ii**, 728–9.

Demarest, J. and Schoch-Ciuffreda, L. (1992) Sex differences in mate preferences across the lifespan. Human Behaviour and Evolution Society 4th Annual Meeting, Albuquerque, USA, *Abstract no. 62*.

Denny, P. and Ashworth, A. (1991) A zinc-finger protein-encoding gene expressed in the post-meiotic phase of spermatogenesis. *Gene*, **106**, 221–7.

de Ruiter, J. R., Scheffrahn, W., Trommelen, G. J. J. M. *et al.* (1992) Male social rank and reproductive success in wild long-tailed macaques, in *Paternity in Primates*, (ed. R. D. Martin, A. F. Dixson and E. J. Wickings), Karger, Basel, pp. 175–91.

Devine, M. C. (1975) Copulatory plugs in snakes: enforced chastity. *Science*, **187**, 844–5.

Dewsbury, D. A. (1982) Ejaculate cost and male choice. *Am. Nat.*, **119**, 601–10.

Dewsbury, D. A. (1984) Sperm competition in muroid rodents, in *Sperm Competition and the Evolution of Animal Mating Systems*, (ed. R. L. Smith), Academic Press, New York, pp. 547–71.

Dewsbury, D. A. (1985) Interactions between males and their sperm during multimale copulatory episodes of deer mice (*Peromyscus maniculatus*). *Anim. Behav.*, **33**, 1266–74.

Dewsbury, D. A. (1988) A test of the role of copulatory plugs in sperm competition in deer mice. (*Peromyscus maniculatus*). *J. Mammal.*, **69**, 854–7.

Dewsbury, D. A. and Hartung, T. G. (1980) Copulatory behaviour and differential reproduction in a two-male, one-female competitive situation. *Anim. Behav.*, **28**, 95–102.

Diagram Group (1981) *Sex: A User's Manual*. Coronet Books, Hodder & Stoughton, London.

Diamond, J. M. (1986) Variation in human testis size. *Nature*, **320**, 488–9.

Diamond, J. M. (1987) Causes of death before birth. *Nature*, **329**, 487–8.

Díaz, S., Aravena, R. Càrdenas, H. *et al.* (1991) Contraceptive efficacy of lactational amenorrhea in urban Chilean women. *Contraception*, **43**, 335–52.

Dickemann, M. (1979) The ecology of mating systems in hypergynous dowry societies. *Soc. Sci. Inf.*, **18**, 163–95.

Diesel, R. (1990) Sperm competition and reproductive success in the decapod *Inachus plalangium* (Majidae): a male ghost spider crab that seals off rivals' sperm. *J. Zool.*, **220**, 213–23.

Dirasse, L. (1978) The socioeconomic position of women in Addis Ababa: the case of prostitution. PhD Dissertation, Boston University.

Dixson, A. F. (1987) Observations on the evolution of the genitalia and copulatory behaviour in male primates. *J. Zool. Lond.*, **213**, 423–44.

Doak, R. H., Hall, E. and Dale, H. F. (1967) Sperm longevity in dogs. *J. Reprod. Fertil.*, **13**, 51–61.

Dobson, H. (1988) Softening and dilating of the uterine cervix, in *Oxford Reviews in Reproductive Biology*, Vol. 10, (ed. J. R. Clarke) Academic Press, New York, pp. 240–6.

Döring, G. K. (1969) The incidence of anovular cycles in women. *J. Reprod. Fertil., Suppl.*, **6**, 77–81.

Dott, H. M. and Walton, A. (1958) Motility and survival of spermatozoa in mixed semen from different bulls. *J. Agric. Sci., Camb.*, **50**, 267–72.

Doty, R. L. (1976) Reproductive endocrine influences upon human nasal chemoreception: a review, in *Mammalian Olfaction, Reproductive Processes, and Behavior*, (ed. R. L. Doty), Academic Press, New York, pp. 295–321.

Doty, R. L. (1978) A review of recent psychophysical studies examining the possibility of chemical communication of sex and reproductive status in humans. *Science*, **196**, 273–86.

Doty, R. L., Ford, M. and Preti, G. (1975) Changes in the intensity and pleasantness of human vaginal odours during the menstrual cycle. *Science*, **190**, 1316–17.

Dreizen, S., Spirikis, C. N. and Stone, R. E. (1967) A comparison of skeletal growth and maturation in undernourished and well nourished girls before and after manarche. *J. Pediatr.*, **70**, 256–63.

Drevius, L. O. (1975) Permeability of the bull sperm membrane, in *The Functional Anatomy of the Spermatozoan*, (ed. B. A. Afzelius), Pergamon Press, Oxford, pp. 373–83.

Duerden, B. I., Reid, T. M. S., Jewsbury, J. M. and Turk, D. C. (1987) *A New Short Textbook of Microbial and Parasitic Infection*. Hodder & Stoughton, London.

Dukelow, W. R. (1978) Ovulation detection and control relative to optimal time of mating in non-human primates. *Symp. Zool. Soc. Lond.*, **43**, 195–206.

Dunbar, R. I. M. (1980) Demographic and life history variables of a population of gelada baboons (*Theropithecus gelada*). *J. Anim. Ecol.*, **49**, 485–506.

Dunham, C., Myers, F., Bernden, N. *et al.* (1991) *Mamatoto: A Celebration of Birth*. The Body Shop/Virago, London.

Dziuk, P. J. (1965) Double mating of rabbits to determine capacitation time. *J. Reprod. Fertil.*, **10**, 389–95.

Eberhard, W. G. (1985) *Sexual Selection and Animal Genitalia*. Harvard University Press, Cambridge.

Eberhard, W. G. (1990) Inadvertent machismo? *Trends Ecol. Evol.*, **5**, 263.

Eberhard, W. G. (1991) Copulatory courtship and cryptic female choice in insects. *Biol. Rev.*, **66**, 1–31.

Eckert, E. D., Bouchard, T. J., Bohlen, J. and Heston, L. L. (1986) Homosexuality in monozygotic twins reared apart. *Br. J. Psychiatry*, **148**, 421–5.

Eckstein, P. and Zuckerman, S. (1956) Morphology of the reproductive tract, in *Marshall's Physiology of Reproduction*, Vol. 1, (ed. A. S. Parkes), Longmans, Green, London, pp. 43–155.

Eddy, E. M. (1988) The spermatozoon, in *Physiology of Reproduction*, Vol. 1, (eds E. Knobil and J. D. Neill). Raven Press, New York, pp. 27–68.

Edwards, J. H. (1957) A critical examination in the reputed primary influence of ABO phenotype on fertility and sex ratio. *Br. J. Prev. Soc. Med.*, **11**, 79–89.

Edwards, M. (1993) Sex in the nineties. *Arena*, no. 38.

Eisenbach, M. and Ralt, D. (1992) Pre-contact mammalian sperm-egg communication and role in fertilization. *Am. J. Physiol.*, **28**, 6–14.

El-Banna, A. A. and Hafez, E. S. E. (1972) The uterine cervix in animals. *Am. J. Obstet. Gynecol.*, **112**, 145–64.

Eliasson, R. and Lindholmer, Ch. (1972) Distribution and properties of spermatozoa in different fractions of split ejaculates. *Fertil. Steril.*, **23**, 252–6.

Eliasson, R. and Lindholmer, Ch. (1973) Effects of human seminal plasma on sperm survival and transport. *INSERM*, **26**, 215–30.

Ellis, L. (1989) *Theories of Rape: Inquiries into the Causes of Sexual Aggression*, Hemisphere Publishing Corporation, New York.

Elstein, M. and Daunter, B. (1976) The structure of cervical mucus, in *The Cervix*, (ed. D. Y. Jordan and M. E. Singer). W.B. Saunders, London.

Elstein, M., Moghissi, K. S. and Borth, R. (1972) *Cervical Mucus in Human Reproduction*, Scriptor, Copenhagen.

Elwood, R. W. (1991) Ethical implications of studies on infanticide and maternal aggression in rodents. *Anim. Behav.*, **42**, 841–50.

Elwood, R. W. (1992) Pup-cannibalism in rodents: causes and consequences, in *Cannibalism: Ecology and Evolution Among Diverse Taxa*, (eds M. Elgar and B. Crespi), Oxford University Press, Oxford.

Ely, J., Alford, P. and Ferrel, R. E. (1991) DNA 'fingerprinting' and the genetic management of a captive chimpanzee population (*Pan troglodytes*). *Am. J. Primat.*, **24**, 39–54.

Emlen, S. T. and Wrege, P. H. (1986) Forced copulations and intra-specific parasitism: two costs of social living in the white-fronted bee-eater. *Ethology*, **71**, 2–29.

Enginsu, M. E., Dumoulin, J. C. M., Pieters, H. E. C. *et al.* (1993) Predictive value of morphologically normal sperm concentration in the medium for in-vitro fertilization. *Int. J. Androl.*, **16**, 113–20.

Englert, Y., Puissant, F., Camus, M. *et al.* (1986) Factors leading to tripronucleate eggs during human in-vitro fertilization. *Human Reprod.*, **1**, 117–19.

Ericsson, R. J., Langevin, C. N. and Nishino, M. (1973) Isolation of fractions rich in human Y sperm. *Nature*, **246**, 421–4.

Essock-Vitale, S. M. and McGuire, M. T. (1985) Women's lives viewed from an evolutionary perspective: I. Sexual histories, reproductive success, and demographic characteristics of a random subsample of American women. *Ethol. Sociobiol.*, **6**, 137–54.

Estep, D. Q. (1988) Copulations by other males shorten the post-ejaculatory intervals of pairs of roof rats, *Rattus rattus*. *Anim. Behav.*, **36**, 299–300.

Estep, D. Q., Gordan, T. P., Wilson, M. E. and Walker, M. L. (1986) Social stimulation and the resumption of copulation in Rhesus (*Macaca mulatta*) and stumptail (*Macaca arctoides*) macaques. *Int. J. Primatol.*, **7**, 507–17.

Evans, E. I. (1933) The transport of spermatozoa in the dog. *Am. J. Physiol.*, **105**, 287–93.

Evans, H. E. (1984) *Insect Biology: A Textbook of Entomology*. Addison-Wesley, London.

Eveleth, P. B. and Tanner, J. M. (1976) *Worldwide Variation in Human Growth*. Cambridge University Press, Cambridge.

Farr, J. A. (1980) The effects of sexual experience and female receptivity on courtship rape decisions in male guppies, *Poecilia reticulata (Pisces: poeciliidae)*. *Anim. Behav.*, **28**, 1195–201.

Farris, E. J. and Murphy, D. P. (1960) The characteristics of the two parts of the partitioned ejaculate and the advantages of its use for intrauterine insemination. *Fertil. Steril.*, **11**, 465–70.

Fausto-Sterling, A. (1992) *Myths of Gender*. Basic Books, New York.

Faux, S. F. and Miller, H. L. (1984) Evolutionary speculations on the oligarchic development of Mormon polygyny. *Ethol. Sociobiol.*, **5**, 15–31.

Fenton, M. B. (1984) The case of vespertilionid and rhinolophid bats, in *Sperm Competition and the Evolution of Animal Mating Systems*, (ed. R. L. Smith), Academic Press, London, pp. 573–87.

Ferin, J., Thomas, K. and Johansson, E. D. B. (1973) Ovulation detection, in *Human Reproduction: Conception and Contraception*, (eds E. S. E. Hafez and T. N. Evans), Harper & Row, New York, pp. 260–83.

Finn, C. A. (1964) Influence of the male on litter size in mice. *J. Reprod. Fertil.*, **7**, 107–11.

Fisher, R. A. (1930) *The Genetical Theory of Natural Selection*. Clarendon Press, Oxford.

Fleagle, J. G. (1988) *Primate Adaption and Evolution*. Academic Press, London.

Flinn, M. V. (1988) Mate guarding in a Caribbean village. *Ethol. Sociobiol.*, **9**, 1–28.

Fooden, J. (1967) Complementary specialization of male and female reproductive structures in the bear macaque, *Macaca arctoides*. *Nature*, **214**, 939–41.

Ford, C. S. and Beach, F. A. (1952) *Patterns of Sexual Behaviour*. Eyre & Spottiswoode, London.

Fortune, R. F. (1963) *The Sorcerers of Dobu*, Dutton: New York.

Fox, W. (1956) Seminal receptacles of snakes. *Anat. Rec.*, **124**, 519–97.

Fox, W. (1963) Special tubules for sperm storage in female lizards. *Nature*, **198**, 500–1.

Fox, C. A., Wolff, H. S. and Baker, J. A. (1970) Measurement of intra-vaginal and intra-uterine pressures during human coitus by radio-telemetry. *J. Reprod. Fertil.*, **22**, 243–51.

France, J. T. (1981) Sperm longevity. *Int. J. Fertil.*, **26**, 143–52.

Francis, C. M. Anthony, E. L. P., Brunton, J. A. and Kunz, T. H. (1994) Lactation in male fruit bats. *Nature*, **367**, 691–2.

Frank, O. and Bongaarts, J. (1991) Behavioural and biological determinants of fertility transition in sub-saharan africa. *Stat. Med.*, **10**, 161–75.

Frazer, J. F. D. (1955) Fetal death in the rat. *J. Embryol. Exp. Morphol.*, **3**, 13–18.

Fredricsson, B. (1978) On the development of different morphologic abnormalities of human sperm. *Andrologia*, **10**, 43–8.

French, F. and Bierman, J. M. (1962) Probabilities of fetal mortality. *Public Health Rep.*, **77**, 835–8.

Frisch, R. E. (1984) Body fat, puberty and fertility. *Biol. Rev.*, **59**, 161–88.

Frisch, R. E. and McArthur, J. W. (1974) Menstrual cycles: fitness as a determinant of minimum weight for height necessary for their maintenance or onset. *Science*, **185**, 949–51.

Frommer, D. J. (1964) Changing age of the menopause. *Br. Med. J.*, **5405**, 349–51.

Fuchs, A. R. (1972) Uterine activity during and after mating in the rabbit. *Fertil. Steril.*, **23**, 915–23.

Fukuda, M., Morales, P. and Overstreet, J. W. (1989) Acrosomal function of human spermatozoa with normal and abnormal head morphology. *Gam. Res.*, **24**, 59–65.

Gaddum-Rosse, P. (1981) Some observations on sperm transport through the uterotubal junction of the rat. *Am J. Anat.*, **160**, 333–41.

Gage, M. J. G. (1991) Risk of sperm competition directly affects ejaculate size in the Mediterranean fruit fly. *Anim. Behav.*, **42**, 1036–7.

Gage, M. J. G. (1992) Removal of rival sperm during copulation in a beetle, *Tenebrio molitor*, **44**, 587–9.

Gage, M. J. G. and Baker, R. R. (1991) Ejaculate size varies with socio-sexual situation in an insect. *Ecol. Entomol.*, **16**, 331–7.

Galbraith, D. A. (1993) Multiple paternity and sperm storage in turtles. *Herpetol. J.*, **3**, 117–23.

Garcia, J. E., Jones, G. S. and Wright, G. L. (1981) Prediction of the time of ovulation. *Fertil. Steril.*, **36**, 308–15.

Garn, S. M. and LaVelle, M. (1983) Reproductive

histories of low weight girls and women. *Am. J. Clin. Nutr.*, **37**, 862–6.
Garton, J. S. (1972) Courtship of the small-mouthed salamander, *Ambystoma texanum*, in southern Illinois. *Herpetologica*, **28**, 145–66.
Gee, H. (1991) The brave new adventure. *Nature*, **354**, 268.
Geer, J. de (1758) in Imms (1951) *A General Textbook of Entomology: Including the Anatomy, Physiology, Development and Classification of Insects*, 8th edn, Methuen, London.
Gerhard, I., Lenard, K., Eggert-Kruse, W. and Runnebaum, B. (1992) Clinical data which influence semen parameters in infertile men. *Hum. Reprod.*, **7**, 830–7.
Giblin, P. T., Poland, M. L., Moghissi, K. S. *et al.* (1988) The effects of stress and characteristic adaptability on semen quality in healthy men. *Fertil. Steril.*, **49**, 127–32.
Gibson, R. M. and Jewell, P. A. (1982) Semen quality, female choice and multiple mating in domestic sheep: a test of Triver's sexual competence hypothesis. *Behaviour*, **80**, 9–31.
Gilbert, L. E. (1976) Postmating female odour in *Heliconius* butterflies: a male-contributed anti-aphrodisiac. *Science*, **193**, 419–20.
Ginsberg, J. R. and Huck, U. W. (1989) Sperm competition in mammals. *Trends Ecol. Evol.*, **4**, 74–9.
Ginsberg, J. R. and Rubenstein, D. I. (1990) Sperm competition and variation in zebra mating behaviour. *Behav. Ecol. Sociobiol.*, **26**, 427–34.
Gledhill, B. L. (1987) Gender preselection by sperm separation, in *New Horizons in Sperm Cell Research*, (ed. H. Mohri), Japan Science Societies Press, Tokyo, pp. 501–10.
Godfray, H. C. J. and Harvey, P. H. (1991) More fecund but not so fit. *Nature*, **354**, 190–1.
Goldenson, R. and Anderson, K. (1986) *Everything you Ever Wanted to Know About Sex – But Never Dared Ask*. Bloomsbury, London.
Goldizen, A. W. (1987) Tamarins and marmosets: communal care of offspring, in *Primate Societies* (ed. B. B. Smuts, D. L. Cheney, R. M. Seyfarth *et al.*), Chicago University Press, Chicago, pp. 34–43.
Goldman, S. E. and Schneider, H. G. (1987) Menstrual synchrony: social and personality factors. *J. Soc. Behav. Personal.*, **2**, 243–50.
Gomendio, M. and Roldan, E. R. S. (1991) Sperm competition influences sperm size in mammals. *Proc. R. Soc. B.*, **243**, 181–5.
Gomendio, M. and Roldan, E. R. S. (1993) Mechanisms of sperm competition: linking physiology and behavioural ecology. *Trends Ecol. Evol.*, **8**, 95–100.
Goodall, J. (1986) *The Chimpanzees of Gombe: Patterns of Behavior*. Harvard University Press, Cambridge, MA.
Gould, S. J. (1987a) Freudian slip. *Nat. Hist.*, **96**, 14–21.
Gould, S. J. (1987b) Stephen Jay Gould replies. *Nat. Hist.*, **96**, 4–6.
Goy, R. W. and Roy, M. (1991) Heterotypical sexual behaviour in female mammals, in *Heterotypical Behaviour in Man and Animals*, (eds M. Haug, P. F. Brian and C. Aron), Chapman & Hall, London, pp. 71–97.
Grafen, A. (1990) Biological signals as handicaps. *J. Theor. Biol.*, **144**, 517–46.
Graham, C. A. and McGrew, W. C. (1980) Menstrual synchrony in female undergraduates living on a coeducational campus. *Psychoneuroendocrinology*, **5**, 245–52.
Grammer, K., Dittami, J. and Fischmann, B. (1993) Changes in female sexual advertisement according to menstrual cycle. International Congress of Ethology. Torremolinos.
Gray, R. H. (1983) The impact of health and nutrition on natural fertility, in *Determinants of Fertility in Developing Countries* Vol. 1, *Supplement and demand for children*, (eds R. A. Bulatao and R. D. Lee), Academic Press, New York, pp. 139–62.
Green, R. (1980) Variant forms of human sexual behaviour, in *Reproduction in Mammals*, Book 8, *Human Sexuality*, (eds C. R. Austin and R. V. Short), Cambridge University Press, Cambridge, pp. 68–97.
Gross, M. R. and Shine, R. (1981) Parental care and mode of fertilization in ectothermic vertebrates. *Evolution*, **35**, 775–93.
Guilloud, N. B. (1968) Personal Communication in Talbert (1977).
Gurman, S. J. (1989) Six of one . . . *Nature*, **342**, 12.
Guyton, A. C. (1991) *Textbook of Medical Physiology*, W. B. Saunders, London.
Gwatkin, R. B. L., Anderson, O. F. and Hutchison, C. F. (1972) Capacitation of hamster spermatozoa *in vitro*: the role of cumulus components. *J. Reprod. Fertil.*, **30**, 389–94.
Gwynne, D. T. (1981) Sexual difference theory: Mormon crickets show role reversal in mate choice. *Science*, **213**, 779–80.
Gyllensten, U. B., Jakobsson, S. and Temrin, H. (1990) No evidence for illegitimate young in monogamous and polygynous warblers. *Nature*, **343**, 168–70.
Hafez, E. S. E. (1971) Reproductive cycles, in *Comparative Reproduction of Nonhuman Primates*, (ed. E. S. E. Hafez), C. C. Thomas, Springfield, Ill.
Hafez, E. S. E. (1972) The comparative anatomy

of the mammalian cervix, in *Biology of the Cervix*, (eds R. J. Blandau and K. S. Moghissi), University of Chicago Press, Chicago, pp. 23–55.

Hafez, E. S. E. (1973) Transport of spermatozoa in the female reproductive tract. *Am. J. Ostet. Gynecol.*, **145**, 703–17.

Hall, K. R. L. and DeVore, I. (1965) Baboon social behavior, in *Primate Behavior*, (ed. I. DeVore), Holt, Rinehardt & Winston, New York.

Halliday, T. R. (1980) *Sexual Strategy*, Oxford University Press, Oxford.

Halliday, T. R. (1983) The study of mate choice, in *Mate Choice*, (ed. P. Bateson), Cambridge, Cambridge University Press, pp. 3–32.

Halliday, T. R. and Verrell, P. A. (1984) Sperm competition in amphibians, in *Sperm Competition and the Evolution of Animal Mating Systems*, (ed. R. L. Smith), Academic Press, London, pp. 487–508.

Hamilton, G. V. (1914) A study of sexual tendencies in monkeys and baboons. *J. Anim. Behav.*, **IV**, 295–318.

Hamilton, G. V. (1929) *A Research in Marriage*, A. C. Boni, New York.

Hamilton, W. D. (1964) The genetical evolution of social behaviour. I,II. *J. Theor. Biol.*, **7**, 1–52.

Hamilton, W. D. (1967) Extraordinary sex ratios. *Science*, **156**, 477–88.

Hamilton W. D. and Zuk, M. (1982) Heritable true fitness and bright birds: a role for parasites? *Science*, **218**, 384–7.

Hamilton, W. J. (1984) Significance of paternal investment by primates to the evolution of male–female associations, in *Primate Paternalism*, (ed. D. M. Taub), Van Nostrand Reinhold, New York.

Hamer, D. H. (1993) Sexual orientation. *Nature*, **365**, 702.

Hamer, D. H., Hu, S., Magnuson, V. L. *et al.* (1993) A linkage between DNA markers on the X chromosome and male sexual orientation. *Science*, **261**, 321–7.

Hammer, C. E., Jennings, L. L. and Sojka, N. J. (1970) Cat (*Feris catus* L.) spermatozoa require capacitation. *J. Reprod Fertil.*, **23**, 477–80.

Hanby, J. P. (1972) The sociosexual nature of mounting and related behaviors in a confined troop of Japanese macaques (*Macaca fuscata*). PhD Thesis, University of Oregon.

Hancock, R. J. T. (1978) Sperm antigens and sperm immunogenicity, in *Spermatozoa, Antibodies and Infertility*, (eds J. Cohen and W. Hendry), Blackwell Scientific Publications, Oxford, pp. 1–28.

Happ, G. M. (1984) Development and reproduction, in *Insect Biology: A Textbook of Entomology*, (ed. H. E. Evans), Addison-Wesley, London, pp. 93–114.

Harcourt, A. H. (1989) Deformed sperm are probably not adaptive. *Anim. Behav.*, **37**, 863–5.

Harcourt, A. H. (1991) Sperm competition and the evolution of nonfertilizing sperm in mammals. *Evolution*, **45**, 314–28.

Harcourt, A. H., Harvey, P. H., Larson, S. G. and Short, R. V. (1981) Testis weight, body weight and breeding systems in primates. *Nature*, **293**, 55–7.

Hardy, J. B. and Mellits, E. D. (1977) Relationship of low birth weight to maternal characteristics of age, parity, education and body size, in *The Epidemiology of Prematurity* (eds D. M. Reed and F. J. Stanley), Urban & Scwarzenberg, Baltimore, pp. 105–17.

Harper, M. J. K. (1988) Gamete and zygote transport, in *The Physiology of Reproduction*, (eds E. Knobil and J. D. Neill), Raven Press, London, pp. 103–35.

Hartl, D. L. and Clark, A. G. (1989) *Principles of Population Genetics*, 2nd edn. Sinauer Associates, Sunderland, MA.

Hartman, C. G. (1924) Observations on the motility of the Opposum genital tract and the vaginal plug. *Anat. Rec.*, **27**, 293–303.

Hartman, C. G. (1957) How do sperms get into the uterus? *Fertil. Steril.*, **8**, 403–27.

Hartmann, P. E., Rattingan, S., Prosser, C. G. *et al.* (1984) Human lactation: back to nature. *Symp. Zool. Soc. Lond.*, **5 1**, 337–68.

Hartung, J. (1985) Matrilineal inheritance: new theory and analysis. *Behav. Brain Sci.*, **8**, 661–88.

Hartz, A. J., Rupley, D. C. and Rimm, A. A. (1984) The association of girth measurement with disease in 32,856 women. *Am. J. Epidemiol.*, **119**, 71–80.

Harvey, C. (1956) The use of partitioned ejaculates in investigating the role of accessory secretions in human semen, in *Studies on Fertility*, Vol. 8. Blackwell, Oxford, pp. 3–10.

Harvey, C. (1960) The speed of human spermatozoa and the effect on it of various dilutents with some preliminary observations on clinical material. *J. Reprod. Fertil.*, **1**, 84–95.

Harvey, P. H. and Harcourt, A. H. (1984) Sperm competition, testis size, and breeding systems in primates, in *Sperm Competition and the Evolution of Animal Mating Systems*, (ed. R. L. Smith), Academic Press, London, pp. 589–600.

Harvey, P. H. and May, R. M. (1989) Out for the sperm count. *Nature.*, **337**, 508–9.

Harvey, P. H., Martin, R. D. and Clutton-Brock, T. H. (1987) Life histories in comparative perspective, in *Primate Societies* (eds B. B. Smuts, D. L. Cheney, R. M. Seyfarth *et al.* Chicago University Press, Chicago, pp. 181–96.

Harvey, S. M. (1987) Female sexual behaviour: fluctuations during the menstrual cycle. *J. Psychosom. Res.*, **31**, 101–10.

Hausfater, G. (1975) Dominance and reproduction in baboons (*Papio cynocephalus*), in *Contributions to Primatology*, Vol. 7, S. Karger, Basel.

Hawker, R. W. (1984) *Notebook on Medical Physiology: Endocrinology*. Churchill Livingstone, London.

Heape, W. (1900) The sexual season of mammals and the relation of prooestrum to menstruation. *Q. J. Microsc. Sci.*, **44**, 1–70.

Hecht, N. B., Bower, P. A., Waters, S. H. *et al.* (1990) Evidence for haploid expression of mouse testicular genes. *Exp. Cell Res.*, **164**, 183–90.

Hendrickx, A. G. and Kraemer, D. C. (1969) Observations on the menstrual cycle optimal mating time and pre-implantation embryos of the baboon *Papio anubis* and *Papio cynocephalus*. *J. Reprod. Fertil., Suppl.*, **6**, 119–28.

Hertig, A. T. and Rock, J. (1959) A series of potentially abortive ova recovered from fertile women prior to the first missed menstrual period. *Am. J. Obstet. Gynecol.*, **58**, 968–73.

Hertig, A. T., Rock, J., Adams, E. C. and Menkin, M. C. (1959) Thirty-four fertilized human ova, good, bad and indifferent, recovered from 210 women of known fertility. *Pediatrics*, **23**, 202–7.

Heske, E. J. and Ostfeld, R. S. (1990) Sexual dimorphism in size, relative size of testes, and mating systems in North American voles. *J. Mammal.*, **71**, 510–19.

Hess, E. A., Ludwick, T., Rickard, H. C. and Ely, F. (1954) Some of the influences of mixed ejaculates upon bovine fertility. *J. Dairy Sci.*, **37**, 649.

Hessell, L. (1992) *Windows on Love: the Ultimate Guide to Sexual Fulfilment*, Crawford House Press, Bathurst, NSW.

Heston, L. L. and Shields, J. (1968) Homosexuality in twins: a family study and a registry study. *Arch. Gen. Psychiatry*, **18**, 149–60.

Heyn, A. (1924) Über sexuelle träume bie frauen. *Arch. Frauenk.*, **60**–5.

Hilgers, T. W. and Bailey, A. J. (1980) Natural family planning. II. Basal body temperature and estimated time of ovulation. *Obstet. Gynecol.*, **55**, 333–9.

Hilgers, T. W., Abraham, G. E. and Cavanagh, D. (1978) Natural family planning I. The peak symptom and estimated time of ovulation. *Obstet. Gynecol.*, **52**, 575–80.

Hill, A., Ward, S., Deino, A. *et al.* (1992) Earliest *Homo*. *Nature*, **355**, 719–22.

Hill, K. and Kaplan, H. (1988) Tradeoffs in male and female reproductive strategies among the Ache: part 2, in *Human Reproductive Behaviour: A Darwinian Perspective*, (eds L. Betzig, M. B. Mulder and P. Turke), Cambridge, Cambridge University Press, pp. 291–306.

Hill, W. C. O. (1953) *Primates: Comparative Anatomy and Taxonomy*, I–*Strepsirhini*, Edinburgh University Press, Edinburgh.

Hill, W. C. O. (1955) *Primates: Comparative Anatomy and Taxonomy*, II–*Haplorhini:Tarsioidea*, Edinburgh University Press, Edinburgh.

Hill, W. C. O. (1957) *Primates: Comparative Anatomy and Taxonomy*, III–*Pithecoidea Platyrrhini*, Edinburgh University Press, Edinburgh.

Hill, W. C. O. (1960) *Primates: Comparative Anatomy and Taxonomy*, IV–*Cebidae Part A*, Edinburgh University Press, Edinburgh.

Hill, W. C. O. (1962) *Primates: Comparative Anatomy and Taxonomy*, V–*Cebidae Part B*, Edinburgh University Press, Edinburgh.

Hill, W. C. O. (1966) *Primates: Comparative Anatomy and Taxonomy*, VI–*Catarrhini Cercopithecoidea*, Edinburgh University Press, Edinburgh.

Hill, W. C. O. (1970a) *Primates: Comparative Anatomy and Taxonomy*, VIII–*Cynopithecinae*, Edinburgh University Press, Edinburgh.

Hill, W. C. O. (1970b) *Sperm Competition and the Evolution of the Reproductive Organs in Primates*. Edinburgh University Press, Edinburgh.

Hill, W. C. O. (1974) *Primates: Comparative Anatomy and Taxonomy*. VII–*Cynopithecinae*. Edinburgh University Press, Edinburgh.

Hill, W. C. O. and Matthews, L. H. (1949) The male external genitalia of the gorilla, with remarks on the os penis of other Hominoidea. *Proc. Zool. Soc. Lond.*, **119**, 363–78.

Hinde, R. A. (1974) *Biological Basis of Human Social Behaviour*. McGraw-Hill, London.

Hiraiwa-Haengawa, M. (1993) Skewed birth sex ratios in primates: should high-ranking mothers have daughters or sons? *Trends Ecol. Evol.*, **8**, 395–9.

Hite, S. (1976) *The Hite Report*. Macmillan, New York.

Hogg, J. T. (1984) Mating in bighorn sheep: multiple creative male strategies. *Science*, **225**, 526–8.

Hogg, J. T., Hass, C. C. and Jenni, D. A. (1992) Sex-biased maternal expenditure in Rocky Mountain bighorn sheep. *Behav. Ecol. Sociobiol.*, **31**, 243–51.

Högland, A. and Odeblad, E. (1977) Sperm penetration in the cervical mucus: a biophysical and

group-theoretical approach, in *The Uterine Cervix in Reproduction*, (eds V. Insler and G. Bettendorf), Georg Theine, Stuttgart, pp. 129–34.
Holstein, A. F. (1975) Morphologische Studien an abnormalen Spermatiden und Spermatozoen des Menschen. *Virchows Arch* [A], **367**, 93–112.
Howell, N. (1979) *Demography of the Dobe !Kung*, Academic Press, New York.
Hrdy, S. B. (1977) *The Langurs of Abu*, Harvard University Press, Cambridge.
Hrdy, S. B. (1979) Infanticide among animals: a review, classification, and examination of the implications for the reproductive strategies of females. *Ethol. Sociobiol.*, **1**, 13–40.
Hrdy, S. B. (1981) *The Woman that Never Evolved*, Harvard University Press, Cambridge.
Hrdy, S. B. and Whitten, P. L. (1987) Patterning of sexual activity, in *Primate Societies*, (eds B. B. Smuts, D. L. Cheney, R. M. Seyfarth *et al.*), Chicago University Press, Chicago, pp. 370–84.
Huang, T. T. F. and Yanagimachi, R. (1984) Fucoidin inhibits attachment of guinea pig spermatozoa to the zona pellucida through binding to the inner acrosomal membrane and equatorial domains. *Exp. Cell Res.*, **153**, 363–73.
Huang, T. T. F., Flemming, A. D. and Yanagimachi, R. (1981) Only acrosome-reacted spermatozoa can bind and penetrate into zona pellucida: a study using guinea pig. *J. Exp. Zool.*, **217**, 286–90.
Hubbard, R. and Wald, E. (1993) *Exploding the Gene Myth*, Beacon, New York.
Huck, U. W. (1982) Pregnancy block in laboratory mice as a function of male social status. *J. Reprod. Fertil.*, **66**, 181–4.
Huck, U. W., Quinn, R. P. and Lisk, R. D. (1985) Determinants of mating success in the golden hamster (*Mesocricetus auratus*) IV. Sperm competition. *Behav. Ecol. Sociobiol.*, **17**, 239–52.
Hull, D. and Johnston, D. I. (1993) *Essential Paediatrics*, Churchill Livingstone, London.
Hunter, F. M., Petrie, M., Otronen, M. *et al.* (1993) Why do females copulate repeatedly with one male? *Trends Ecol. Evol.*, **8**, 21–6.
Hunter, R. H. F. (1973) Transport, migration and survival of spermatozoa in the female genital tract: species with intra-uterine deposition of semen. *INSERM*, **26**, 309–42.
Hunter, R. H. F. (1981) Sperm transport and reservoirs in the pig oviduct in relation to time of ovulation. *J. Reprod. Fertil.*, **63**, 109–17.
Hunter, R. H. F. (1987a) Human fertilization in vivo, with special reference to progression storage and release of competent sperm. *J. Reprod. Fertil.*, **72**, 203–11.
Hunter, R. H. F. (1987b) Peri-ovulatory physiology of the oviduct with special reference to progression storage and capacitation of spermatozoa, in *New Horizons in Sperm Cell Research*, (ed. H. Mohri), Japan Science Societies Press, Tokyo, pp. 31–45.
Hunter, R. H. F. (1988) *The Fallopian Tubes*. Springer-Verlag, Berlin.
Hunter, R. H. F. (1991) Oviduct function in pigs with particular reference to the pathological condition of polyspermy. *Mol. Reprod. Dev.*, **29**, 385–91.
Huntingdon, E. (1938) *The Season of Birth*, John Wiley, New York.
Huxley, J. S. (1912) A 'disharmony' in the reproductive habits of the wild duck (*Anas boschas* L.) *Biol. Zentralbl*, , **32**, 621–3.
Imms, A. D. (1951) *A General Textbook of Entomology: Including the Anatomy, Physiology, Development and Classification of Insects*, 8th edn, Methuen, London.
Inoue, M., Takenaka, A., Tanaka, S. *et al.* (1990) Paternity discrimination in a Japanese macaque group by DNA fingerprinting. *Primates*, **31**, 563–70.
Inoue, M., Mitsunaga, F., Ohsawa, H. *et al.* (1991) Male mating behaviour and paternity discrimination by DNA fingerprinting in a Japanese macaque group. *Folia Primatol.*, **56**, 202–10.
Inoue, M., Mitsunaga, F. Ohsawa, H. *et al.* (1992) Paternity testing in captive Japanese macaques (*Macaca fuscata*) using DNA fingerprinting, in *Paternity in Primates*, (eds R. D. Martin, A. F. Dixson and E. J. Wickings), Karger, Basel, pp. 141–54.
Insler, V., Glezerman, M., Zeidel, L. *et al.* (1980) Sperm storage in the human cervix: a quantative study. *Fertil. Steril.*, **33**, 288–93.
Jaffe, S. B., Jewelewicz, R., Wahl, E. and Khatamee, M. A. (1991) A controlled study for gender selection. *Fertil. Steril.*, **56**, 254–8.
James, W. H. (1971) The distribution of coitus within the human intermenstruum. *J. Biosoc. Sci.*, **3**, 159–71.
Jamieson, I. G. (1989) Levels of analysis or analyses at the same level. *Anim. Behav.*, **37**, 696–7.
Jarett, L. R. (1984) Psychosocial and biological influences on menstruation: synchrony, cycle length, and regularity. *Psychoneuroendocrinology*, **9**, 21–8.
Jarvis, J. U. M. (1981) Eusociality in a mammal: cooperative breeding in naked mole rat colonies. *Science*, **212**, 571–3.
Jaszczak, S. and Hafez, E. S. E. (1973) Physiopathology of sperm transport through the human vagina and cervix. *INSERM*, **26**, 557–74.

Jerison, H. J. (1973) *Evolution of the Brain and Intelligence*. Academic Press, New York.

Jöchle, W. (1973) Coitus induced ovulation. *Contraception*, **7**, 523–64.

Jöchle, W. (1975) Current research in coitus-induced ovulation: a review. *J. Reprod. Fertil., Suppl.* **22**, 165–207.

Johnson, A. M., Wadsworth, J., Field, J. *et al.* (1990) Surveying sexual attitudes. *Nature*, **343**, 109.

Johnson, A. M., Wadsworth, J., Wellings, K. *et al.* (1992) Sexual lifestyles and HIV risk. *Nature*, **360**, 410–12.

Johnson, A. M., Wadsworth, J., Wellings, K. *et al.* (1994) *Sexual Attitudes and Lifestyles*. Blackwell, London.

Johnson, L., Petty, C. S. and Neaves, W. B. (1980) A comparative study of daily sperm production and testicular composition in humans and rats. *Biol. Reprod.*, **22**, 1233–43.

Johnson, M. H. and Everitt, B. J. (1988) *Essential Reproduction*, 3rd edn, Blackwell Scientific Publications, London.

Joly, D., Cariou, M.-L., Lachaise, D. and David, J. R. (1989) Variation of sperm length and heteromorphism in drosophilid species. *Genet. Sel. Evol.*, **21**, 283–93.

Jones, J. W. (1959) *The Salmon*. Collins, London.

Jonkel, C. J. and Cowan, I. McT. (1971) The black bear in the spruce-fir forest. *Wildl. Monogr.*, **27**, 5–57.

Junod, H. A. (1962) *The Life of a South African Tribe*. University Books, New Hyde Park.

Kallmann, F. J. (1952a) Twin and sibship study of overt male homosexuality. *Am. J. Hum. Genet.*, **4**, 136–46.

Kallmann, F. J. (1952b) Comparative twin study on the genetic aspects of male homosexuality. *J. Nerv. Ment. Dis.*, **115**, 283–98.

Kanagawa, H. and Hafez, E. S. E. (1975) The penis, in *Scanning Electron Microscopic Atlas of Mammalian Reproduction*, (ed. E. S. E. Hafez), Georg Theime, Stuttgart, pp. 106–10.

Kaneko, S., Oshio, S., Kobayashi, T. *et al.* (1986) Effects of acrosin, acrosin inhibitors and anti-acrosin antibodies on the motility of human sperm. *Jap. J. Fertil. Steril.*, **31**, 136–42.

Katamee, M. A. (1988) Infertility: a preventable epidemic. *Int. J. Fertil.*, **33**, 246–51.

Katz, D. F. and Overstreet, J. W. (1982) The mechanisms and analysis of sperm migration through cervical mucus, in *Mucus in Health and Disease* II, (eds E. Chantler and M. Elstein), Plenum, New York, pp. 319–30.

Katz, D. F. and Yanagimachi, R. (1980) Movement characteristics of hamster and guinea pig spermatozoa within the oviduct. *Biol. Reprod.*, **22**, 759–64.

Katz, D. F., Morales, P., Samuels, S. J. and Overstreet, J. W. (1990). Mechanisms of filtration of morphologically abnormal human sperm by cervical mucus. *Fertil. Steril.*, **54**, 512–16.

Keeton, W. T. (1980) *Biological Science*, 3rd edn. W. W. Norton, New York.

Kelly, J. V. (1962) Myometrial participation in human sperm transport. *Fertil. Steril.*, **13**, 84–92.

Kempenaers, B., Verheyen, G. R., Van den Broeck, M. *et al.* (1992) Extra-pair paternity results from female preference for high quality males in the blue tit. *Nature*, **357**, 494–6.

Kempf, E. J. (1917) The social and sexual behaviour of infra-human primates with some comparable facts in human behaviour. *Psychoanal. Rev.*, **4**, 127–54.

Kenagy, G. J. and Trombulak, S. C. (1986) Size and function of mammalian testes in relation to body size. *J. Mammal.*, **67**, 1–22.

Kennedy, C. R. (1975) *Ecological Animal Parasitology*, Blackwell, London.

Kenneth, J. H. and Richie, G. R. (1953) *Gestation Periods: A Table and Bibliography*, Commonwealth Agricultural Bureau, Slough, Bucks.

Kesserü, E. (1984) Sexual intercourse enhances the success of artificial insemination. *Int. J. Fertil.*, **29**, 143–5.

Keverne, E. B. (1976) Reply to Goldfood *et al.* *Horm. Behav.*, **7**, 369–72.

Khatamee, M. A. (1988) Infertility: a preventable epidemic. *Int. J. Fertil.*, **33**, 246–51.

Kierzenbaum, A. T. and Tres, L. L. (1975) Structural and transcriptional features of the mouse spermatid genome. *J. Cell Biol.*, **65**, 258–70.

Kierzenbaum, A. T. and Tres, L. L. (1978) RNA transcription and chromatin structure during meiotic and post-meiotic stages of spermatogenesis. *Fed Proc. Fed. Am. Soc. Exp. Biol.*, **37**, 2512–16.

Kiltie, R. A. (1982) On the significance of menstrual synchrony in closely associated women. *Am. Nat.*, **119**, 415–19.

Kim, D. H. and Lee, H. Y. (1982) Clinical investigations of testicular size. *J. Korean Med. Assoc.*, **25**, 135–43.

King, A. S. (1981) Phallus, in *Forms and Function in Birds*. Vol. 2, (eds A. S. King and J. McLelland), Academic Press, London.

King, R. C. (1970) Ovarian development in *Drosophila melanogaster*. Academic Press, New York.

Kinsey, A. C., Pomeroy, W. B. and Martin, C. E. (1948) *Sexual Behaviour in the Human Male*, W. B. Saunders, Philadelphia.

Kinsey, A. C., Pomeroy, W. B., Martin, C. E. and Gebhard, P. H. (1953) *Sexual Behaviour in the Human Female*, W. B. Saunders, Philadelphia.

Kinzey, W. G. (1974) Male reproductive systems and spermatogenesis. *Comparative Reproduction of Nonhuman Primates*, Academic Press, London, pp. 85–114.

Kloek, J. (1961) The smell of some steroid sex hormones and their metabolites: reflections and experiments concerning the significance of smell for the mutual relation of the sexes. *Psychiatr. Neurol. Neurochir.*, **64**, 309–44.

Knight, C. (1991) *Blood Relations – Menstruation and the Origins of Culture.* Yale University Press, Boston.

Knowlton, N. (1974) A note on the evolution of gamete dimorphism. *J. Theor. Biol.*, **46**, 283–5.

Knowlton, N. (1979) Reproductive synchrony, parental investment, and the evolutionary dynamics of sexual selection. *Anim. Behav.*, **27**, 1022–33.

Koch, J. U. (1980) Sperm migration in the human female genital tract with and without intrauterine devices, in *Proceedings of 4th International Meeting on the Control of Fertility*, Geneva. pp. 33–60.

Koenig, W. D., Mumme, R. L. and Pitelka, F. A. (1984) The breeding system of the acorn woodpecker in central coastal California. *Z. Tierpsychol.*, **65**, 289–308.

Kolodny, R. C., Jacobs, L. S. and Daughaday, W. H. (1972) Mammary stimulation causes prolactin secretion in non-lactating women. *Nature*, **238**, 284–6.

Konner, M. J. (1976) Maternal care, infant behavior, and development among the !Kung, in *Kalahari Hunter–Gatherers: Studies of the !Kung San and their Neighbours* (eds R. B. Lee and I. DeVore), Harvard University Press, Cambridge, MA.

Koprowski, J. L. (1992) Removal of copulatory plugs by female tree squirrels. *J. Mammal.*, **73**, 572–6.

Krebs, J. R. and Davies, N. B. (1993) *An Introduction to Behavioural Ecology*, 3rd edn, Blackwell Scientific Publications, Oxford.

Kremer, J. and Jeger, S. (1988) Sperm–cervical interaction, in particular in the presence of anti-spermatozoal antibodies. *Hum. Reprod.*, **3**, 69–73.

Kroeks, M. V. A. M. and Kremer, J. (1977) The pH in the lower third of the genital tract, in *The Uterine Cervix in Reproduction*, (eds V. Insler and G. Bettendorf), Thieme, Stuttgart, pp. 109–18.

Kruger, T. F., Menkveld, R., Stander, F. S. H. *et al.* (1986) Sperm morphologic features as a prognostic factor in in vitro fertilization. *Fertil. Steril.*, **46**, 1118–23.

Krzanowska, H. (1974) The passage of abnormal spermatozoa through the uterotubal junction of the mouse. *J. Reprod. Fertil.*, **38**, 81–90.

Kummer, H. (1986) *Social Organisation of Hamadryas Baboons*, Chicago University Press, Chicago.

Kusan, F., Flemming, A. D. and Seidel, G. (1984) Successful fertilization *in vitro* of fresh intact oocytes by perivitelline (acrosome reacted) spermatozoa in the rabbit. *Fertil. Steril.*, **41**, 766–70.

Kvist, U. (1991) Can disturbances of the ejaculatory sequence contribute to male infertility? *Int. J. Androl.*, **14**, 389–93.

Labov, J. B. (1980) Factors influencing infanticidal behavior in wild male house mice (*Mus musculus*). *Behav. Ecol. Sociobiol.*, **6**, 297–303.

Lack, D. (1954) *The Natural Regulation of Animal Numbers*, Oxford University Press, Oxford.

Lamb, M. E. (1984) Observational studies of father–child relationships in humans, in *Primate Paternalism*, (ed. D. M. Taub), Van Nostrand Reinhold, New York.

Langford, B. B. and Howarth, B. (1974) A trypsin-like enzyme in acrosomal extracts of chicken, turkey and quail spermatozoa. *Poult. Sci.*, **53**, 834–7.

Lassalle, B., Courtot, A. M. and Testart, J. (1987) In vitro fertilization of hamster and human oocyte by microinjection of human sperm. *Gam. Res.*, **16**, 69–78.

Last, R. J. (1978) *Anatomy: Regional and Applied*, Churchill Livingstone, London.

Laufer, N., Segal, S., Ron, M. and Grover, N. B. (1978) Size and size distribution of subfertile human spermatozoa. *Fertil. Steril.*, **30**, 188–91.

Lawrence, S. E. (1992) Sexual cannibalism in the praying mantid, *Mantis religiosa*, a field study. *Anim. Behav.*, **43**, 569–85.

Laws-King, A., Trounson, A., Sathananthan, H. and Kola, I. (1987) Fertilization of human oocytes by microinjection of a single sperm under the zona pellucida. *Fertil. Steril.*, **48**, 637–42.

Le Boeuf, B. J. (1972) Sexual behaviour in the northern elephant seal, *Mirounga angustirostris. Behaviour*, **41**, 1–26.

Lechtig, A. and Klein, R. E. (1981) Pre-natal nutrition and birth weight: is there a causal association? in *Maternal Nutrition in Pregnancy – Eating for Two?* (ed. J. Dobbing), Academic Press, New York, pp. 131–74.

Lee, R. B. and DeVore, I. (1968) Problems in the study of hunters and gatherers in *Man the*

Hunter, (eds R. B. Lee and I. DeVore), Aldine, Chicago, pp. 3–12.
Lee, S. (1988) Sperm preparation for assisted conception. *Conceive*, **12**, 4–6.
Lee, S. van der and Boot, L. M. (1955) Spontaneous pseudopregnancy in mice. *Acta Physiol. Pharmacol. Nèerland.*, **4**, 442–4.
Leidl, W., Stolla, R. and Botzenhardt, A. (1972) The DNA content of morphologically normal and abnormal sperm cells, in *Male Infertility*, University of Pittsburgh Press, Pittsburgh, pp. 241–4.
Leighton, D. R. (1987) Gibbons: territoriality and monogamy, in *Primate Societies*, (eds B. B. Smuts, D. L. Cheney, R. M. Seyfarth, *et al.*.), Chicago University Press, Chicago, pp. 135–45.
Leikkola, A. (1955) Seminal Fluids; Composition in Barren Marriages. PhD Thesis, University of Turku, Finland.
LeVay, S. (1991) A difference in hypothalamic structure between heterosexual–homosexual men. *Science*, **253**, 1034–6.
LeVay, S. (1993) *The Sexual Brain*. MIT Press, Boston.
Levin, M. P. (1973) Preferential mating and the maintenance of the sex-linked dimorphism in *Papilio glaucous*: evidence from laboratory matings. *Evolution*, **27**, 257–64.
Levin, R. J. (1981) The female orgasm – a current appraisal. *J. Psychosom. Res.*, **25**, 119–33.
Liang, Z. G., Kamada, M. and Koide, S. S. (1991) Structural identity of immunoglobulin-binding factor and prostatic secretory protein of human seminal plasma. *Biochem. Biophys. Res. Commun.*, **180**, 356–9.
Lifjeld, J. T. and Robertson, R. J. (1992) Female control of extra-pair fertilization in tree swallows. *Behav. Ecol. Sociobiol.*, **31**, 89–96.
Lightcap, J. L., Kurland, J. A. and Burgess, R. L. (1982) Child abuse: a test of some predictions from evolutionary theory. *Ethol. Sociobiol.*, **3**, 61–7.
Lindholmer, Ch. (1973) Survival of human sperm in different fractions of split ejaculates. *Fertil. Steril.*, **24**, 521–6.
Linkie, D. M. (1982) The physiology of the menstrual cycle, in *Behaviour and the Menstrual Cycle*, (ed. R. C. Friedman), Marcel Dekker, New York.
Lisk, R. D., Reuter, L. A. and Raub, J. A. (1983) Effects of grouping on sexual receptivity in female hamsters. *J. Exp. Zool.*, **189**, 1–6.
Littleton, J. M. (1981) Biochemistry of alcohol tolerance and dependence, in *Handbook of Biological Psychiatry* Part IV: *Brain mechanisms and abnormal behaviour – Chemistry*, (ed. H. M. Van Praag) Marcel Dekker, New York, pp. 234–75.
Lloyd, R. and Coulam, C. B. (1989) The accuracy of urinary luteinizing hormone testing in predicting ovulation. *Am. J. Obstet. Gynecol.*, **160**, 1370–5.
Lobban, C. F. (1972) *Law and Anthropology in the Sudan (An Analysis of Homicide Cases in Sudan)*. African Studies Seminar Series No. 13, Sudan Research Unit, Khartoum University.
LoPiccolo, J. and Lobitz, W. J. (1972) The role of masturbation in the treatment of orgasmic dysfunction. *Arch. Sex. Behav.*, **2**, 163–71.
Lovejoy, C. O. (1981) The origin of man. *Science*, **211**, 341–50.
Lundquist, F. (1949) Aspects of the biochemistry of human semen. *Acta Physiol. Scand.*, Suppl. **66**, 1–45.
Luterman, M., Twamoto, T. and Gagnon, C. (1991) Origin of the human seminal plasma motility inhibitor within the reproductive tract. *Int. J. Androl.*, **14**, 91–8.
Macfarlane, A. and Mugford, M. (1984) *Birth Counts: Statistics of Pregnancy and Childbirth*, HMSO, London.
MacGillivray, I. and Campbell, D. M. (1978) The physical characteristics and adaptations of women with twin pregnancies, in *Twin Research: Clinical Studies*, (ed. W. E. Nance), Alan R. Liss, New York, pp. 81–6.
Macintyre, S. and Sooman, A. (1992) Non-paternity and prenatal genetic screening. *Lancet*, **338**, 839.
MacKinnon, J. (1978) *The Ape Within Us*. Collins, London.
MacLeod, J. (1951) Effect of chicken pox and of pneumonia on semen quality. *Fertil. Steril.*, **2**, 523–33.
MacLeod, J. (1953) The male factor in fertility–infertility. *Fertil. Steril.*, **4**, 10–33.
MacLeod, J. (1970) The significance of deviations in human sperm morphology, in *The Human Testis*, (eds E. Rosemburg and C. A. Paulsen) Plenum Press, London, pp. 481–94.
Maddox, J. (1989) Sexual behaviour unsurveyed. *Nature*, **341**, 181.
Maddox, J. (1991) Is homosexuality hard-wired? *Nature*, **353**, 13.
Madsen, T., Shine, R., Loman, J. and Håkansson, T. (1992) Why do female adders copulate so frequently? *Nature*, **355**, 440–1.
Magnusson, W. E. (1979) Production of an embryo by an *Acrochordus javanicus* isolated for 7 years. *Copeia*, **1979**, 744–5.
Magnusson, W. E., Vliet, K. A., Pooley, A. C. and Whitaker, R. (1989) Reproduction, in

Crocodiles and Alligators, (ed. C. A. Ross), Merehurst Press, London, pp. 118–35.

Mahadevan, M. M., Trouson, A. O., Wood, C. and Leeton, J. F. (1987) Effects of oocyte quality and sperm characteristics on the number of spermatozoa bound to the zona pellucida of human oocytes inseminated *in vitro*. *J. In Vitro Fert. Embryo Transf.*, **4**, 223–7.

Mahmoud, S-u-H. (1985) Health development planning. *Research Methods for Health Development*, World Health Organization, Geneva.

Malinowski, B. (1929) *The Sexual Life of Savages in North-Western Melanesia*, Routledge, London.

Mandal, A. and Bhattacharyya, A. K. (1985) Studies on the coagulational characteristics of human ejaculates. *Andrologia*, **17**, 80–6.

Mann, T. (1960) Serotonin (5-hydroxytryptamine) in the male reproductive tract of the spiny dogfish. *Nature*, **188**, 941.

Mann, T. (1975) Spermatophores, in *The Biology of the Male Gamete*, (eds J. G. Duckett and P. A. Racey), *Biol. J. Linn. Soc., Suppl.* **1**, 417–25.

Mann, T. and Lutwak-Mann, C. (1981) *Male Reproductive Function and Semen*, Springer-Verlag, New York.

Mann, T., Short, R. V., Walton, A. *et al.* (1957) The 'tail-end sample' of stallion semen. *J. Agric. Sci.*, **49**, 301–6.

Margulis, L. (1981) *Symbiosis in Cell Evolution: Life and its Environment on the Early Earth*, W. H. Freeman, San Francisco.

Marmar, J. L., Praiss, D. E. and Debenedictis, T. J. (1979) Statistical comparison of the parameters of semen versus the fractions of the split ejaculate. *Fertil. Steril.*, **30**, 439–43.

Marmar, J. L., Praiss, D. E. and Debenedictis, T. J. (1979) An estimate of the fertility potential of the fractions of the split ejaculate in terms of motile sperm count. *Fertil. Steril.*, **32**, 202–5.

Marsden, H. M. and Bronson, F. H. (1965) The synchrony of estrus in mice: relative roles of the male and female environments. *J. Endrocrinol.*, **32**, 313–19.

Marshall, D. S. (1971) Sexual behaviour on Mangaia, in *Human Sexual Behaviour: Variations in the Ethnographic Spectrum*, (eds D. S. Marshall and R. C. Suggs), Basic Books, London, pp. 103–63.

Martan, J. and Shepherd, B. A. (1976) The role of the copulatory plug in reproduction of the guinea pig. *J. Exp. Zool.*, **196**, 79–84.

Martin, P. A. and Dzuik, P. J. (1977) Assessment of relative fertility of males (cockerels and boars) by competitive mating. *J. Reprod. Fertil.*, **49**, 323–9.

Martin, P. A., Reimers, T. J., Lodge, J. R. and Dzuik, P. J. (1974) The effect of ratios and numbers of spermatozoa mixed from two males on proportions of offspring. *J. Reprod. Fertil.*, **39**, 251–8.

Martin, R. D. (1993) Primate origins: plugging the gaps. *Nature*, **363**, 223–34.

Martinet, L. and Raynaud, F. (1974) Survie prolongée des spermatozoides dans l'uterus de la Hase: explication de la superfoetation. *Proceedings Symposium on Sperm Transport, Survival and Fertilising Ability in Vertebrates*, (eds E. S. E. Hafez and C. G. Thibault), INSERM, Paris.

Martinez, A. R., Bernardus, R. E., Vermeiden, J. P. W. and Shoemaker, J. (1992) Reliability of home urinary LH tests for timing of insemination: a consumer's study. *Hum. Reprod.*, **7**, 751–3.

Masters, W. H. and Johnson, V. E. (1966) *Human Sexual Response*. J. & A. Churchill, London.

Mather, K. (1974) Statistical appendix (Wallace, 1974). *Heredity*, **33**, 428–9.

Matson, P. L., Troup, S. A., Lowe, B. *et al.* (1989) Fertilization of human oocytes *in vitro* by spermatozoa from oligozoospermic and normospermic men. *Int. J. Androl.*, **12**, 117–23.

Matthews, M. K. and Adler, N. T. (1977) Systematic interrelationship of mating, vaginal plug position, and sperm transport in the rat. *Physiol. Behav.*, **20**, 303–9.

Mattner, P. E. (1968) The distribution of spermatozoa and leucocytes in the female genital tract in goats and cattle. *J. Reprod. Fertil.*, **17**, 253–61.

Mattner, P. E. (1969) Phagocytosis of spermatozoa by leucocytes in bovine cervical mucus in vitro. *J. Reprod. Fertil.*, **20**, 133–4.

Mayer, F. (1993) Genetic population structure of the noctule bat *Nyctalus noctula*. Symposium Abstract in: *Recent Advances in Bat Biology*. Zoological Society, London.

Maynard Smith, J. (1964) Group selection and kin selection. *Nature*, **201**, 1145–7.

Maynard Smith, J. (1976) Sexual selection and the handicap principle. *J. Theor. Biol.*, **57**, 239–42.

Maynard Smith, J. (1978) *The Evolution of Sex*, Cambridge University Press, Cambridge.

Mazer, C. and Ziserman, A. J. (1932) Pseudomenstruation in the human female. *Am. J. Surg.*, **18**, 332–7.

McCance, R. A., Luff, M. C. and Widdowson, E. E. (1937) Physical and emotional periodicity in women. *J. Hyg.*, **37**, 571–611.

McCann, T. S. (1981) Aggression and sexual activity of southern elephant seals. *J. Zool.*, **195**, 295–310.

McClintock, M. K. (1971) Menstrual synchrony and suppression. *Nature*, **229**, 229–45.

McGregor, L., Flaherty, S. P. and Breed, W. G. (1989) Structure of the zona pellucida and cumulus oophorus in three species of native Australian rodents. *Gam. Res.*, **23**, 279–87.

McKinlay, S., Jeffreys, M. and Thompson, B. (1972) An investigation of the age at menopause. *J. Biosoc. Sci.*, **4**, 161–73.

McKinney, F., Cheng, K. M. and Bruggers, D. J. (1984) Sperm competition in apparently monogamous birds, in *Sperm Competition and the Evolution of Animal Mating Systems*, (ed. R. L. Smith), Academic Press, London, pp. 523–40.

Meddis, R. (1984) *Statistics Using Ranks: a Unified Approach*, Blackwell, Oxford.

Meikle, D. B., Tilford, B. L. and Vessey, S. H. (1984) Dominance rank, secondary sex ratio and reproduction of offspring in polygynous primates. *Am. Nat.*, **124**, 172–88.

Mènard, N. Scheffrahn, W., Vallet, D. *et al.* (1992) Application of blood protein electrophoresis and DNA fingerprinting to the analysis of paternity and social characteristics of wild Barbary macaques, in *Paternity in Primates*, (eds R. D. Martin, A. F. Dixson and E. J. Wickings), Karger, Basel, pp. 155–74.

Metcalf, M. G. and Mackenzie, J. A. (1980) Incidence of ovulation in young women. *J. Biosoc. Sci.*, **12**, 345–52.

Meves, F. (1902) Uber oligopyrene und apyrene Spermien und uber ihre Entstehung, nach Beobachtungen an *Paludina* und *Pygaera. Arch. Mikrosk. Anat. Entwicklungsgesch.*, **61**. 1–84.

Michael, R. P. and Keverne, E. B. (1970) Primate sex pheromones of vaginal origin. *Nature*, **225**, 84–5.

Milligan, D., Drife, J. O. and Short, R. V. (1975) Changes in breast volume during normal menstrual cycle and after oral contraceptives. *Br. Med. J.*, **iv**, 813–14.

Mills, M. E. (1992) Gender differences predicted by evolutionary theory are reflected in survey responses of college students. Human Behaviour and Evolution Society 4th Annual Meeting, Albuquerque, USA, *Abstract no. 64.*

Milton, K. (1985) Urine washing behaviour in the woolly spider monkey. *Z. Tierpsychol.*, **67**, 154–60.

Mjelstad, H. (1991) Displaying intensity and sperm quality in the capercaillie *Tetrao urogallus. Fauna norv. Ser. C. Cinclus.*, **14**, 93–4.

Mock, D. W. and Fukioka, M. (1990) Monogamy and long-term pair bonding in vertebrates. *Trends Ecol. Evol.*, **5**, 39–42.

Moghissi, K. S. (1972) The function of the cervix in fertility. *Fertil. Steril.*, **23**, 295–306.

Moghissi, K. S. (1977) Sperm migration through the human cervix, in *The Uterine Cervix in Reproduction* (eds V. Insler and G. Bettendorf), Theime, Stuttgart, pp. 146–65.

Moghissi, K. S. and Syner, F. N. (1976) The effects of seminal protease on sperm migration through the cervical mucus. *Int. J. Fertil.*, **21**, 246–53.

Mohnot, S. M. (1984) Langur interactions around Jodhpur (*Presbytis entellus*), in *Current Primate Researches*, (eds M. L. Roonwal, S. M. Mohnot and N. S. Rathore), University Press, Jodhpur.

Møller, A. P. (1988a) Testis size, ejaculate quality and sperm competition in birds. *Biol. J. Linn. Soc.*, **33**, 273–83.

Møller, A. P. (1988b) Ejaculate quality, testis size and sperm competition in primates, *J. Hum. Evol.*, **17**, 479–88.

Møller, A. P. (1988c) Ejaculate quality, testis size and sperm production in mammals. *Funct. Ecol.*, **3**, 91–6.

Møller, A. P, (in prep) Why are sperm so small? Sperm size, sperm abnormality and selfish DNA. *Proc. R. Soc. Lond. B.*

Moore, C. R. (1926) The biology of the mammalian testis and scrotum. *Q. Rev. Biol.*, **1**, 4–50.

Morin, L. P. (1986) Environment and hamster reproduction: responses to phase-specific starvation during estrous cycle. *Am. J. Physiol.*, **251**, R663–R669.

Morris, D. (1967) *The Naked Ape*, Cape, London.

Morris, D. (1972) The reproductive behaviour of the ten-spined stickleback, in *Patterns of Reproductive Behaviour*, (ed. D. Morris), Panther Books, London.

Morris, M. (1993) Telling tails explains the discrepancy in sexual partner reports. *Nature*, **365**, 437–40.

Morris, N. M. and Udry, J. R. (1970) Variations in pedometer activity during the menstrual cycle. *Obstet. Gynecol.*, **35**, 199–201.

Morris, N. M., Underwood, L. E. and Easterling, W. (1976) Temporal relationship between basal body temperature, nadir, luteinizing hormone surge in normal women. *Fertil. Steril.*, **27**, 780–3.

Morroll, D. R., Critchlow, J. D., Matson, P. L. and Lieberman, B. A. (1992) The use of cryopreserved aged human oocytes in a test of the fertilizing capacity of human spermatozoa. *Hum. Reprod.*, **7**, 671–6.

Morroll, D. R., Lieberman, B. A. and Matson. P. L. (1993) Use of human zonae from cryopreserved oocytes in a test to assess the binding capacity of human spermatozoa. *Int. J. Androl.*, **16**, 97–103.

Mortimer, D., Leslie, E. E., Kelly, R. W. and Templeton, A. A. (1982) Morphological selection of human spermatozoa in vivo and in vitro. *J. Reprod. Fertil.*, **64**, 391–9.

Mortimer, D. (1983) Sperm transport in the human female reproductive tract, in *Oxford Reviews of Reproductive Biology*, Vol. 5, (ed. C. A. Finn); Clarendon Press, Oxford, pp. 20–7.

Morton, D. B. and Glover, T. D. (1974a) Sperm transport in the female rabbit: the role of the cervix. *J. Reprod. Fertil.*, **38**, 131–8.

Morton, D. B. and Glover, T. D. (1974b) Sperm transport in the female rabbit: the effect of inseminate volume and sperm density. *J. Reprod. Fertil.*, **38**, 139–46.

Mosig, D. W. and Dewsbury, D. A. (1970) Plug fate in the copulatory behaviour of rats. *Psycho. Sci.*, **20**, 315–16.

Mueller, U. (1993) Social status and sex. *Nature*, **363**, 490.

Münster, K., Schmidt, L. and Helm, P. (1992) Length and variation in the menstrual cycle – a cross-sectional study from a Danish county. *Br. J. Obstet. Gynaecol.*, **99**, 422–9.

Murdock, G. P. (1967) *Culture and Society*. University of Pittsburgh Press, Pittsburgh.

Murphy, F. K. and Patamasucon, P. (1984) Congenital syphilis, in *Sexually Transmitted Diseases*, (eds K. K. Holmes, P. Mardh, P. F. Sparling and P. J. Wiesner), McGraw-Hill, New York, pp. 352–74.

Nadler, R. D. and Collins, D.C. (1991) Copulatory frequency, urinary pregnanediol and fertility in great apes. *Am. J. Primat.*, **24**, 167–79.

Nagae, T., Yanagimachi, R., Strivastava, P. N. and Yanagimachi, H. (1986) Acrosome reaction in human spermatozoa. *Fertil. Steril.*, **45**, 701–7.

Nagai, T., Niwa, K. and Iritani, A. (1984) Effect of sperm concentration during preincubation in a defined medium on fertilization *in vitro* of pig follicular oocytes. *J. Reprod. Fertil.*, **70**, 271–5.

Neaves, W. B., Johnson, L., Porter, J. C. *et al.* (1984) Leydig cell numbers, daily sperm production, and serum gonadotropin levels in aging men. *J. Clin. Endocrinol. Metab.*, **59**, 756–63.

Neel, J. V. and Weiss, K. M. (1975) The genetic structure of a tribal population, the Yanomama Indians. XIII. Biodemographic studies. *Am. J. Phys. Anthropol.*, **42**, 25–51.

Newton, M. E., Wood, R. J. and Southern, D. J. (1976) A cytogenetic analysis of meiotic drive in the mosquito, *Aedes aegypti* (L.) *Genetica*, **46**, 256–60.

Nishida, T. and Hiraiwa-Hasegawa, M. (1987) Chimpanzees and bonobos: cooperative relationships among males, in *Primate Societies* (eds B. B. Smuts, D. L. Cheney, R. M. Seyfarth *et al.*), Chicago University Press, Chicago, pp. 165–78.

Niwa, K., Ohara, K., Hoshi, Y. and Iritani, A. (1985) Early events of *in vitro* fertilization of cat eggs by epididymal spermatozoa. *J. Reprod. Fertil.*, **74**, 657–60.

Novacek, M. J. (1992) Mammalian phylogeny: shaking the tree. *Nature*, **356**, 121–5.

Okamura, F. and Nishiyama, H. (1978) The passage of spermatoza through the vitelline membrane in the domestic fowl, *Gallus gallus*. *Cell Tissue Res.*, **188**, 497–508.

Orbach, J., Miller, M., Billimoria, A. and Solhkhah, N. (1967) Spontaneous seminal ejaculation and genital grooming in rats. *Brain Res.*, **5**, 520–3.

Osol, A. (ed.) (1972) *Blakiston's Gould Medical Dictionary*, 3rd edn. McGraw-Hill, New York.

Overstreet, J. W. (1983) Transport of gametes in the reproductive tract of the female mammal, in *Mechanisms and Control of Animal Fertilization*, (ed. J. F. Hartmann), Academic Press, London, pp. 499–543.

Overstreet, J. W. and Cooper, G. W. (1978) Sperm transport in the reproductive tract of the female rabbit. I. The rapid transit phase of transport. *Biol. Reprod.*, **19**, 101–14.

Overstreet, J. W., Katz, D. F. and Johnson, L. L. (1980) Motility of rabbit spermatozoa in the secretions of the oviduct. *Biol. Reprod.*, **22**, 1083–8.

Packer, C. (1977) Reciprocal altruism in *Papio anubis*. *Nature*, **265**, 441–3.

Page, A. (1993) Chimpanzee contraception. *Focus*, June, 35.

Pandya, I. J. and Cohen, J. (1985) The leukocytic reaction of the human uterine cervix to spermatozoa. *Fertil. Steril.*, **43**, 417–21.

Pant, K. P. and Mukherjee, D. P. (1972) The effects of seasons on the sperm dimensions of the buffalo bulls. *J. Reprod. Fertil.*, **29**, 425–9.

Parker, G. A. (1970a) Sperm competition and its evolutionary consequences in the insects. *Biol. Rev.*, **45**, 525–67.

Parker, G. A. (1970b) The reproductive behaviour and the nature of sexual selection in *Scatophaga stercoraria* (L.) (Diptera: Scatophagidae). V. The female's behaviour at the oviposition site. *Behaviour*, **37**, 140–68.

Parker, G. A. (1972) Reproductive behaviour of *Sepsis cynipsea* (L.) (Diptera: Sepsidae): I. A preliminary analysis of the reproductive strategy and its associated behaviour patterns. *Behaviour*, **41**, 172–206.

Parker, G. A. (1982) Why are there so many tiny sperm? Sperm competition and the maintenance of two sexes. *J. Theor. Biol.*, **96**, 281–94.

Parker, G. A. (1984) Sperm competition and the evolution of animal mating strategies, in *Sperm Competition and the Evolution of Animal Mating*

Systems, (ed. R. L. Smith), Academic Press, London, pp. 1–60.

Parker, G. A. (1990a) Sperm competition: games, raffles and roles. *Proc. R. Soc. Lond. B.*, **242**, 120–6.

Parker, G. A. (1990b) Sperm competition: sneaks and extra-pair copulations. *Proc. R. Soc. Lond. B.*, **242**, 127–33.

Parker, G. A. (1992) Snakes and female sexuality. *Nature*, **355**, 395–6.

Parker, G. A. (1993) Sperm competition games: sperm size and sperm number under adult control. *Proc. R. Soc. Lond. B.*, **253**, 245–54.

Parker, G. A. and Begon, M. (1993) Sperm competition games: sperm size and sperm number under gametic control. *Proc. R. Soc. Lond. B.*, **253**, 255–62.

Parker, G. A., Baker, R. R. and Smith, V. G. F. (1972) The origin and evolution of gamete dimorphism and the male:female phenomenon. *J. Theor. Biol.*, **36**, 529–53.

Parker, G. A., Simmons, L. W. and Kirk, H. (1990) Analysing sperm competition data: simple models for predicting mechanisms. *Behav. Ecol. Sociobiol.*, **27**, 55–65.

Parkes, A. S. (1969) Multiple births in man. *J. Reprod. Fertil. Suppl.*, **6**, 105–16.

Parrish, J. J. and Foote, R. H. (1985) Fertility differences among male rabbits determined by heterospermic insemination of fluorochrome-labelled spermatozoa. *Biol. Reprod.*, **33**, 940–9.

Parsons, L. and Sommers, S. C. (1978) *Gynecology*, W. B. Saunders, Philadelphia.

Partridge, L. and Harvey, P. H. (1992) What the sperm count costs. *Nature*, **360**, 415–58.

Passingham, R. E. (1982) *The Human Primate*, Freeman, San Francisco.

Paul, A. and Kuester, J. (1990) Adaptive significance of sex ratio adjustment in semifree-ranging Barbary macaques (*Macaca sylvanus*) at Salem. *Behav. Ecol. Sociobiol.*, **27**, 287–93.

Payne, R. B. and Kahrs, A. J. (1961) Competitive efficiency of turkey sperm. *Poult. Sci.*, **40**, 1598–604.

Perloff, W. H. and Steinburger, E. (1964) In vivo survival of spermatozoa in cervical mucus. *Am. J. Obstet. Gynecol.*, **88**, 439–42.

Perry, J. (1954a) Fecundity and embryonic mortality in pigs. *J. Embryol. Exp. Morphol*, **2**, 308–13.

Perry, J. S. (1954b) *The Ovarian Cycle of Mammals*. Oliver & Boyd, Edinburgh.

Petrie, M. (1992) Copulation frequency in birds: why do females copulate more than once with the same male? *Anim. Behav.*, **44**, 790–2.

Philipp, E. E. (1973) Discussion: moral, social, and ethical issues, in *Law and Ethics of A.I.D. and Embryo Transfer. Ciba Foundation Symposium* 17, (eds G. E. W. Wolstenholme and D. W. Fitzsimons), Associated Scientific, London, pp. 663–6.

Polakoski, K. L. K. and Siegel, M. S. (1986) The proacrosin–acrosin system, in *Andrology, Male Fertility and Sterility*, (ed. J. D. Paulsen, A. Negor-Vilar and L. Martini), Academic Press, Orlando, FL, pp. 359–75.

Poland, M. L., Giblin, P. T., Ager, J. W. and Moghissi, K. S. (1986) Effect of stress on semen quality in semen donors. *Int. J. Fertil.*, **31**, 229–31.

Pommerenke, W. T. (1946) Cyclic changes in physical and chemical properties of cervical mucus. *Am. J. Obstet. Gynecol.*, **52**, 1023–8.

Pond, C. M. (1984) Physiological and ecological importance of energy storage in the evolution of lactation: evidence for a common pattern of anatomical organization of adipose tissue in mammals, in *Physiological Strategies in Lactation*. Symposium Zoological Society of London no. 51, (eds M. Peaker, R. G. Vernon and C. H. Knight), Academic Press, London, pp. 1–32.

Pond, C. M. (1992a) An evolutionary and functional view of mammalian adipose tissue. *Proc. Nutr. Soc.*, **51**, 367–77.

Pond, C. M. (1992b) The structure and function of adipose tissue in humans, with comments on the evolutionary origin and physiological consequences of sex differences. *Coll. Anthropol.*, **16**, 135–43.

Portmann, A. (1952) *Animal Forms and Patterns*. Faber & Faber, London.

Potts, M., Diggory, P. and Peel, J. (1977) *Abortion*, Cambridge University Press, Cambridge.

Pough, F. H., Heiser, J. B. and McFarland, W. N. (1990) *Vertebrate Life*, Macmillan, New York.

Prentice, A. M. and Whitehead, R. G. (1987) The energetics of human reproduction. *Symp. Zool. Soc. Lond.*, no. 57, 275–304.

Preti, G., Cutler, W. B., Garcia, C. T. *et al.* (1986) Human axillary secretions influence women's menstrual cycles: the role of donor extract of females. *Horm. Behav.*, **20**, 474–82.

Profet, M. (1993) Menstruation as a defence against pathogens transported by sperm. *Q. Rev. Biol.*, **68**, 335–86.

Propst, F., Rosenberg, M. P. and Vande Woude, G. F. (1988) Proto-oncogene expression in germ cell development. *Trends Genet.*, **4**, 183–7.

Pursel, V. G. and Johnson, L. A. (1974) Glutaraldehyde fixation of boar spermatozoa for acrosome evaluation. *Theorio.*, **1**, 63–9.

Pyburn, W. F. (1970) Breeding behaviour of the

leaf-frogs, *Phyllomedusa callidryas*, and *Phyllomedusa dacnicolor* in Mexico. *Copeia*, **2**, 209–18.
Quadagno, D. M., Shubeita, H. E., Deck, J. and Fancoeur, D. (1981) Influence of male social contacts, exercise and all-female living conditions on the menstrual cycle. *Psychoneuroendocrinology*, **6**, 239–44.
Quagliarello, J. and Arny, M. (1986) Inaccuracy of basal body temperature charts in predicting urinary luteinizing hormone surges. *Fertil. Steril.*, **45**, 334–7.
Quinlivan, W. L. G. and Sullivan, H. (1977) The immunological effects of husband's semen on donor spermatozoa during mixed insemination. *Fertil. Steril.*, **28**, 448–50.
Quinn, P. and Marrs, R. P. (1991) Effect of concentration and total number of spermatozoa on fertilization of human oocytes *in vitro*. *Hum. Reprod.*, **6**, Suppl. 1, Abst 238.
Rabenold, P. P., Rabenold, K. N., Piper, W. H. *et al.* (1990) Shared paternity revealed by genetic analysis in cooperative breeding tropical wrens. *Nature*, **348**, 538–42.
Raboch, J. and Faltus, F. (1991) Sexuality of women with anorexia nervosa. *Acta Psychiatr. Scand.*, **84**, 9–11.
Raboch, J. and Skachova, J. (1965) The pH of human ejaculate. *Fertil. Steril.*, **16**. 252–6.
Racey, P. A. (1973) The viability of spermatozoa after prolonged storage by male and female European bats. *Period. Biol.*, **75**, 201–6.
Racey, P. A. (1979) The prolonged storage and survival of spermatozoa in Chiroptera. *J. Reprod. Fertil.*, **56**. 391–6.
Rajasekarasetty, M. R. (1954) Studies on a new type of genetically-determined quasisterility in the house mouse. *Fertil. Steril.*, **5**, 68–97.
Ralt, D., Goldenberg, M., Fettorolf, P. *et al.* (1991) Sperm attraction to a follicular factor(s) correlates with human egg fertilizability. *Proc. Natl. Acad. Sci. USA*, **88**, 2840–4.
Rebuffé-Scrive, M. (1987) Regional adipose tissue metabolism in men and in women during menstrual cycle, pregnancy, lactation and menopause. *Int. J. Obes.*, **11**, 347–55.
Ridley, M. (1978) Paternal care. *Anim. Behav.*, **26**. 904–32.
Ridley, M. (1989) The incidence of sperm displacement in insects: four conjectures, one corroboration. *Biol. J. Linn. Soc.*, **38**, 349–67.
Robertson, D. H. H., McMillan, A. and Young, H. (1989) *Clinical Practice in Sexually Transmissible Diseases*, Churchill Livingstone, London.
Robey, B., Rutstein, S. O. and Morris, L. (1993) The fertility decline in developing countries. *Sci. Am.* December, 30–7.
Robl, J. M. and Dzuik, P. J. (1988) Comparison of heterospermic and homospermic inseminations as a measure of male fertility. *J. Exp. Zool.*, **245**, 97–101.
Roche, J. F., Dzuik, P. J. and Lodge, J. R. (1968) Competition between fresh and aged spermatozoa in fertilizing rabbit eggs. *J. Reprod. Fertil.*, **16**, 155–7.
Rodger, J. C. and Bedford, J. M. (1982) Conception in marsupials 2: Separation of sperm pairs and sperm–egg interaction in the opossum, *Didelphis virginiana*, *J. Reprod. Fertil.*, **64**, 171–9.
Rodman, P. S. and Mitani, J. C. (1987) Orangutans: sexual dimorphism in a solitary species, in *Primate Societies*, (eds B. B. Smuts, D. L. Cheney, R. M. Seyfarth *et al.*,) Chicago University Press, Chicago, pp. 146–54.
Roetzer, J. (1977) *Fine Points of the Symptom-Thermic Method of Natural Family Planning*. The Human Life Center, Collegeville, MN.
Rogel, M. J. (1978) A critical evaluation of the possibility of higher primate reproductive and sexual pheromones. *Psychol. Bull.*, **85**, 810–30.
Rogers, A. R. (1990) The evolutionary economics of reproduction. *Ethol. Sociobiol.*, **2**, 54–68.
Rogers, A. R. (1993) Why menopause? *Evol. Ecol.*, **7**, 406–20.
Roldan, E. R. S., Gomendio, M. and Vitullo, A. D. (1992) The evolution of eutherian spermatozoa and underlying selective forces: female selection and sperm competition. *Biol. Rev.*, **67**, 551–93.
Roosen-Runge, E. C. (1969) Comparative aspects of spermatogenesis. *Biol. Reprod.*, **1**, 24–39.
Rosenblatt, J. S. and Schneirla, T. C. (1962) The behaviour of cats, in *The Behaviour of Domestic Animals*, 1st edn. (ed. E. S. E. Hafez), Williams and Wilkins, Baltimore, pp. 453–88.
Rossman, I. D. and Bartelmez, G. W. (1946) Delayed ovulation, a significant factor in the variability of the menstrual cycle. *Am J. Obstet. Gynecol.*, **52**, 28–33.
Rothschild, M. L. (1991) Arrangement of sperm within the spermatheca of fleas, with remarks on sperm displacement. *Biol. J. Linn. Soc.*, **43**, 313–23.
Roughgarden, J. (1991) The evolution of sex. *Am. Nat.*, **138**, 934–53.
Roussel, J. D. and Austin, C. R. (1967) Enzymic liquefaction of primate semen. *Int. J. Fertil.*, **12**. 288–90.
Rowell, T. E. (1972) Female reproduction cycles and social behaviour in primates. *Adv. Study Behav.*, **4**, 69–105.
Rowland, W. J. (1979) Stealing fertilizations in the fourspine stickleback, *Apeltes quadracus*. *Am. Nat.*, **114**. 602–4.

Rubenstein, B. B., Strauss, H., Lazarus, M. L. and Hankin, H. (1951) Sperm survival in women. *Fertil. Steril.*, **2**, 15–19.

Rushton, J. P. (1988) Race differences in behaviour: a review and evolutionary analysis. *Person. Individ. Diff.*, **9**, 1009–24.

Rushton, J. P. and Bogaert, A. F. (1987) Race differences in sexual behavior: testing an evolutionary hypothesis. *J. Res. Person.*, **21**, 529–51.

Russel, D. E. (1984) *Sexual Exploitation*, Sage, Beverley Hills, CA.

Russell, M. J., Switz, G. M. and Thompson, K. (1980) Olfactory influences on the human menstrual cycle. *Pharmacol. Biochem. Behav.*, **13**, 737–8.

Russell, R. J. H. and Wells, P. A. (1987) Estimating paternity confidence. *Ethol. Sociobiol.*, **8**, 215–20.

Ryder, N. B. (1973) Contraceptive failure in the United States. *Fam. Plan. Perspec.*, **5**, 133–42.

Sagiroglu, N. and Sagiroglu, E. (1970) Biological mode of action of the Lippes Loop in intrauterine contraception. *Am. J. Obstet. Gynecol.*, **106**, 506–15.

Saling, P. M. and Storey, B. T. (1979) Mouse gamete interaction during fertilization *in vitro*: chlortetracycline as a fluorescent probe for the mouse sperm acrosome reactions. *J. Cell. Biol.*, **83**, 544–55.

Salisbury, G. W. and Baker, F. N. (1966) Nuclear morphology of spermatozoa from inbred and line-cross Hereford bulls. *J. Anim. Sci.*, **25**, 476–9.

Salmon, D., Seger, J. and Salmon, C. (1980) Expected and observed proportion of subjects excluded from paternity by blood phenotypes of a child and its mother in a sample of 171 families. *Am. J. Hum. Genet.*, **32**, 432–44.

Sauther, M. L. (1991) Reproductive behaviour of free-ranging *Lemur Catta* at Beza Mahafaly special reserve, Madagascar. *Am. J. Phys. Anthropol.*, **84**, 463–7.

Schacht, L. E. and Gershowitz, H. (1963) Frequency of extra-marital children as determined by blood groups, in *Proceedings of the Second International Congress on Human Genetics*, (ed. L. Gedda), G. Mendel, Rome, pp. 894–7.

Schaller, G. B. (1963) *The Mountain Gorilla: Ecology and Behavior*, Chicago University Press, Chicago.

Schaller, G. B. (1972) *The Serengeti Lion*, Chicago University Press, Chicago.

Schank, R. C. (1990) What is AI, anyway? in *The Foundations of Artificial Intelligence*, (eds D. Partridge and Y. Wilks), Cambridge University Press, Cambridge, pp. 3–13.

Schill, W. -B., Preissier, G., Dittmann, B. and Müller, W. P. (1979) Effect of pancreatic kallikrein, sperm acrosin and high molecular weight (HMW) kininogen on cervical mucus penetration ability of seminal plasma-free human spermatozoa, in *Kinins – II. Systemic protease and cellular function*, (eds S. Fujii, H. Moriya and T. Suzuki), Plenum Press, New York.

Schill, W.-B., Töpfer-Peterson, E. and Heissler, E. (1988) The sperm acrosome: functional and clinical aspects. *Hum. Reprod.*, **3**, 139–45.

Schirren, C. G., Holstein, A. F. and Schirren, C. (1971) Über die Morphogenese rundöpfiger Spermatozoen des Menschen. *Andrologia*, **3**, 117–25.

Schneider, J. E. and Wade, G. M. (1989) Availability of metabolic fuels control estrous cyclicity of Syrian hamsters. *Science*, **244**, 1526–8.

Schonfield, W. A. (1943) Primary and secondary sexual characteristics. Study of their development in males from birth through maturity with biometric study of penis and testes. *Am. J. Dis. Child.*, **65**. 535–49.

Schulze-Hagen, K., Swatschek, I., Dyrcz, A. and Wink, M. (1993) *J. Orinithol.*, **134**, 145–54.

Schwagmeyer, P. L. and Foltz, D. W. (1990) Factors affecting the outcome of sperm competition in thirteen-lined ground squirrels. *Anim. Behav.*, **39**. 156–62.

Scott, L. M. (1984) Reproductive behavior of adolescent female baboons (*Papio anubis*) in Kenya, in *Female Primates: Studies by Women Primatologists*, (ed. M. F. Small), Alan R. Liss, New York.

Sekoni, V. O. and Gustafsson, B. K. (1987) Seasonal variations in the incidence of sperm morphology abnormalities in dairy bulls regularly used for artificial insemination. *Br. Vet. J.*, **143**, 312–17.

Sekulic, R. (1982) Behavior and ranging patterns of a solitary female red howler (*Alouatta seniculus*). *Folia Primatol.*, **38**, 217–32.

Setchell, B. P. (1978) *The Mammalian Testis*, Paul Elek, London.

Settlage, D. S. F., Motoshima, M. and Tredway, D. R. (1973) Sperm transport from the external cervical os to the fallopian tubes in women. *INSERM*, **26**, 201–18.

Seuánez, H. N. (1980) Chromosomes and the spermatozoa of the African great apes. *J. Reprod. Fertil. Suppl.*, **28**, 91–104.

Seuánez, H. N., Carothers, A. D., Martin, D. E. and Short, R.V. (1977) Morphological abnormalities in the spermatozoa of man and apes. *Nature*, **270**, 345–7.

Shalash, M. R. (1981) Seasonal variation in the semen characters of buffalo. *Ind. J. Vet. Sci.*, **24**, 478–9.

Shalgi, R., Kaplan, R., Nebel, L. and Kraicer, P. F. (1981) The male factor in fertilization of rat eggs *in vitro*. *J. Exp. Zool.*, **217**, 399–402.
Shanson, D. C. (1989) *Microbiology in Medical Practice*, Wright, London.
Shapiro, S., Levine, H. S. and Abramowicz, M. (1971) Factors associated with early and late fetal loss. *Adv. Planned Parenthood*, **6**, 45–50.
Sharma, O. P. and Hays, R. L. (1975) Heterospermic insemination and its effect on the offspring ratio in rats. *J. Reprod. Fertil.*, **45**. 533–5.
Sharman, G. B. (1976) Evolution of viviparity in mammals, in *The Evolution of Reproduction*, (eds C. R. Austin and R. V. Short), Cambridge University Press, Cambridge, pp. 32–71.
Shearer, L. (1978) Sex sensation. in 'Intelligence Report', *Parade Magazine*, Sept. 10.
Sherman, P. W. (1988) The level of analysis. *Anim. Behav.*, **36**, 616–19.
Sherman, P. W. (1989a) Mate guarding as paternity insurance in Idaho ground squirrels. *Nature*, **338**, 418–20.
Sherman, P. W. (1989b) The clitoris debate and the levels of analysis. *Anim. Behav.*, **37**, 697–8.
Shields, W. M. and Shields, L. M. (1983) Forcible rape: an evolutionary perspective. *Ethol. Sociobiol.*, **4**, 125–36.
Short, R. V. (1976) The evolution of human reproduction. *Proc. R. Soc., B*, **195**, 3–24.
Short, R. V. (1979) Sexual selection and its component parts, somatic and genital selection, as illustrated by man and the great apes. *Adv. Study Behav.*, **9**, 131–58.
Short, R. V. (1980) The origins of human sexuality, in *Reproduction in Mammals, Book 8, Human Sexuality*, (eds C. R. Austin and R. V. Short), Cambridge University Press, Cambridge, pp. 1–33.
Short, R. V. (1981) Sexual selection in man and the great apes, in, *Reproductive Biology of the Great Apes*, (ed. C. E. Graham), Academic Press, London, pp. 319–41.
Short, R. V. (1984) Testis size, ovulation rate and breast cancer, in *One Medicine*, (eds O. A. Ryder and M. L. Byrd), Springer-Verlag, New York, pp. 32–44.
Siddiquey, A. K. S. (1982) Effectiveness of spermatozoa in the mouse and the human. PhD thesis, University of Birmingham, UK.
Siddiquey, A. K. S. and Cohen, J. (1982) *In vitro* fertilization in the mouse and the relevance of different sperm-egg concentrations and volumes. *J. Reprod. Fertil.*, **66**, 237–42.
Silberglied, R. E., Shepherd, J. G. and Dickinson, J. L. (1984) Eunuchs: the role of apyrene sperm in Lepidoptera? *Am. Nat.*, **123**, 255–65.
Sillén-Tullberg, B. and Møller, A. P. (1993) The relationship between concealed ovulation and mating system in anthropoid primates: a phylogenetic analysis. *Am. Nat.*, **141**, 1–25.
Silk, J. B. (1983) Sex ratio in macaques. *Am. Nat.*, **121**, 56–66.
Silverstone, P. A. (1976) A revision of the poison arrow frogs of the genus *Phyllobates* Bibron in Sagra (family Dendrobatidae). Natural History Museum Los Angeles Co., *Sci. Bull.*, **27**, 1–53.
Simmons, L. W., Craig, M., Llorens, T. *et al.* (1993) Bushcricket spermatophores vary in accord with sperm competition and parental investment theory. *Proc. R. Soc. Lond. B*, **251**, 183–6.
Simpson, J. L., Golbus, M. S., Martin, A. O. and Sarto, G. E. (1982) *Genetics in Obstetrics and Gynaecology*, Grune & Stratton, New York.
Simpson, M. J. A. and Simpson, A. E. (1982) Sex ratio variation in macaques. *Nature*, **300**, 440–1.
Singer, I. (1973) Fertility and the female orgasm, in *The Goals of Human Sexuality*, (ed. I. Singer), Wildwood House, London, pp. 159–97.
Singer, S. (1985) *Human Genetics: an Introduction to the Principles of Heredity*, 2nd edn. W. H. Freeman, New York.
Singh, D. (1993) Body shape and women's attractiveness: the critical role of waist-to-hip ratio. *Hum. Nat.*, **4**. 297–321.
Singh, G. (1992) Ultrastructural features of round-headed human spermatozoa. *Int. J. Fertil.*, **37**, 99–102.
Sivinski, J. (1980) Sexual selection and insect sperm. *Florida Entomol.*, **63**, 99–111.
Slattery, M., Overall, J. C. and Abbott, T. M. (1989) Sexual activity, contraception, genital infection and cervical cancer: support for a sexually transmitted disease hypothesis. *Am. J. Epidemiol.*, **130**, 248–58.
Small, M. F. (1988) Female primate sexual behaviour and conception: are there really sperm to spare? *Curr. Anthropol.*, **29**, 81–99.
Small, M. F. (1989) Aberrant sperm and the evolution of human mating patterns. *Anim. Behav.*, **38**, 544–6.
Smallcombe, A. and Tyler, K. R. (1980) Semen-elicited accumulation of antibodies and leucocytes in the rabbit female tract. *Experientia*, **36**, 88–89.
Smith, R. L. (1984a) *Sperm Competition and the Evolution of Animal Mating Systems*, Academic Press, London.
Smith, R. L. (1984b) Human sperm competition, in *Sperm Competition and the Evolution of Animal Mating Systems*, (ed. R. L. Smith), Academic Press, London, pp. 601–60.
Smith, T. T. and Yanagimachi, R. (1991) Attachment and release of spermatozoa from the caudal

isthmus of the hamster oviduct. *J. Reprod. Fertil.*, **91**, 567–73.

Smuts, B. B. (1985) *Sex and Friendship in Baboons*. Aldine, Hawthorne, NY.

Smuts, B.B. (1987) Gender, aggression and influence, in *Primate Societies*, (eds B. B. Smuts, D. L. Cheney, R. M. Seyfarth *et al.*), University of Chicago Press, London, pp. 400–12.

Snaith, L. and Williamson, M. (1947) Pregnancy after the menopause. *J. Obstet. Gynaecol. Br. Emp.*, **54**, 496–8.

Snell, R. S. (1986) *Clinical Anatomy for Medical Students*. Little, Brown, Boston.

Sonnenschein, M. and Håuser, C. L. (1991) Evolution of sperm dimorphism within lepidopteran suborders, in *Comparative Spermatology* II, (ed. B. Baccetti), Raven Press, New York, pp. 1011–15.

Speroff, L., Glass, R. H. and Kase, N. G. (1989) *Clinical Gynecology, Endocrinology and Infertility*, Williams & Wilkins, London.

Špinka, M. (1988) Different outcomes of sperm competition in right and left sides of the female reproductive tract revealed by thymidine ^{3}H-labelled spermatozoa in the rat. *Gam. Res.*, **21**, 313–21.

Spira, A. (1984) Seasonal variations of sperm characteristics. *Arch. Androl.*, **12** (Suppl.), 23–38.

Stavy, M. and Terkel, J. (1992) Interbirth interval and duration of pregnancy in hares. *J. Reprod. Fertil.*, **95**, 609–15.

Steen, E. B. and Price, J. H. (1977) *Human Sex and Sexuality*, Wiley, New York.

Steptoe, P. C. and Edwards, R. G. (1976) Reimplantation of a human embryo with subsequent tubal pregnancy. *Lancet*, **i**, 880–2.

Steptoe, P. C. and Edwards, R. G. (1978) Birth after the reimplantation of a human embryo. *Lancet*, **ii**, 366.

Stewart, K. J. and Harcourt, A. H. (1987) Gorillas: variations in female relationships, in, *Primate Societies*, (eds B. B. Smuts, D. L. Cheney, R. M. Seyfarth *et al.*), Chicago University Press, Chicago, pp. 155–64.

Stoddart, D. M. (1990) *The Scented Ape*, Cambridge University Press, Cambridge.

Struhsaker, T. T. and Leland, L. (1987) Colobines: infanticide by adult males. in *Primate Societies*, (eds B. B. Smuts, D. L. Cheney, R. M. Seyfarth *et al.*), Chicago University Press, Chicago, pp. 83–97.

Strum, S. C. (1981) Processes and products of change: baboon predatory behavior at Gilgil, Kenya, in *Omnivorous primates: Gathering and Hunting in Human Evolution*, (eds R. S. O. Harding and G. Teleki), Columbia University Press, New York.

Summers-Smith, J. D. (1955) The communal display of the house sparrow, *Passer domesticus*. *Ibis*, **96**, 116–28.

Sumption, L. J. (1961) Multiple sire mating in swine. *J. Agric. Sci.*, **56**, 31–7.

Suominen, J. and Vierula, M. (1993) Semen quality of Finnish men. *Br. Med. J.*, **306**, 1579.

Svärd, L. (1985) Paternal investment in a monandrous butterfly, *Pararge aegeria*. *Oikos*, **45**, 66–70.

Svärd, L. and Wiklund, C. (1989) Mass production rate of ejaculates in relation to monandry–polyandry in butterflies. *Behav. Ecol. Sociobiol.*, **24**, 395–402.

Swartz, M. N. (1984) Neurosyphilis, in *Sexually Transmitted Diseases*, (eds K. K. Holmes, P. Mardh, P. F. Sparling, and P. J. Wiesner), McGraw-Hill, New York, pp. 318–34.

Symonds, E. M. (1992) *Essential Obstetrics and Gynaecology*, Churchill Livingstone, London.

Symons, D. (1979) *The Evolution of Human Sexuality*, Oxford University Press, Oxford.

Symons, D. and Ellis, B. (1989) Human male–female differences in sexual desire, in *The Sociobiology of Sexual and Reproductive Strategies*, (eds A. E. Rasa, C. Vogel and E. Voland), Chapman & Hall, London, pp. 131–47.

Takahata, Y. (1982) The socio-sexual behavior of Japanese monkeys. *Z. Tierpsychol.*, **59**, 89–108.

Talbert, G. B. (1977) The reproductive system, in *Handbook of the Biology of Aging*, (ed. C. E. Finch and L. Hayflick), Van Nostrand Reinhold, New York, pp. 318–56.

Talbot, P. and DiCarlantonio, G. (1984) Ultrastructure of opossum oocyte investing coats and their sensitivity to trypsin and hyaluronidase. *Dev. Biol.*, **103**, 159–67.

Talbot, P., DiCarlantonio, G., Zao, P. *et al.* (1985) Motile cells lacking hyalurodidase can penetrate the hamster oocyte cumulus complex. *Dev. Biol.*, **108**, 387–98.

Tanner, J. M. (1962) *Growth in Adolescence*, 2nd edn, Blackwell, Oxford.

Tanner, R. E. S. (1970) *Homicide in Uganda, 1964. Crime in East Africa.* Scandinavian Institute of African Studies, Uppsala.

Taub, D. M. (1980) Female choice and mating strategies among wild Barbary macaques (*Macaca sylvanus* L.), in *The Macaques: Studies in Ecology, Behavior, and Evolution*, (ed. D. G. Lindburg), Van Nostrand Reinhold, New York.

Tauber, P. F. and Zaneveld, L. J. D. (1976) Coagulation and liquefication of human semen, in *Human Semen and Fertility Regulation in Men*, (ed. E. S. E. Hafez), C. V. Mosby, St Louis, pp. 153–67.

Tauber, P. F., Zaneveld, L. J. D., Propping, D. and Schumacher, G. F. B. (1976) Components of human split ejaculates I. Spermatozoa, fructose, immunoglobulins, albumin, lactoferrin, transferrin and other plasma proteins. *J. Reprod. Fertil.*, **43**, 249–67.

Taylor, N. J. (1982) Investigation of sperm-induced cervical leucocytosis by a double mating study in rabbits. *J. Reprod. Fertil.*, **66**, 157–62.

Terasaki, P. I., Gjertson, D., Bernoco, D. *et al.* (1978) Twins with two different fathers identified by HLA. *N. Engl. J. Med.*, **209**, 590–2.

Thadani, V. M. (1982) Mice produced from eggs fertilized *in vitro* at a very low sperm:egg ratio. *J. Exp. Zool.*, **219**, 277–83.

Thomas, K., De Hertogh, R., Pizzaro, M. *et al.* (1973) Plasma LH-HCG, 17β-extradiol, estrone, and progesterone monitoring around ovulation and subsequent nidation. *Int. J. Fertil.*, **18**, 65–70.

Thompson, L. A., Tomlinson, L. J., Barratt, C. L. R. *et al.* (1992) Positive immunoselection – a method of isolating leukocytes from leukocytic-reacted human cervical mucus samples. *Am. J. Reprod. Immunol.*, **38**, 92–104.

Thomson, J. A., Lincoln, P. J. and Mortimer, P. (1993) Paternity by a seemingly infertile vasectomised man. *Br. Med. J.*, **307**, 299–300.

Thornhill, R. (1976) Sexual selection and nuptual feeding behaviour in *Bittacus apicalis*. *Am. Nat.*, **110**, 529–48.

Thornhill, R. (1980) Rape in *Panorpa* scorpionflies and a general rape hypothesis. *Anim. Behav.*, **28**, 52–9.

Thornhill, R. (1981) *Panorpa* (Mecoptera: Panorpidae) scorpionflies: systems for understanding resource-defence polygyny and alternative male reproductive efforts. *Annu. Rev. Ecol. Syst.*, **12**, 355–86.

Thornhill, R. (1983) Cryptic female choice and its implications in the scorpionfly *Harpobittacus nigriceps*. *Am. Nat.*, **122**, 765–88.

Thornhill, R. and Thornhill, N. W. (1983) Human rape: an evolutionary analysis. *Ethol. Sociobiol.*, **4**, 137–73.

Thorpe, W. H. (1974) *Animal Nature and Human Nature*, Methuen, London.

Tietze, C. (1957) Reproductive span and rate of reproduction among Hutterite women. *Fertil. Steril.*, **8**, 89–97.

Tilbrook, A. J. and Pearce, D. T. (1986) Patterns of loss of spermatozoa from the vagina of the ewe. *Aust. J. Biol. Sci.*, **39**, 295–303.

Tokarz, R. R. and Slowinski, J. B. (1990) Alternation of hemipenis use as a behavioural means of increasing sperm transfer in the lizard *Anolis sagrei*. *Anim. Behav.*, **40**, 374–9.

Tomlinson, M. J., Barratt, C. L. R., Bolton, A. E. and Cooke, I. D. (1990) Seminal leucocytes and semen quality. *J. Reprod. Fertil. Abstract Series no. 5*, **61**, pp. 35.

Tomlinson, M. J., White, A., Barratt, C. L. R. *et al.* (1992) The removal of morphologically abnormal sperm forms by phagocytes: a positive role for seminal leukocytes. *Hum. Reprod.*, **7**, 517–22.

Torrey, T. W. (1962) *Morphogenesis of the Vertebrates*, Wiley, New York.

Tortora, G. J. and Anagnostakos, N. P. (1984) *Principles of Anatomy and Physiology*, Harper & Row, New York.

Trivers, R. L. (1971) The evolution of reciprocal altruism. *Q. Rev. Biol.*, **46**, 35–57.

Trivers, R. L. (1972) Parental investment and sexual selection, in *Sexual Selection and the Descent of Man*, (ed. B. Campbell), Aldine, London, pp. 139–79.

Trivers, R. L. and Willard, D. E. (1973) Natural selection of paternal ability to vary the sex ratio. *Science*, **179**, 90–2.

Trounson, A. A. (1984) In vitro fertilization problems of the future. *Br. J. Hosp. Med.*, **7**, 104–10.

Tsunoda, Y. and Chang, M. C. (1975) Penetration of mouse eggs *in vitro*, optimal sperm concentration and minimal number of spermatozoa. *J. Reprod. Fertil.*, **44**, 139–42.

Turner, C. L. (1947) Viviparity in teleost fishes. *Sci. Monthly*, **65**, 508–18.

Tutin, C. E. G. (1979) Mating patterns and reproductive strategies in a community of wild chimpanzees (*Pan troglodytes schweinfurthii*). *Behav. Ecol. Sociobiol.*, **6**, 29–38.

Tutin, C. E. G. and McGinnis, R. P. (1981) Sexuality of the chimpanzee in the wild, in *Reproductive Biology of the Great Apes: Comparative and Biomedical Perspectives*, (ed. C. E. Graham) Academic Press, New York, pp. 239–64.

Tyler, J. P. P. and Crockett, N. G. (1982) Comparison of the morphology of vital and dead human spermatozoa. *J. Reprod. Fertil.*, **6**, 667–70.

Tyler, K. R. (1977) Histological changes in the cervix of the rabbit after coitus. *J. Reprod. Fertil.*, **49**, 41–5.

Tyler, K. R. (1978) Studies on the interactions between spermatozoa immunoglobulins and the female reproductive tract in the rabbit and human. PhD thesis, University of Birmingham, UK.

Udry, J. R. and Morris, N. M. (1968) Distribution

of coitus in the menstrual cycle. *Nature*, **220**, 593–6.

Udry, J. R. and Morris, N. M. (1970) The influence of contraceptive pills on the distribution of sexual activity in the menstrual cycle. *Nature*, **226**, 502–3.

Udry, J. R. and Morris, N. M. (1977) The distribution of events in the human menstrual cycle. *J. Reprod. Fertil.*, **51**, 419–25.

United Nations (1987) Contraceptive practice, in *Fertility Behaviour in the Context of Development: Evidence from the World Fertility Survey*, (UN no. 100), pp. 129–65.

Vandenburgh, J. G., Alitselt, J. M. and Lombardi, J. R. (1975) Partial isolation of a pheromone accelerating puberty in female mice. *J. Reprod. Fertil.*, **43**, 515–23.

Vander Vliet, W. L. and Hafez, E. S. E. (1974) Survival and aging of spermatozoa: a review. *Am. J. Obstet. Gynecol.*, **118**, 1006–105.

Van Steirteghem, A. C., Liu, J., Joris, H. *et al.* (1993) Higher success rate by intracytoplasmic sperm injection than by subzonal insemination. Report of a second series of 300 consecutive treatment cycles. *Hum. Reprod.*, **8**, 1055–60.

Van Tienhoven, A. (1968) *Reproductive Physiology of Vertebrates*. W. B. Saunders, London.

Van Voorhies, W. A. (1992) Production of sperm reduces nematode lifespan. *Nature*, **360**, 456–8.

Van Wagenen, G. (1970) Menopause in a subhuman primate. (*Macaca mulatta*). *Anat. Rec.*, **166**, 392.

Veith, J. L., Buck, M., Getzlap, S. *et al.* (1983) Exposure to men influences the occurrence of ovulation in women. *Physiol. Behav.*, **31**, 313–15.

Vermesh, M., Kletzky, O. A., Davajan, V. and Israel, R. (1987) Monitoring techniques to predict and detect ovulation. *Fertil. Steril.*, **477**, 259–64.

Verrell, P. A. (1982) The sexual behaviour of the red-spotted newt, *Notophthalamus viridescens* (Amphibia, Urodela, Salamandridae). *Anim. Behav.*, **30**, 1224–36.

Vessey, M., Lawless, M. and Yeates, D. (1982) Efficacy of different contraceptive methods. *Lancet*, **i**, 841–2.

Villavaso, E. J. (1975) Functions of the spermathecal muscle of the boll weevil, *Anthomus grandis*. *J. Insect Physiol.*, **21**, 1275–8.

Voland, E. (1984) Human sex-ratio manipulation: historical data from a German parish. *J. Hum. Evol.*, **13**, 99–107.

Voland, E. (1988) Differential infant and child mortality in evolutionary perspective: data from late 17th to 19th century Ostfriesland (Germany), in *Human Reproductive Behaviour: A Darwinian Perspective*, (eds L. Betzig, M. B. Mulder and P. Turke), Cambridge University Press, Cambridge, pp. 253–62.

von Helversen, D. and von Helversen, O. (1991) Pre-mating sperm removal in the bushcricket *Metaplastes ornatus* Ramme 1931 (Orthoptera, Tettigonoidea, Phaneropteridae). *Behav. Ecol. Sociobiol.*, **28**, 391–6.

Voss, R. (1979) Male accessory glands and the evolution of copulatory plugs in rodents. Occasional papers of the Museum of Zoology University of Michigan, no. 689.

Waage, J. K. (1979) Dual function of the damselfly penis: sperm removal and sperm transfer. *Science, NY*, **203**, 916–18.

Wakerley, J. B., Clarke, G. and Summerlee, A. J. S. (1988) Milk ejection and its control, in *The Physiology of Reproduction*, (eds E. Knobil and J. D. Neill), Raven Press, London, pp. 2283–322.

Wallace, H. (1974) Chiasmata have no effect on fertility. *Heredity*, **33**, 423–9.

Wallach, S. J. R. and Hart, B. L. (1983) The role of the striated penile muscles in seminal plug dislodgement and deposition. *Physiol. Behav.*, **31**, 815–21.

Wallis, J. (1985) Synchrony of estrous swelling in captive group-living chimpanzees (*Pan troglodytes*). *Int. J. Primatol.*, **6**, 335–50.

Webb, J. E., Wallwork, J. A. and Elgood, J. H. (1977) *Guide to Living Mammals*, Macmillan, London.

Weingold, A. B. (1990) Abnormal bleeding, in *Principles and Practice of Clinical Gynecology*, (eds K. N. G. Weingold and D. M. Gershenson), Churchill Livingstone, New York.

Weissenberg, S. (1922) Das geschlechtsleben de russichen studentinnen (Schbankov Study 1906). *Ztschr. Sexual Wissensch.*, **11**, 7–14.

Weller, A. and Weller, L. (1992) Menstrual synchrony in female couples. *Psychoneuroendocrinology*, **17**, 171–7.

Weller, A. and Weller, L. (1993) Menstrual synchrony between mothers and daughters and between roommates. *Physiol. Behav.*, **53**, 943–9.

Wells, K. D. (1981) Paternal behaviour of male and female frogs, in *Natural Selection and Social Behaviour*, (ed. R. D. Alexander and D. W. Tinkle), Chiron Press, New York, pp. 184–97.

West, M. M. and Konner, M. J. (1976) The role of the father: an anthropological perspective, in *The Role of the Father in Child Development*, (ed. M. E. Lamb), Plenum Press, New York.

Weygoldt, P. (1980) Complex brood care and reproductive behavior in captive poison-arrow frogs, *Dendrobates pumilia*. *Behav. Ecol. Sociobiol.*, **7**, 329–32.

Whitten, P. L. (1983) Diet and dominance among female vervet monkeys (*Cercopithecus aethiops*). *Am. J. Primatol.*, **5**, 139–59.

Whitten, P. L. (1987) Infants and adult males, in *Primate Societies*, (eds B. B. Smuts, D. L. Cheney, R. M. Seyfarth *et al.*.), Chicago University Press, Chicago, pp. 343–57.

Whitten, W. K. (1959) Occurrence of anoestrus in mice caged in groups. *J. Endocrinol.*, **18**, 102–7.

Whitten, W. K. (1966) Pheromones and mammalian reproduction. *Adv. Reprod. Physiol.*, **1**, 155–77.

WHO (1990) Global estimates for health situation assessment and projections. WHO, Geneva.

Wickler, W. (1967) Socio-sexual signals and their intra-specific imitation among primates, in *Primate Ethology*, (ed. D. Morris), Weidenfeld & Nicolson, London.

Wildt, D. E., Bush, M., Howard, J. G. *et al.* (1983) Unique seminal quality in the South African cheetah and a comparative evaluation in the domestic cat. *Biol. Reprod.*, **29**, 1019–25.

Wildt, D. E., Howard, J. G., Hall, L. L. and Bush, M. (1986) Reproductive physiology of the clouded leopard: 1. Electroejaculates contain high proportions of pleiomorphic spermatozoa throughout the year. *Biol. Reprod.*, **34**, 937–47.

Wildt, D. E., Bush, M., Goodrowe, K. L. *et al.* (1987a) Reproductive and genetic consequences of founding isolated lion populations. *Nature*, **329**, 328–31.

Wildt, D. E., O'Brien, S. J., Howard, J. G. *et al.* (1987b) Similarity in ejaculate-endocrine characteristics in captive versus free-ranging cheetahs of two subspecies. *Biol. Reprod.*, **36**, 351–60.

Wilson, E. O. (1971) *The Insect Societies*, Belknap Press, Cambridge, USA.

Wilson, E. O. (1978) *On Human Nature*, Harvard University Press, Cambridge, MA.

Wilson, H. C. (1987) Female axillary secretions influence women's menstrual cycles: a critique. *Horm. Behav.*, **21**, 536–46.

Wilson, H. C. (1992) A critical review of menstrual synchrony research. *Psychoneuroendocrinology*, **17**, 565–91.

Wilson, H. C., Kiefhaber, S. H. and Gravel, V. (1991) Two studies of menstrual synchrony: negative results. *Psychoneuroendocrinology*, **16**, 353–9.

Wilson, W. and Durrenberger, R. (1982) Comparison of rape and attempted rape victims. *Psychol. Rep.*, **50**, 198.

Winokur, G., Guze, S. B. and Pfeiffer, E. (1959) Nocturnal orgasm in women. *Arch. Gen. Psychiatry.*, **1**, 76–80.

Wolf, L. L. (1975) 'Prostitution' behaviour in a tropical hummingbird. *Condor*, **77**, 140–4.

Wolf, D. P., Byrd, W., Dandekar, P. and Quigley, M. M. (1984) Sperm concentration and the fertilization of human eggs in vitro. *Biol. Reprod.*, **31**, 837–48.

Wolfe, L. (1981) *The Cosmo Report*. Arbor House, New York.

Wood, B. A. (1978) *Human Evolution*, Chapman & Hall, London.

Wood, B. A. (1992) Origin and evolution of the genus *Homo*. *Nature*, **355**, 783–90.

Wood, B. A. (1993) Four legs good, two legs better. *Nature*, **363**, 587–8.

Wood, J. L., Johnson, P. L. and Campbell, K. L. (1985) Demographic and endocrinological aspects of low natural fertility in highland New Guinea. *J. Biosoc. Sci.*, **17**, 57–79.

Wood, J. W. (1989) Sperm longevity. *Oxford Rev. Reprod. Biol.*, **11**, 61–109.

Wood, P. D. P., Foulkes, J. A., Shaw, R. C. and Melrose, D. R. (1986) Semen assessment fertility and the selection of Hereford bulls for use in AI. *J. Reprod. Fertil.*, **76**, 783–95.

Wood, R. J. (1961) Biological and genetical studies on sex ratio in DDT resistant and susceptible strains of *Aedes aegypti* Linn. *Genet. Agr.*, **13**, 287–307.

Wood, R. J. and Newton, M. E. (1976) Meiotic drive and sex ratio distortion in the mosquito *Aedes aegypti*. *Proceedings of the XV International Congress Entomology*, 97–105.

Wrangham, R. W. (1979) On the evolution of ape social systems. *Soc. Sci. Inf.*, **18**, 334–68.

Wright, R. M., John, F. M., Klotz, K. *et al.* (1990) Cloning and sequencing of genes coding for the human intra-acrosomal antigen SP10. *Biol. Reprod.*, **42**, 693–701.

Yamada, K., Stevenson, A. F. G. and Mettler, L. (1988) Fertilization through spermatozoal microinjection; significance of acrosome reaction. *Hum. Reprod.*, **3**, 657–61.

Yanagimachi, R. (1970) The movement of golden hamster spermatozoa before and after capacitation. *J. Reprod. Fertil.*, **23**, 193–6.

Yanagimachi, R. (1988) Mammalian fertilization, in *Physiology of Reproduction* Vol. 1, (eds E. Knobil and J. D. Neill), Raven Press, New York, pp. 135–85.

Yanagimachi, R. and Mahi, C. A. (1976) The sperm acrosome reaction and fertilization in the guinea pig: a study *in vivo*. *J. Reprod. Fertil.*, **46**, 49–54.

Zaadstra, B. M., Seidell, J. C., Van Noord, P. A. H. *et al.* (1993) Fat and female fecundity: prospective study of effect of body fat distribution on conception rates. *Br. Med. J.*, **306**, 484–7.

Zahavi, A. (1975) Mate selection – a selection for a handicap. *J. Theor. Biol.*, **53**, 205–14.
Zavos, P. M. and Goodpasture, J. C. (1989) Clinical improvements of specific seminal deficiencies via intercourse with a seminal collection device versus masturbation. *Fertil. Steril.*, **55**, 190–3.

Index